AF323501

Digital Optical Measurement Techniques and Applications

Digital Optical Measurement Techniques and Applications

Pramod Rastogi

Editor

ARTECH HOUSE

BOSTON | LONDON
artechhouse.com

Library of Congress Cataloging-in-Publication Data
A catalog record for this book is available from the U.S. Library of Congress

British Library Cataloguing in Publication Data
A catalog record for this book is available from the British Library.

ISBN-13: 978-1-60807-806-6

Cover design by John Gomes

© 2015 Artech House
685 Canton St.
Norwood, MA

For color versions of figures in Chapter 8, please go to www.artechhouse.com/static/Downloads/RastogiColorFigs.zip.

10 9 8 7 6 5 4 3 2 1

Contents

CHAPTER 3

Phase-Shifting Interferometry 89

CHAPTER 4

Systematic Approach to Digital Holographic Imaging and Its Applications in Interferometry 129

CHAPTER 7

CHAPTER 8

CHAPTER 9

Digital Particle Image Velocimetry   347

CHAPTER 10

Optical Fiber Sensors   377

CHAPTER 11

Optical Coherence Tomography: Principles, Implementation, and Applications in Ophthalmology

Foreword

Optics has been effectively utilized to address diverse applications in a variety of fields in science and engineering through its multiple arms, such as optical testing, optical metrology, and image processing. Many of these applications lie in far-flung fields such as experimental mechanics, fracture mechanics, materials, biology, and medicine. Since the introduction of the laser in 1960, different lasers with extended wavelengths and high powers have become available. As a by-product of enhanced coherence made available by lasers, speckles continue to be perceived as noise in coherent optical imaging. Speckles occur by the scattering of coherent light from a rough surface or by its transmission through an optically rough medium, having thickness variations sufficient to randomize the transmitted wave, and play an important role in nondestructive testing. Digital information processing, together with the developments in electro-optical detection, is instrumental to the success of optical measurement techniques and their applications in science and engineering. Technology has changed dramatically since the time when early versions of these techniques were conceived. CCD and CMOS detectors with large numbers of pixels, extended color sensitivity, and directly computer-addressed pixels form the basic elements of digital information processing systems. Storage capacity and the capability of digital computers have also continued to increase exponentially with respect to the computational power per unit of cost, allowing high data manipulation.

Optical metrology being contactless does not damage the object surface. The information about the object is carried by the transmitted or scattered wave without affecting the state of the object significantly. The requirements with respect to the precision, the measurement speed, and the gradual shift toward applications involving micro- and nanotechnology are also rapidly growing. Novel developments of new photoelectric receivers, with ever-improving spatial resolution and shorter response time, should be helpful in handling a large amount of data or in extending the wavelength range into the infrared and ultraviolet regions. This should in addition require appropriate image processing tools. The enormous progress that digital information processing has witnessed since the beginning of this century has paved the way to handling this large amount of data flow.

Digital image processing in photography aids in rendering the scene with the highest fidelity possible, taking care of the band-limited nature of the electro-optical recording. Imaging technology has advanced in step with the image processing hardware in recent years. Digital processing techniques using almost-real-time contrast enhancement and blur reduction schemes have led to enhancing the subjective image quality. As well, edge enhancement, image sharpening, and histogram equalization have contributed to improving subjective image perception. However, for an objective image enhancement, image processing needs to be extended, as discussed in Chapter 1, by applying Fourier transform techniques with frequency domain analysis. For improving the image quality, the modulation transfer function has been used in order to restore the image quality quantitatively.

Optical testing, discussed in Chapter 2, is essential in high-precision manufacturing, especially when producing high-precision optical components and systems. Interferometric testing of optical surfaces and systems has been known for many years; Newton rings are still referred to in shop testing of surfaces and optical systems. In interferometric testing of aspherical surfaces, computer-generated holograms are useful to generate an aspheric reference wave. High-precision and complex manufacturing machines can only be effective when contactless testing is realized in a very short time or in quasi-real time. Contactless optical testing of aspheric surfaces in industry was for many years and still is a subject of research for single elements and low f-numbers when computer-generated holograms are too expensive or their generation takes too long. The metrology of freeform surfaces is even more demanding.

Digital fringe analysis, discussed in Chapter 3, and fringe interpretation thereof is a compulsory need of the day. Phase-shifting techniques and Fourier transform methods are frequently used for the automated fringe analysis in interferometry, holographic interferometry, and speckle pattern interferometry, as well as in photoelasticity for stress analysis and in fringe projection for shape comparison and measurement. Instead of using point-wise strain gauge, optical techniques allow noncontact full-field displacement and strain analysis. Digital holographic and speckle pattern interferometry, which are the subjects of Chapters 4 and 5, are a well-established class of whole-field interferometric techniques for the measurement of in-plane and out-of-plane displacements, deformations, strains, slopes, curvatures, shapes, and vibrations. Digital information processing with automated fringe analysis is discussed together with digital holographic imaging and digital speckle pattern interferometry.

In Chapter 6, which discusses digital image correlation, recording and processing of object information is required before and after object deformation as is the case in holographic and speckle pattern interferometry. The technique has gained popularity in its use for planar objects and in-plane deformation measurement. A binocular stereo vision system combining the projection of computer generated speckle patterns with a digital projector was developed for curved surfaces and for 3-D deformation measurement. With an improved stereo vision calibration, 3-D digital image correlation is bound to gain in popularity for applications in the measurement of deformation of macroscopic objects and structures. Fringe analysis is well established in digital fringe projection profilometry, which is the discussed in Chapter 7, and has been widely applied for shape measurement where robustness, measurement speed, and automated analysis are essential.

Digital photoelasticity, detailed in Chapter 8, is related to common path interferometry with low coherence requirements. It is an excellent tool for visualizing and quantifying stress fields and is based on the birefringence that is temporarily induced in the object by imposing it under stress. Whole-field automated photoelasticity has become possible with the development of phase-shifting-based techniques. From the early use of digital hardware to the use of phase-shifting or color image processing to bring in automation to conventional photoelasticity has necessarily required the use of digital information processing. Experimental analysis has always been considered as essential for overcoming the problems and uncertainties associated with numerical modeling, and which, in turn, is favorable for industrial applications due to the low costs involved in image acquisition.

Laser-based fluid velocity measurement techniques can be divided into point-measuring (e.g., laser Doppler anemometry) and field-measuring techniques (e.g., laser speckle velocimetry, better known as particle image velocimetry, discussed in Chapter 9). The recording of 3-D location of markers in fluid multiple times leads to the reconstruction of 3-D fields. However, instead of tracking individual markers the velocity components in the flow are determined using digital processing of a finite 3-D integration volume.

Optical fiber sensors, the subject of Chapter 10, are finding use worldwide from monitoring stresses in airplane wings to medical applications. Although the basic principle of these sensors is well established, the implementation of the optical fiber sensor systems and networks requires considerable exploitation of digital techniques.

Optical coherence tomography, discussed in Chapter 11, is a variation of low coherence interferometry and has found an important niche in ophthalmology. The method provides noninvasive tomography of in vivo human tissues. The purpose of tomography is to obtain a cross-sectional slice of the object. The rapid development of optical coherence tomography to a standard method, in particular in ophthalmology, where it was invented originally, is exceptional.

The reader will find this book well organized and well written; with many built-in didactic features, such as solved exercises, case studies, and text boxes, and with appropriate references to the past work. Overall a valuable resource, the book provides a state-of-the-art as well as a conceptual perspective on the subject and should be a standard text on digital optical measurements and their applications for years to come.

Hans J. Tiziani
Berneck, Switzerland
April 2015

Preface

Thirty-five years ago, in its early phases of development, digital processing was only a small part of a few optical imaging and measurement systems. Nevertheless, the introduction of electronic imaging devices such as charge-coupled devices were already setting alight a hope for the future by their perceived potential at creating human-friendly and economically driven solutions for the existing optical measurement techniques that would enable them to surmount the obstacles and to respond to the demands of technology in the twenty-first century.

Digital equivalents of optical measurement systems have been devised by integrating image processing technologies into them, such as the relevant tools within digital image and signal processing. Far from just translating the strengths of the analog methods into their digital counterparts, the integration process has brought to light capabilities that could never exist in their respective analog forms, in addition to enhancement in efficiencies and accuracies.

This marriage of analog methods with digital data processing has been one of the most vaunted success stories of the last couple of decades. Their combined synergy has not only given rise to the ability to deal with more complex problems but has also created mechanisms for coming up with innovative solutions by working in ways that were previously impossible to imagine. This synergy has also provided a robust means of controlling the processes involved and in managing the information and its flow.

A body and soul unison of digital information processing and analysis with optical metrology has engendered the possibility of a significant increase in the functional range, resulting in challenging research avenues to explore both for the present and the future. The rise in the complexity reflects an important demand for the need to increase the fundamental understanding and the necessary skills in digital data processing and optical metrology for our science and engineering students. This book has been developed with this view in mind, and as a follow up to *Optical Measurement Techniques and Applications* (Artech House, 1997).

The ever-deepening imprint of pluridisciplinarity that continues to color the horizons of progress in most fields has been slowly but steadily marking a shift in book publishing by bringing together groups of experts in diverse fields with the objective of addressing a wide breadth of subjects in a style befitting or closer to a textbook as compared to a reference work. This book, an example of this shift, is intended to fulfill the following expectations: The topics covered here have been selected and presented in a way that the reader can understand their underlying physics and functionality before embarking on a unifying approach to the subject. The book lays out these elements in a pedagogical style and provides a good understanding of the state-of-the-art developments in the field. Solved exercises, applications, case studies, and boxes containing background material have been employed throughout to help expand the understanding on the subject.

I would like to take this opportunity for thanking very warmly all the authors contributing to this book for their deep sense of collegiality and for their promptness in responding to my calls.

Pramod Rastogi
April 2015

CHAPTER 1

Digital Image Processing

Richard H. Vollmerhausen

1.1 Introduction

A tourist visiting the Grand Canyon would probably bring a color digital camera. On the other hand, uncooled thermal technology is popular for a soldier's rifle sight. Thermal imagers are heavier and more expensive, but they provide target acquisition capability at night. The choice of technology depends on the application and expected environment. However, the color camera and the thermal rifle sight have something in common. Namely, these cameras will be used by someone, somewhere, at some time in the future, to do an unspecified visual task. Whether the camera is in a cell phone or an expensive military weapon system, it must provide good imagery of a variety of scenes.

The goal of this chapter is to describe digital image processing techniques that aid in rendering the scene with the highest fidelity possible. We want to transpose radiometric variations across the scene to luminance variations across the display with minimal error given the band limited nature of electro-optical (EO) imagers. To that end, this chapter describes image deblurring and contrast enhancement techniques.

Digital restoration improves scene rendering by reducing blur. Diffraction, optical aberrations, defocus, vibration, detector cross-talk, and many other factors cause blur in the image. Digital processing can reduce the blur and enhance image fidelity.

There are occasions, however, when scene detail cannot be properly displayed because of wide variations in scene radiance. For example, part of the scene might be in sunlight and part in deep shade. Another example occurs with infrared imagery, where atmospheric path radiance might create a large variation in intensity across the scene. On those occasions, imagery benefits from contrast enhancement techniques that remove the wide swings in radiance. In other words, on occasion, we must forgo absolute fidelity in rendering the scene because of limitations in the display and eye.

There is extensive literature on edge enhancement, image sharpening, and many variations of histogram equalization methods. Those methods can be applied iteratively to a picture to improve subjective image quality. That is, apply the process (e.g., apply a sharp filter), judge whether you like the result, and if not try something else. However, those methods are suitable for Photoshop but not camera design, because they distort as many scene details as they enhance. Those methods are not described here because they are not suitable for incorporation into camera electronics.

The material in the chapter assumes that the reader has a basic understanding of digital signal processing and wishes to apply digital techniques to image enhancement. Also, Fourier transforms are used to model the imaging process, but the explanation of theory is very brief. The material in the chapter assumes that the reader has a working level understanding of computer processing and familiarity with common signal analysis practices.

Figure 1.1 illustrates the components of an electro-optical (EO) imager. The lens forms an image of the scene on a two-dimensional (2D) array of photo-detectors called the focal plane array (FPA). Photo-electrons are generated in proportion to image intensity at the location of each detector. Therefore, the FPA provides a 2D array of intensity samples that represent the imaged scene. The FPA samples are digitized and stored for digital processing. Digital processing can enhance contrast, compress dynamic range for viewing, restore some of the blur caused by the optics, and provide other enhancements to the image. A 2D array of display pixels forms an image for viewing.

Although the focus of this chapter is the *digital filter* box in Figure 1.1, understanding the relationship among the optical image, the digital signal processing, and the displayed image is crucial, because the function of the digital processor is to correct the blur and contrast problems that originate in the analog signal portion of the imager. Section 1.2.1 explains how Fourier transforms are used to analyze electro-optical (EO) imagers. Section 1.2.2 provides the theoretical basis for the deconvolution procedures described in Section 1.2.3. Section 1.2.2 also describes the transfer function attributes that lead to good imaging performance.

Section 1.3 describes digital deconvolution filters that remove image blur. That section explains windowing and discusses the circumstances under which windowing improves restoration results. Several image restoration examples are presented. Section 1.4 describes the benefits and implementation of LACE. The fundamental differences between reflective and thermal imagery are discussed in order to illustrate the benefits of LACE to both types of imagery. Again, several examples are presented to illustrate the benefits of LACE. Conclusions are in Section 1.5.

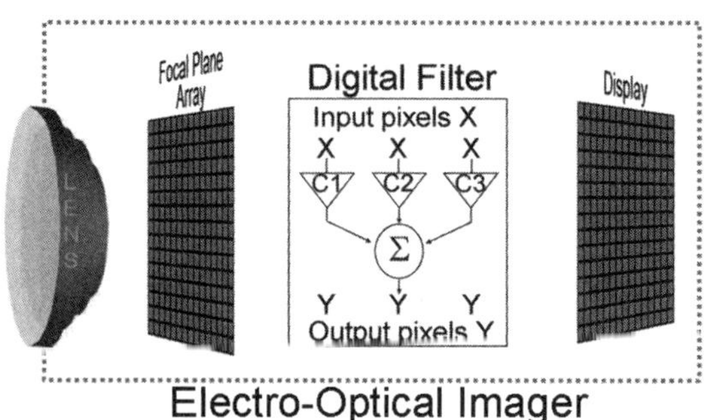

Figure 1.1   The component parts of an electro-optical imager.

1.2   Fourier Optical Analysis

Fourier analysis is used to evaluate how well the imager replicates scene details on the display. The EO imager is treated a spatial filter of scene details. Each stage of the imaging process, optical image formation, photo-detection, digital processing, and display either removes or enhances scene details by blurring or sharpening the image. Each imaging stage has a transfer function that quantitatively describes its spatial filtering properties. The transfer function of the imager is the product of all the component filters.

Figure 1.2 illustrates how an image is formed in an EO camera. The lens forms an image of the scene on the FPA. The image on the FPA is blurred compared to the scene because the optics are not perfect. Optical blurs occur due to diffraction and optical aberrations. In the figure, a portion of the optical image is magnified in order to clearly show the image blur. The squares represent active detector areas. Each detector integrates the photo-electrons generated by the light falling on it. That is, all of the photo-electrons generated by the light falling on a single square are summed. Spatially integrating over the detector area causes an additional blur beyond that generated in the optics. The photo-electrons generated in each photo-detector are samples of the image intensity at the location of the detector.

The FPA samples are digitized and stored for processing. The FPA intensity values can be manipulated in a digital processor to increase resolution, improve contrast, or filter out noise. However, in Figure 1.2, the FPA intensity values are passed directly to the display. The display is a 2D array of pixels that generate light in proportion to the FPA pixel values. The viewer therefore sees a representation of the scene on the display.

Photo-detection is not a perfect process. In virtually all cases, each detector in the FPA has slightly different gain and dark current, and these FPA nonuniformities result in fixed-pattern noise in the image. Also, shot noise corrupts each detector

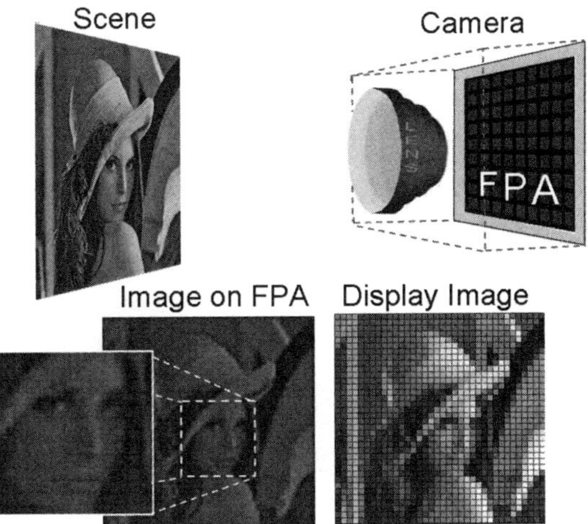

Figure 1.2   Illustration of image formation in a camera. The squares represent detectors on the FPA.

sample. For square-law detectors where the signal is proportional to light intensity, shot noise equals the square root of the number of signal photo-electrons. Additional noise is added because of noise in the read-out electronics. The analog to digital converter also adds quantization noise.

In addition to noise, aliasing can corrupt the image. The displayed image in Figure 1.2 is not very good. Part of the problem is blur, but a large part of the problem is that the scene is poorly sampled because of the large and widely spaced detectors on the FPA. The widely spaced detectors cause aliasing. A further problem with the displayed image in Figure 1.2 is that the display pixels are large and sharply demarcated. Aliasing is caused by sampling, but poor engineering of the display can greatly exacerbate the negative effects of aliasing.

Section 1.2.1 describes the frequency domain transfer response of a well-sampled imager; the transfer response quantifies imager system blur. Section 1.2.2 discusses the transfer function attributes that lead to good imager performance. Section 1.2.3 describes the response function of a sampled imager; the response function quantifies both the transfer response and aliasing in the imager.

See Vollmerhausen (2010) for a description of thermal and reflective imaging and a discussion of the primary causes of camera blur. See Goodman (1996) for a thorough treatment of Fourier optics and discussion of the analysis of two-dimensional systems. Caniou (1999) provides a thorough and clear discussion of imager noise sources.

1.2.1   Frequency Domain Transfer Response

Each point in the scene generates a blurred spot on the FPA called the point spread function (psf). The photo-electrons generated per square centimeter (cm^2) in each detector is proportional to the irradiance in watts per cm^2 on that detector. In this chapter, only incoherent imaging is considered, and the phase of the light waves arriving from each point in the scene varies randomly. The image of the scene is the sum of all the psf intensities.

The psf includes blur due to the optics and blur due to spatial integration of light over the detector active area. In most optical systems, the psf is not constant over the entire field of view because optical aberrations vary with field angle. The optical blur is generally smaller at the center of an image than it is at the edge. However, the image plane can be subdivided into regions within which the optical blur is approximately constant. A region of the image with approximately constant blur is called an *isoplanatic patch*.

The image within an isoplanatic patch can be represented as a convolution of the psf over the scene. A convolution is illustrated in Figure 1.3. If $h_{op}(x,y)$ represents the spatial shape (the intensity distribution) of the psf, then $h_{op}(x - x', y - y')$ represents a psf at location (x',y') in the image plane. The psf quantifies the way that the light from each point in the scene is distributed over an isoplanatic patch of the FPA. The units of x and y are milliradians (mrad), and (x,y) defines a point in a Cartesian coordinate system. We assume a one-to-one mapping of the scene onto the FPA and from the FPA to the display, such that if $scn(x,y)$ describes the radiant intensity of the scene, then $img(x,y)$ describes the brightness of the displayed image.

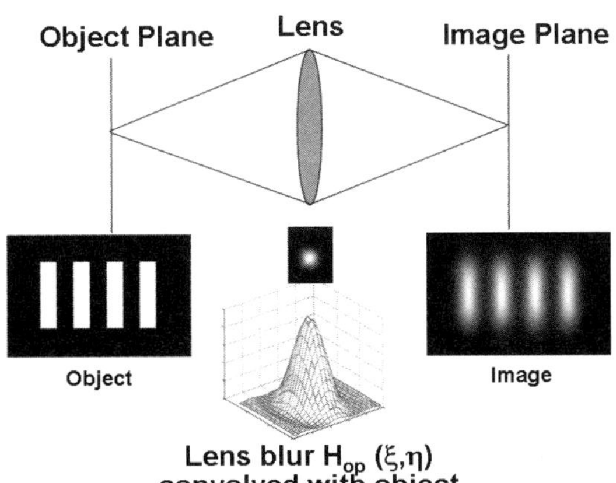

Figure 1.3   Convolving optical blur over the scene causes a blurred image on the FPA.

BOX 1.1

Ignoring sampling for the moment, the image on the FPA $i_{\mathrm{FPA}}(x,y)$ is found by convolving the psf over the scene.

$$i_{FPA}(x,y) = \int\limits_{-\infty}^{\infty} \int\limits_{-\infty}^{\infty} s_{cn}(x',y')h_{op}(x-x',y-y')dx'dy' \tag{1.1}$$

The image on the display is found by convolving the detector blur $h_{\mathrm{det}}(x,y)$, the display blur pixel(x,y), and the digital blur $h_{\mathrm{dig}}(x,y)$ over the FPA image

$$i_{mg}(x,y) = s_{cn}(x,y) * h_{op}(x-x',y-y') * h_{det}(x-x',y-y') \tag{1.2}$$
$$* h_{dig}(x-x',y-y') * p_{ix}(x-x',y-y')$$

where (*) represents convolution. Any additional blurs such as stabilization and eye can be handled the same way

$$i_{mg}(x,y) = s_{cn}(x,y) * h_{op}(x-x',y-y') * h_{det}(x-x',y-y')$$
$$* p_{ix}(x-x',y-y') * h_{dig}(x-x',y-y') * s_{tab}(x-x',y-y') \tag{1.3}$$
$$* e_{ye}(x-x',y-y').$$

Equation (1.3) involves a lot of convolutions all of which are double integrals. For this reason, imaging system analysis is usually performed in the frequency domain. A convolution in the spatial domain equates to a multiplication in the frequency domain, and frequency domain analysis avoids much of the computational difficulty of (1.3).

A Fourier transform (FT) finds the sine wave content of an image. Spatial sine waves are shown in Figure 1.4. The sine wave at the top of the figure is higher frequency than the sine waves at the bottom. The lowest sine wave in the figure has a different phase than the middle because it has the same period but the peak is offset in space. Note that sine waves can be oriented at any angle. Also note that a single

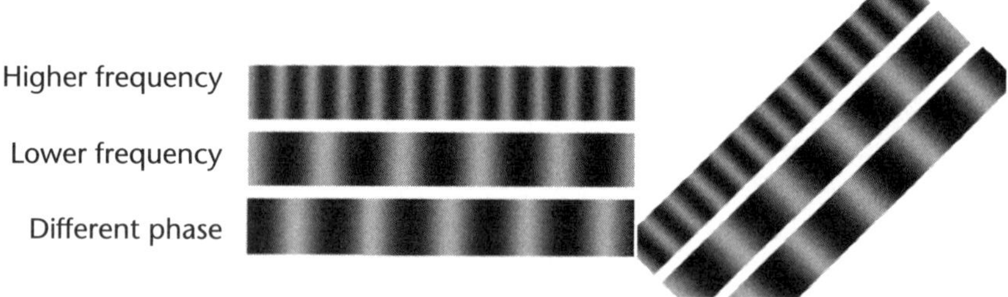

Figure 1.4 Illustration of spatial sine waves.

frequency is a sine wave that extends throughout space, and that the truncated sine waves shown in the figure do not represent single frequencies.

Any image can be represented by the sum of sine waves. Figure 1.5 illustrates how sine waves can be summed to form a three-bar pattern. Individual sine waves are shown at left. Note that these sine waves have different intensity amplitudes and phases as well as different periods. To the right in the figure, the sine waves are progressively summed. When only a few, low frequency sine waves are summed, the bar pattern is blurry, and it becomes progressively sharper as more sine waves are added.

A FT from spatial coordinates (x,y) in milliradians into frequency domain coordinates (ξ,η) in cycles per milliradian (mrad^{-1}) is given by (1.4), and the inverse transform is given in (1.5). We capitalize the first letter of each spatial domain function to represent its FT.

$$S_{cn}(\xi,\eta) = \int\limits_{-\infty}^{\infty} \int\limits_{-\infty}^{\infty} s_{cn}(x,y)e^{-2\pi i(x\xi+y\eta)}\,dxdy \tag{1.4}$$

$$s_{cn}(x,y) = \int\limits_{-\infty}^{\infty} \int\limits_{-\infty}^{\infty} S_{cn}(\xi,\eta)e^{2\pi i(x\xi+y\eta)}\,d\xi d\eta \tag{1.5}$$

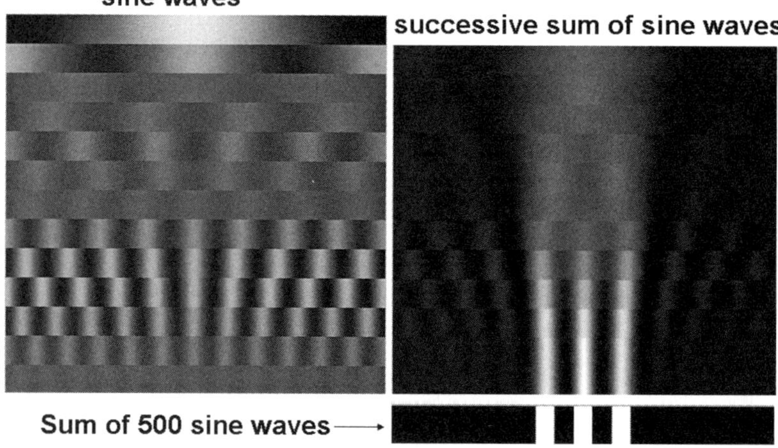

Figure 1.5 The image of a three-bar pattern can be obtained by summing sine waves of various frequencies, phases, and amplitudes.

In order to simplify analysis, we normalize each FT such that H_{op}, H_{det}, H_{dig}, etc., represent spatial filters that operate on the scene frequency content. So, for example,

$$H_{op}(\xi, \eta) = \frac{\displaystyle\int_{-\infty}^{\infty}\int_{-\infty}^{\infty} h_{op}(x,y)e^{-2\pi i(x\xi + y\eta)}\,dx\,dy}{\displaystyle\int_{-\infty}^{\infty}\int_{-\infty}^{\infty} h_{op}(x,y)\,dx\,dy} \tag{1.6}$$

H_{op} is referred to as the optical transfer function (OTF) of the imager, and the magnitude |OTF| is the optical modulation transfer function (MTF). Each image plane in the camera (FPA, digital processing, and display) has a transfer function and MTF. Then the system transfer function $H(\xi,\eta)$ is

$$H(\xi, \eta) = H_{op}(\xi, \eta)H_{det}(\xi, \eta)H_{dig}(\xi, \eta)P_{ix}(\xi, \eta) \tag{1.7}$$

and the FT of the image is in modulation on the display

$$I_{mg}(\xi, \eta) = S_{cn}(\xi, \eta)H(\xi, \eta) \tag{1.8}$$

The calculation of scene and display modulation or modulation contrast is illustrated in Figure 1.6. At the top is an oscilloscope trace of the intensity of the sine wave shown at the bottom. Modulation contrast C_{mod} is given by the equation

$$C_{mod} = \frac{\text{max} - \text{min}}{\text{max} + \text{min}} \tag{1.9}$$

Figure 1.7 shows that frequency domain analysis arrives at the same blurred image as the spatial convolution illustrated in Figure 1.3. Instead of convolving the lens blur over the scene, the FT of the scene is multiplied by the OTF to get the FT of the image on the FPA. While it is true that using frequency domain analyses requires performing a double integral on $s_{cn}(x,y)$ to get $S_{cn}(\xi,\eta)$ and on $I_{mg}(\xi,\eta)$ to get the final image $i_{mg}(x,y)$, the intermediate step is a multiplication by the lens OTF. The steps illustrated in Figure 1.7 are actually more complicated than the convolution

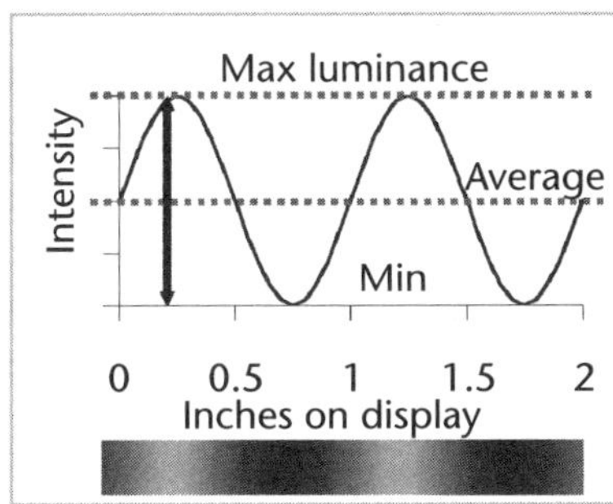

Figure 1.6   The intensity trace of a sine wave is used to calculate modulation.

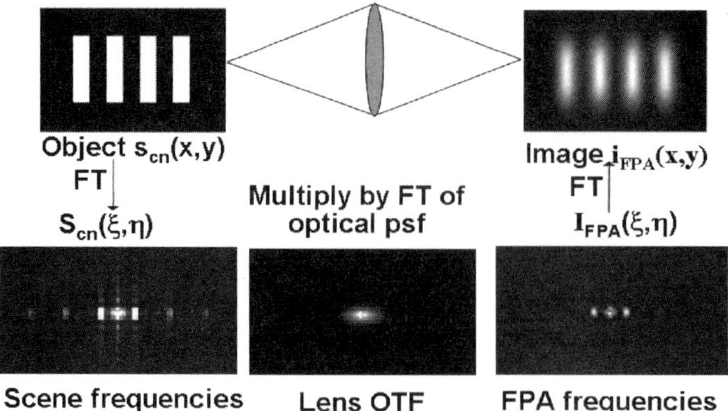

Figure 1.7   Using FT analysis to arrive at the same blurred image as shown in Figure 1.3.

process illustrated in Figure 1.3. However, all of the intermediate steps in (1.3) are now multiplications rather than double integrals, and frequency domain analysis is easier when the entire signal chain is considered.

Figure 1.8 summarizes the analysis technique. A camera images a point in the scene onto the display. The point is blurred on the display because the system transfer function is not perfect, and all frequencies that constitute a point do not pass through the camera. Envision two sine waves in the scene; these are shown at the bottom of Figure 1.8. The sine wave with a small spatial period is more blurred and has reduced contrast on the display than the sine wave with a large spatial period. The failure of the camera to pass all scene spatial frequencies at full modulation creates the image blur. We find the modulation on the display at every spatial frequency by multiplying scene modulation by the system transfer function.

Equations (1.1) through (1.8) apply to well-corrected optical systems where the blur is constant over some substantial region in the field of view (FOV). Also, to this point, we have ignored sampling. Equation (1.8) treats the camera as a filter of scene content, and no frequencies can appear on the display that do not originate in the scene. That is not true in practice, and it is a very important flaw in the non-sampled camera model.

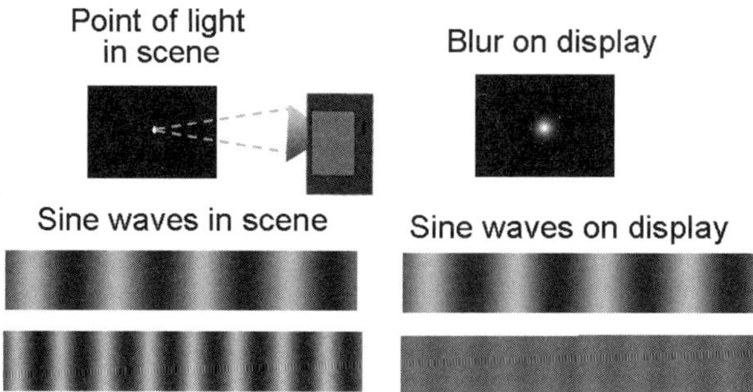

Figure 1.8   The camera lowers the modulation of high frequency sine waves, and this causes image blur. Display modulation is found by multiplying scene modulation by the system transfer function.

However, a camera's transfer function provides some important insights into expected performance. The next section discusses the features that are desired in a transfer function.

EXERCISE 1.1

A thermal imager has the component MTF shown in the figure. The scene is a full modulation sine wave like that shown in Figure 1.6. What is the display modulation at 5 cycles per milliradian? At 10 cycles per milliradian?

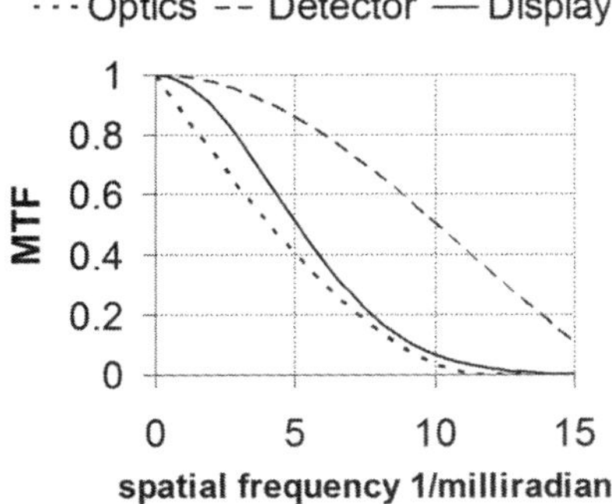

Figure Ex. 1.1

Solution

At 5 cycles per milliradian, multiply optics MTF (0.4) by display MTF (0.5) and then detector MTF (0.85) to get display modulation of 0.17. At 10 cycles per milliradian, the display MTF is approximately 0.001.

1.2.2　Desirable Features in a Camera Transfer Function

The camera should have as big a spatial bandwidth as possible in order to minimize the error between the scene and the rendered image. In other words, the system MTF should be maintained as near to one as possible throughout as wide a spatial frequency passband as possible.

BOX 1.2

Not boosting high frequencies might seem like a mistake; is it not true that high frequencies should be boosted more than low frequencies? The answer to that question is "no." High frequency boost helps scene rendering when it overcomes modulation loss in the imager. That is, high frequency boosts helps when it creates a flat, wide imager transfer function. We want to transpose radiometric variations across the scene to luminance variations across the display with the highest fidelity possible given the band limited nature of EO imagers. The least error between scene and display is achieved by making the MTF flat throughout the passband of the imager.

The justification for requiring a wide, flat passband harks back to our assumption that the camera will be used by someone, somewhere, to do some unspecified visual task. Many scenes must be considered, and the FT of "the target" is not known. Even in the case of searching for one specific object, range and aspect are not known, and these determine the target FT.

Two-dimensional plots are sometimes hard to interpret and do not accommodate comparisons on the same coordinate axes. For those reasons, one-dimensional (1D) plots of MTF are often used. The 1D plots show the intersection of a 2D surface with either the horizontal or vertical (MTF, spatial frequency) plane. Figure 1.9 provides an example. A 2D Gaussian MTF is shown at left; the intersection of the MTF surface with the (MTF, horizontal frequency) plane is shown to the right.

Although 1D MTF plots are common, and we shall use them in this chapter, spatial bandwidth W is found using a 2D integral:

$$W = \int_0^{\eta_{max}} \int_0^{\xi_{max}} H(\xi,\eta)d\xi d\eta \qquad (1.10)$$

Spatial bandwidth is the volume under the surface defined by $H(\xi,\eta)$. Multiplying the areas under the 1D horizontal and vertical MTF curves only works if $H(\xi,\eta)$ is separable in Cartesian coordinates. That is, the 1D plots only represent the system if $H(\xi,\eta)$ equals $H_\xi(\xi)$ multiplied by $H_\eta(\eta)$. Note that the separable assumption is almost never true for either scenes or transfer functions, and the 1D representations must be used with caution.

The following examples illustrate the nature of a good transfer function. These examples assume a thermal imager operating at 10 microns wavelength with a 5 degree field of view (FOV) and 512 by 512 pixels. Figure 1.10 shows two apertures with the same collecting areas; one is circular and the other is a ring. The ring aperture has a center obstruction that is the same size as the circular aperture. The circular aperture has a 3-millimeter (mm) diameter; resolution through both apertures is diffraction limited.

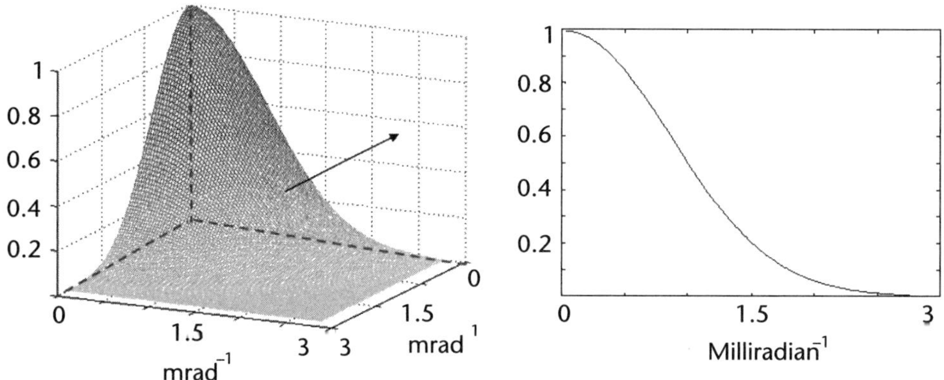

Figure 1.9 Illustration of 1D MTF showing the intersection of a Gaussian surface with the (MTF, horizontal frequency) plane.

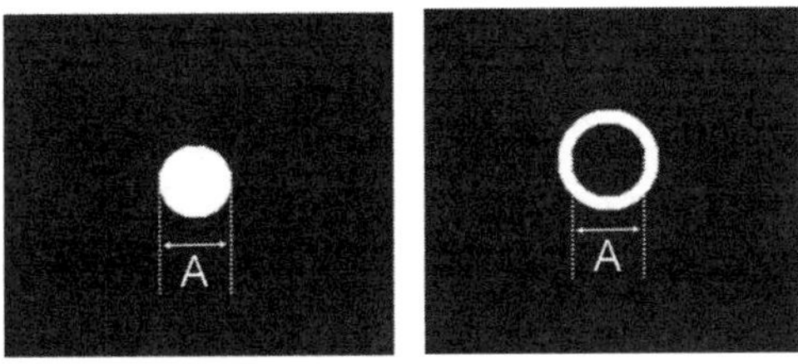

Figure 1.10   Circular aperture to the left has the same area as the central obstruction of the ring aperture to the right. Both apertures have the same collecting area.

Figure 1.11 shows the MTF for both apertures. The circular MTF is to the left, the ring MTF in the middle, and 1D plots of both to the right. The ring has a larger diameter than the circle, so diffraction cutoff is at a higher frequency. However, modulation transfer at low frequencies is poorer with the ring than the circle.

Figure 1.12 shows the MTF in Figure 1.11 applied to a thermal image of a tank. Both the root-mean-square (RMS) error and absolute error are about equal for the circular and ring apertures, even though the ring aperture provides a poor image. RMS error is a measure of energy transport, not image fidelity. In terms of RMS error, the higher frequency cutoff of the ring aperture compensates for the poor modulation transfer at lower frequencies, but image fidelity is hurt by the irregular band pass characteristics of the ring aperture. RMS error can be useful, however,

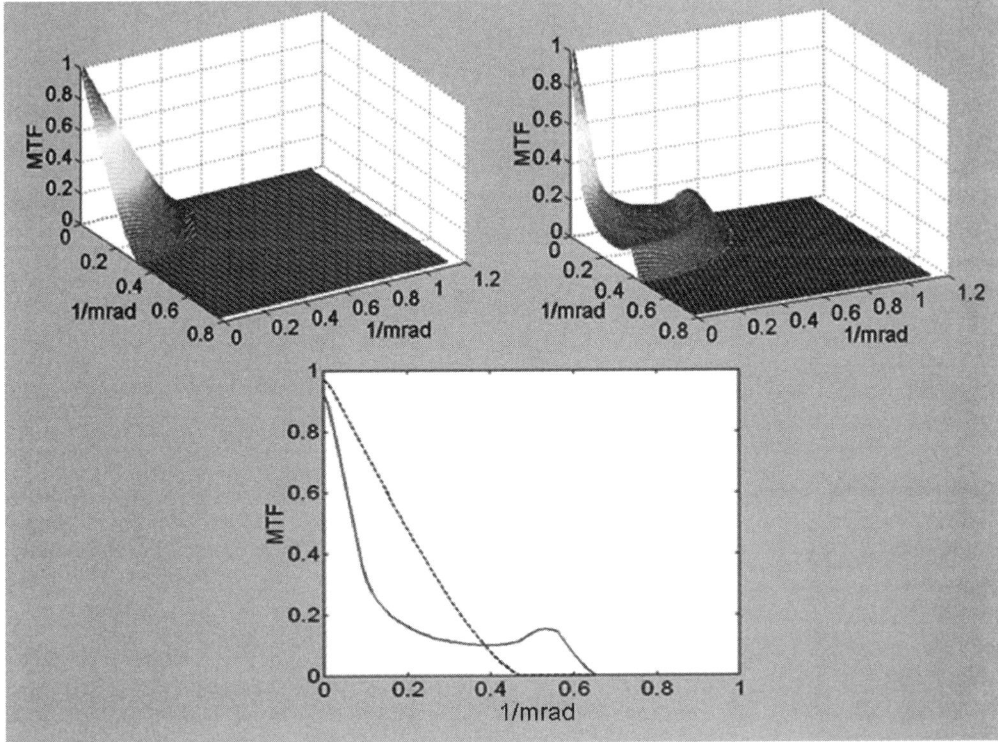

Figure 1.11   MTF of circular (left) and ring (right) apertures. A comparison of the two MTF is on the bottom.

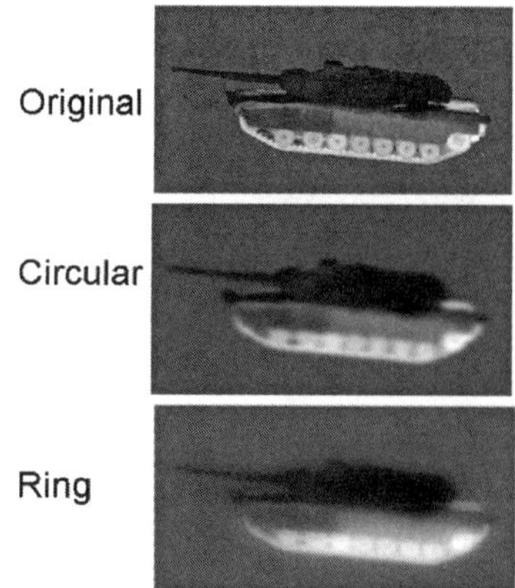

Figure 1.12   Images formed by applying the circular and ring MTF to the original image.

in quantifying the benefit of image restoration, provided that the comparison is not between different imagers.

As long as the noise is very low, both images can be digitally enhanced using Weiner restoration filters as described later in this chapter. The boost and restored MTF are shown in Figure 1.13; the restored images are shown in Figure 1.14. The total error with boost compared to without boost drops by a factor of forty for the ring MTF and a factor of twenty for the circular MTF. Most importantly, the ring aperture now provides a good image. The ring aperture only provides a better image than the circular aperture after the transfer response has been flattened using a digital restoration filter.

The clean boost shown in Figure 1.14 is only possible because of very low noise. An example with noisy imagery will illustrate the difficulty associated with restoring the ring aperture diffraction MTF. In the next example, the standard deviation of the detector noise is equal to the standard deviation of the target vehicle thermal contrast. The noisy images are shown at the top of Figure 1.15, and Weiner restored images are shown at the bottom of the figure. The various MTF are shown in Figure 1.16. The restored ring MTF shown in Figure 1.16 is not flat, and the restored

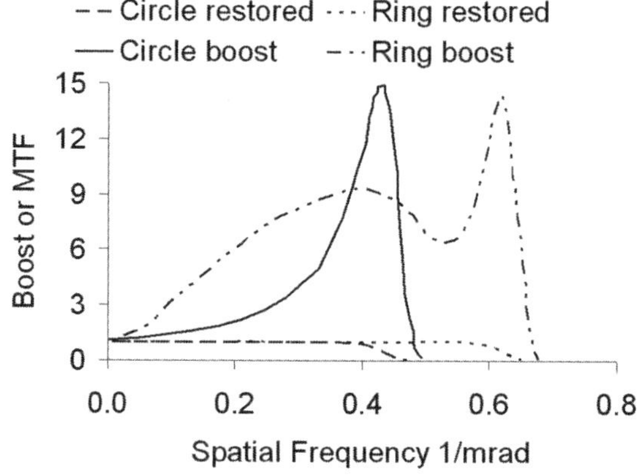

Figure 1.13   Boost and restored MTF for circular and ring apertures.

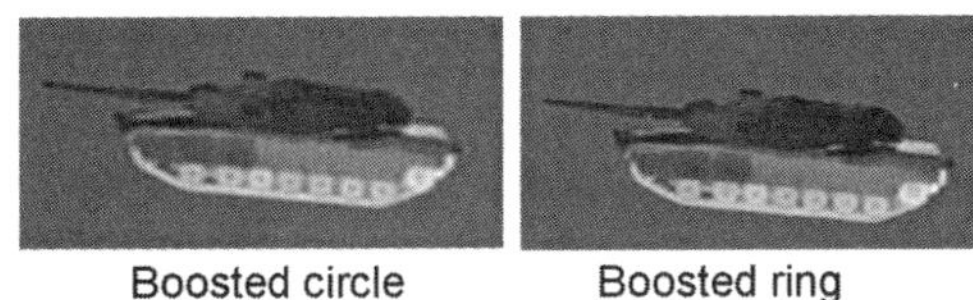

Figure 1.14 Restored images. The restored ring MTF provides a better image because the diffraction cutoff is higher, and the lower frequency passband has been boosted.

tank image at the bottom right of Figure 1.15 is quite poor when compared to the circular aperture image at the bottom left.

But there is at least one other option, and that is to apply a noise-suppression filter before restoration. Image blur occurs before the addition of detector noise, so that blurring affects the noise more than scene details. The noise is suppressed using a Gaussian blur prior to full restore using the restoration MTF shown in Figure 1.13. As seen in Figure 1.17, the resulting transfer response is not flat, but it has the Gaussian shape of the noise filter. The Weiner restored (Figure 1.16) and the

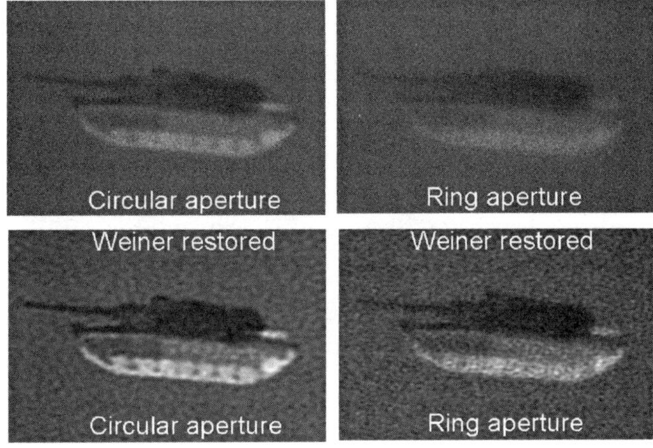

Figure 1.15 Thermal images of tank with detector noise added. The pictures at the bottom are restored using a Weiner filter.

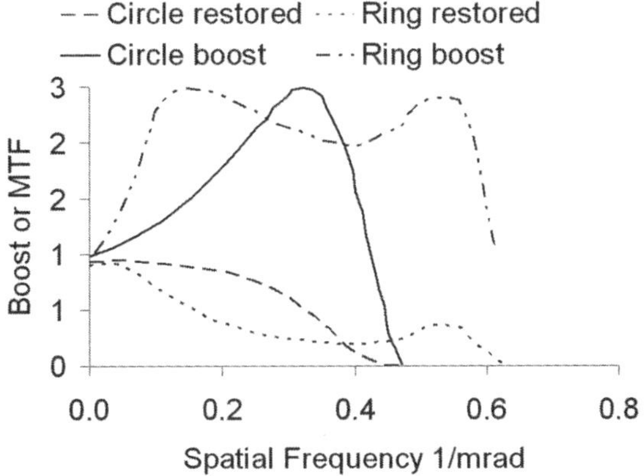

Figure 1.16 Restoration MTF used to restore noisy images.

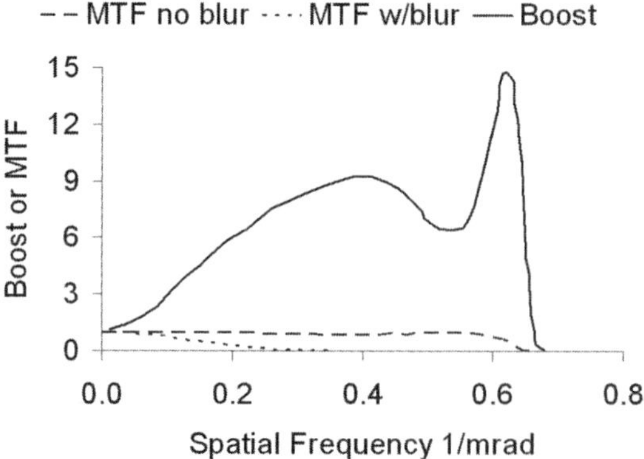

Figure 1.17 MTF for restoration of ring aperture with pre-blur to remove noise.

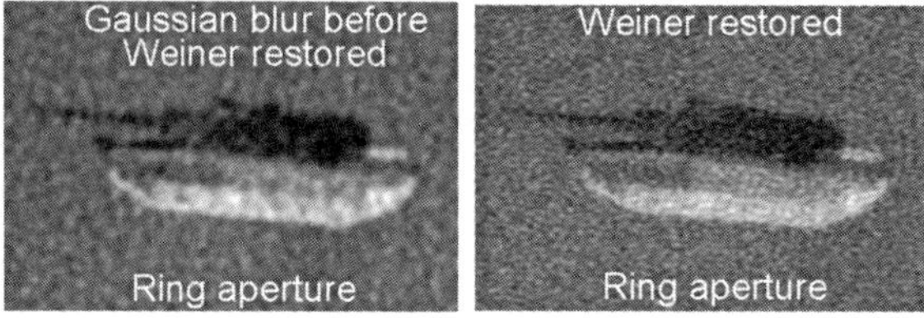

Figure 1.18 Comparison of restoration with preblur (at left) to restoration without preblur (at right).

preblurred and then Weiner restored (Figure 1.17) images are both shown in Figure 1.18. To our eyes, the preblur helped a little bit, but certainly not a lot.

The point of this example is that the higher diffraction cutoff of the ring aperture only benefits image quality when the band pass is flattened, and flattening the passband can be difficult in practice. The reduced modulation transfer at low spatial frequencies results in degraded imagery when noise is introduced.

In summary, extending the high frequency response of an imager is, of course, beneficial. However, extending the high frequency response at the expense of reduced modulation transfer at mid or low spatial frequency is generally not a good trade. Since display modulation cannot exceed one, boosting high spatial frequencies beyond the modulation lost in the optics is counter-productive, because the result is reduced display modulation at lower spatial frequencies. For general imaging purposes, the goal is to accurately render the scene, and that requires a system with MTF equal to one over as large a spatial frequency interval as possible.

EXERCISE 1.2

MTF boost is added to the thermal camera in Exercise 1.1. When the MTF in Exercise 1.1 are multiplied by the boost MTF, the transfer MTF in this figure results. Is display modulation at 5 cycles per milliradian greater than one? Has the scene-to-display contrast transfer at 2.5 cycles per milliradian been improved?

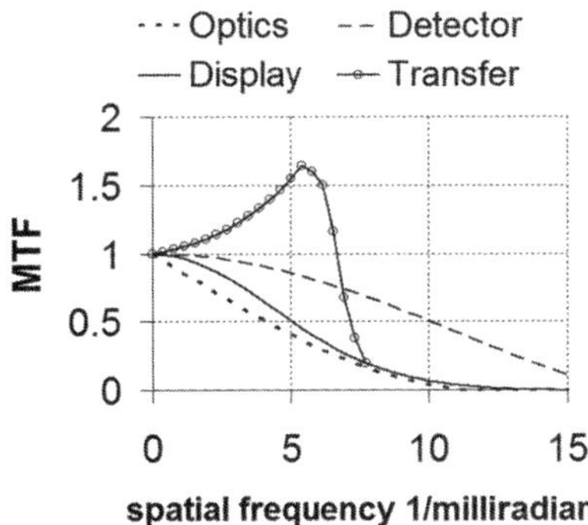

Figure Ex. 1.2

Solution

Display modulation = (max − min)/(max + min) and cannot be bigger than 1.0. The correct transfer response is normalized to a peak of 1.0. All spatial frequencies above approximately 2.3 cycles per milliradian have better modulation than in Exercise 1.1; the modulation at lower spatial frequencies degrades.

1.2.3   Sampling and Image Reconstruction

All EO imagers are sampled in at least one direction, and almost all modern imagers are sampled both horizontally and vertically. Since digital processing acts upon the samples, we must modify (1.1) through (1.8) to include the effects of sampling. This section discusses the sampling and image reconstruction processes. See Vollmerhausen (2010).

In Sections 2.1 and 2.2, the EO system is considered linear and shift invariant (LSI). LSI systems are well-behaved. For one thing, no frequency content appears on the display that did not originate in the scene. Spatial frequencies might be reduced, amplified, or shifted in phase by an LSI imager, but LSI systems are strictly filters or amplifiers and do not generate spurious frequency content. Sampled imagers are not LSI, and they do generate spurious signal called aliasing.

Linearity assumes that superposition holds; that is, if input A yields output A and input B yields output B then input $A + B$ yields output A + output B. To the same, reasonable approximation as any EO imager, sampled imagers are linear. However, shift invariance assumes that every point in the scene is treated alike by the camera, and sampled imagers are not shift invariant.

Figure 1.2 illustrates that all points in the imaged scene are not treated alike. The scene is focused by a lens on the FPA. Light falling on the active area of each detector generates photo-electrons. All of the light falling on a single detector is summed, and this does two things. First, the summing is a blur beyond that caused by diffraction and aberrations in the optics. It is a blur because signals from adjacent points in the scene are combined. Second, the photo-signal now represents an area and not a point. This is not just blur, because we associate the photo-signal from any angle within the detector instantaneous field of view with a discrete detector position. Light from closely adjacent points in the scene might fall on two different detectors, whereas light from two more distant points might fall on the same detector. Some light falls between detectors and generates no photo-signal at all. All points in the scene are not treated alike, and the sensor is not shift invariant.

BOX 1.3

The displayed image of a sampled psf is found by sampling the pre-sample blur (the optical blur h_{op} convolved with the detector blur h_{det}) and then convolving the samples with the display and digital processing blurs. The pre-sample blur h_{pre} includes all pre-sample blurs encompassing diffraction, optical aberrations, summing over the detector area, and any other blur that occurs before photo-detection.

$$r_{sp}(x,y) = \left[h_{pre}(x,y) \sum_{m=-\infty}^{m=\infty} \sum_{n=-\infty}^{n=\infty} \delta(x - nX, y - mY) \right] * p_{ix}(x,y) * h_{dig}(x,y) \qquad (1.11)$$

where X and Y are horizontal and vertical detector pitch in mrad and n and m are integers. A multiplication in space is a convolution in frequency, so the transfer function for a sampled imager $R_{sp}(\xi,\eta)$ is

$$R_{sp}(\xi,\eta) = P_{ix}(\xi,\eta) H_{dig}(\xi,\eta) \sum_{m=-\infty}^{m=\infty} \sum_{n=-\infty}^{n=\infty} H_{pre}\left(\xi - \frac{n}{X}, \eta - \frac{m}{Y} \right) \qquad (1.12)$$

Figure 1.19 uses 1D charts to illustrate what happens. Sampling replicates the presample MTF at all multiples of the sample frequency. The product of display and digital filter MTF forms the post-MTF. The presampled replica centered at zero frequency and multiplied by the post MTF is the transfer function $H(\xi,\eta)$ (see Figure 1.20). The sampled imager transfer function is shown in (1.13), and that equation is exactly the same as (1.7). The other presample replicas multiplied by the post MTF constitute aliasing $A_1(\xi,\eta)$; see (1.14) and Figure 1.20. The transfer function and aliasing taken together constitute the sampled imager response function $R_{sp}(\xi,\eta)$.

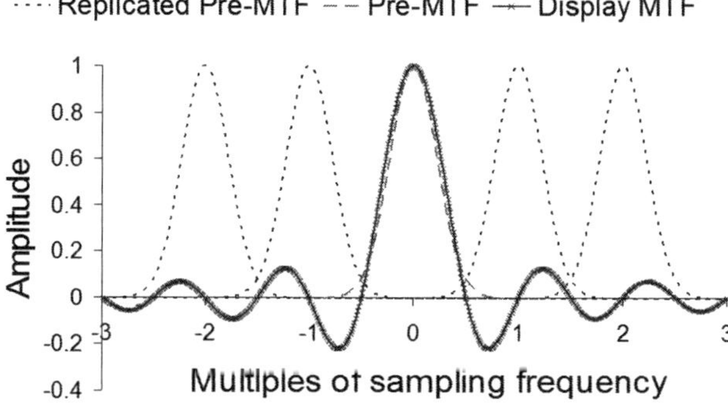

Figure 1.19 Sampling replicates pre-MTF at all multiples of the sample frequency. Some of the replicas are shown. Also shown is the MTF of a square display pixel that has the same size as the sample spacing.

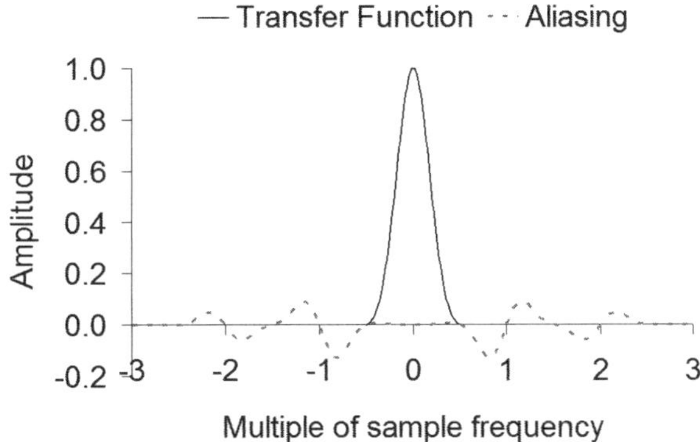

Figure 1.20 The transfer function (solid line) is obtained by multiplying the presample MTF by the pixel MTF. Aliasing (dotted line) is the product of the sampling replicas with the pixel MTF.

$$H(\xi,\eta) = H_{pre}(\xi,\eta)H_{dig}(\xi,\eta)P_{ix}(\xi,\eta) \tag{1.13}$$

$$A_l(\xi,\eta) = P_{ix}(\xi,\eta)H_{dig}(\xi,\eta)\sum_{\text{all }m\neq0}\sum_{\text{all }n\neq0} H_{pre}\left(\xi - \frac{n}{X},\eta - \frac{m}{Y}\right) \tag{1.14}$$

The signal at frequencies beyond the half-sample rate should not be ignored. That frequency content represents poor interpolation between the samples. If the pixel is large enough to see frequency content just below the half-sample rate, then the spurious content just above the half-sample rate is also visible. The pictures of Lena in Figure 1.21 illustrate that using multiple display pixels per sensor sample greatly improves image quality. Image quality improves when the aliasing content above the half-sample rate is removed by display interpolation.

Figure 1.22 compares the image frequency spectrum when using one square display pixel versus two horizontally and two vertically. The interpolations are bilinear and a seven coefficient interpolation kernel. Interpolation does an excellent job of removing out-of-band visible aliasing. (Out-of-band means that the spatial frequency content is above the half-sample frequency.) Using multiple pixels per sensor sample reconstructs spatial frequencies below the half-sample rate and filters out frequencies above the half-sample rate. The large, sharply demarcated pixels on the left in the figure interrupt the eye's ability to integrate the underlying scene.

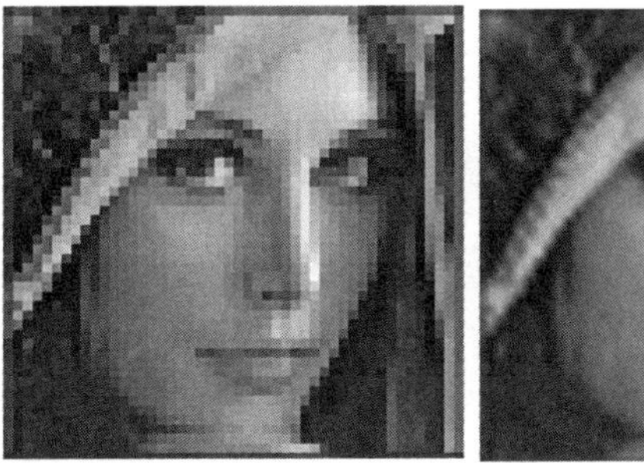

Figure 1.21 Picture to left uses one pixel per imager sample, whereas picture to right interpolates samples and uses multiple display pixels per sample.

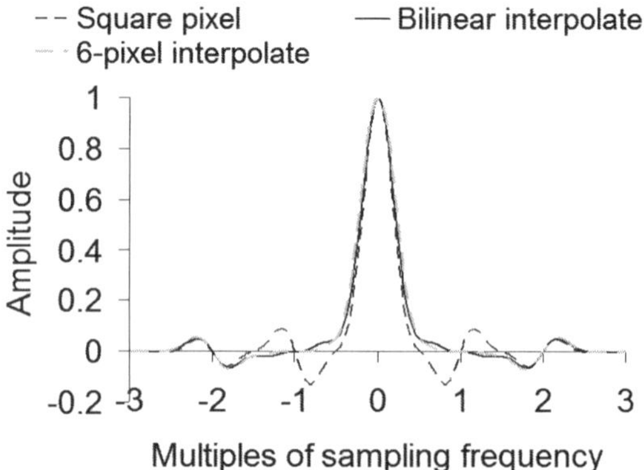

Figure 1.22 Image spectrum using one square pixel per image sample (dashed line). The bilinear (solid line) and seven coefficient interpolations (dotted line) use square pixels half the sample size. The large, sharply demarcated pixels (dashed line) generate visible edges that make it difficult for the eye to integrate the underlying imagery.

Aliasing beyond the half-sample rate does not contribute to scene comprehension and actually tends to hide useful details.

Figure 1.23 illustrates a different aliasing problem. At the top, as the distance between the clock face and the imager increases, the clock shrinks in image space and the frequency spectrum in mrad^{-1} spreads out. This figure illustrates that an object that is adequately sampled at close range might become poorly sampled at longer range. The clock numerals are adequately sampled at close range and become poorly sample as range increases.

The transfer response and aliasing of the camera used to image the clock are shown in Figure 1.24. Note that, because the MTF replicas substantially overlap the transfer response, there is substantial aliasing at spatial frequencies below the half-sample rate. Like detector noise, aliasing is generated within the imager and corrupts the signal. Unlike detector noise, the strength and frequency spectrum of aliasing depends on target structure, target range, and target contrast. The effect of in-band aliasing depends on range and on the particular visual task being performed.

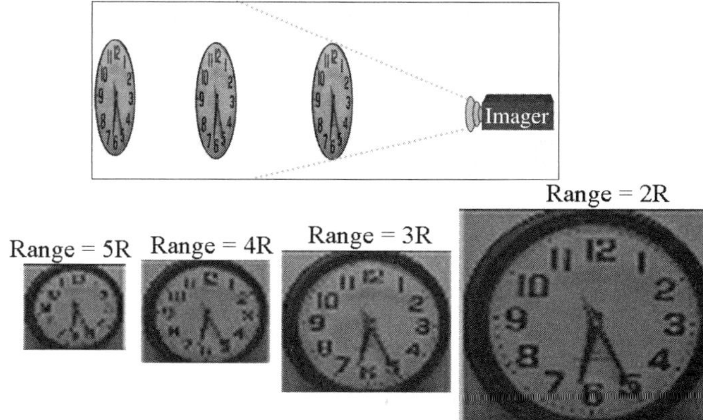

Figure 1.23 At the top, the clock face shrinks in image space as range increases. At bottom, the numerals become poorly sampled as range increases because they become smaller on the display and there are fewer samples per numeral.

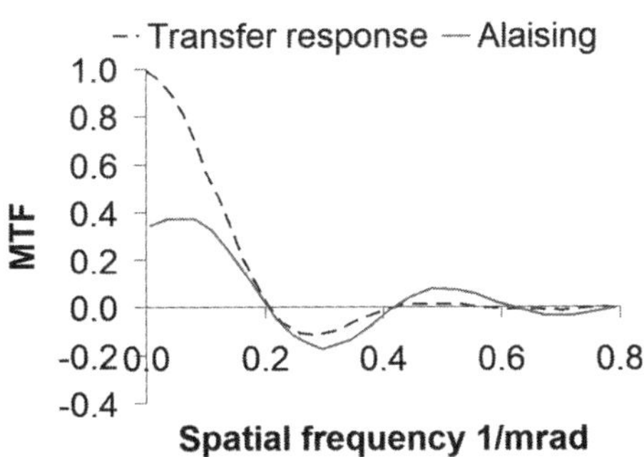

Figure 1.24   Transfer response and aliasing of camera used to photograph clock face.

Nothing in this section is meant to suggest to the reader that optical blur should be added to correct under-sampling of the image. Other than large amounts of noise, nothing hurts an image worse than blur. Image aliasing can be corrected by adding detectors to the FPA, and interpolation aliasing problems can be corrected by digital processing and adding display pixels. Of course, these approaches to correcting under-sampling might be impractical. Nonetheless, it is hard to imagine a practical situation where adding optical blur improves subjective image quality or helps any kind of visual task performance. The aliasing shown in Figure 1.24 can be removed by degrading presample MTF. However, the result would be a blurrier image at close range, and at long range the numerals would simply be too blurred to read. At long range, the numerals are under-sampled, and adding blur removes aliasing but does not correct the sampling problem.

There are two parts to the Sampling Theorem. First, sample the entire image at more than twice the highest spatial frequency present. Second, reconstruct the image by generating and summing sinc waves centered at each sample and with the sample amplitude. However, any band-limited signal cannot be spatially limited, and any spatially limited signal cannot be band-limited. Perfect reconstruction requires either sampling at an infinite rate or sampling an infinite interval. There will always be at least some in-band aliasing. (In-band aliasing causes corruption at spatial frequencies less than the half-sample frequency.) If in-band aliasing is significant, it tends to have greater impact on visualizing small or distant objects.

Also, we do not reconstruct using sinc waves but rather display pixels. Sinc waves have no frequency content beyond the half-sample rate, but display pixels have whatever shape the manufacturer selected, and the frequency spectrum is the Fourier transform of the pixel shape. The idea that frequencies above the half-sample rate do not exist in the image is simply not true, and the frequency content can be found using (1.14). Since the out-of-band frequencies are caused by the display pixels, that aliasing tends to have spatial structure that obstructs and even hides scene detail. The out-of-band aliasing is suppressed by using digital interpolation and two display pixels per sensor sample.

EXERCISE 1.3

The figure shows the component MTF and aliasing in a thermal camera. Explain how there can be aliasing beyond the half-sample frequency.

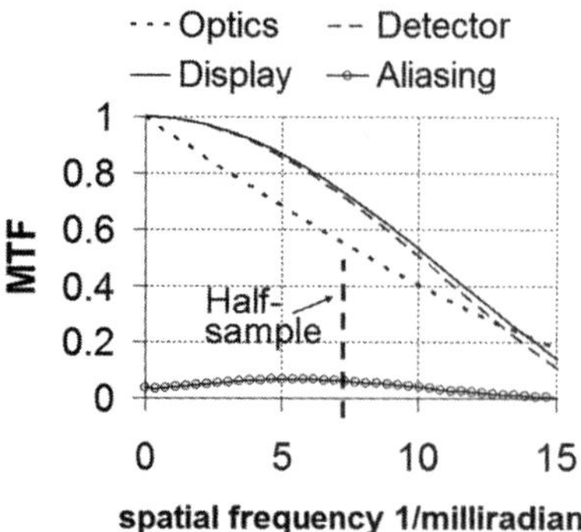

Figure Ex. 1.3

Solution

The Sampling Theorem reconstructs the sampled signal using sinc waves [sine(x)/(x)]; sinc waves have no frequency content beyond the half-sample frequency. However, display pixels do not have a sinc shape. Using 1D for simplicity, the frequency content of the image FT(x) is

$$\text{FT}(\xi) = \sum_{\text{all integer } m} I_m P_{ix}(\xi) e^{-2\pi m X \xi}$$

where I_m is pixel intensity, P_{ix} is the FT of the pixel shape, and the exponential places pixels at spatial intervals of X. That is, the FT of the displayed image is the sum of the FT of all of the pixels, each with its own intensity. We cannot draw conclusions based on the Sampling Theorem unless we actually *implement* the Sampling Theorem.

1.3   Digital Restoration Filters

Section 1.2 established the goal of image restoration. Namely, the transfer function should be as wide and flat as possible in order to render the scene with the best fidelity possible. All EO imagers are limited by diffraction even if optical aberrations are perfectly corrected.

Figure 1.11 shows the diffraction MTF of the apertures in Figure 1.10. Figure 1.12 shows the blurred imagery that results from diffraction. However, while it is true that no frequencies above the diffraction cutoff are available in the image, the modulation amplitude at frequencies below the diffraction cutoff can be amplified. Restoring the MTF of all frequencies below diffraction cutoff to a value of one results in the imagery in Figure 1.14. Clearly, boosting the high spatial frequencies results in much better imagery. We do not want to over-amplify any frequency, however, because the goal is to match the FT of the image to the FT of the scene.

This section describes a procedure for creating finite impulse response (FIR) deconvolution kernels. The kernels are obtained by windowing Wiener type inverse filters (WTIF). Implementing a high-gain WTIF in either the space or frequency domains can lead to artifacts that corrupt the image. Creating a FIR filter by windowing an infinite-duration, idealized filter is a well-known technique for designing digital filters; in this section, that technique is applied to image restoration. See Vollmerhausen (October 2010), Mersereau (1981), and Fiasconaro (1979).

In this section, imagery is assumed to be relatively noise-free. Assuming that the noise is small compared to signal minimizes the impact of the scene power spectrum on restoration filter design. Under low-noise conditions, the restoration filter is essentially an inverse filter that compensates for imager blur. Kernel values depend primarily on imager characteristics and less on the expected scene. Another simplifying assumption is that aliasing is unimportant.

Section 1.3.1 describes the generation of deconvolution kernels and discusses the benefit of limiting kernel size by using windows. In Section 3.4, fast Fourier transforms (FFT) are used to apply the deconvolution kernel.

1.3.1   Generation of Deconvolution Kernels

This section describes how to generate restoration kernels and discusses the use of windows to limit deconvolution kernel size while flattening the modulation passband.

Weiner type restoration filters are defined by (1.15). The derivation of the Weiner equation by Goldman (1981) is straightforward but provides details lacking elsewhere; also, see Jannson (1997) for a discussion of Weiner restoration and many other deconvolution methods. These filters restore an expected scene representing a broad average; all spatial frequencies are equally represented in the expected scene. Under low noise conditions, Weiner filters create a square band pass applied to the scene.

BOX 1.4

The scene power spectral density (psd) ϕ_s is taken as uniform over the band limit of the imager. The *cc* symbol denotes the complex conjugate.

$$R(\xi,\eta) = \frac{H^{cc}(\xi,\eta)\phi_s(\xi,\eta)}{H^2(\xi,\eta)\phi_s(\xi,\eta) + H^2_{post}(\xi,\eta)\phi_n(\xi,\eta)} \text{ for } \xi < \Omega_x \ \& \ \eta < \Omega_y \quad (1.15)$$

$$= 0 \text{ for } \xi \geq \Omega_x \text{ or } \eta \geq \Omega_y$$

where

$\quad\quad R(\xi,\eta) =$ the Wiener boost filter

$\quad\quad\quad \xi =$ horizontal spatial frequency in cycles per pixel (pixel^{-1})

$\quad\quad\quad \eta =$ vertical spatial frequency in pixel^{-1}

$\quad \Omega_x$ and $\Omega_y =$ horizontal and vertical band limits in pixel^{-1}

$\quad\quad H(\xi,\eta) =$ MTF from scene through the display

$\quad H_{post}(\xi,\eta) =$ MTF from noise insertion through the display

$\quad\quad\quad\quad \phi_n =$ noise power spectral density (the square of a square-law detector signal)

$\quad\quad\quad\quad \phi_s =$ scene power spectral density.

The Weiner theory assumes that the camera noise is Gaussian normal, is evenly distributed over frequency, is time varying but statistically stationary, and is independent of the scene. In that case, the image noise spectrum is set by camera post-MTF including the digital filter, display MTF, and the MTF of the eye. The Weiner theory further assumes that ϕ_n and ϕ_s are statistically independent.

Quite often, a camera engineer has little or no control over display MTF, image size, and viewing distance. Those factors establish $H_{\text{post}}(\xi,\eta)$ and are important to a Weiner optimization when the imagery is noisy. Further, the Weiner assumptions about ϕ_n and ϕ_s appear to only apply to thermal imagery where detector noise is caused by background flux. That is, the large background flux creates a random noise pattern over the field of view that is independent of the relatively small variations in thermal contrast that represent a typical thermal image. Under most contrast conditions, however, the noise in reflective imagers is proportional to the square root of signal intensity in each pixel; that means that ϕ_n and ϕ_s are not independent. Given the practical and theoretical problems with applying Weiner theory, we choose a more empirical approach and use the following form of the restoration filter

$$R(\xi,\eta) = \frac{H^{cc}(\xi,\eta)}{H^2(\xi,\eta) + \alpha^2} \tag{1.16}$$

where α is established by trial and error. See Vollmerhausen (October 2010) for a more detailed description of the derivation that follows.

To simplify the discussion, one-dimensional (1D) kernels are used except for an example at the end of the section. Two-dimensional equations and graphs make it difficult to convey concepts. In 1D, the psf of the deconvolution filter is $r(x)$. Note that, in this section, we are using pixel dimension as sample spacing and pixel^{-1} as frequency, so the sample spacing is one.

Since $R(\xi)$ is band limited at some spatial frequency less than 0.5 pixel^{-1}, (1.17) expresses $r(x)$ using the Sampling Theorem

$$r(x) = \sum_{j=-\infty}^{\infty} r_j \frac{\sin(\pi x - j\pi)}{(\pi x - j\pi)} \tag{1.17}$$

where the r_j are given by

$$r_j = \frac{1}{2\pi} \int_{-0.5}^{0.5} R(\xi)e^{i2\pi\xi j}d\xi \tag{1.18}$$

Equations (1.17) and (1.18) provide a theoretical solution that is not helpful in practice because the sum is infinite. We can limit the sum by multiplying $r(x)$ by a window function $w(x)$ yielding a new function $r'(x)$ with transform $R'(\xi)$. Since a multiplication in space is a convolution in frequency,

$$R'(\xi) = R(\xi) * W(\xi) \tag{1.19}$$

where $W(\xi)$ is the Fourier transform of $w(x)$. $R'(\xi)$ only approximates $R(\xi)$. The restored image is generated by convolving the blurred image digital values I_m at location m with a sampled $r(x)$. The restored image is sampled at the same locations as the original image. By the nature of the sinc function, the sampled value at each restored image location R_m is

$$R_m \approx \sum_{j=-K}^{K} r_j \Gamma_j I_{m+j} \qquad (1.20)$$

where Γ_j are window values. R_m is an approximation because we have truncated the restoration kernel to 2K+1 coefficients.

$R(\xi)$, of course, depends on the value of α. A large value of α limits the amount of blur restoration but also minimizes image artifacts. The point is that both α and $W(\xi)$ can be used to optimize restoration for any desired kernel size. Simply truncating the kernel at the desired size and using α to minimize artifacts is one approach. In that case, $w(x)$ is a rectangle (rect) function. Selecting $w(x)$ other than a rect function provides additional flexibility in designing the restoration kernel.

The utility of using a window function in addition to varying α is illustrated using the Gaussian blur defined by the equation

$$r_j = e^{-(j/8)^2} \qquad (1.21)$$

Figure 1.25 shows restored MTF for values of α from 0.004 increasing by a factor of two to 1.024. These curves are the product of the blur MTF by the restoration MTF. Each restoration kernel has 511 coefficients. These restoration kernels are not practical even if the processing power is available to apply this length of convolution because the edge effects corrupt much of the image. For this example, if the image is 512 by 512 pixels, then only a 2 by 2 pixel portion would be uncorrupted by edge effects. The restoration kernel must be shorter in practice.

Figure 1.26 shows restoration results for the same Gaussian blur but limiting kernel size to 31 coefficients. Multiplying the 511 coefficient array by a spatial rect function 31 pixels wide convolves the MTF $R(\xi)$ with a sinc wave function that is fairly wide in the frequency domain and that results in a badly degraded restoration. Of course, some blurs do not require a large kernel for good restoration, and truncating those kernels does not result in poor restorations. Nonetheless, for this example and in the general case, applying a rect $w(x)$ results in significant image artifacts unless a large α is used, and a large α blurs the image.

We now explore the benefits of using windows other than the rect function. In other words, the 511 coefficient restoration kernel is multiplied by something other than zeros and ones. For the moment, the alternative $w(x)$ are referred to as

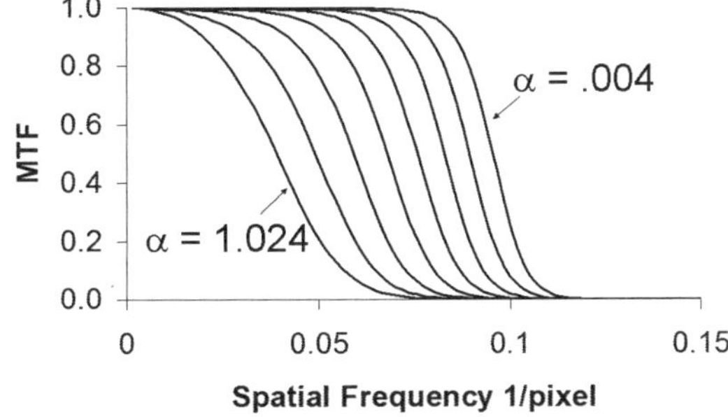

Figure 1.25 Each curve represents the Gaussian blur MTF multiplied by a restoration MTF calculated using (1.16) through (1.20). The parameter α is varied from 0.004 to 1.024 increasing by a factor of two. Notice that restoration becomes less and less effective as the value of α increases.

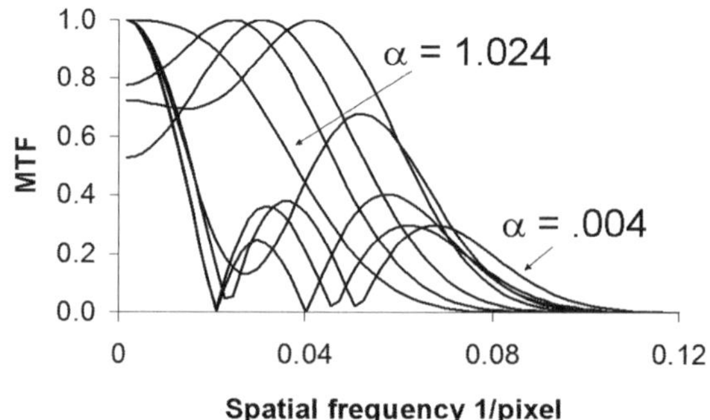

Figure 1.26 Each curve represents the Gaussian blur MTF multiplied by a restoration MTF calculated using (1.16) through (1.20). The curves in Figure 1.25 use a 511 coefficient restoration kernel, and these curves use a 31 coefficient restoration kernel. The parameter α is varied from 0.004 to 1.024 increasing by a factor of two. Notice that restoration with these kernels creates bad image artifacts unless α is very large and high spatial frequencies are not enhanced.

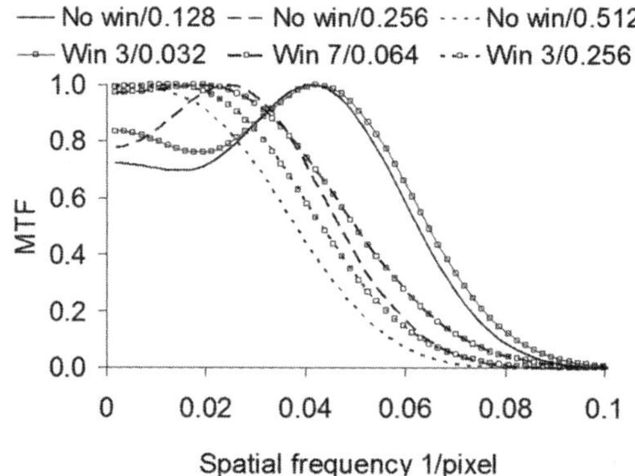

Figure 1.27 Comparison of restored MTF with and without windowing. The nature of Win 3 and Win 7 is explained in the next section. The numbers after the slash (/) are α. The alternative windows allow the use of smaller α while still achieving a flat passband along with a higher frequency cutoff.

windows 1 through 7. Suitable window functions are described in Section 1.3.2. Figure 1.27 compares the restored MTF for three truncated kernels and three kernels using selected windows. The window examples are selected to have the same or greater cutoff frequency with a flatter passband.

Note that use of the window functions tends to flatten the MTF for each α, such that effective restoration is achieved at greater cutoff frequencies when using the alternative windows. Put another way, ripples in the restoration MTF cause image artifacts, and one significant cause of those ripples is truncation of the restoration kernel.

Figures 1.28 through 1.31 show original pictures, blurred pictures, pictures restored using a rect window and an α of 0.128, and pictures restored using Win 3 and an α of 0.032. The restored MTF are the right-hand curves in Figure 1.27. On the average over all of the pictures, using the Win 3 window results in 40 per cent less root-mean-square error. Also, use of the alternative window to create the restoration kernel results in subjectively more pleasing imagery.

Figures 1.32 and 1.33 show restorations using a K equal to 8 (17 coefficient kernels) and α equal to 0.002. A rect window is used in Figure 1.32 and a Win 2

Figure 1.28 Original pictures.

Figure 1.29 The pictures in Figure 1.28 blurred with the kernel given by (1.21).

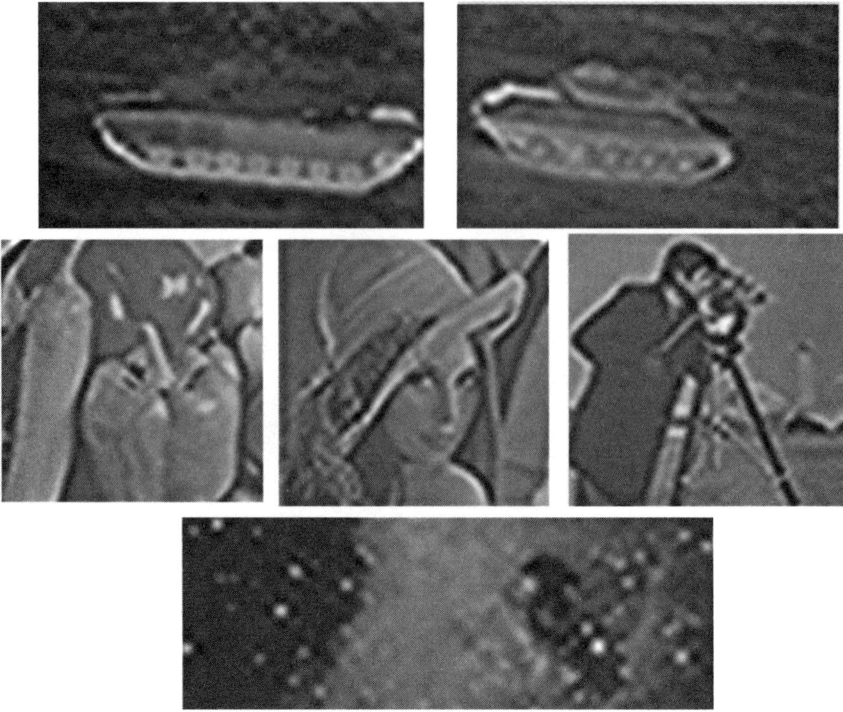

Figure 1.30 Pictures in Figure 1.29 restored using (1.16) through (1.20) to generate a 31 coefficient kernel ($K = 15$) and a rect function as $w(x)$.

Figure 1.31 Pictures in Figure 1.29 restored using (1.16) through (1.20) to generate a 31 coefficient kernel ($K = 15$) and one of the windows described in the next section as $w(x)$.

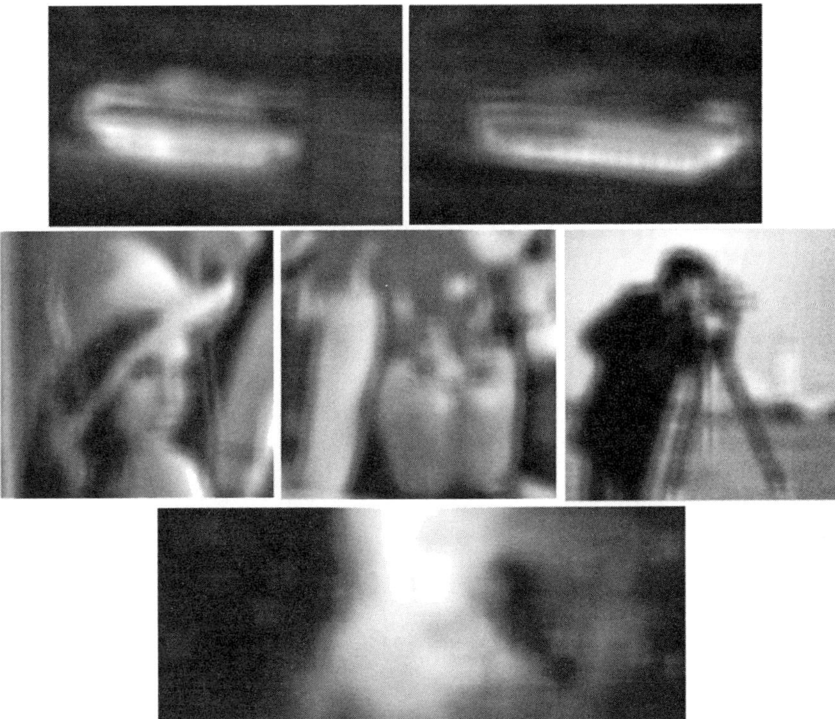

Figure 1.32 The blurred images in Figure 1.29 restored using K equal to 8, α equal to 0.002, and a rect window. Truncating the window to 17 coefficients causes stark artifacts.

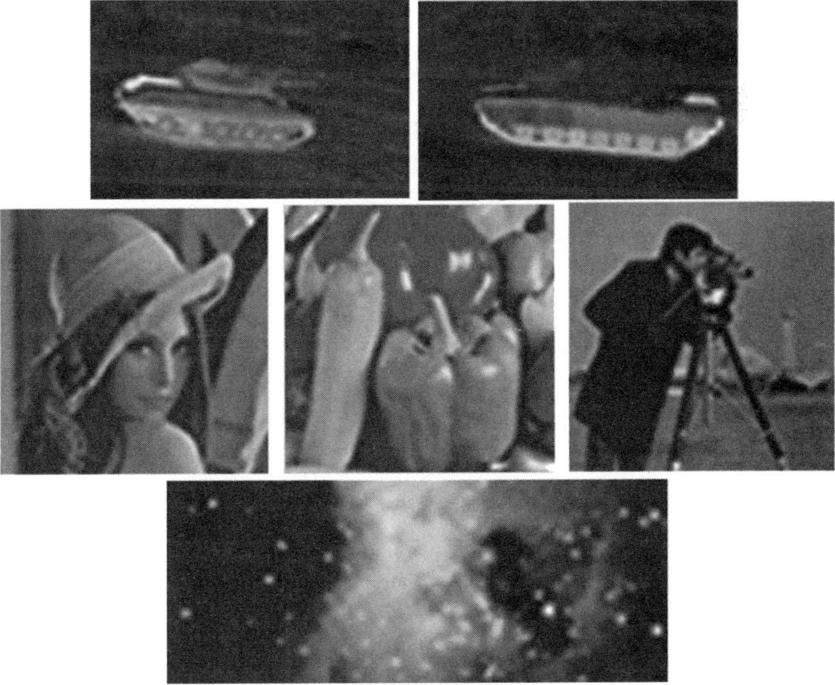

Figure 1.33 The blurred images in Figure 1.29 restored using K equal to 8, α equal to 0.002, and a nonrect window. The nonrect window eliminates the artifacts seen in Figure 1.32.

window in Figure 1.33. Figure 1.32 illustrates the stark image artifacts caused by truncating a deconvolution kernel, and Figure 1.33 shows the benefit of using a nonrect function window to limit kernel size.

Some restoration kernels are small by nature, and the FIR design technique using nonrect windows is not needed. At other times, however, using the windows described in the next section can significantly improve restoration results. FIR restoration kernels can be optimized by varying Γ_j in (1.20) as well as by varying α in (1.16).

EXERCISE 1.4

A Gaussian blur has a half-width to $1/e$ of 4 pixels. Find the central 61 coefficients of the restoration kernel when α equals 0.004, the window is a rect function, and use a 511 element discrete Fourier transform.

Solution

- Create a 511 element array (A_{rray}) with the Gaussian blur at the center. The amplitude A_j of the jth array element is

$$A_j = e^{-(256-j)^2/16}$$

- Take the DFT of the array (DFT_Array) and normalize the peak amplitude to 1.
- Find the DFT of the restoration kernel

$$\text{DFT\_Restore} = \frac{\text{conjugate(DFT\_Array)}}{\text{conjugate(DFT\_Array) .* (DFT\_Array)} + (0.004)^2}$$

where .* means to multiply the arrays element by element.
- The result is shown in the figure.

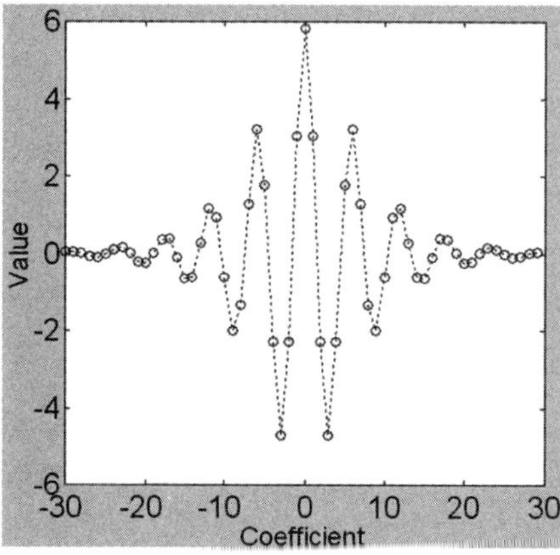

Figure Ex. 1.4

1.3.2   Useful Window Functions

Prolate spheroidal wave functions (PSWF) are selected as window functions because they concentrate the largest possible fraction of band limited energy onto the minimum spatial interval.

BOX 1.5

Consider a *psf* with shape $f(x)$ and Fourier transform $F(\xi)$. Define θ and β

$$\theta^2 = \int_{-X}^{X} f^2(x)dx \bigg/ \int_{-\infty}^{\infty} f^2(x)dx \tag{1.22}$$

$$\beta^2 = \int_{-\Omega}^{\Omega} F^2(\xi)d\xi \bigg/ \int_{-\infty}^{\infty} F^2(\xi)d\xi \tag{1.23}$$

where X is a given size in pixels. If either θ or β is specified, the other must remain below a certain value depending on the product ΩX. For each choice of θ or β, the product $\theta\beta$ is maximized by making $f(x)$ a selected PSWF Ψ. Papoulis (1962) provides concise descriptions of the relevant properties of these functions.

The Ψ are functions of $2\pi\Omega X$. Each Ψ must be scaled spatially to have the same band limit Ω pixel^{-1} as the imagery. Define a new set of functions ψ (see Vollmerhausen, October, 2010)

$$\psi(\gamma, x) = \psi(\gamma, 2\pi\Omega x/\gamma) \tag{1.24}$$

The Fourier transform $\Gamma(\gamma, \xi)$ of $\psi(\gamma, x)$ is given by

$$\Gamma(\gamma, \xi) = \psi(\gamma, \gamma\xi/2\pi\Omega)/\psi(\gamma, 0), \quad |\xi| < \Omega \tag{1.25}$$

$$\Gamma(\gamma, \xi) = 0, \quad |\xi| \geq \Omega$$

Notice that the Fourier transform is a truncated version of the wave function itself.

The Γ are the window functions, and the Ψ are the Fourier transforms of the window functions. This is because the window function must be spatially limited. So the window functions are $\Gamma(\gamma, x)$ where x is in pixels.

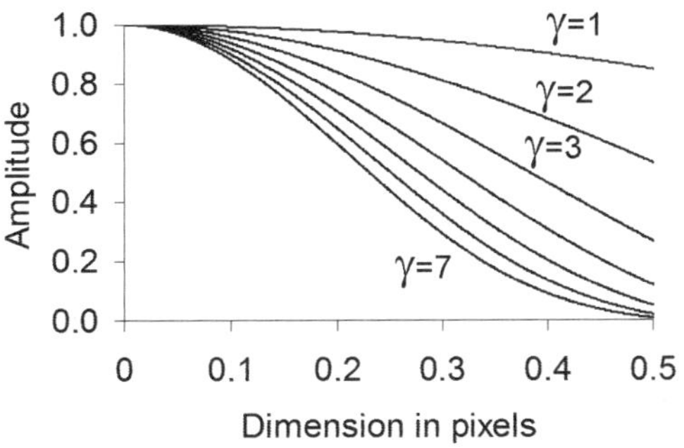

Figure 1.34   PSWF windows. The windows drop to zero amplitude at 0.5.

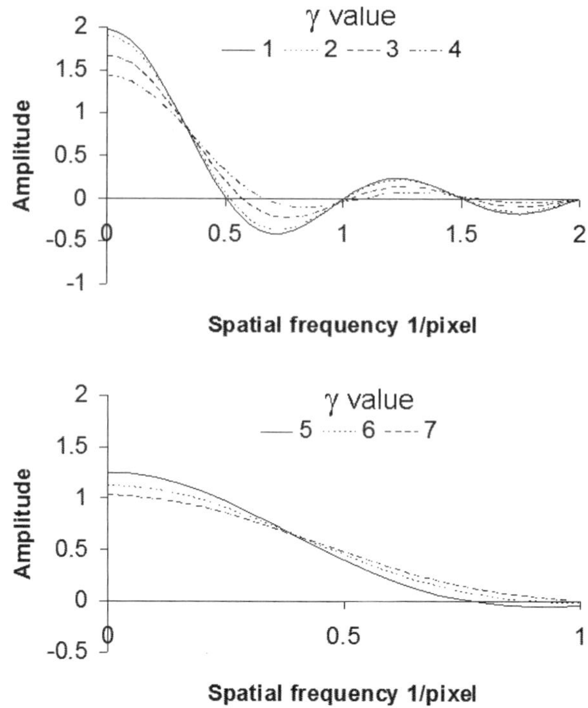

Figure 1.35　Fourier transforms of windows shown in Figure 1.34. Note that the FT are constrained in the frequency domain even though the windows themselves are zero beyond 0.5 pixel.

The window functions Γ for γ one to seven are shown in Figure 1.34. The Fourier transforms are shown in Figure 1.35. Since Γ is spatially limited, the Fourier transform cannot be band limited. However, notice that there is little content beyond the half-sample frequency. When Γ is scaled to create bigger windows with more kernel elements, the Fourier transform of Γ is even more constrained in the frequency domain. PSWF are selected for the windowing application because they provide low-amplitude side-lobes for a given size of central blur. PSWF windows are constrained in space but also effectively band limited.

1.3.3　Restoration Using Fast Fourier Transforms

Most optical blurs are not separable in Cartesian coordinates, and restoration cannot be applied as sequential convolutions of horizontal and vertical kernels. Convolving an N by N separable kernel with an image requires $2N$ multiplications and $2N - 2$ sums per restored pixel. Applying an N by N nonseparable kernel requires N^2 multiplications and $N^2 - 1$ sums per pixel. The 31 coefficient kernel used as an example in the last section would require 961 multiplies and 960 sums for every restored pixel if applied as a 2D kernel. Using an FFT to apply the restoration filter is computationally much more efficient.

Within the valid region of a convolution, using the discrete transform provides the same output values as (1.20).

That is true even though convolutions by multiplying DFT transforms are cyclic; the cyclic nature of a discrete transform result in different edge effects. The FFT

Figure 1.36   Red, green, and blue psf when the camera is focused at 15 feet and the light source is at 4 feet.

can be used to apply a restoration kernel. One must be cautious, however, about limiting the *kernel size* when using discrete transforms to apply restoration.

In the following, *image_DFT* is the DFT of the blurred image, and *blur_DFT* is the complex conjugate of the DFT of the camera *psf*. Since this is a color example, each step applies to one color. That is, both the *image_DFT* and the *blur_DFT* are different for each color, and the restoration is done separately for each color slice. The *blur_DFT* is found by photographing a (essentially) point source of light in a dark box and taking the discrete transform. The red, green, and blue blurs are shown in Figure 1.36.

Once the DFT of the blur is measured or predicted:

- Find the Wiener filter R using the formula

$$R = \frac{blur\_DFT}{\left|blur\_DFT\right|^2 + a^2}$$ (1.26)

- Take the inverse DFT of R.
- Apply (multiply by) a window centered on the kernel.
- Take the DFT of the windowed kernel to get R'.
- The restored image is the inverse transform of the product of R' multiplied by *image_DFT*.

In Figure 1.37, Phyllis is standing 4 feet from the camera, which is 15 feet from the bookcase in the background. In (a) at the top, the books are in focus but Phyllis's

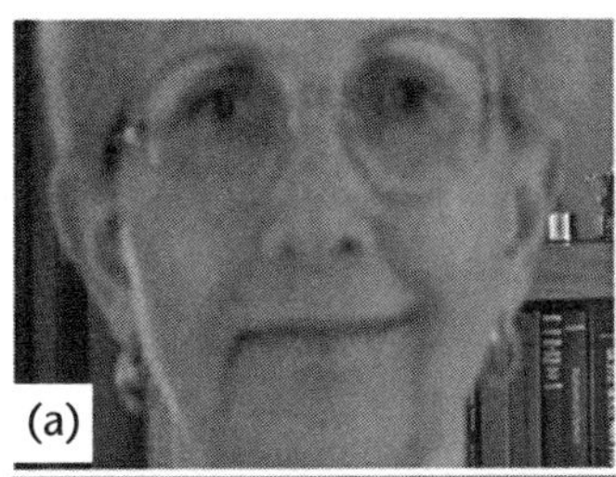

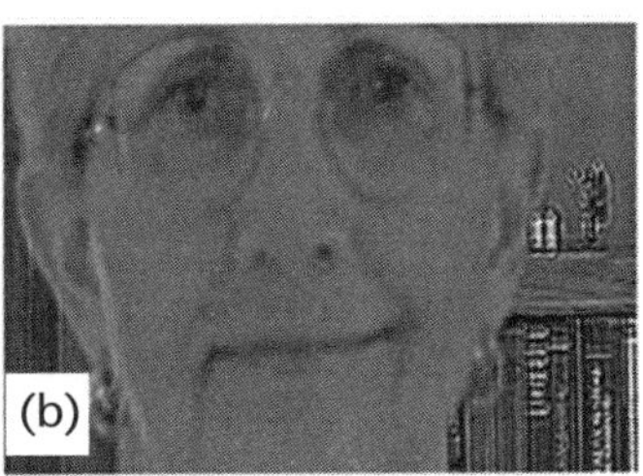

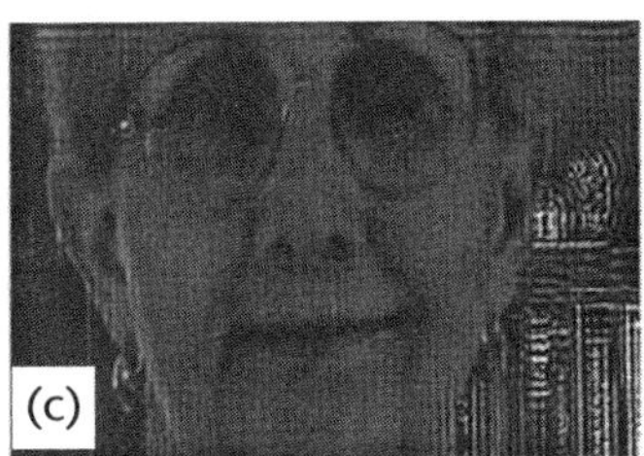

Figure 1.37   The face in (a) is blurred because the camera is focused on the books in the background. In (b), the focus blur is corrected by applying restoration using the procedure described earlier. In (c), the procedure is used without multiplying by a window, and edge effects corrupt the whole picture.

face is not. This is because the camera is focused at 15 feet. Image (b) is restored using this procedure and a 131 element PSWF/7 window. Note that the face, glasses, and earrings are in focus, but the books are now corrupted. Restoration has corrected focus to 4 feet, and the books are at 15 feet. Picture (c) is restored without using a window. The lack of a window results in image artifacts.

The point of this procedure is not so much the application of a nonrect window, although the benefits of a good window function still apply. *Rather, the point is that a window is necessary to ensure size of the valid region of the convolution.*

EXERCISE 1.5

Using numerical integration, find the difference between the magnitude of the FT of the analog Gaussian and the magnitude of the FT of the Gaussian samples used in the Exercise 1.4. Is there much difference? What does this indicate?

Solution

The FT of the analog Gaussian is

$$FT_{\text{Gaus}}(\xi) = \int_{-\infty}^{\infty} e^{-x^2/16} e^{-2\pi i x \xi} dx$$

where x is noninteger dimension in pixel space with x equal to 0 at pixel 256.

One sample is an impulse function; the FT of the Gaussian samples is the sum of the FT of all the impulse functions. The FT of the samples of the Gaussian is

$$FT_{\text{samps}}(\xi) = \sum_{j=-255}^{255} A_j e^{-2\pi i j \xi}$$

where A_j is given in Exercise 1.6.

The normalized magnitudes of both FT are plotted in the figure. One curve is visible because the two curves overlap; there is no visible difference. This indicates that the Gaussian is well sampled, and that using (1.17) and (1.18) does not create errors in this particular case.

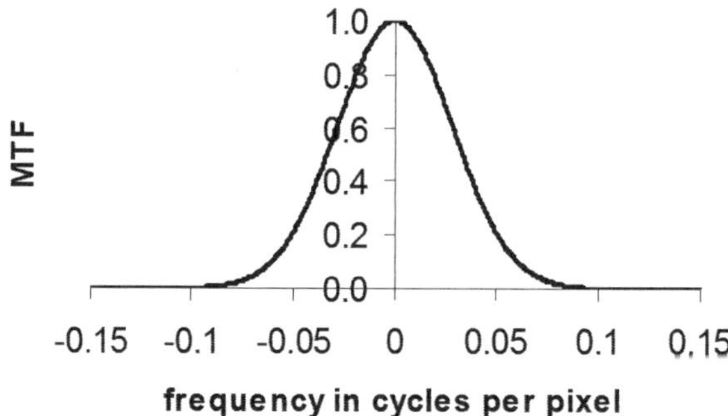

Figure Ex. 1.5

EXERCISE 1.6

For the Gaussian blur defined in Exercise 1.4, find the 61 element restoration kernel using

$$FT_{\text{samps}}(\xi) = \sum_{j=-255}^{255} A_j e^{-2\pi i j \xi}$$

and plot the coefficient values along with the Exercise 1.4 result. Are they the same? If not, why not?

Solution

FT_{samp} replaces DFT_Array in Exercise 1.4; otherwise, the procedure for finding the deconvolution coefficients is the same. The coefficient values for the two kernels are plotted in the figure; the circles are DFT values, and the (x) are FT_{samp} values. The deconvolution kernels are not the same. In the figure, the two deconvolution kernels appear more different than they are, because the sum of the coefficients for each kernel sums to one. In other words, the large swings in amplitude cancel. Nonetheless, the two procedures provide different answers.

A DFT finds frequency content at discrete locations in the frequency domain, and the value at each discrete frequency does not represent an accurate average over the nearby frequency interval. Also, a DFT assumes periodicity, but that is not a problem in this particular case because the Gaussian is at the center of the array.

Experience suggests that the DFT approach provides reasonable approximations to the FT_{samp} formula. Nonetheless, the kernels are somewhat different, and it is always safest to use the FT_{samp} formula to calculate deconvolution kernel coefficient values.

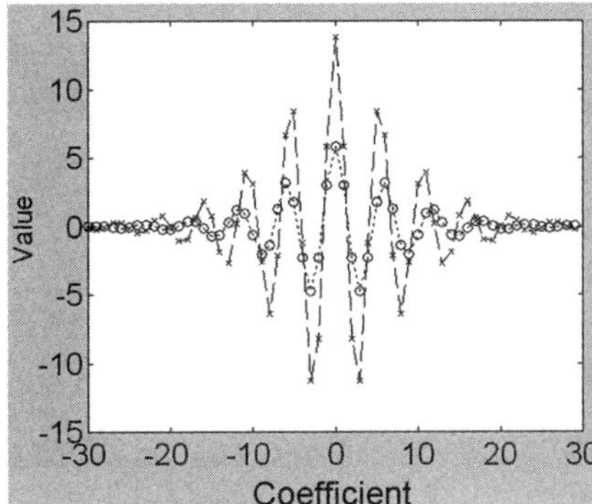

Figure Ex. 1.6

1.4   Local Area Contrast Enhancement

Local area contrast enhancement (LACE) has multiple uses. It adjusts display contrast and brightness automatically, relieving the operator of that tedious task. It adapts the wide dynamic range of cameras to the lower dynamic range of displays. Also, the contrast of each area of the image is adjusted (more or less) independently, so that scene details are more visible in both light and dark areas. This section starts with a description of the need for LACE; LACE algorithms for thermal and reflective imagers are discussed in separate subsections.

Note that LACE solves a display problem; LACE only works if sufficient gray scale is provided by the camera. LACE becomes a camera issue when a thermal or reflective imager produces real time video, because the display must be used real time without the need for the operator to continually adjust display contrast and brightness.

The dynamic range of modern cameras is 10 or 12 bits, meaning the camera can render up to 4,000 shades of gray. However, the smallest contrast distinguishable by the eye is one part in 500, and that contrast threshold is only achieved at spatial frequencies near 0.25 cycles per milliradian. Lower or higher spatial frequencies require higher contrast to be visible. Further, modern displays are not capable of rendering small changes in contrast. Typical digital display interfaces limit even large area contrast rendition to seven or eight bits, providing at most 256 shades of gray. Cameras provide more shades of gray than can be rendered on a display or interpreted by the eye.

Further, improving scene contrast improves viewing even when the entire image is encompassed within the dynamic range of the display. Consider Figure 1.38. The scene in (a) is lit by a street light not far from the camera. The nearby ground is well lit, but illumination on the bridge is poor. Applying LACE results in the picture in Figure 1.36(b); details in the distant scene are much easier to see.

Thermal imagery provides excellent examples of the need for contrast enhancement, because everything in the terrestrial environment radiates heat. Figure 1.39 is a thermal image of a tank. Figure 1.39(a) shows scene radiance displayed with an absolute intensity scale; each pixel intensity represents scene radiance at that point. The scene is very low contrast. For example, although the tank engine is hot, it radiates only 5% more energy in the 8 to 12 micron spectral band than the terrain background. See (c) for a plot of the radiance along the dashed line in (a) and (b). In Figure 1.39(b), only differences in contrast are displayed. That is, the minimum intensity pixel in (a) establishes the black level in (b).

The uniform intensity pedestal in (a) lowers display contrast and obscures scene details. Removing the pedestal improves image contrast. In (b), the mapping of scene radiance to display luminance is no longer one to one, but the image is

Figure 1.38   Applying LACE to (a) on left results in picture (b) to right. The original image in (a) is a simulation with eight bits per color.

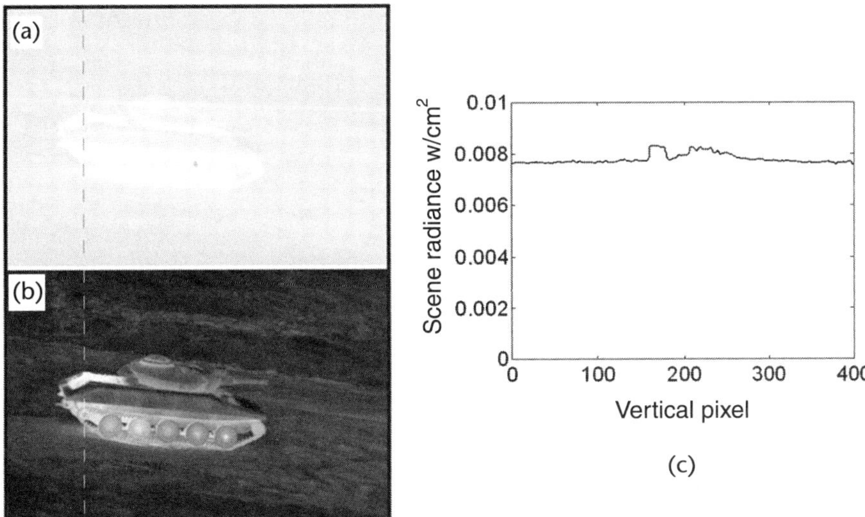

Figure 1.39 Thermal image of a tank. Absolute radiance is displayed in (a) at the top left, and only differences within the scene are imaged in (b) at the bottom left. To the right in (c), radiance along the dashed line is plotted to show that even hot areas on the tank are low contrast.

better in the sense that the observer has better vision of scene details. Removing the thermal pedestal compensates for limitations in the display and eye. Absolute intensity information is lost in order to enhance viewing of the thermal contrast (that is, radiance differences) within the scene.

In Figure 1.40(a), atmospheric path radiance causes a large increase in the thermal pedestal from the bottom to the top of the picture. The display controls are set such that no area is saturated, and that lowers the apparent contrast of the distant target. In Figure 1.40(b), a high pass filter is applied to the scene. The need to apply the filter is out of our control, because both path radiance and display and eye contrast limitations are physical realities. Nonetheless, some amount of information about the target vehicles has been lost by applying the high pass filter.

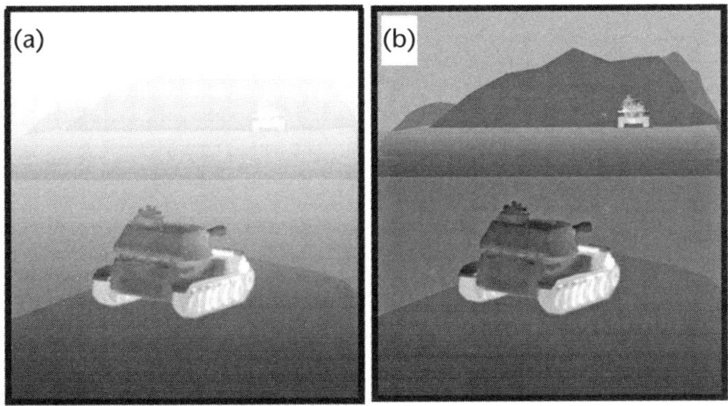

Figure 1.40 In (a) at left, atmospheric path radiance creates a large signal swing from the bottom to the top of the picture. Imager gain is reduced in order to prevent saturation. In (b) at right, a high pass filter removes the low-frequency radiance component. Increased imager gain enhances the view of the distant target.

However, the advantages of contrast enhancement greatly outweigh the effect of the lost information content.

Different LACE algorithms are used for reflective and thermal imagers. Intensity in a thermal image is caused by radiated heat. Intensity differences in a reflective imager are caused by both varying illumination and variations in scene reflectance. A portion of a reflective image might be dark because it is in shadow, or it might be dark because the surface is not reflective. Thermal and reflective imagers require different approaches to LACE.

1.4.1 Local Area Contrast Enhancement Algorithms for Thermal Imagers

Thermal terrain signatures vary widely from day to night and depending on weather, but they also vary on a given day within a fraction of a mile from a single location. Perhaps the sun preferentially heats a rise in the road more than a shaded, grassy glen. Thermal signatures can vary widely when looking different directions from one vantage point. Early thermal imagers had manual gain and level controls, and they required continual adjustment.

Another problem is that the hump in the road might be in the same field of view as the grass field. That would not be a problem if we were interested in hills and grass, but the thermal signature of a man might be small compared to the road versus grass. Modern thermal imagers must handle two problems. First, we cannot expect the user to continually fuss with gain and level. Second, regardless of the wide dynamic range of the thermal image, we want to see the road, the grass field, and the man.

Quite often, the dynamic range of the camera output exceeds the dynamic range of the display. And even when the dynamic range of the camera and display are compatible, the thermal pedestal illustrated by Figure 1.39(a) tends to move up and down, necessitating control adjustments. LACE overcomes both of these display problems.

The most widely accepted solution is to high-pass filter the thermal image. The following is a computationally manageable procedure.

- Divide the image into N by N blocks where N might be as few as 3 or as many as 50 or so. Take the mean of each block. The median or minimum or some other statistic can be substituted for the mean. Note that this procedure is equivalent to convolving an N by N kernel of ones and then down-sampling.
- Use bilinear or preferably bicubic interpolation to generate a full-size image from the block averages.
- Subtract most (perhaps three-quarters) of the interpolated low pass image from the original; that will remove gradual variations of thermal contrast.
- Restore the video black level and gain by setting the minimum and maximum image intensities based on the standard deviation. Perhaps the minimum is set to the mean minus two or three standard deviations, and the maximum intensity is set to the mean plus two or three standard deviations.

Using a block size of three by three, this procedure is an approximation to convolving with a kernel

$$\begin{pmatrix} -1 & -1 & -1 \\ -1 & 8 & -1 \\ -1 & -1 & -1 \end{pmatrix} = \begin{pmatrix} -1 & -1 & -1 \\ -1 & -1 & -1 \\ -1 & -1 & -1 \end{pmatrix} + \begin{pmatrix} 0 & 0 & 0 \\ 0 & 9 & 0 \\ 0 & 0 & 0 \end{pmatrix} \qquad (1.27)$$

where (1.27) subtracts all of the low frequencies. Three-quarters of the low frequencies are subtracted by the kernel

$$\begin{pmatrix} -1 & -1 & -1 \\ -1 & 11 & -1 \\ -1 & -1 & -1 \end{pmatrix} \qquad (1.28)$$

Generally, a three by three kernel is too small because too much of the image frequency content is removed; the block averaging procedure is computationally more efficient.

Note that, if a kernel is used—(1.28) rather than the block averaging procedure—then LACE might provide the opportunity to add boost. Both the block procedure and kernels like (1.28) tend to generate image artifacts. Deconvolution requires more processing power but provides a flat passband.

In Figure 1.41(a), simulation is used to create a scene with known thermal contrast characteristics. Atmospheric path radiance increases by three times the average

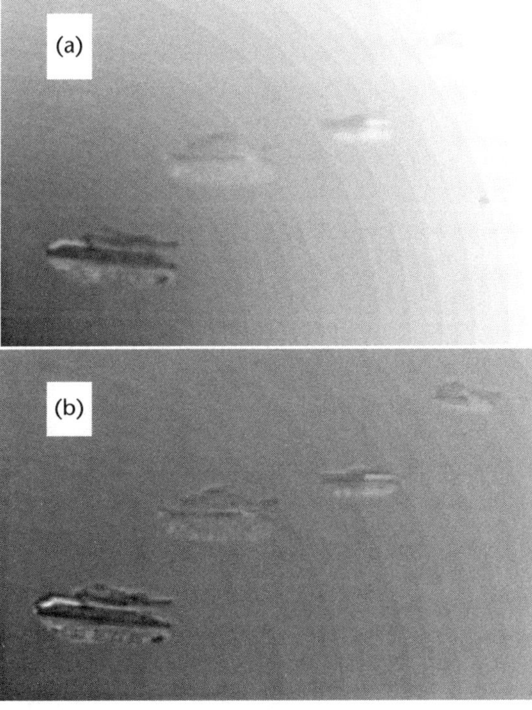

Figure 1.41   At top in (a), atmospheric path radiance causes a large thermal ramp from the bottom-left to the top-right. In (b), the image is enhanced by subtracting a low pass created by convolving a five by five pixel kernel over the original and subtracting the low pass image. The approximate procedure described earlier is used with a multiplication factor of 0.75. The contrast of all the tracked vehicles is enhanced, plus the high frequencies are boosted, and that decreases the blur.

tracked-vehicle contrast from the bottom-left corner to the top-right corner of the field of view. Atmospheric radiance increases with path length, which is longer at the top-right of the image. Also, atmospheric transmission to the most distant vehicle is 0.1. A Gaussian blur is used to represent camera MTF.

In the absence of LACE, the display contrast must be reduced to avoid saturation, and the distant target vehicle is displayed at low contrast in Figure 1.41(a). Applying a high pass filter to the image removes the gradual radiance change caused by path radiance. With the large signal swing removed, the display gain can be increased to render the distant vehicle better. The result of LACE using a five by five block size is shown in Figure 1.41(b).

The frequency boost provided by subtracting a low passed image does not flatten the transfer response in the same manner as a restoration filter. Restoration is better if the blurs are known and constant over the field of view. However, as long as boost is not excessive, LACE partially compensates for camera blur and improves both contrast and MTF performance. LACE reduces blur while improving display contrast and eliminates the need for constant control adjustments.

Figures 1.42 through 1.44 show low resolution, uncooled thermal images. These figures illustrate the benefit of LACE when the input images do not have a large

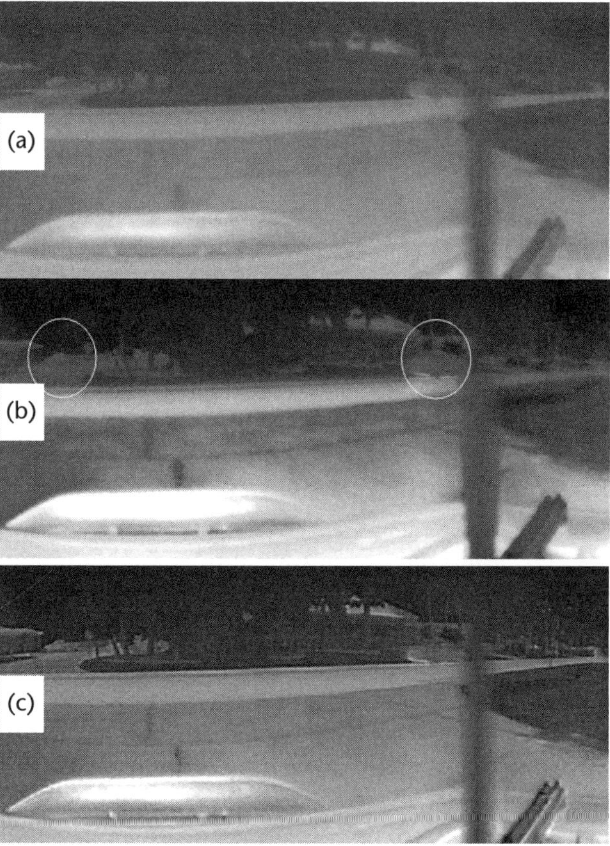

Figure 1.42 Uncooled thermal image in (a) at top is histogram equalized in (b) and LACE is applied in (c). The circles in (b) indicate places where histogram equalization has reduced the contrast of scene details.

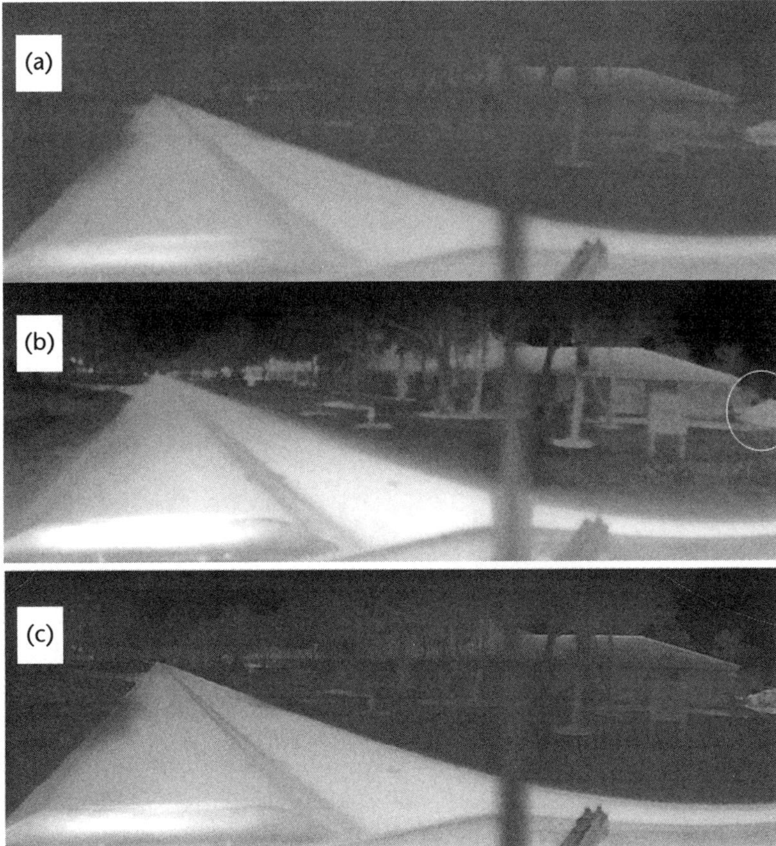

Figure 1.43 Uncooled thermal image in (a) at top is histogram equalized in (b) and LACE is applied in (c). The circles in (b) indicate places where histogram equalization has reduced the contrast of scene details.

dynamic range. The originals are in (a) in all the figures. The pictures in (b) show histogram equalization, and the (c) pictures are enhanced with the same LACE algorithm used in Figure 1.41(b).

Note the cars inside the circles in the (b) pictures. Histogram equalization results in high-contrast images that we subjectively judge as "good." However, histogram equalization changes grayscale content such that some image features actually lose contrast. Broadening grayscale to enhance display contrast is consistent with our goal of accurately rendering scene detail; changing the relative intensities of pixel content is not consistent with our goal. Histogram equalization is a good Photoshop tool, but it is not appropriate for real-time camera processing, because histogram equalization cannot be depended upon to render scene detail accurately.

Figure 1.45 compares the histogram of the three images in Figure 1.44. Note that LACE (the bottom histogram) expands the contrast of the original image (the top histogram). The thermal pedestal is removed, and camera gray shades are expanded to fill the entire dynamic range of the display. Histogram equalization, however, remaps the camera grayscale to take full advantage of the display grayscale. Remapping the camera grayscale does provide a higher contrast image, and

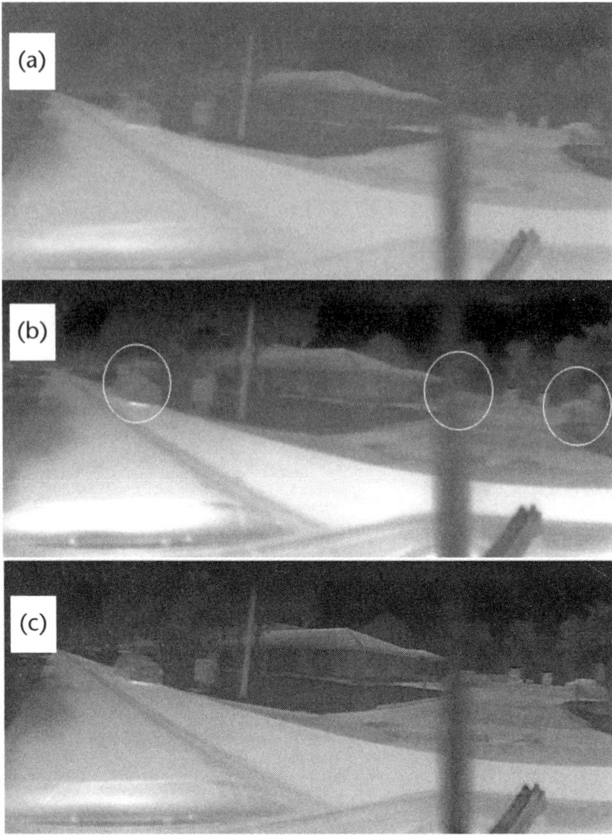

Figure 1.44 Uncooled thermal image in (a) at top is histogram equalized in (b) and LACE is applied in (c). The circles in (b) indicate places where histogram equalization has reduced the contrast of scene details.

that is subjectively pleasing. However, as highlighted by the circles in Figures 1.42(b) through 1.44(b), histogram equalization actually degrades some scene details.

Many variations of histogram equalization have been proposed to correct the problem in the basic algorithm. Pace (2008) uses a Laplacian pyramid to decompose the image and applies histogram equalization to each level of filter. Unlike a typical histogram, her implementation will remove the thermal ramp that causes a problem in Figure 1.41(a). A further advantage of her technique is that frequency boost can be tailored during image reconstruction. From a theoretical standpoint, however, our goal is to accurately render a diversity of scenes. High pass LACE filters out unwanted signal but passes most of the scene frequency content to the display. It is hard to see how any algorithm that remaps camera grayscale to create a uniform display grayscale could accurately render a large variety of scenes.

1.4.2 Local Area Contrast Enhancement Algorithms for Reflective Imagers

Modern reflective cameras are either charge coupled device (CCD) or complimentary metal oxide semiconductor (CMOS) technology. Except for some special applications, these devices provide at most 10 bit imagery with 1,000 shades of gray. Some

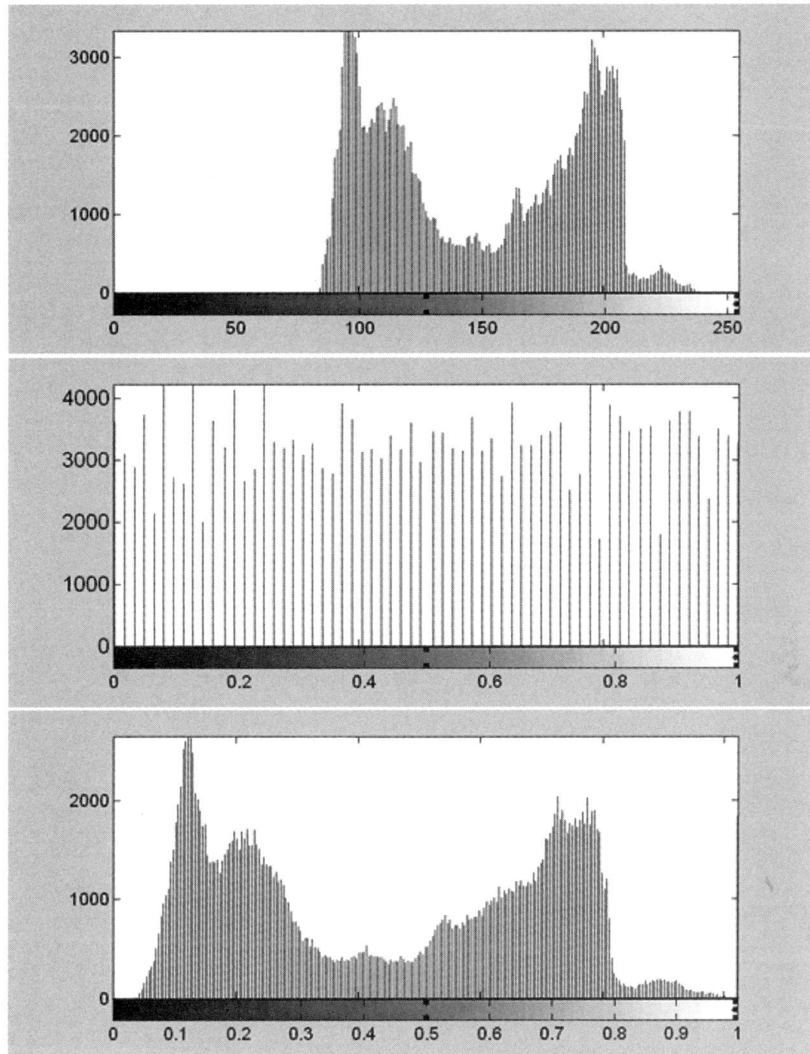

Figure 1.45   Histograms of the three pictures in Figure 1.44.

medical displays are capable of displaying 1,000 shades of gray, but most displays do not. Unfortunately, the high-pass algorithm used with thermal imagers cannot be applied to reflective imagers. We do not know a priori whether large variations in signal result from illumination or from the reflectance characteristics of the scene. Reflective cameras require a different LACE technique than thermal imagers.

For real-time imaging, camera manufacturers generally use a nonlinear mapping of the camera output to the display. The concept is well known, but the details are generally proprietary. See, for example, the white paper describing Panasonic cameras by Behar (2009). We present a generic mapping and apply it to some examples.

Figure 1.46 shows a curve traditionally associated with "the film look." Camera output on the low end receives extra emphasis so that dark areas are visible. Camera output at the high end does not saturate but is deemphasized. Only the middle grayscale is linear. The nonlinear mapping is useful because it allows simultaneous

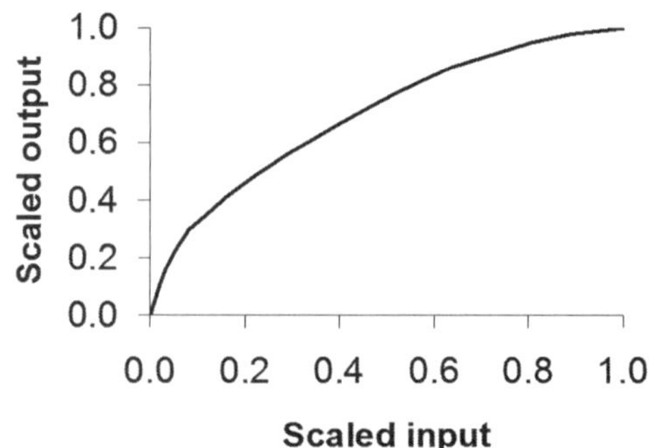

Figure 1.46 Curve that maps camera output to display input. Both scales are normalized. The output would normally be 7 or 8 bits to the display, and the input would normally be 8 or 10 bits from the camera.

rendering of both shadowed areas and bright highlights. Of course, this particular curve is a generic example.

To use the graph in Figure 1.46, the minimum and maximum camera output establish the "scaled input" zero and one, respectively. By camera output, we mean the actual data from the scene, but filtered in time over a few frames. It is likely that best results are obtained by letting some pixels saturate on both the low and high ends. The seven-, eight-, or ten-bit digital display input establish the "scaled output" zero and one.

Figures 1.38, 1.47, and 1.48 show examples of the benefit of nonlinear mapping. In Figure 1.38(a), a streetlight illuminates the nearby scene, leaving the bridge in the dark. In (b), the dark areas are lightened by applying the camera-to-display mapping shown in Figure 1.46. In this case, the original color picture has eight bits per color and so does the corrected image. Each color is enhanced separately.

The picture in Figure 1.47(a) also has eight bits per color, and the cameraman picture in Figure 1.48(a) is eight bit monochrome. Figures 1.47(b) and 1.48(b) illustrate the effectiveness of LACE. In Figure 1.47(b), the shadowed areas and the bright areas are simultaneously rendered by the LACE. In Figure 1.48(b), shading in the cameraman's coat is visible, whereas the coat is solid black in Figure 1.48(a). These three figures demonstrate the utility of a nonlinear mapping from the camera to the display.

Figure 1.47 In (a), the sunlit backyard is well lit, but the doors and chairs are in deep shadow. In (b), details in the shadowed areas are visible.

Figure 1.48   The picture to the left (a) is the original available on the Internet. The nonlinear mapping was applied to get (b) on right. Note the coat details visible in (b) but not in (a).

The use of a nonlinear display mapping will, to some small extent, corrupt scene details. The case here is the same as with thermal imagers; we accept some degree of imperfection to overcome display limitations.

1.5   Conclusions

Imaging technology has advanced at a rapid rate in recent years, and image processing hardware has kept pace. Computers, digital signal processors, field programmable gate arrays, and other digital hardware provide the processing power needed to enhance the imagery. This chapter described digital processing techniques for real-time resolution and contrast enhancement.

Section 1.2 described imager components and explained how to analyze imager blur using MTF. MTF quantifies how well the spatial frequency content in the scene is transferred to the display. If the display brightness and contrast are properly adjusted, then modulation in the scene multiplied by the MTF of the system (the transfer response) yields the modulation on the display. In a very literal sense, MTF tells us how well-modulated the display signal will be.

Section 1.2 also discussed sampling. Sampling at or above the Nyquist frequency is only half of the Sampling Theorem; the other half is proper reconstruction. Since display pixels do not reconstruct in accordance with the Sampling Theorem, Section 1.2 describes how to calculate the frequency content of the displayed image. If the camera engineer wants frequencies near the half-sample rate to be visible, then two display pixels per camera pixel should be used to display the images. Two display pixels are needed to reconstruct in-band information near the Nyquist frequency, and two display pixels also avoids out-of-band aliasing caused by visible display pixels.

Section 1.2 also explained that scene information is best rendered by a broad imager bandwidth. The point spread function is minimized by broad bandwidth. The camera and display can be thought of as a communication channel. With a temporal filter, each frequency increment conveys the same amount of information. In the spatial case, (1.10) tells us that image bandwidth is an area in the spatial frequency domain, and high frequencies cover a bigger area than low spatial frequencies. Nonetheless, the same logic applies to imagery as to a communication channel. Image fidelity is optimized by maintaining unity modulation transfer throughout the camera passband. Excessive peaking of high frequencies degrades bandwidth. Peaking might help subjective image quality, but it degrades the accurate rendering of scene detail.

Section 1.3 explains how to generate deconvolution kernels and discusses the benefits of using a nonrect window when limiting kernel size. Large deconvolution kernels require powerful digital processors, and large kernels also reduce image size because of extended edge effects. However, truncating a deconvolution kernel is equivalent to multiplying by a rect function, and a multiplication in space is a convolution in frequency. Multiplying by a rect function causes rippling and spreading of the deconvolution MTF. Prolate spheroidal wave functions make good window functions, because they tend to be limited in both the spatial and frequency domains. Using a nonrect window to limit kernel size can be an effective deconvolution design tool.

LACE is effective for both thermal and reflective imagers, but the digital algorithms are quite different. For both types of imagers, LACE maps the large camera dynamic range into the more limited dynamic range of the display. However, with a thermal imager, using a spatial filter to remove a large but gradual change in radiance over the field of view helps with viewing scene details. With a reflective camera, a large swing in radiance might be the result of luminance difference; perhaps one part of the field of view is in sunlight and other parts in shade. With thermal cameras, spatial filtering is used to remove low spatial frequencies and enhance the contrast of scene details. Reflected light cameras use a nonlinear display mapping to brighten shadowed areas and subdue bright areas.

References

[1]   Joseph Caniou, *Passive Infrared Detection*, Boston: Kluwer Academic Publishers, 1999.

[2]   J. G. Fiasconaro, "Two-dimensional Nonrecursive Filters," Chapter 3 in *Picture Processing and Digital Filtering*, New York: Springer-Verlag, 1979.

[3]   Stanford Goldman, *Information Theory*, New York: Dover Publications, 1981.

[4]   Joseph W. Goodman, *Introduction to Fourier Optics*, Boston: McGraw Hill, 1996.

[5]   Peter A. Jansson, *Deconvolution of Images and Spectra*, San Diego: Academic Press, 1997.

[6]   R. M. Mersereau, "The Design of Nonrecursive Filters Using Windows," Section 2.2 in *Two-dimensional Digital Signal Processing I*, New York: Springer-Verlag, 1981.

[7]   Athanasios Papoulis, *The Fourier Integral and Its Applications*, New York: McGraw Hill, 1962.

[8]   Richard Vollmerhausen, Don Reago, and Ronald Driggers, *Analysis and Evaluation of Sampled Imaging Systems*, SPIE Tutorial Series, Bellingham, WA, 2010.

[9]   Richard Vollmerhausen, "Design of Finite Impulse Response Deconvolution Filters," *Applied Optics*, Vol. 49, 2010, pp. 5814–5827.

[10] John M. Wozencraft and Irwin Mark Jacobs, *Principles of Communication Engineering*, New York: John Wiley & Sons, 1965.

CHAPTER 2

Interferometric Testing of Optical Surfaces

Jim Burge, Chunyu Zhao, and Ping Zhou

2.1 Fundamentals of Interferometry

Interferometers use wave properties of light to measure the shape of surfaces that transmit or reflect the light. The precision of the interferometer comes from the small wavelength of light, less than 1 micrometer, and the ability to collect and process images in a computer. The accuracy of interferometers comes from the design, the use of precision components, and the ability to perform in situ calibrations. This section summarizes the fundamental physics of interferometry and discusses the applications for measurement systems. Consult *Basics of Interferometry* [1] for further reading on this subject.

2.1.1 Interference Made Possible by the Wave Nature of Light

Light is a transverse electro-magnetic wave, as shown in Figure 2.1. For interferometry, the amplitude, phase, and direction of oscillation (polarization) all matter. This section provides some definitions and background on the fundamentals of the nature of light.

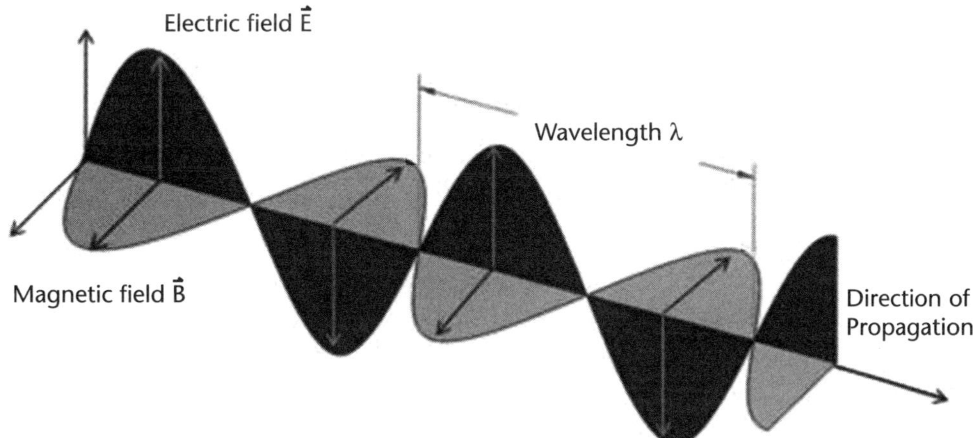

Figure 2.1 Light waves consist of electric and magnetic fields that vary sinusoidally in directions that are perpendicular to the direction of propagation.

Light, such as that created by a laser, can be defined as waves of electromagnetic energy that propagate in free space with no attenuation or change in frequency. The electromagnetic waves that define the light used for interferometers can be thought of as originating with an oscillation in the electric field with typical frequency f is approximately 5×10^{14} Hz. The electric field actually induces a changing magnetic field in the perpendicular direction, which in turn reinduces an electric field oscillation one wavelength away from the initial field. These self-propagating undulations travel at the universal speed of light $c = 3 \times 10^8$ m/s in vacuum. The waves are slowed down to a speed of c/n in a medium with index of refraction n. The frequency f is always conserved as the light travels from one medium to the next. The value of the wavelength λ is calculated as

$$\lambda \cdot f = \frac{c}{n} \tag{2.1}$$

Light is made of transverse field variations, with the electric and magnetic fields pointing in directions perpendicular to the direction of propagation and perpendicular to each other, as shown in Figure 2.1. We define a state of polarization according to the direction of the electric field since most materials interact more strongly to an electric field than a magnetic field. Light with vertical polarization will have the electric field oscillating in the vertical direction (and the magnetic field in the horizontal direction). Light polarized in the 45° direction can be decomposed as equal amounts of horizontal and vertical polarization that are in phase. Light with circular polarization has constant 90° phase shift between the two. Unpolarized light has equal contributions of both vertical and horizontal directions with a randomly varying phase between them.

Electromagnetic waves used for interferometry have wavelength defined by the source. Most interferometers use laser sources that are highly coherent, meaning that the wavelength range is very small and the phase distribution across the beam of light is well defined. Electromagnetic waves with longer wavelength, such as radio waves, utilize frequencies that are low enough that the field variations can be measured directly using an antenna. But the high frequencies of visible wavelength light are not directly observable. Instead, the energy of the light is captured with solid-state detectors. The light detection is then proportional to the average energy captured within an exposure time, rather being proportional to the field strength. The fields vary sinusoidally about a zero mean. The intensity or energy density of the light is proportional to the square of the field. So if the field strength of the light is doubled, then the intensity and the measured brightness will increase by a factor of four.

As an electromagnetic wave, light exhibits the effect of interference. When two beams of light are combined, the resulting distribution is defined by the sum of the *fields* that define the light. If the two beams of light are in phase with each other, meaning that the sinusoidal waves line up with such as shown in Figure 2.2, then the fields add constructively yielding a new wave with increased intensity. If one beam is shifted by a half wavelength, it is out of phase, and the fields add destructively where one cancels the other and the resulting intensity is decreased.

Interferometers use a single source of light that is split into two portions using some type of beamsplitter that reflects some of the light and passes the rest. The light is then recombined to create the interference. The two portions travel different

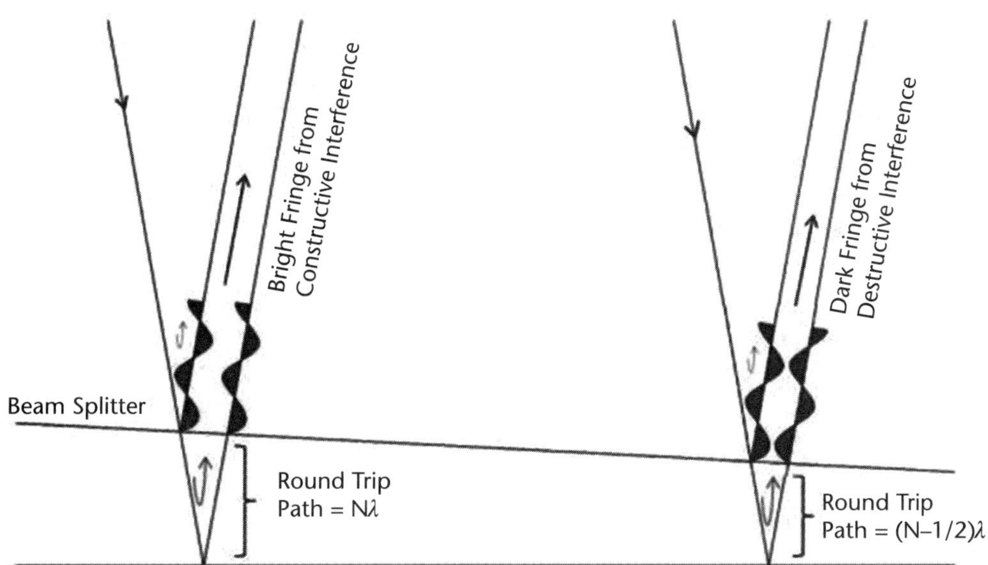

Figure 2.2   Light reflected from two difference surfaces is recombined to create interference. Bright fringes are formed when the light is combined in phase, with optical path difference of an integral number of wavelengths and dark fringes resulting from the light being out of phase.

paths before being recombined. We define the optical path length (OPL) for a ray of light as the product of its geometric path length and the refractive index it travels in. We define the optical path difference (OPD) as the difference in the OPL between the two beams before they are recombined. The light will interfere constructively (creating a bright fringe) in regions where the two beams are in phase and will interfere destructively (creating a dark fringe) in areas where the two beams are out of phase. The resulting fringe pattern shows the distribution of light where one full fringe (bright region to bright region) represents an OPD of exactly one wavelength. This effect is shown in Figure 2.2.

The simplest type of interferometric measurement compares a flat or curved surface with a mating master surface that is made onto a piece of glass called a test plate. The test plate is placed against the surface under test and illuminated with monochromatic light such as that from a mercury lamp. Fringes are formed from the interference between the light reflected from the master surface and that reflected from the surface being measured. Each fringe represents exactly one wave of optical path difference, which would be caused by a difference in the surfaces of one-half of a wavelength, as shown in Figure 2.3. The OPD at near-normal incidence is twice the surface distortion. A depression in the surface that is one micron deep would require light to travel one micron further to get to the bottom of the depression. Upon reflection, it would travel an additional micron to catch up with the light reflected from the nominal surface. Succinctly, *the variation in separation between surfaces = λ/2 × number of fringes.*

The resulting interferogram is interpreted as a contour map of the surface with the fringes spaced at half-wave contours. An accurate measurement of the surface under test can be performed by evaluating the fringe pattern and using an accurately shaped master test plate. Further details on the use of test plates for measurement of optical surfaces are provided in the authoritative book *Optical Shop Testing* [4].

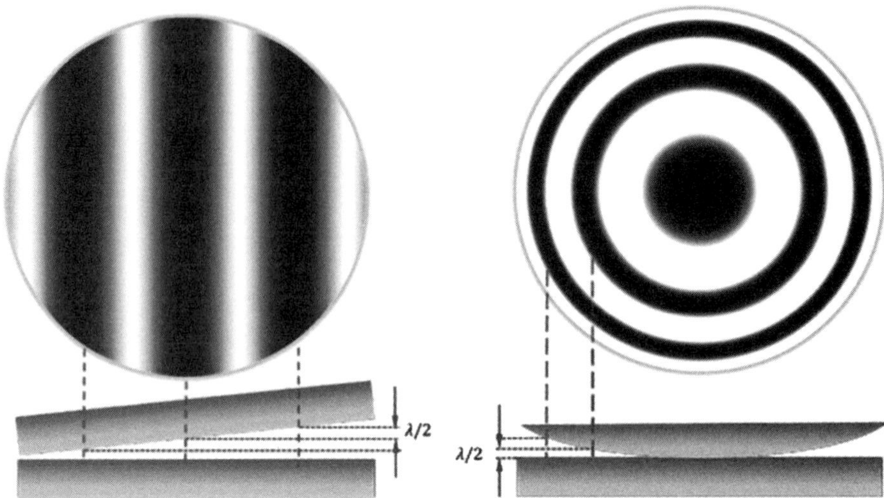

Figure 2.3 Two polished surfaces in contact form fringes of interference with each fringe defining round trip path length difference of one wavelength of light. When one surface is a master flat, the other can be measured by the fringe pattern where each fringe represents surface contours of one half of a wavelength.

EXERCISE 2.1

Two flat glass plates that are 100 mm square across are stacked, but a thin gold foil separates the plates at one edge. When illuminated with green light from a mercury lamp, wavelength 0.546 μm, and viewed in reflection, 10 straight fringes are seen as shown:

    a.   Calculate the thickness of the gold foil.

    b.   Calculate the angle between the two glass plates.

Solution

    a.   Each fringe corresponds to $\lambda/2$ or 0.273 μm contours. So 10 fringes is equivalent with $10 \times 0.273 = 2.7$ μm. The foil is about 2.7 μm thick.

    b.   Using the small angle approximation, the angle between the plates is 2.7 μm/100mm, or 27 μrad.

Figure Ex 2.1

The formation of good interference fringes from two beams of light requires the following properties, explained in detail next:

A. Intensity matching of the two beams;
B. Temporal coherence, defined by the narrowness of the wavelength range;
C. Spatial coherence, defined by the effective size of the light source;
D. Polarization matching of the two beams;
E. Vibration of the fringes must not be excessive during an exposure.

A. Contrast

The contrast or fringe visibility describes how well the fringes show up. Contrast is quantified as the amplitude of a sinusoidal function divided by its DC offset, as shown in Figure 2.4.

For ideal contrast of 1, the two beams will have matching intensity, allowing full cancellation for the dark fringes and full field addition for the bright fringes, giving them twice the brightness of the incoherent sum of the two beams.

In general, the intensity variation of fringes formed by two mutually coherent beams of light is described as

$$I(x,y) = I_A + I_B + 2\sqrt{I_A I_B}\,\cos\big(\phi(x,y)\big) \tag{2.2}$$

where I_A and I_B are the intensities of the two beams, and $\Phi(x,y)$ is the phase difference between them.

The maximum and minimum intensities are

$$I_{\max} = I_A + I_B + 2\sqrt{I_A I_B}$$
$$I_{\min} = I_A + I_B - 2\sqrt{I_A I_B} \tag{2.3}$$

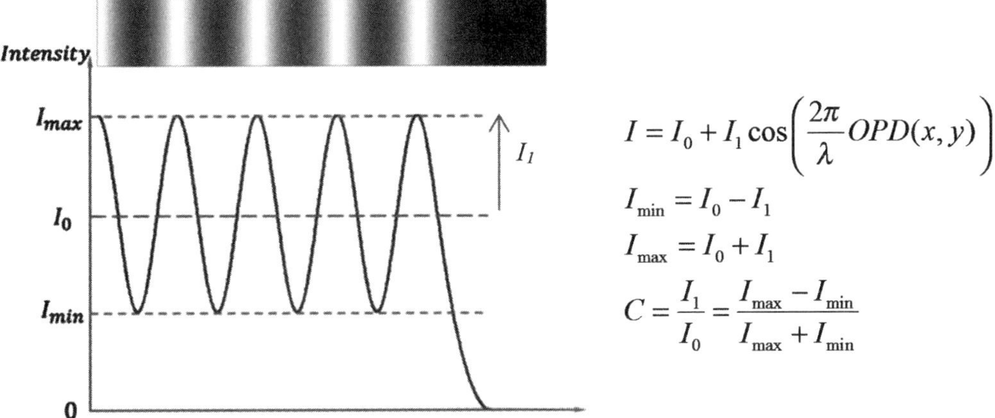

Figure 2.4   The fringe contrast, or visibility, is calculated as the sinusoidal amplitude divided by the DC offset.

The contrast or fringe visibility is calculated as the sinusoidal intensity magnitude divided by its DC offset,

$$C = \frac{I_{max} - I_{min}}{I_{max} + I_{min}} = \frac{2\sqrt{I_A I_B}}{I_A + I_B} \tag{2.4}$$

B. Temporal Coherence

Temporal coherence is required for interference such that the field variation maintains its sinusoidal behavior over the time that it takes for the light to travel the optical path difference. The coherence time τ relates to the coherence length L_C by the speed of the light. If the OPD is greater than the coherence length, the field summation will fluctuate at very high frequencies, which will wash out contrast of any interference patterns. This condition generally requires that the light for both beams come from the same source and that the light is monochromatic. This relationship is easy to understand by converting the optical path difference to units of waves and using the fact that light at different frequencies cannot create stable interference patterns. We can define a fringe pattern in the light intensity for each wavelength. A continuous spectrum from $\lambda - \Delta\lambda/2$ to $\lambda + \Delta\lambda/2$ will create a continuous distribution of fringes with phase proportional to the wavelength. We define the coherence length (or time) for such a distribution such that the phase shift for the full range of wavelengths will be less than 90° as the light propagates over a distance equal to the coherence length. If the optical path difference is less than this coherence length, then all of the light contributes to the same set of sinusoidal fringes. Otherwise, the different fringe sets from the full range of wavelengths will be sufficiently out of phase that the fringes will wash out.

Mathematically, the coherence length L_C, coherence time τ, and bandwidth $\Delta\lambda$ are related as

$$L_c = c\tau = \frac{\lambda^2}{2\Delta\lambda} \tag{2.5}$$

Laser light is usually highly monochromatic and can exhibit coherence length of many meters.

EXERCISE 2.2

Some lasers operate in multiple modes which output several specific wavelengths. Consider a 30 cm HeNe laser that emits two modes of equal intensity. The nominal wavelength is 632.8 nm, and the two modes are different in frequency by 500 MHz.

a. Calculate the nominal frequency of the electromagnetic wave.

b. Calculate the wavelength difference between the two modes.

c. Determine the condition when the fringe contrast is maximum and minimum.

d. Calculate the optical path difference and cavity length for minimum and maximum fringe contrast.

Solution

a. $c = f \cdot \lambda$, $f = c/\lambda = (3E8 \text{ m/s})/(632.8E^{-9} \text{ m}) = 4.74E14 \text{ Hz}$

b. Start from the simple relationship between wavelength and frequency and take differentials.

$$c = f \cdot \lambda$$

$$\Delta c = 0 = f \cdot \Delta\lambda + \Delta f \cdot \lambda$$

$$\frac{\Delta f}{f} = -\frac{\Delta\lambda}{\lambda}$$

$$|\Delta\lambda| = \frac{\Delta f}{f}\lambda = \frac{500E6}{4.74E14}(632.8 \text{ nm}) = 6.7E - 4\text{nm}$$

c. The resulting interferogram is made from the sum of two interferograms, one for each wavelength. Maximum contrast of 1 occurs when the fringe patterns coincide, which occurs when the optical path difference OPD is an integral number of waves. The patterns wash out resulting in contrast of zero when the patterns are shifted by one half fringe, which occurs when the OPD is $(N + 1/2)$ waves where N is an integer.

d. For maximum contrast, find the OPD (which has units of length) that satisfies this relationship by evaluating the change in the number of wavelengths defined by the OPD as the value of the wavelength is changed. When this change is an integer k, the two fringe patterns overlap and the fringe contrast is optimal.

$$\frac{d}{d\lambda}\left(\frac{\text{OPD}}{\lambda}\right)\Delta\lambda = k$$

$$\text{OPD} \cdot \frac{\Delta\lambda}{\lambda^2} = k$$

$$\text{OPD} = \frac{\lambda^2}{\Delta\lambda}k = \frac{(632.8E - 9\text{m})^2}{(6.7E - 13\text{m})}k = 0.6\text{m} \times k$$

Since the OPD is twice the cavity length, the contrast will be optimal when the cavity length is 0, 0.3m, 0.6m, 0.9m, ...

Using the same analysis, but evaluating the condition when the change in the number of waves in the OPD equals $(k + 1/2)$ where k is an integer, find contrast minimum at

$$\text{OPD} = 0.6 \text{ m} \times (k + 1/2)$$

where the cavity length is 0.15, 0.45, 0.75, ... meters

C. Spatial Coherence

Spatial coherence describes the degree to which we can treat the wavefront as a simple phase distribution in space, such as that created by a single source. In general, we can treat the light as a superposition of waves from different source points that don't interfere with each other. As long as the path difference of all of these waves is within ±0.25 waves, then the resulting fringes will be in phase. A point source, which can be created by focusing a laser through a small pinhole, has high spatial coherence, allowing it to readily form interference fringes. However, the use of such a source also allows interference from light scattered from surface defects or dust. Some interferometers improve this effect using a small source that preserves enough spatial coherence for the measurements but decreases the noise from the scattered light.

D. Polarization Matching of the Two Beams

Since the interference uses the superposition of the electric fields, light with horizontal polarization will not interfere with vertical polarization light. Also, light with left-hand circular polarization will not interfere with right-hand circular light. The addition of a polarizer in the combined beam always passes the matching component of both beams, which allows interference. The intensity and phase of each component can be calculated using a matrix method called Jones calculus.

E. Vibration

Interferometers are quite sensitive to vibration. If a surface under test is vibrating with amplitude of only 0.3 microns, then the wavefront phase will be changing by twice this. For an interferometer using a HeNe laser source with 0.633 μm wavelength, the wavefront phase will vary by a full wave, causing the fringes to move by a full period. If the exposure that captures the interferogram is longer than the period of the vibration, then the fringe will be washed out. Interferometric measurements are commonly performed on isolated tables to minimize the coupling of ground vibration into the system.

2.1.2   Types of Interferometers for Surface Measurement

Fringe patterns can be captured as interferograms that give information about the surface under test. In fact most spherical surfaces are measured by a trained optician who views the interferogram from a test plate and subjectively judges the irregularity in the fringe pattern. A well-corrected optical surface with "a quarter wave" surface irregularity will be judged to have less than one-half fringe deviation from the expected shape. (Remember that the reflected wavefront irregularity has twice that of the surface.) An example interferogram of a half-fringe surface is shown in Figure 2.5.

Phase shifting interferometry (PSI) allows measurements that are much more precise and objective. In PSI systems, the relative phase between the reference and test beams is shifted while the interferograms are captured for processing. Typically, five exposures are taken with 90° phase shift between them. The set of five values allows the phase to be calculated for each pixel in the focal plane using an equation such as

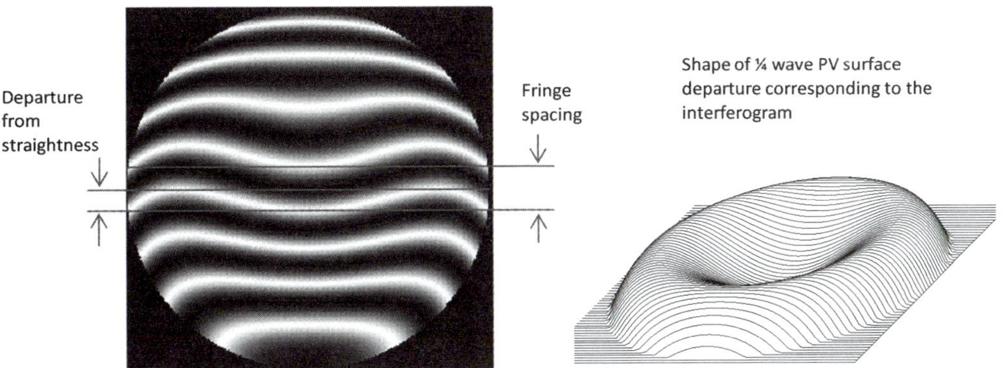

Figure 2.5   Test plate fringes showing one-half fringe of irregularity, which is the deviation from straightness in units of fringes. This is due to a quarter wave of surface irregularity as shown to the right.

$$\phi = \tan^{-1}\left[\frac{2(I_2 - I_4)}{2I_3 - I_5 - I_1}\right] \tag{2.6}$$

where I_1, I_2, … are the measured intensities for each pixel for the five frames with 90° phase shift between them. Equation (2.6) can determine phase within $\pm\pi$. The 2π ambiguities are resolved using the continuity of the surface. As long as the interferometer reference is accurate, the pixel-to-pixel phase variations of the light will be proportional to the irregularity of the surface under test.

Three common interferometer configurations that use a collimated laser source and the phase shifting technique are shown in Figure 2.6. The Mach-Zehnder interferometer uses two different beamsplitters—one to separate the two beams and the second to recombine them. The Twyman Green interferometer uses a single beamsplitter twice. The outgoing light is split into a test and reference beam. The same beamsplitter recombines them to create the interference. The Fizeau interferometer uses a master surface as the only beam splitter. The reference beam is created directly from reflection from this surface. The test beam transmits out through this surface, reflects off the surface under test, and then transmits back through this surface, where it is recombined with the reference beam.

2.2   Commercial Interferometers

2.2.1   Overview of Commercially Available Interferometers

Most commercial interferometers use phase shifting techniques and can provide high-speed and high-accuracy measurements. There are several types of phase-shifting interferometers (PSI) on the market. Depending on the application, one type could work better over another. A few common commercial PSI designs are discussed next.

Temporal Phase Shifting Interferometer

This is the most common interferometer, which introduces phase shift by translating the reference optic with a piezoelectric transducer (PZT). Three or more

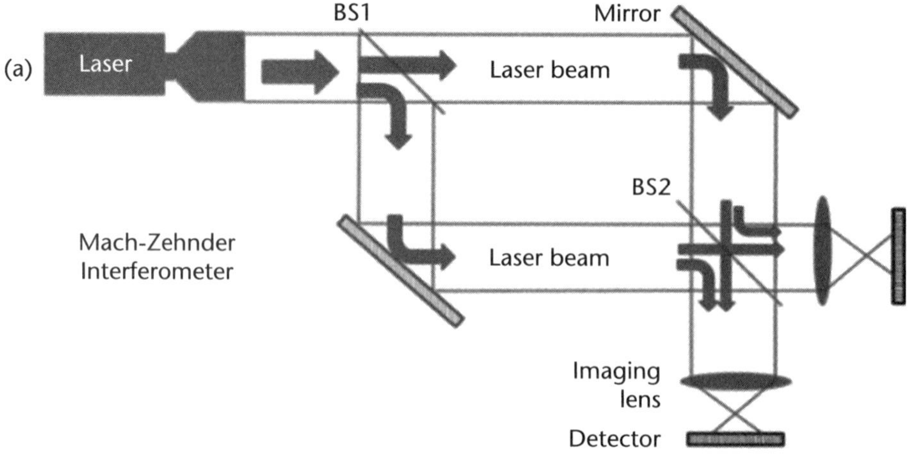

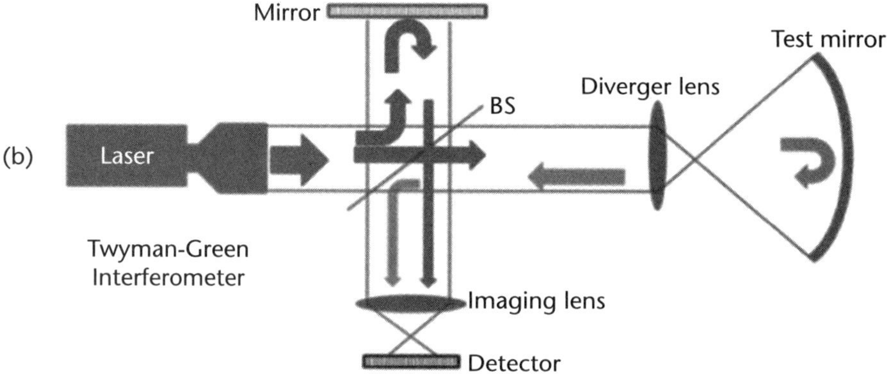

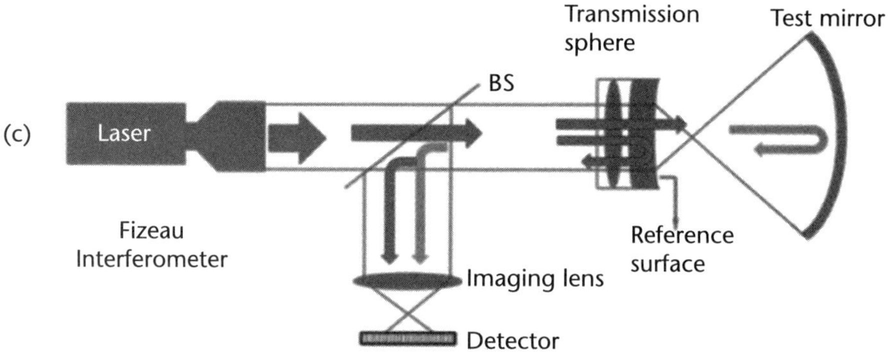

Figure 2.6 Common interferometer configurations: (a) Mach-Zehnder, (b) Twyman-Green, and (c) Fizeau.

phase-shifted frames of interferometric data are taken sequentially. This method, so-called temporal phase shifting, is very sensitive to mechanical vibration of system. Vibration causes the phase shifts between the data frames to be different from what is desired and therefore introduces errors in the measurement. However, it has been used and is still being used in many applications when the test system is relatively

Figure 2.7   Zygo Verifire XPZ interferometer (left) and 4D AccuFiz interferometer (right). (Courtesy of Zygo Corporation and 4D Technology.)

stable and the air turbulence is well controlled. The AccuFiz interferometer from 4D Technology and the Verifire XPZ from Zygo (shown in Figure 2.7) are examples of this conventional type of interferometer.

Vibration Insensitive/Dynamic Interferometer

This type of interferometers overcomes the difficulty of the temporal PSI, and all the phase-shifted data is collected in a single exposure in order to minimize the impact from vibration and air turbulence. It is also called spatial phase-shifting interferometer, or dynamic PSI.

There are several methods to take multiple frame data in one shot. One method uses three CCD cameras to obtain three phase-shifted interferograms simultaneously, such as the ESDI Dimetior interferometer. The calibration and alignment of the three cameras is critical to this type of dynamical PSI.

Spatial carrier technique is another approach. In this method, a series of tilt fringes are purposely introduced to the interferogram with known spatial frequency and orientation. Fourier methods are then used to extract the phase from a single shot interferogram [8]. Both Zygo and 4D Technology have interferometers using this spatial carrier technique.

Another dynamic interferometer that uses a pixelated phase mask was developed by 4D Technology. The phase-mask is a micropolarizer array and that imparts a discrete phase shift on each pixel and makes use of the polarization properties of the source [5].

Low Coherence Noise Interferometer

Commercial interferometers typically use lasers as a light source to achieve high-contrast fringes. However, lasers cause coherent noise that comes from surface defects or dust particles. Stray light from these artifacts adds coherently with the test and reference beams of the interferometer, creating spurious fringes in the interferogram, and introducing phase errors in the measurement. These coherent noises can be reduced by decreasing the source coherence. One way to reduce it is to use a polychromatic source. This solution requires the interferometer to match the path length for the test and reference beams. The FizCam 2000 developed by 4D Technology uses this type of short-coherence light source.

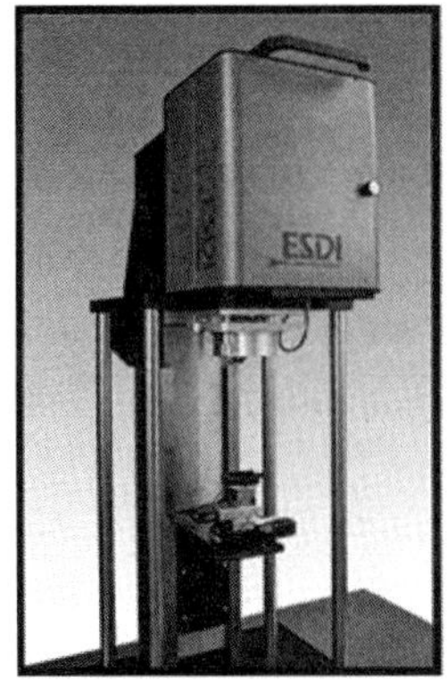

Figure 2.8 ESDI Dimetor interferometer (left) and 4D phaseCam (right). (Courtesy of Engineering Synthesis Design Inc. and 4D Technology.)

Extended source can also reduce the coherence of the system. A classic way to create an extended source consists of sending the laser beam on a spinning ground glass diffuser. The distance between the point source and the diffuser defines the size of extended source. However, this type of extended source reduces fringe contrast. A ring source approach, like the Ring of Fire developed by Zygo, maintains the fringe contrast while reducing the coherent noise [3].

Path Matching Interferometer

This interferometer greatly extends the capability of traditional interferometry for measuring planar optics. Traditional interferometers use long coherence laser sources. Interference occurs between reflections from any surfaces in the beam path; therefore, multiple sets of fringes are observed at the same time. When this happens, it is difficult or impossible to extract information related to any one surface.

Figure 2.9(a) shows an example of testing a window with a standard interferometer. Two sets of fringes are observed. One is the interference between the reference and one surface of the window; the other is interference between the two surfaces of the window. To isolate and measure the front surface, one may put some scattering coating, such as Vaseline, on the back surface to minimize its reflection. However this is not always practical due to the surface contamination. Figure 2.9(b) shows an interferogram of the surface shown in Figure 9(a) but taken with a path matching interferometer (4D FizCam 2000).

The path matching interferometer has a short coherence length, and an "optical path matching" mechanism within the interferometer selects only fringes that occur between the desired surfaces. With this system, one can select the interference between certain surfaces. For example, one can measure the interference between the front and back surfaces of a window and measure the quality of the window itself.

Stitching Interferometer

Interferometers are usually equipped to easily measure flats and spherical surfaces. Aspheric surfaces are usually measured interferometrically with the addition of null

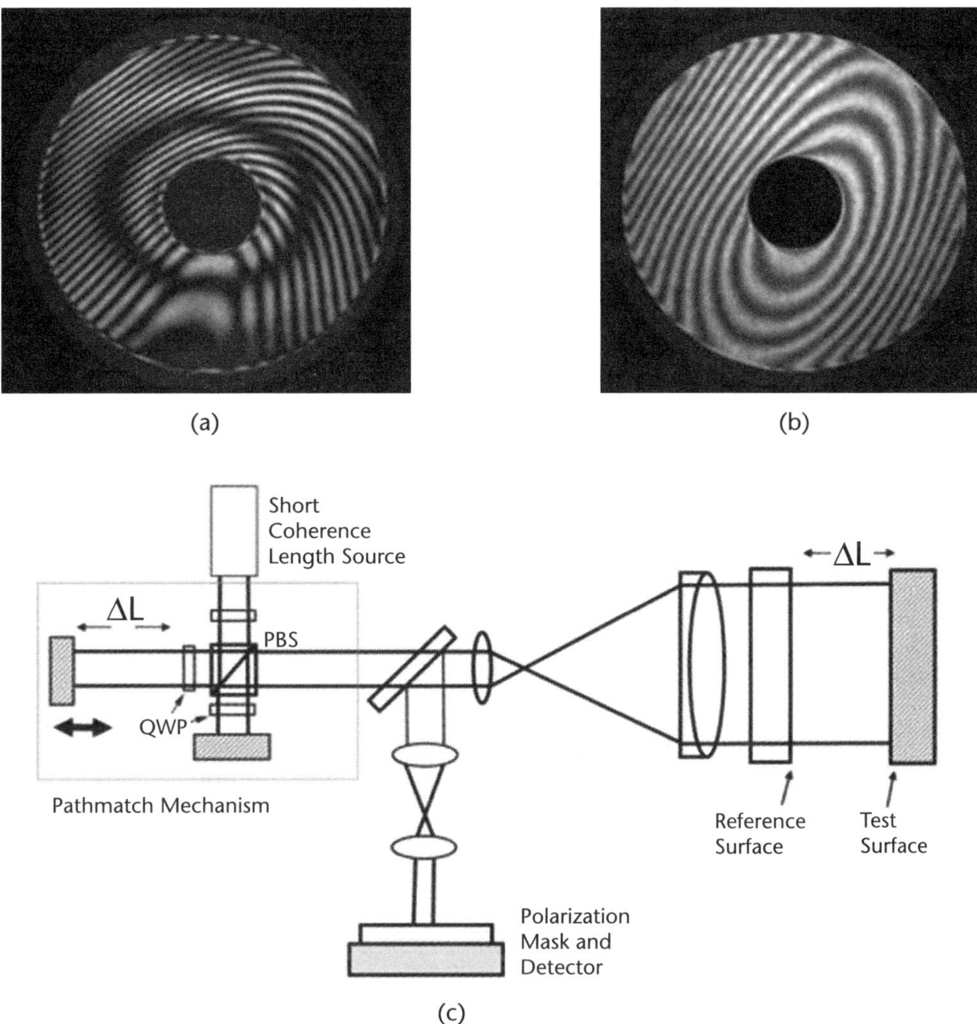

Figure 2.9   (a) A window measured with a standard interferometer; (b) the same window measured with a 4D FizCam 2000 dynamic interferometer; (c) the 4D FizCam 2000 path matching mechanism. (Courtesy of 4D Technology.)

optics, such as computer-generated holograms (CGHs). However for each different aspheric shape, a specific null optic has to be made for the measurement.

To get around this, some special interferometers have scanning system. They can measure subaperture of a large flat, sphere, or asphere without a null optic and then stitch the subaperture measurements to have a high-resolution 3D surface. Zygo's Verifire Asphere and QED's subaperture stitching interferometer, shown in Figure 2.10, are of this type. The interferometer has multiple axes of motorized stages, and offers automated alignment, acquisition, and analysis of asphere. To improve aspheric capability and measurement speed, QED developed the variable optical null (VON) technology. It consists of a pair of prisms that can be rotated with respect to one another [6]. Tilt of the prism pair introduces coma and astigmatism that can compensate the aberrations from subapertures of an asphere.

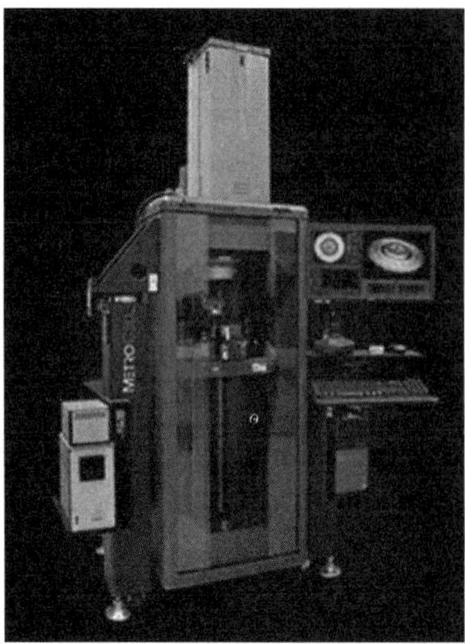

Figure 2.10 Zygo Verifire Asphere (left) and QED aspheric stitching interferometer (ASI) (right). (Courtesy of Zygo Corporation and QED Technologies Inc.)

Sub-Nyquist Interferometer

Compared to conventional interferometers, this interferometer has a much wider dynamic range. It extends the resolution of a sampled system to frequencies beyond its Nyquist frequency [Grievenkamp, 1987]. While the dense fringes from the aspherical surface cannot be resolved by the detector in the interferometer, the regular sampling of the detector creates an aliased pattern that can be processed to calculate the phase. The technique is based on the concept of sub-Nyquist sampling and is used in conjunction with phase-shifting interferometry. The measurement precision inherent to phase shifting interferometry is maintained, and the measurement range is greatly improved. ESDI Intellium Asphere, shown in Figure 2.11, uses this technology, and it can measure not only aspheric surfaces but also spherical and flat surfaces.

Figure 2.11 ESDI Intellium Asphere is a sub-Nyquist interferometer. (Courtesy of Engineering Synthesis Design Inc.)

2.2.2   Additional Features and Accessories

A well-equipped interferometer lab needs a wide variety of accessories to expand or enhance the capabilities of the interferometer.

Transmission Flat/Spheres

Transmission flats are used with an interferometer system for measuring a planar surface. They are available in different size, such as 4 inch and 6 inch. Standard transmission flats have a reflectivity of 4%. This enables users to measure majority of surfaces. For high reflective surfaces, either an attenuation filter (e.g., pellicle) or a Dynaflect coated transmission flat may be used to reduce the intensity of the test beam returning to the interferometer. Dynaflect flats have a reflectivity of 20% and can accommodate surfaces ranging from 4% to 99% reflectivity.

Transmission spheres, such as those shown in Figure 2.12, are used when measuring curved surfaces or diverging/converging wavefronts. They are available in a variety of F-numbers. (The F-number is the ratio of the focal length of the objective to its aperture diameter.) The quality of the transmission flat/sphere is very high, and the PV error is usually between $\lambda/40$ and $\lambda/10$.

Beam Expanders

A beam expander is designed to increase or decrease the diameter of a collimated input beam. It is convenient to have when one needs to measure a surface with either a large or small aperture. Some typical beam expanders are shown in Figure 2.13.

Reference Optics

Reference optics are often used with a transmission sphere when measuring the transmitted wavefront of spherical optics. They are usually a fast concave sphere

Figure 2.12   Transmission sphere and diverging lenses. (Courtesy of Zygo Corporation and 4D Technology.)

Figure 2.13　Beam expander or aperture converter. (Courtesy of Zygo Corporation and 4D Technology.)

uncoated or with high-reflectivity coating. A high-quality ball, such a CaliBall [9], can also be used as reference optics to retro-reflect the light back into the interferometer. Patches of the CaliBall are selected randomly and measured. The phase maps are averaged to obtain an estimate of the transmission sphere because errors in the CaliBall average to zero in the limit of infinite measurements.

Mechanical Mounts

A set of good mechanical mounts is essential for an interferometer lab. A 5-axis mount allows adjusting tip/tilt, translation x/y, and defocus. A self-centering element holder, as shown in Figure 2.14, allows holding optics with various sizes.

2.2.3　Radius of Curvature Measurement

Interferometers can also be used to measure the radius of curvature of a spherical surface. It requires a linear rail with an encoder. The test surface needs to be measured at both cat's eye and confocal positions, as shown in Figure 2.15. The cat's eye position means the test surface is at the focus of the light from the interferometer, while the confocal position means the center of curvature of the test surface is at the focus of the interferometer. The distance that the optic travels between these

Figure 2.14　Self-centering element holder.

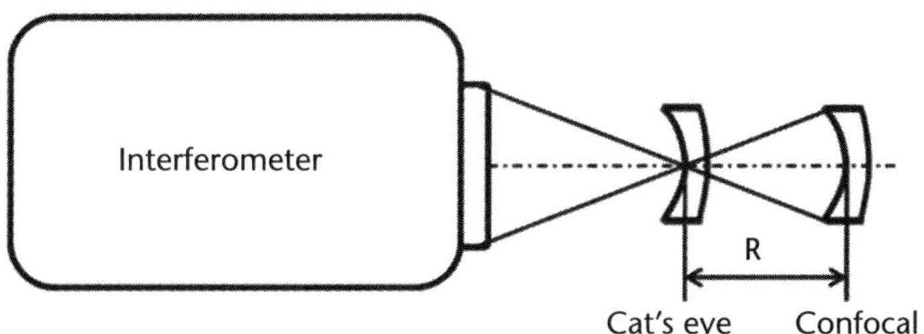

Figure 2.15   Radius measurement geometry, showing cat's eye and confocal measurement positions.

two positions is the radius of curvature of the surface. The accuracy of the radius measurement depends on the precision of the rail and the indicator. Typically, one can achieve an accuracy of a few tens of microns in radius.

2.3   Practical Issues for Measuring Flats and Spheres with Commercial Interferometers

2.3.1   Set Up Hardware and Alignment

The general procedure to test a flat/sphere is described next.

To Test a Flat

- Insert transmission flat and align it to the interferometer so that the reference surface of the transmission flat is precisely perpendicular to the optical axis of the interferometer. Many interferometers have alignment mode. Turn the tip and tilt knobs of the transmission flat until the return spot that represents the reflection off the transmission flat superimposes on the alignment target (crosshair center).
- Place the test flat closer to the interferometer to minimize the environmental effects. Then adjust tip/tilt of the flat to null the fringes.

To Test a Sphere

- Select the transmission sphere that is fast enough to fill aperture of the sphere. Insert and align it to the interferometer in the alignment mode, similar to the alignment of the transmission flat.
- Fine alignment of transmission sphere is optional. You can achieve a better alignment of transmission sphere by placing the test optic in the beam at the cat's eye position first. The cat's eye position means the reflecting surface is at the focus of the light from the interferometer. Note that tip and tilt adjustment of test optic at the cat's eye reflection will not change the fringe pattern. Adjust the tip/tilt knob of the transmission optic to remove tilt fringes in the interferogram. Cat's eye tilt fringes are due only to tip/tilt misalignment of the transmission optics.

- Approximately position the test sphere based on its radius of curvature. The center of curvature should be at the focus of the light from the interferometer. Adjust the tip/tilt/focus of the test sphere until the return spot is on top of the alignment target center in the alignment mode. To test a concave surface, it is often helpful to place a pinhole aperture near the focus of the transmission sphere. Adjust the distance between the test surface and interferometer, so that the reflection from the test surface on the pinhole is focused. Adjust tip/tilt to bring the reflected spot back through the pinhole to the interferometer. Then you should be able to see an interferogram.

For Both Flat and Sphere (Fine Alignment)

- Switch to the fringe view mode and adjust the test optic until the fringes are nulled.
- Adjust the zoom to maximize the image size and adjust the imaging focus to make the edge of aperture look sharp in the interferogram. You can also place a piece of paper right in front of the test optic, and look at the edge of the paper to check the imaging focus.

Some typical interferometer setups are given in Table 2.1.

Table 2.1 Configurations for Common Measurements with an Interferometer*

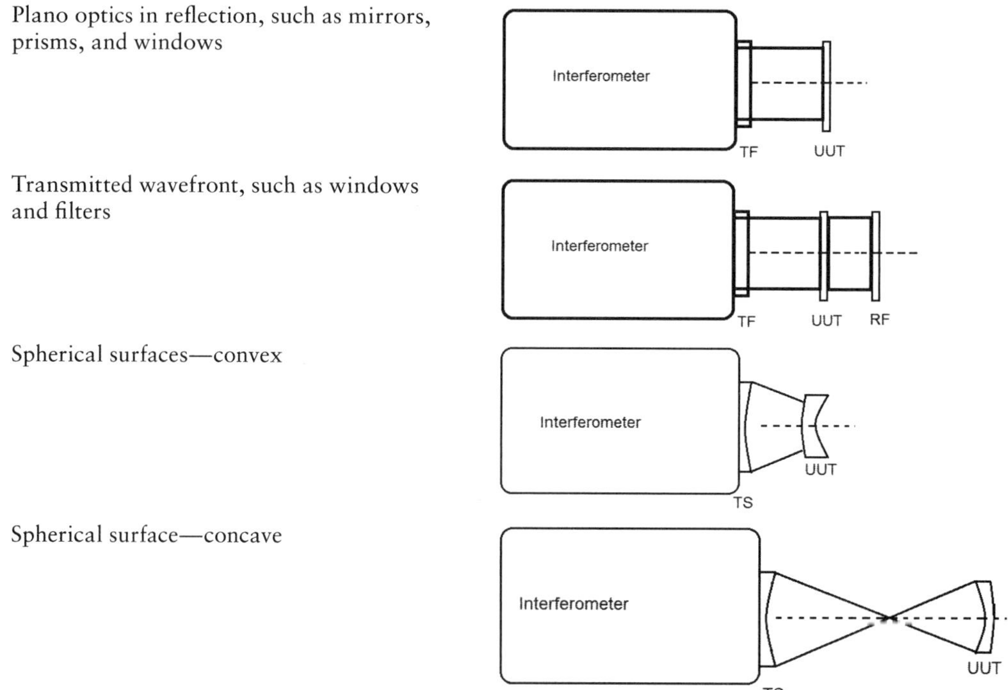

Table 2.1   *(Cont.)*

Lens/system performance method 1

Lens/system performance method 2

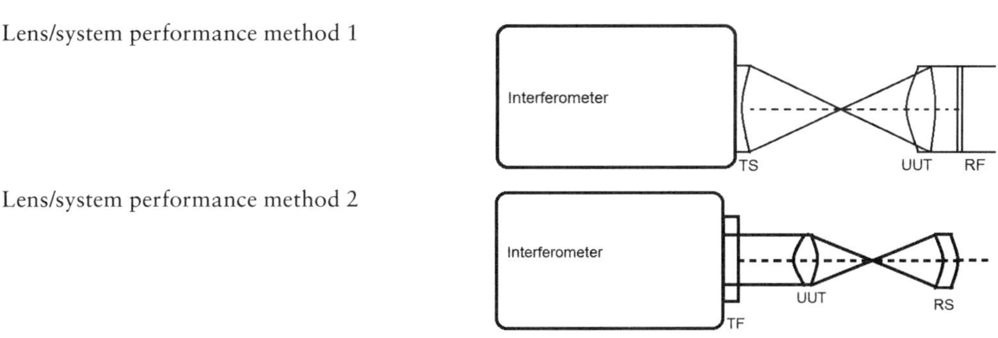

Only the test arm is shown in the figures. TF: transmission flat; RF: return flat; TS: transmission sphere; RS: return sphere; UUT: unit under test.

EXERCISE 2.3

When measuring the transmitted wavefront of a window, the return flat has to be placed close to the window. Can you explain why?

Solution

To correctly measure the surface, an interferometer has to focus on it. When the surface is not in focus, it will cause two issues. One is edge diffraction—the edge of the test surface has diffraction "ripples." The other is smoothing of the high spatial frequency features on the test surface.

When measuring a window in transmission, the test beam sees the windows twice: before and after reflection off the return flat (RF). If we unfold the light path about the RF, the window and its mirror image are located at both sides of the RF. The interferometer cannot focus on the window and its mirror image simultaneously. The window and its mirror image can both be in focus as long as we place the return flat close to window.

2.3.2   Software Setting

It is important to set up the software configuration correctly to get meaningful results.

Map Orientation

Depending on the test geometry, the interferogram may be flipped horizontally or vertically or both. It is helpful to stick a piece of paper or pen in front of the test optic and take a measurement. By looking at the shadow created by the paper or pen, one can easily figure out the orientation of the map and edit it in the software.

Sign of the Phase

The interferometer provides a 2D map. It is important to know whether the high shown in the map is actually high, instead of low. An easy way to check that is to introduce some tilt or power fringes along a known direction and then use the measured phased map (without removing tilt and power) to find the sign of the interferometer.

Wedge Factor

An interferometer measures the wavefront difference between the test and reference beam. The direct information from an interferogram has no indication as to whether the wavefront has traveled through a window, reflected from a surface, or at what angle it has reflected from a surface. The interferogram wedge factor (or scale factor) specifies how this input wavefront error is scaled to properly represent the physical parameters to be measured. The common scale factor is 0.5 because one often is interested in error in the surface, rather than the wavefront. Figure 2.16 shows an example of test configuration. The wavefront error the interferometer measures from the test surface is $W = 2 \times 2S\cos\theta$, where S indicates the surface error and θ is the incident angle on the surface under test. If one is interested in the surface error of the test optic, then the wedge factor should be $1/4\cos\theta$.

Mask

Before executing a measurement, one should define a mask so only the data of interest are included in the calculation. Many measurement results, such as RMS, PV, and Zernike coefficients, depend on the mask.

Term Removal

It is very common to remove the piston, tilt, and power from the measurement because these are often alignment errors and not properties of the surface.

2.3.3 Calibration

There are many sources of errors in measured wavefront data, such as stray reflections, quantization errors, detector nonlinearity, frequency and intensity instability

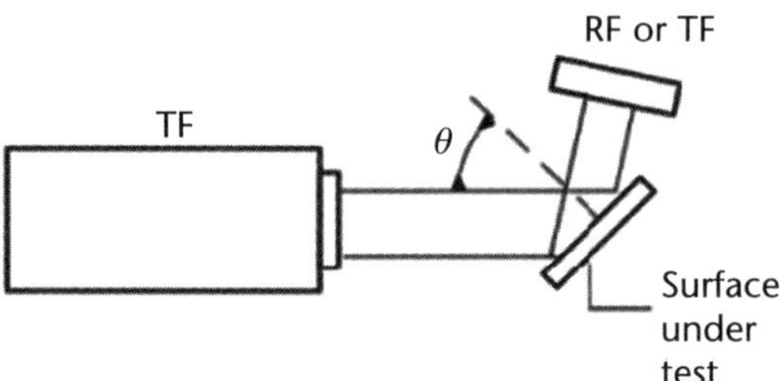

Figure 2.16 The interferometer wedge factor is set to scale the data appropriately. The interferometer measures reflected wavefront variations. The test geometry dictates the relationship between this wavefront and the surface under test.

in the light source, and incorrect phase shifts between data frames. For commercial interferometers, most of these errors have been either minimized or calibrated.

Fringe Print-Through Error

Fringe print-through error is a common error that appears in the phase map as shown in Figure 2.17. The print-through error is an error in the shape measurement that is correlated with the interference fringe pattern, and it often has twice the original fringe frequency. The cause of fringe print-through errors are vibration, saturated light intensity, or incorrect phase shifter calibration. Vibration causes the phase shift between frames that are different from the designed value. Saturated light intensity makes it difficult to find the correct fringe center. Incorrect phase shift calibration between data frames also introduces fringe print-through errors. An example of phase shift calibration for a five-step measurement is given in Figure 2.18. In general, adding more phase steps reduces this error but increases the likelihood of measurement errors from the environment. One should identify the cause of the print-through errors and then address it accordingly.

In most interferometer software, the phase shifter can be calibrated. For example, one can adjust the PZT slope (voltage/sec) so that the phase shift is exactly 90 degree, as shown in Figure 2.18.

Transmission Sphere Calibration

In interferometric testing, the measured optical path difference (OPD) is the difference between the reference and the test beams. For high accuracy measurement, errors in transmission optics are on the order of the magnitude of the test optic; therefore, it must be calibrated to achieve the best estimate of the quality of the test optic. One calibration method is the random ball test, where randomly selected patches of a high-quality spherical ball are measured and the results are averaged to obtain an estimate of the reference surface. Errors in the ball average to zero in the limit of infinite measurements. Practical implementation of this test can use the commercially available CaliBall and the kinematic mount shown in Figure 2.19(a).

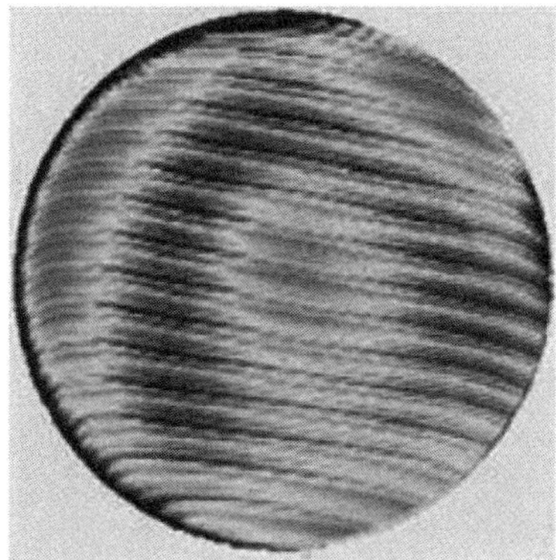

Figure 2.17   The phase map is an image that depicts the measured surface shape errors. This maps shows an example of the fringe print-through errors in the phase map.

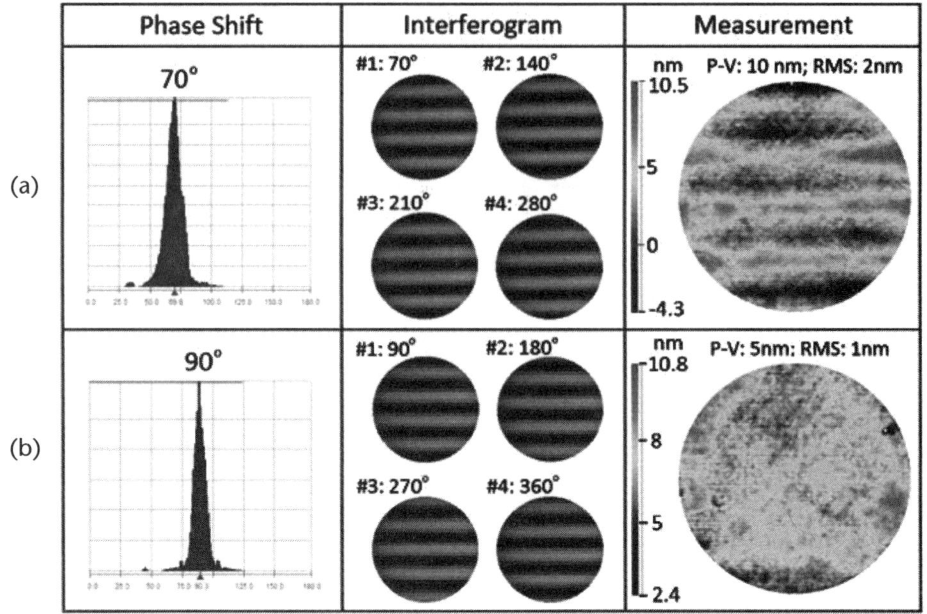

Figure 2.18 Incorrect phase shift calibration causes fringe print-through error; the required phase shift is 90°: (a) the phase shift is incorrect; (b) the phase shift is correctly calibrated.

Figure 2.19(b) gives an example of the calibrated reference surface for a transmission sphere.

There are other absolute tests, such as three-flat/sphere [7] and three position tests [2]. Those techniques generally require critical alignment of the test surface and an extremely good rotation stage for 180° rotation.

2.3.4 Processing Data

Zernike Fit

Zernike polynomials are often used to describe wavefront data since they are of the same form as the types of aberrations observed in optical tests. These polynomials are a complete set and orthogonal over a unit circle. There are two common

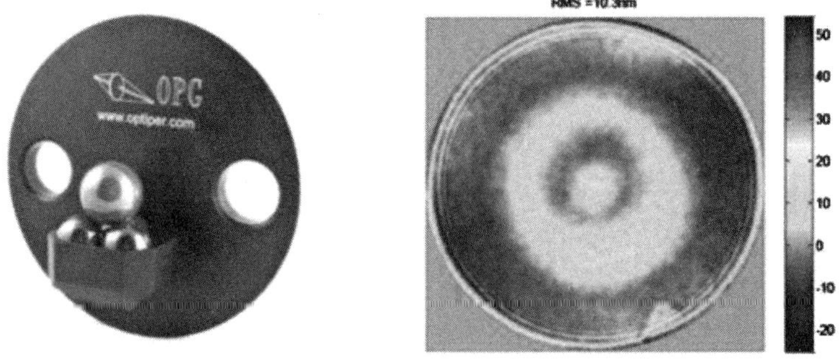

Figure 2.19 CaliBall and its mount (left); calibrated reference surface error (right). (Courtesy of Optical Perspectives Group.)

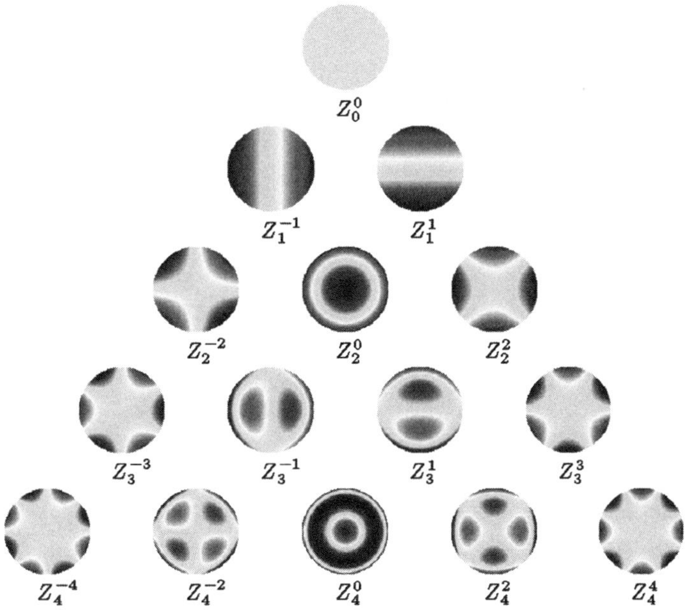

Figure 2.20   Zernike polynomial wavefront maps.

definitions for Zernike polynomials: one is Zernike fringe polynomials; the other is Zernike standard polynomials. They have different numbering and normalization. Figure 2.20 shows the lowest order Zernike polynomials.

Any wavefront can be described as a sum of Zernike polynomials and a residual that describes all remaining features:

$$W(x,y) = \sum_j C_j Z_j(x,y) + W_{resid}(x,y) \tag{2.7}$$

Figure 2.21 shows a wavefront that is described by 37 terms of Zernike standard polynomials and the residual.

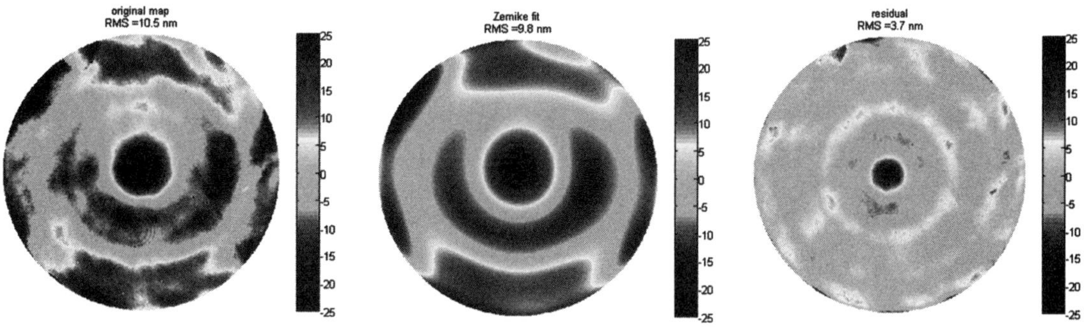

Figure 2.21   Original wavefront (left), 37 standard Zernike fit of the original wavefront (middle), and the residual (right).

Removing Alignment Terms from the Data

Misalignment of the test causes specific measurement errors that appear as tilt and power in the data. Small rotations of a flat or sphere by an angle α from ideal will cause a surface measurement with linear variation and a peak-to-valley (PV) variation of αD where D is the diameter of the measured region. The decenter of a sphere is equivalent with a small amount of tilt.

We define the R-number (R_n) as the ratio of the radius of curvature R to the diameter D. Decentering a spherical surface by an amount δx is equivalent to tilting the surface by an amount $\alpha = \delta x/R$. The measured surface variation will have PV tilt of $\delta x/R_n$.

Axial displacement of a flat does not affect the measurement. But axial displacement δz of a spherical surface will cause power, or quadratic variation, in the measurement. The PV power in the surface measurement due to this displacement is computed using the relation $\delta z/SR_n^2$.

Since tilt and power are created by alignment of the test and are not intrinsic to the surface being measured, these terms are frequently removed from the data.

2.3.5 Estimation of Measurement Uncertainties

Errors in interferometric measurement can be categorized into random error, retrace errors, and imaging distortion. Retrace error is negligible as long as the measurement is a null. Once there are many fringes in the measurement, the test beam and reference beam have a large mismatch. Therefore, the two beams are no longer common path, and this introduces errors.

Imaging distortion typically occurs when using a fast transmission sphere or null optics. It will cause two problems: one is that the surface defects appear shifted, and the other is that lower order alignment errors appear as high order wavefront errors. Imaging distortion can be corrected by remapping to remove the distortion and is discussed in Section 2.5.5.

Random errors usually dominate most measurements of flats and spheres. Sources for this error include laser source instability, mechanical vibrations, air turbulence, and detector noise. Averaging a number of measurements reduces the random error and improves measurement accuracy. The variation in the data can be used to estimate the noise in individual maps and the magnitude of the random error in the averaged map.

Since the true surface figure of the test optic is unknown, an average of many measurements is used instead as an estimate of the surface figure. The difference between a single measurement and the average map provides an estimate of the measurement noise.

An example of estimating the measurement noise for a reference sphere with a radius of curvature of 16m is provided here. For this example, the radius of curvature of the mirror is so long that the air turbulence causes a large amount of noise. Figure 2.22(a) shows a map of a single interferometric surface measurement with an RMS surface error of 47.7 nm. The astigmatism shown in the picture is mainly due to the turbulence. The average of 70 measurements, $\overline{W}$, shown in Figure 2.22(b), is more rotationally symmetric, and the RMS error is reduced to 28.9 nm.

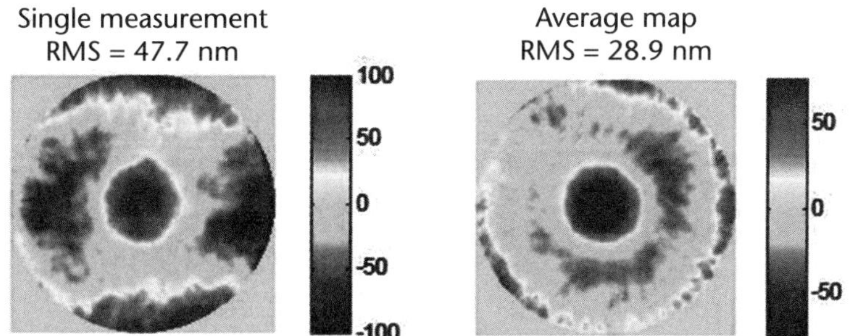

Figure 2.22   Map of (a) a single measurement and (b) an average of 70 measurements.

The average of 70 measurements is certainly less noisy, and the noise in the average map can be estimated using the method described next.

From the 70 measurements, $N(N < 70$) maps are chosen at random and averaged. This average map, denoted as $\overline{W_N}$, is subtracted from $\overline{W}$, and the RMS of the average is calculated as $\mathrm{RMS}(\overline{W_N} - \overline{W})$. This process is repeated several times, and the mean and the standard deviation of $\mathrm{RMS}(\overline{W_N} - \overline{W})$ are calculated. This is then repeated for various values of N (from 1 to 39 in this case), and the results are plotted in Figure 2.23(a). The curve follows the mean, and the error bars represent the standard deviation. The mean of $\mathrm{RMS}(\overline{W_N} - \overline{W})$ becomes significantly smaller as the number of averaged measurements increases. This plot gives the magnitude of residual noise in the average of N maps. For example, the noise is about 10 nm rms when 15 measurements are averaged. Figure 2.23(b) is the same curve plotted in a log-log scale. A straight line with a slope of -0.51 is fitted to the data, which means that the noise in the measurements is random since it is dropping off by $1/\sqrt{N}$. By extrapolating, the noise in the average of 100 maps is reduced to about 3.5 nm.

A plot like Figure 2.23 is very useful, because it can be used to estimate the residual error in an average map and the number of measurements needed to achieve the desired accuracy. As conditions improve, the slope of the log-log plot remains

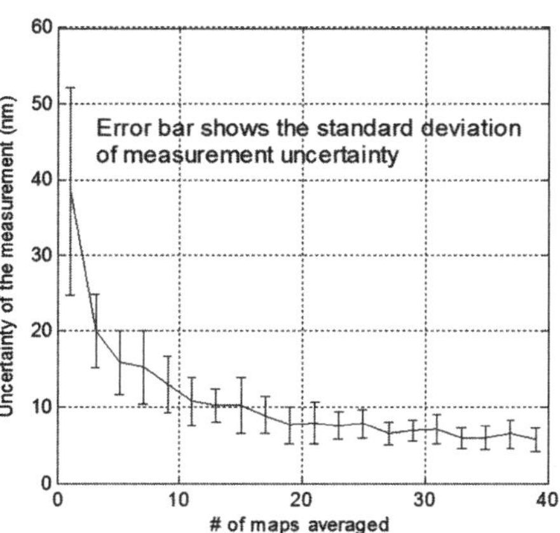

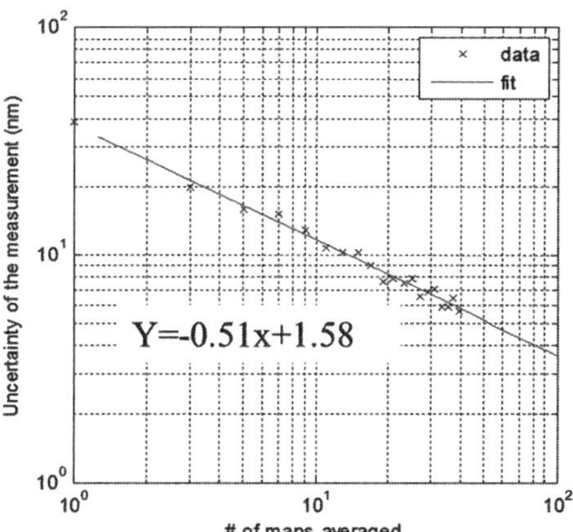

Figure 2.23   Measurement noise as a function of the number of measurements averaged: normal scale (left); log scale (right) showing that the noise is dropping inversely with the square root of the number of measurements in the average.

0.5, but the entire curve shifts down. The y-intercept of this curve represents the quality of the testing environment and the accuracy of a single measurement.

EXERCISE 2.4

What is the reason that we do not use the standard deviation of the RMS value of each individual map to estimate the random error?

Solution

Two wavefront maps can have the same RMS values but completely different shapes. If we use only RMS value to estimate the random error, we will not correctly estimate the random error in the measurements. The random errors are determined using direct pixel-to-pixel subtraction of individual maps from an average of many maps.

2.4 Measurement of Aspherics

2.4.1 Definition and Types of Aspherics

An asphere is a surface that is neither flat nor spherical. Some typical asphere types are described next.

Conic Surface

A conic surface is described by the following sag equation:

$$Z(r) = \frac{r^2/R}{1 + \sqrt{1 - (1 + k)r^2 / R^2}} \tag{2.8}$$

where

$r = \sqrt{x^2 + y^2}$: the radial coordinate

R: radius of curvature at the vertex of the surface

k: conic constant

$K = 0$: sphere

$k = -1$: parabola

$k < -1$: hyperbola

$-1 < k < 0$: prolate ellipsoid

$k > 0$: oblate ellipsoid.

An off-axis segment of a conic surface is described by an additional parameter: off-axis distance, d, which is defined as the radial position of the segment's center in the parent coordinate.

A conic surface has two foci, which are conjugates to each other.

Conic Surface Plus Additional Polynomial Terms

This type of surface is described by the sag equation:

$$Z(r) = \frac{r^2/R}{1 + \sqrt{1 - (1+k)r^2/R^2}} + \sum_{n=2}^{m} A_{2n} r^{2n} \tag{2.9}$$

where A_{2n} is the coefficient for the polynomial term r^{2n}.

Again an off-axis segment of such a surface is described by an additional parameter: off-axis distance d.

Toroidal and Atoroidal Surface

The shape of a toroidal surface can be thought of geometrically as the surface obtained by rotating an arc of a circle about an axis. Depending on the radius of the arc, the surface has the shape of a donut, or a barrel, as seen in Figure 2.24. A cylinder surface is a special type of toroidal surface. If the rotating arc is part of a noncircular curve, then an atoroidal surface results.

Freeform Surface

In optical applications, a freeform surface can be loosely defined as a smooth surface of irregular shape. Figure 2.25 shows the surface shape of a progressive addition eyeglass lens, which is a freeform. This surface cannot be described by any of the previous equations.

2.4.2   Configurations for Null Testing Aspheric Surfaces

A null test is an interferometric test of an aspheric surface where null fringes are obtained when the surface under test has the exact prescription. Basically, additional optics are included to allow the interferometer's measurement of its spherical wavefront to provide the data on the aspheric surface.

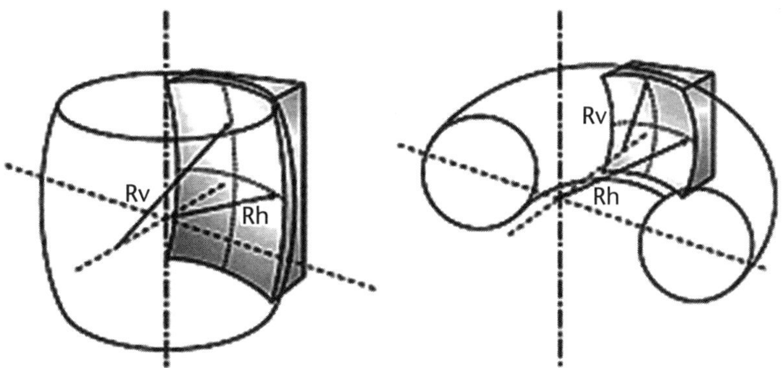

Figure 2.24 Toroidal surfaces of barrel type (left) and donut type (right).

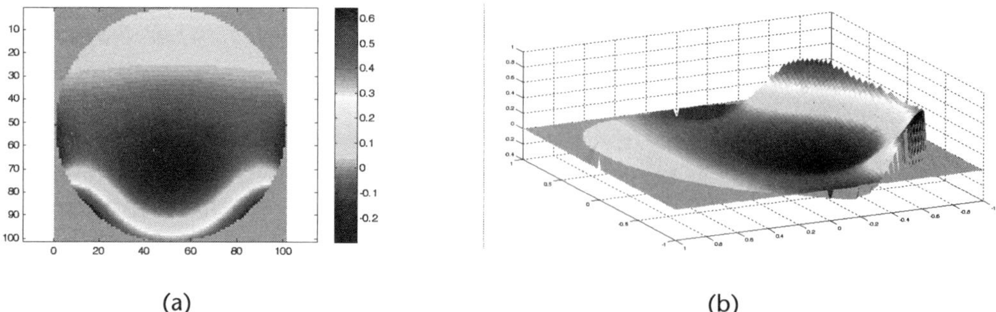

(a)   (b)

Figure 2.25 Shape of a freeform surface of a progressive lens: (a) 2-D plot; (b) 3-D plot.

There are two primary types of interferometric null tests of the aspheric surfaces, the conjugate test and the null lens test.

Conjugate Test

The conic surface has two conjugate foci, which is taken advantage of in the null test of the surface. Some well-known techniques include autocollimating test for a paraboloid or an off-axis segment of it, and the Hindle test of hyperboloid.

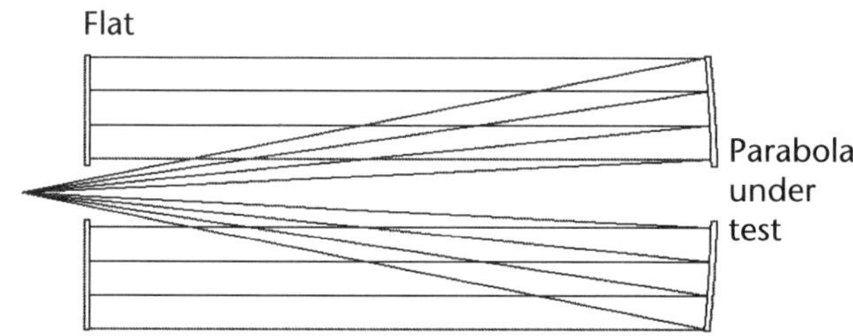

Figure 2.26 Autocollimation test of a paraboloid (a parabola of revolution).

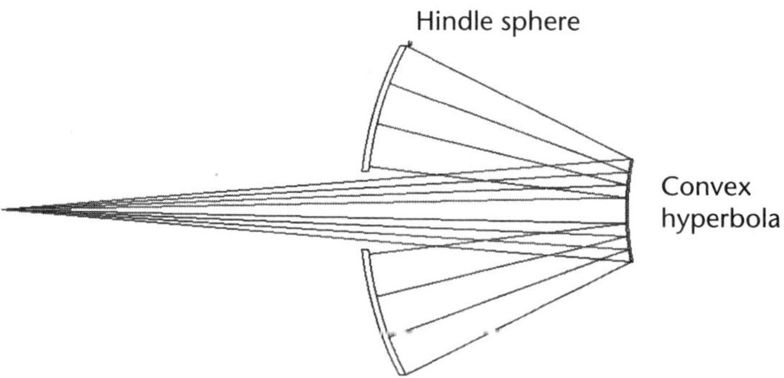

Figure 2.27 Layout of a Hindle test of hyperbola.

Null Lens Test

In this type of null test, a null lens is used to create an aspheric wavefront that matches the perfect aspheric surface. Depending on the way the aspheric wavefront is created, there are different types of null lenses:

a. Conventional refractive or reflective null: an example of Offner null lens is shown in Figure 2.28, where two lenses—relay and field lenses—are used to convert a spherical wavefront to a wavefront of parabolic shape that matches the ideal surface under test.

b. Computer generated holograms (CGH): CGH changes wavefront by diffraction. Figure 2.29 shows a test of a general aspheric surface with a CGH as null lens. See details about the CGH in Section 2.5.

c. Combination of refractive, reflective, or diffractive nulls: in some special cases (e.g., when the surface to test has huge aspheric departure—a few mm PV), combinations of different types of null lenses are required. Figure 2.30 shows the null test of the 8.4m segment of the Giant Magellan Telescope's primary mirror (14 mm PV aspheric departure) where the null lens consists of three components: a 3.75m diameter large fold sphere, a 0.75m diameter small fold sphere, and a 6-inch CGH.

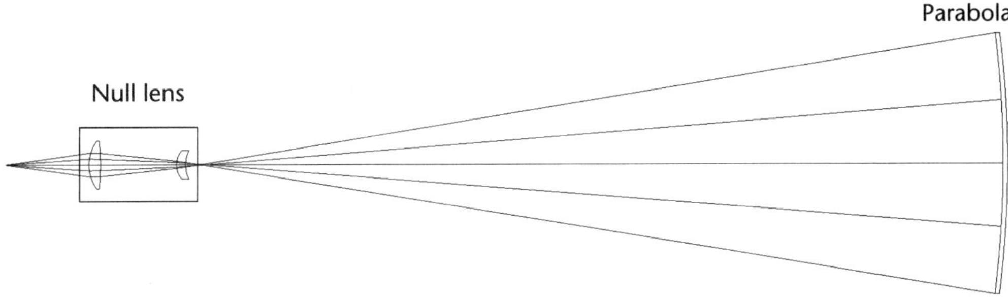

Figure 2.28 Testing an F/1.75 parabolic surface with an Offner null lens. (Interferometer and transmission sphere are not shown.)

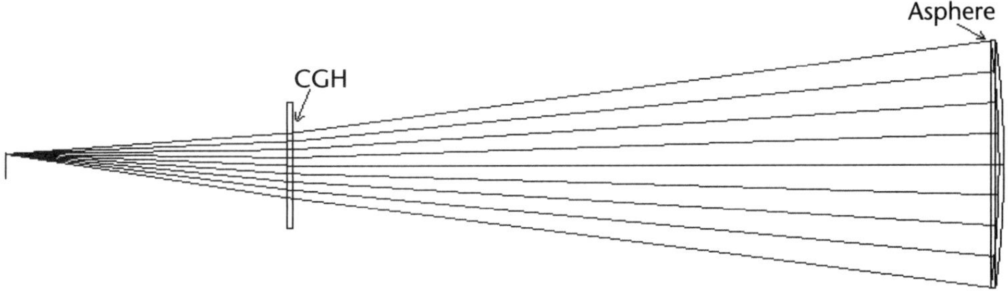

Figure 2.29 Layout of a CGH test of a general aspheric surface. (Interferometer and transmission sphere are not shown.)

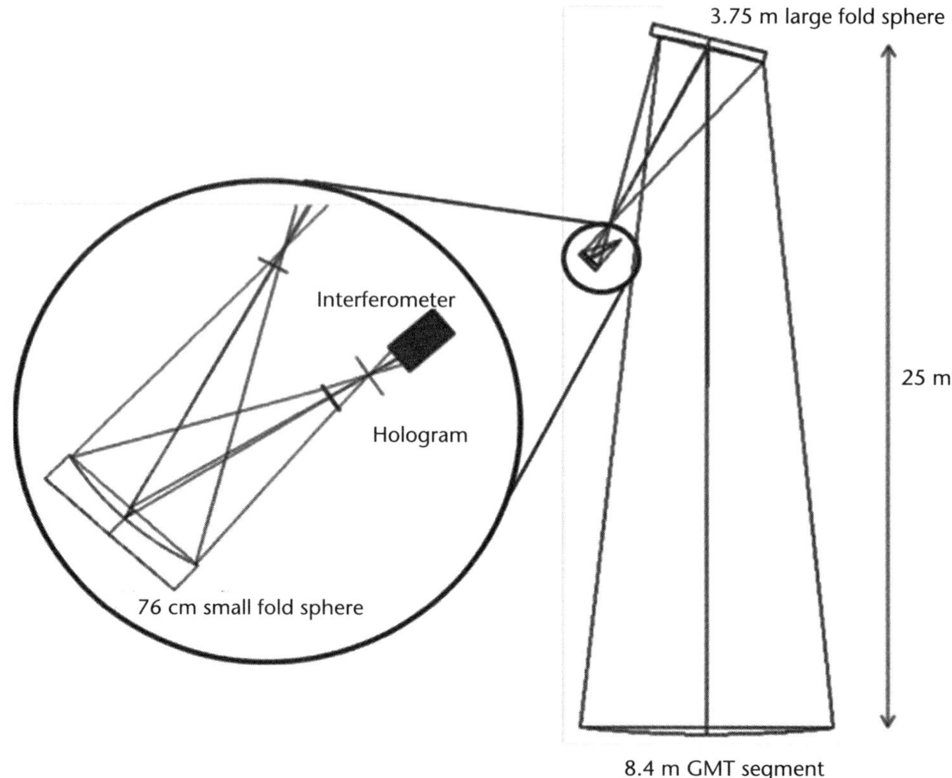

Figure 2.30 Layout of optical test for the Giant Magellan Telescope off-axis primary segment.

2.5 Measurements of Aspherics Using Computer Generated Holograms

The computer generated hologram (CGH) is a diffractive null lens for measuring an aspheric optical surface. It is used together with a laser interferometer, similar to its traditional refractive or reflective counterpart. A spherical or flat incident wavefront is converted by the CGH to an aspheric wavefront that matches the shape of the ideal aspheric surface under test, resulting in a null test of the surface, as shown in Figure 2.31.

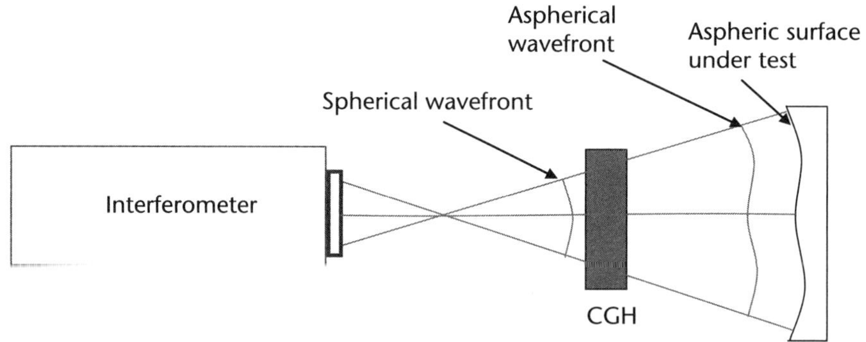

Figure 2.31 Illustration of an interferometric setup for testing an aspheric surface with a CGH.

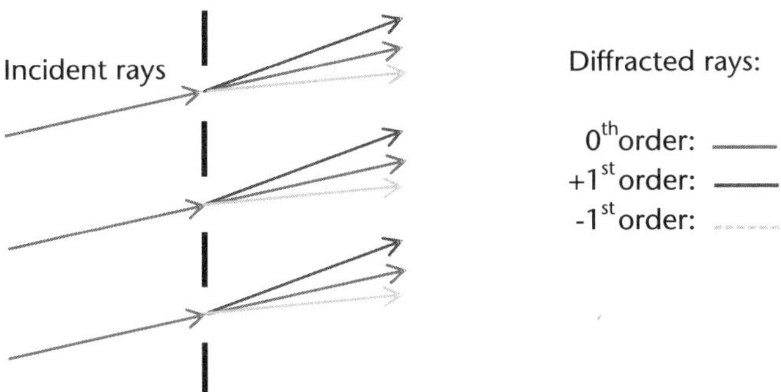

Figure 2.32 Illustration of CGH redirecting rays via diffraction.

The CGH changes the wavefront, or redirects rays, via diffraction like a grating, as illustrated in Figure 2.32.

For a majority of aspheric surfaces, a single CGH is adequate to measure the surface, and the CGH can be fabricated accurately at a reasonable cost using semiconductor photomask fabrication technology. In comparison, a refractive or reflective null lens usually requires two or more components. Each component of the null lens must be fabricated accurately, and then all the components need be assembled accurately. As such, it often costs more and takes longer to make a refractive or reflective null corrector than the CGH.

2.5.1   Principles Behind CGH Interferometry

Mathematical Representation of CGH

A CGH, or a diffractive optical element in general, is mathematically described by a phase function. When a ray hits a point on a CGH surface, it picks up the phase at that point. Then, the ray would change direction according to the general vector form of Snell's Law:

$$\left(n_1 \hat{r}_1 - n_2 \hat{r}_2 + \nabla\Phi \right) \times \hat{n} = 0 \tag{2.10}$$

where, as illustrated in Figure 2.33, Φ is the phase function of a diffractive surface; n_1 and n_2 are refractive indices before and after the surface, respectively, $\hat{r}_1$ and $\hat{r}_2$ are the unit vectors of the ray before and after the surface; and $\hat{n}$ is a vector along the normal of the surface.

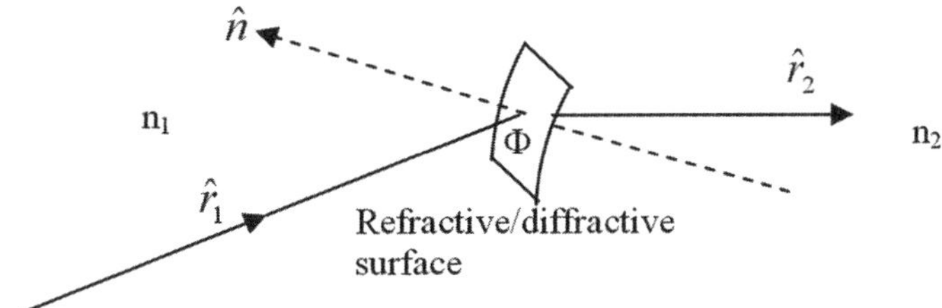

Figure 2.33 Illustration of a ray being refracted/diffracted at a diffractive surface defined by a phase function Φ.

CGH Types and Diffraction Efficiency

The CGH for optical testing consists of line patterns made onto an optical surface that is most often flat. The lines are curved and unequally spaced, looking similar to interference fringes; therefore, the CGH lines are often called fringes. In any local area though, the CGH can be approximated as a linear grating. The CGH is usually binary. There are two types of binary CGHs:

1. Amplitude CGH, where the intensity of the laser beam is modulated;
2. Phase CGH, where phase of the laser beam is modulated.

The phase CGH has higher diffraction efficiency than the amplitude CGH. When testing an uncoated surface in double pass setup, a phase CGH is required to get fringes of good contrast. The phase CGH costs more than the amplitude CGH because additional steps are required for fabrication.

Chang and Burge developed a parametric model for the CGH [10]. Figure 2.35 depicts a cross-sectional view of a binary grating. It is defined by period S and the etching depth t. The duty-cycle of the grating is defined as $D = b/S$, where b is width of grating line, n is the refractive index of the grating, and A_0 and A_1 correspond to the amplitudes of the output wavefront from the space and line of the grating, respectively. The phase step ϕ represents the phase difference between these two areas, which equals $2\pi(n - 1)t/\lambda$ for a grating used in transmission.

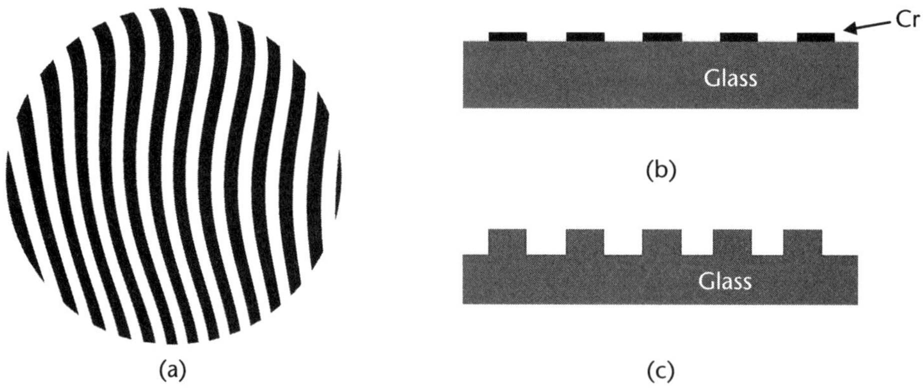

Figure 2.34   Illustration of (a) CGH lines or fringes, (b) amplitude CGH defined by Cr lines on glass surface, and (c) phase CGH defined by grooves etched into glass.

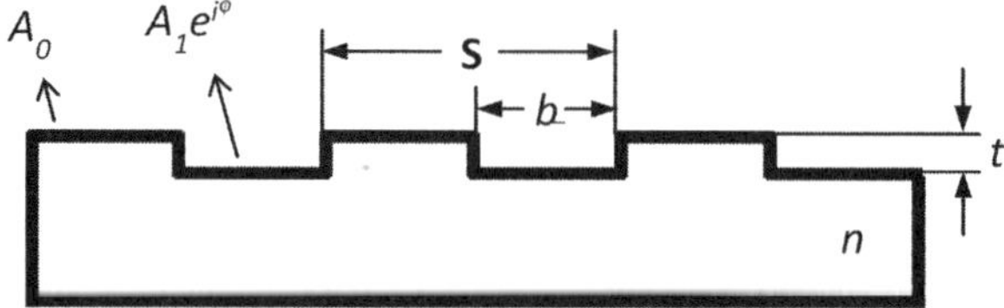

Figure 2.35   Parametric model for the shape of the wavefront immediately after the CGH. Most of the CGHs use 50% duty cycle (i.e., $D = 0.5$).

Using scalar diffraction theory and the linear grating approximation, equations for calculating diffraction efficiency can be derived.

For 0th diffraction order:

$$\eta_{m=0} = A_0^2(1 - D)^2 + A_1^2 D^2 + 2A_0 A_1 D(1 - D)\cos(\phi) \tag{2.11}$$

For non-0th order:

$$\eta_{m\neq 0} = \left[A_0^2 + A_1^2 - 2A_0 A_1 \cos(\phi) \right] D^2 \operatorname{sinc}^2(mD) \tag{2.12}$$

For 50% duty cycle, the amplitude CGH has diffraction efficiency of 25% in 0th order, and 10% in the ±1st orders. For 50% duty cycle and a half wave phase step ($\phi = \pi$), the phase CGH has diffraction efficiency of 0% in the 0th order and 40% in the ±1st orders. For order m where $|m| > 1$, the diffraction efficiency is $\eta_m = \eta_1/m^2$.

EXERCISE 2.5

Question: Estimate the fringe contrast of the test of an aspheric surface with a Fizeau interferometer and a CGH, assuming the null pattern (1st diffraction order is used) is phase etched with 50% duty cycle and half wave phase step, the aspheric surface is made on a BK7 substrate, and the laser wavelength is 632.8 nm.

Solution

Assume the total laser energy incident at the reference surface of the transmission sphere 1, the reference beam intensity right after the reference surface is then 0.04. The test beam goes through the reference surface (96% transmission single pass) and the CGH (40% transmission single pass) twice, and reflected by the test surface once (4% reflection.) So the test beam intensity is 0.96^2 * 0.40^2 * 0.04 = 0.0059 right after it passes the reference surface in the return path. Then the fringe contrast is 2 * sqrt(0.04 * 0.0059)/(0.04 + 0.0059) = 66.9%.

Benefits of Using CGHs—Great Accuracy and Flexibility

There exist a few machines in the world that were developed specifically for writing symmetric CGH patterns, but the majority of the CGH are written with electron-beam or laser writers developed for making photomask for semiconductor industry. The writing accuracy of either type of machines, defined by the placement accuracy of the pattern lines, is extremely high (~10 nm rms over 100 mm diameter area), which enables accurate control of the aspheric wavefront shape produced by the CGH.

The error of the wavefront a CGH produces has direct relationship with the writing error as follows:

$$\Delta W(x,y) = \vec{\varepsilon}(x,y) \cdot \vec{S}(x,y) \tag{2.13}$$

where

ΔW is wavefront error

$\bar{\varepsilon}$ is the local writing error

$\vec{S}$ is the local phase slope

For a given CGH phase function $\Phi(x, y)$, $\vec{S}(x,y) = -\nabla\Phi(x,y)$.

Another CGH advantage is its flexibility—it is able to produce virtually any wavefront shape, which means it is able to measure any surface shape, limited only by the minimum writable features. Traditional refractive or reflective nulls have very limited design degrees of freedom because each surface of each element is usually sphere. For nonsymmetric surfaces, it is extremely difficult to design and make a traditional null lens. As mentioned earlier, the CGH is approximately a linear grating in any small local area. The direction and pitch of the linear grating determines how a ray is redirected. Think of CGH as an ensemble of a great number of linear gratings, which affords it the flexibility to change a wavefront any way one wants.

Furthermore, multiple patterns are often made on the same substrate, each of which performs a specific function. For example, the major pattern produces the aspheric wavefront, an alignment pattern used in reflection allows accurate alignment of the CGH with regard to the interferometer, and some fiducial projection patterns projects focused spots around the surface under test to allow rough alignment of the test surface.

Limitations for the Use of CGHs—Substrate Quality and Ghost Orders

Specifically built CGH writers allow high-precision substrates but only write circular patterns. While the photomask writers are able to write arbitrary patterns, they only allow standard substrates. For the latter, the limitation of the overall CGH accuracy is the quality of the substrate. The standard photomask substrates have a larger form factor than what is acceptable in optics industry. The quality of the substrate surface and the transmitted wavefront is not ideal either. For demanding applications where nonsymmetrical patterns are required, custom-made high-quality substrates are necessary.

Redirecting rays via diffraction also brings about a disadvantage—the unwanted diffraction orders. The unwanted diffraction orders form ghost fringes, which reduce measurement accuracy if not fully blocked; they also reduce the number of photons of the wanted order, which may degrade the fringe contrast. In a lot of cases,

Table 2.2 Dimension Specifications for Selected Standard Photomask Substrates

Type	Edge Length Min. (mm)	Edge Length Max. (mm)	Thickness Min. (mm)	Thickness Max. (mm)
4009	100.8	101.6	2.2	2.4
5009	126.2	127.0	2.2	2.4
6012	151.6	152.4	2.95	3.15
6025	151.6	152.4	6.25	6.45
7012	177.0	177.8	2.95	3.15

these disadvantages can be are overcome: separation of diffraction orders are made possible by adding carriers—tilt carrier for lateral separation or power carrier for longitudinal separation, and phase CGH is used to increase diffraction efficiency of the wanted order (usually 1st order). In the CGH design process, a lot of effort is spent on either totally eliminating the ghost fringes from the unwanted orders or reducing them to an acceptable level.

2.5.2   CGH Design

The typical CGH test involves an interferometer, a CGH, and the surface under test. The first goal of the design process is to find a working test layout (i.e., where each element is with regard to each other in the test). The design is done in an optical design code, such as Zemax or Code V, where the interferometer is represented by a focus point or a collimated beam assuming a transmission sphere (TS) or transmission flat (TF) is used, respectively. The CGH design is done by purely geometric ray tracing despite the fact that the CGH is a diffractive element. The reason for this is because a CGH for optical testing is represented mathematically by a phase function. The phase function is the outcome of the design process besides the working test layout.

Before design of a CGH, gather the following information:

1. Prescription of the surface to be tested;
2. The interferometer the CGH is to be used with, especially the laser wavelength, the aperture size, and the available transmission spheres or diverger lenses;
3. The CGH substrate material, dimensions, and maximal patternable area;
4. Required alignment features;
5. Required measurement accuracy.

Note: Laser wavelength must be accurately known before a CGH is designed! Be very careful using diode laser whose wavelength may shift.

To begin the design process, set up a single pass ray tracing model starting from the test surface that has the same shape as the wavefront that the CGH needs to create for the null test of the surface. A trick is to set the index of refraction for the medium right before the surface to be 0, then according to the Snell's law, the

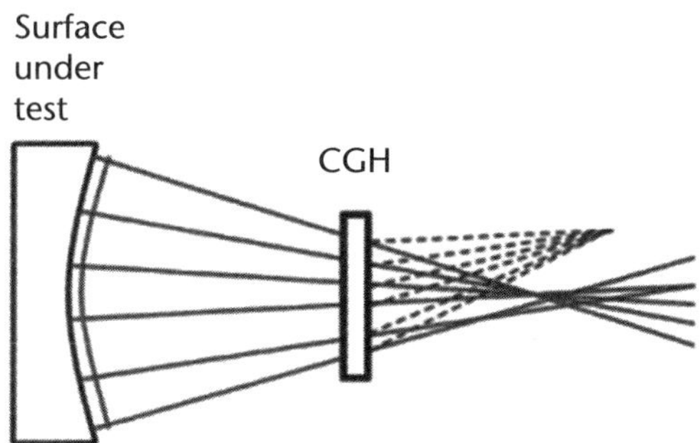

Figure 2.36   Single pass ray tracing model for determining a CGH test layout.

rays coming out of the surface are perpendicular to the surface. By definition, the wavefront defined by these rays coincides with the surface.

We need to find where the CGH and focus need to be with respect to the surface under test such that

1. The size of the null pattern is appropriate (i.e., together with other patterns fit to the available patternable area of the substrate);
2. Ghosts from other orders are absent or acceptably small;
3. The line spacings of the patterns are big enough such that they are writable, and the wavefront the pattern produces is accurate enough knowing the certain writing accuracy.

Once a working test layout is found, the CGH phase function for this layout is uniquely defined. We can use the built-in polynomial types of surface of an optical design code to represent the phase function, set the coefficients of the polynomials as variables, and optimize the system to produce the wanted results. In some special cases, this approach fails to converge, but we can still numerically calculate the phase values at a grid of points, then fit the phase values at any arbitrary point using cubic spline fitting. Zemax has a built-in surface type—grid phase—that does just that.

After the null pattern phase function is determined, make sure to set up a double pass model to perform the ghost analysis. If the ghost is not totally eliminated or reduced to an acceptable level, go back to the single pass model to find another layout. Iterate this process until a working layout is obtained. Then, start with the double pass model and design other patterns to be made on the same substrate (e.g., alignment pattern, the fiducial projection patterns, and so on).

2.5.3 CGH Fabrication

CGH phase must be converted to fringe position data before it can be fabricated. For symmetric CGH to be fabricated with a circular laser writer, this process is straightforward, as illustrated in Figure 2.37. To fabricate a CGH with the photomask fabrication technology, a smooth fringe must be approximated by polygons because only polygons can be written by the machine. This process is called digitization and is illustrated in Figure 2.38. After digitization, the data is usually encoded into an intermediate exchange format (e.g., GDSII). The writing machine reads in the encoded data and further converts it to a machine specific format; this process is called fracturing. Afterward, the patterns are written (and etched) following the processes illustrated in Figure 2.39.

To make sure a CGH is designed correctly for a specific surface and fabricated accurately, the following quality assurance checks are recommended:

1. The CGH is designed for testing the specific surface.
2. Models for different patterns (i.e., null, alignment, and fiducial patterns) are consistent.
3. CGH fabrication data consistent with the design data.
4. Patterns are placed accurately.

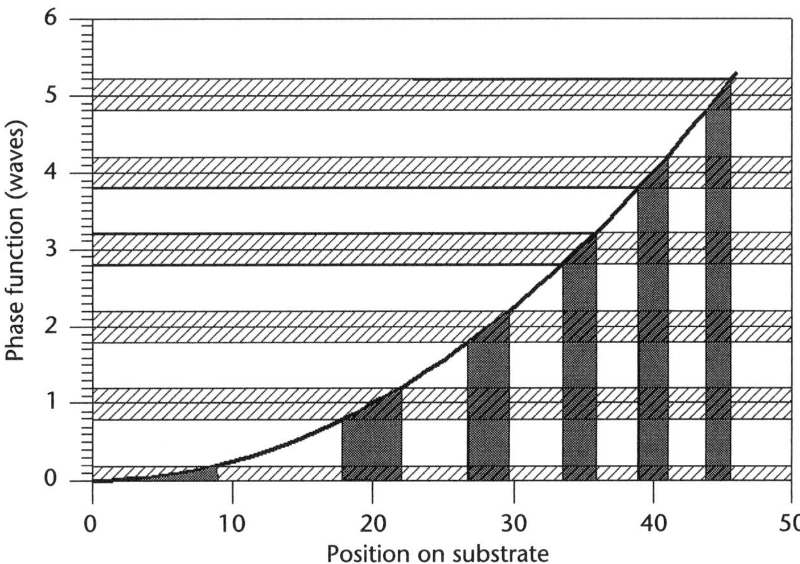

Figure 2.37 Generation of the fringe positions from the phase data for a symmetric CGH to be fabricated by a circular laser writer.

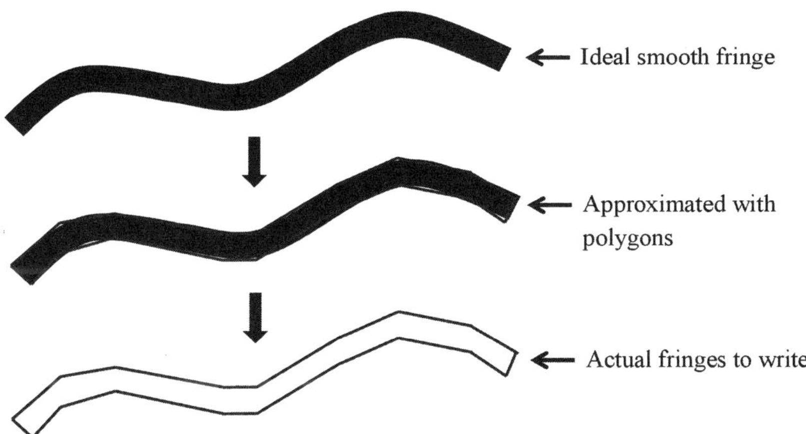

Figure 2.38 Illustration of the digitization process for fabrication of a general CGH.

2.5.4   Alignment of CGH Test

For a typical CGH test consisting of an interferometer, a CGH, and the surface under test, the CGH and the aspheric surface need to be aligned to the interferometer, which can be made possible or aided by making the following patterns together with the null pattern on the same substrate:

1. CGH alignment pattern: for aligning the CGH as the name suggests. It's a Cr pattern retro-reflecting the incident laser beam to form fringes, the number of which is minimized by adjusting the CGH with regard to the interferometer to get the CGH aligned.

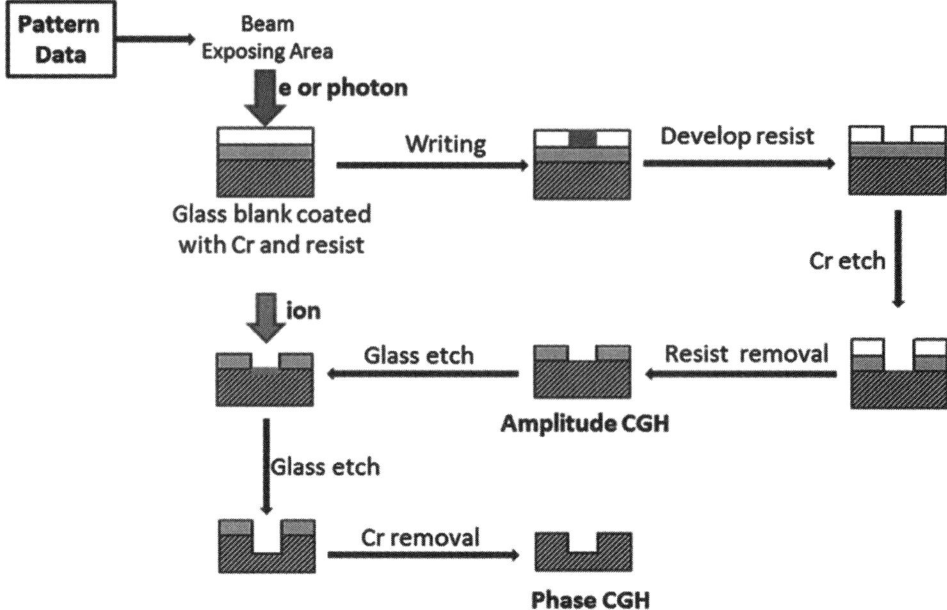

Figure 2.39   Illustration of the fabrication process of both the amplitude and phase types of CGH.

2. Fiducial projection pattern: for aligning the aspheric surface. Focused spots are projected to particular positions of the aspheric surface (e.g., the center or the edge for the coarse alignment of the surface).

Figure 2.40 shows a scaled fringe plot of a CGH with multiple patterns and the actual picture of the CGH. The ray trace model of the test is shown in Figure 2.41.

2.5.5   Data Reduction

One important characteristic of the CGH testing is the image distortion of the aspheric surface, which is manifested by the mapping distortion between the aspheric surface and the CGH. Therefore, morphing is necessary to get the correctly scaled

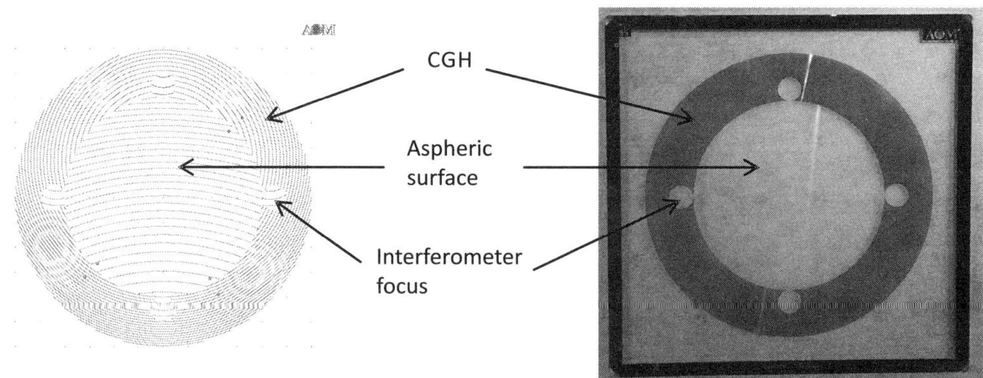

Figure 2.40   (a) Scaled fringe plots of all the CGH patterns; (b) actual picture of the CGH.

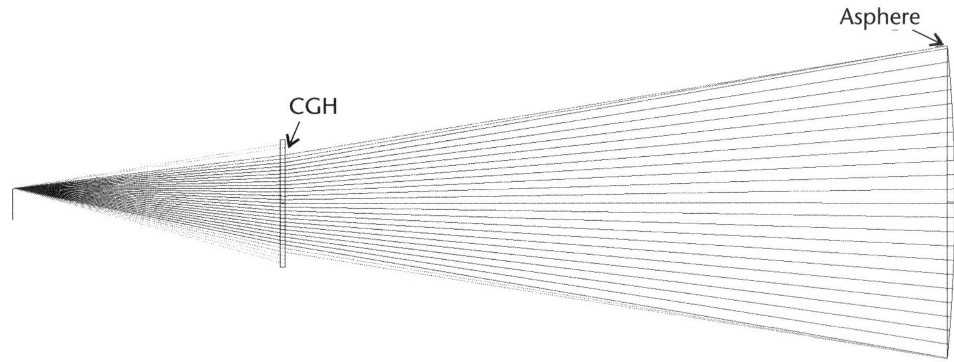

Figure 2.41   Schematic of the CGH test layout including CGH alignment pattern (green), null pattern (blue), and focused spot projection patterns (red and orange).

measurement map of the surface. After morphing, all the analysis for an ordinary interferometric measurement can be performed on the morphed map (e.g., Zernike decomposition, removal of alignment terms, or any combination of Zernike terms). Note that the alignment terms such as tip/tilt/focus can be removed only from the morphed map.

For the test of an off-axis parabola (OAP), the image distortion is illustrated in Figure 2.42 (a) and (b). Figure 2.42(a) shows the beam footprint diagram at the

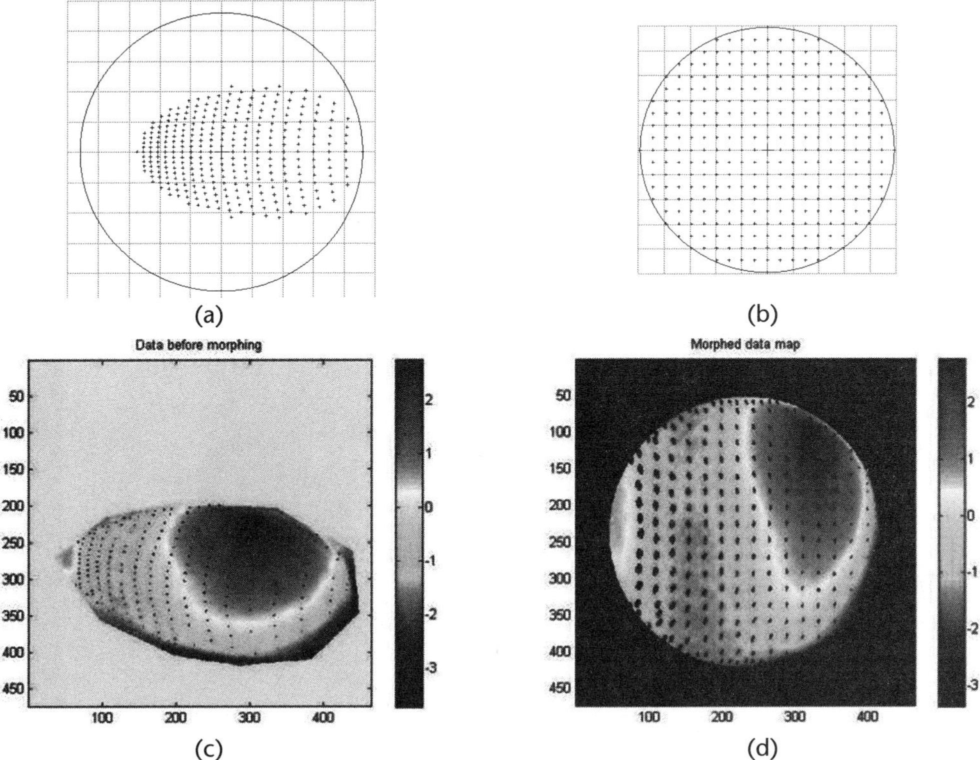

Figure 2.42   (a) Positions of the points on the CGH surface, corresponding to (b) the points on the aspheric surface under test; (c) fiducial points' positions overlaid on the raw measurement map; and (d) the morphed measurement map with the fiducials.

CGH surface, while Figure 2.42(b) shows the beam footprint at the OAP surface. The raw surface measurement map and the morphed map are shown in Figure 2.42(c) and (d).

The image distortion, or mapping relation, is described mathematically as the following:

$$X = \sum u_{ij} x^i y^j$$
$$Y = \sum v_{ij} x^i y^j$$

(2.14)

where

(X,Y) is the normalized pixel coordinate of a point in the distorted measurement map prior to morphing

(x,y) is the normalized coordinate of the corresponding point on the test surface

$(u_{i,j}, v_{i,j})$ are the coefficients of the polynomials

The mapping relations can be obtained by putting some fiducial marks, or a mask as shown in Figure 2.43, on the test surface and then measuring both their physical positions (X,Y) on the aspheric surface and their corresponding pixel positions (x,y) in the raw measurement map. With these numbers, we can fit the mapping relations to get $(u_{i,j}, v_{i,j})$, which are then applied to any point on a regular grid on the aspheric surface to get its corresponding position on the measurements, therefore the surface error at this point. In doing so, we get a morphed measurement map of the surface. Figure 2.42(c) shows a raw measurement map with the fiducials positions overlapped,

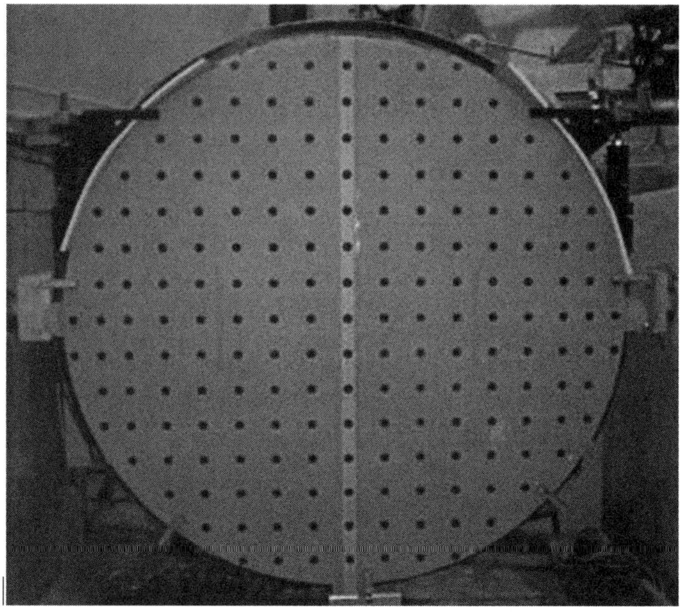

Figure 2.43 A mask of holes is put on the test surface to get the mapping relations of the surface and its image in the CGH test.

and Figure 2.42(d) shows the morphed surface map, again with (morphed) fiducials overlapped. The regular distributions of the morphed fiducials indicate good quality of the morphing. Figure 2.43 shows a mask of holes is put on the test surface to get the mapping relations of the surface and its image in the CGH test.

2.5.6   Error Analysis

As any optical component, a CGH cannot be made perfectly. Therefore, the aspheric wavefront a CGH produces is not perfect, which contributes to the uncertainty of the aspheric surface measurement. The uncertainty of the wavefront produced by the CGH consists of contributions from the following sources:

1. Design error: the residual wavefront error in the design of the CGH phase function;
2. Encoding error: error introduced by approximating smooth CGH lines with polygons;
3. Fabrication error: represented by the line placement error, the line width error, uniformity error of the etching width and depth for the phase CGH, and so on;
4. Substrate error:
   a. Surface error when the CGH is used in reflection, or the transmitted wavefront error when used in transmission;
   b. Uncertainty in knowledge of the index of refraction, substrate thickness, and wedge;
   c. Meniscus error (i.e., both surfaces are not flat but have the same radius of curvature).

Substrate quality is the major limitation of the CGH measurement accuracy of an aspheric surface. The transmitted wavefront error can be calibrated after patterning using the 0th order. The calibration steps are illustrated in Table 2.3. The calibration setup consists of a Fizeau interferometer, a transmission flat (TF), and a reference flat (RF). Two measurements are made in sequence: the first with the CGH in the Fizeau cavity, and the second without for calibrating the Fizeau cavity error. The difference map was taken between the two measurements to yield the absolute transmitted wavefront error map that can be used to back out the CGH substrate error from the aspheric surface measurement.

Table 2.3   Illustration of the CGH Substrate TWE Calibration Procedure

Step 1: Measurement of the CGH TWE at the 0th diffraction order	Step 2: Calibration measurement of empty cavity
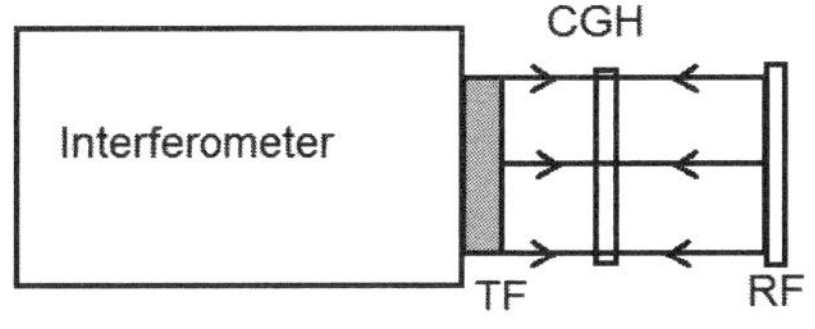	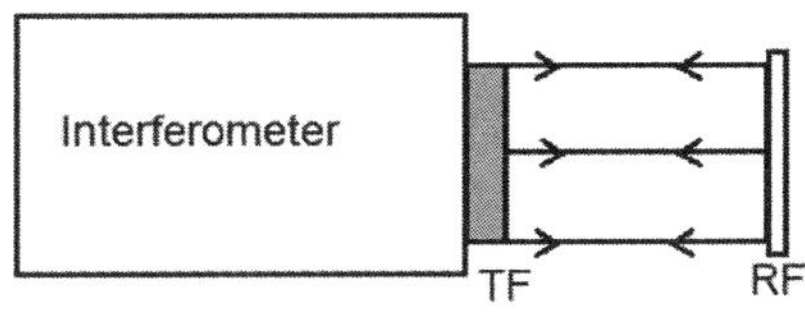

For amplitude CGH, the calibration captures the full intrinsic substrate error and can be used directly. However, the calibration of the phase CGH is more complicated. On the one hand, the 0th order measurement calibration captures the full effect of etching depth variation on the wavefront (for 50% duty cycle CGH); on the other hand, it also captures the effect of the duty cycle variation on the 0th order beam. Note that the non-0th order test beam wavefront is not affected by duty cycle variation, yet such calibrated TWE would add the effects of duty cycle variation to the 0th order substrate measurement. The duty cycle variations are typically small, which adds only a small contribution to the measurement error. When necessary, the effect of the duty cycle variations can be measured and compensated.

An example error analysis is given in Table 2.4.

Besides the intrinsic error of the CGH component, alignment of the CGH also introduces wavefront error, which can be evaluated in optical design software, such as Zemax. The wavefront error produced by the alignment pattern, a tiny fraction of a wave, has an insignificant contribution to the overall CGH alignment error. The mechanical adjustment resolution and accuracy also contributes to the CGH alignment error. The biggest contributor is usually the substrate quality. Alignment pattern is used in reflection, yet the substrate surface is far from perfectly flat. For accurate alignment of the CGH, the substrate surface must be calibrated as well.

Some Convenient Equations, Rules of Thumb, and Numbers Related to the CGHs

- CGH line pitch = $\lambda/\delta\theta$, where λ is the laser wavelength and $\delta\theta$ is ray angle change by the CGH;
- Fringe contrast = $2 * \mathrm{sqrt}(I_A * I_B)/(I_A + I_B)$, where I_A and I_B are intensities of the two beams being combined;
- Amplitude CGH, 50% duty cycle: 1st order diffraction efficiency: ~10%;
- Phase CGH, 50% duty cycle, half wave phase step: 1st order diffraction efficiency: ~40%;

Table 2.4 An Example of Complete CGH Error Analysis

Error Source		Single Pass WFE (nm rms)
Design residual		0.0
Laser wavelength uncertainty of 0.002 nm		0.2
Fabrication	Encoding error	0.5
	Patterning error	0.7
	Etching error	2.2
Substrate errors	CGH substrate refractive index error	0.0
	CGH substrate thickness error	0.0
	CGH substrate wedge error	0.1
	CGH substrate deformation error (the same deformation on both faces, that is not seen in the transmission measurement)	0.1
	Substrate TWE calibration residual	1.0
RSS		2.6

- For both types of CGH, the mth order has diffraction efficiency $1/m^2$ as large as the 1st order;
- Dark Cr reflectivity: ~20% @633 nm;
- Bright Cr reflectivity: ~50% @633 nm;
- Index of refraction of fused silica: 1.457017@633 nm.

References

[1]   Hariharan, P., and P. Hariharan, *Basics of Interferometry*, Second Edition, Elsevier, 2007.

[2]   Jensen, A. E., "Absolute Calibration Method for Laser Twyman-Green Wave-Front Testing Interferometers," *J. Opt. Soc. Am.* Vol. 63, 1973, p. 1313A.

[3]   Kuchel, M., "Spatial Coherence in Interferometry. Zygo's New Method to Reduce Intrinsic Noise in Interferometers," *Zygo Technical Report*, 2004.

[4]   Malacara, D. (ed.), *Optical Shop Testing*, Third Edition, Wiley, 2007.

[5]   Millerd, J., B. Neal, J. Hays, M. North-Morris, M. Novak, et al., "Pixelated Phase-Mask Dynamic Interferometer," *Proc. SPIE*, Vol. 5531, 2004, pp. 304–314.

[6]   Murphy, P., et al., "Measurement of High-Departure Aspheric Surfaces Using Subaperture Stitching with Variable Null Optics," *Proc. SPIE*, Vol. 7426, 2009.

[7]   Schulz, G., and J. Schwider, "Interferometric Testing of Smooth Surfaces," in *Progress in Optics*, Vol. 13, 1976, pp. 93–167.

[8]   Takeda, M., "Temporal versus Spatial Carrier Techniques for Heterodyne Interferometry," *Proc. SPIE*, Vol. 813, 1987, pp. 329–330.

[9]   http://www.optiper.com/products/caliball-i-ii.

[10]   Chang, Y., Zhou, P., and Burge, J. H., "Analysis of Phase Sensitivity for Binary Computer-Generated Holograms," *Appl. Opt.*, Vol. 45, 2006, pp. 4223–4234.

[11]   Zhao, C., and Burge, J. H., "Optical Testing with Computer Generated Holograms: Comprehensive Error Analysis," *Proc. SPIE*, 2013, pp. 8838–88380H.

CHAPTER 3

Phase-Shifting Interferometry

Lionel R. Watkins

Interferometers have been used for optical testing since at least the early 1900s and at that time were entirely reliant on the analysis of static interferograms, as the early papers by Kingslake attest [1]. This analysis generally required one to find fringe centers, interpolate to a regular grid, and necessitated a trade-off between precision and number of data points. From the early 1970s, the availability of computers and digital imaging revolutionized the field and led to the rapid development of phase-shifting interferometry. The key to this expansion was the ability to electronically record a series of interferograms, each of which corresponded to a different value of the reference phase, and which thereby enabled the phase map (representing, for example, wavefront distortion) to be recovered on a pixel-by-pixel basis. Phase-shifting interferometry significantly improved the convenience, speed, and accuracy with which measurements could be made.

This chapter provides an introduction to some of the principles behind phase-shifting interferometry and has two major parts. In the first, we begin with fundamental concepts around interference, two-beam interferometers, and the definition of optical intensity before describing some of the most commonly used phase-shifting algorithms. Of course, each algorithm has its relative strengths and weaknesses and although there are many ways in which their performance may be characterised, we conclude this part by describing a particular elegant method based on the characteristic polynomial. The second part is devoted to techniques for phase stepping and their practical applications. Topics covered include mechanical approaches based on piezoelectric transducers, rotating gratings, and translating glass wedges. Several important recent developments are based on various implementations of the geometric phase, sometimes called polarization modulation in the literature, and so we devote a section to a reasonably detailed discussion of this phenomenon. We conclude by describing techniques that allow all the frames in a phase stepping interferometer to be captured simultaneously, an ability that significantly reduces these instruments' sensitivity to vibration and air turbulence.

3.1 The Interferometer Equation

The interferometer equation describes how the intensity at the output of an interferometer varies as a function of the optical path length and is the fundamental starting point for all phase stepping methods, so we begin with its derivation here.

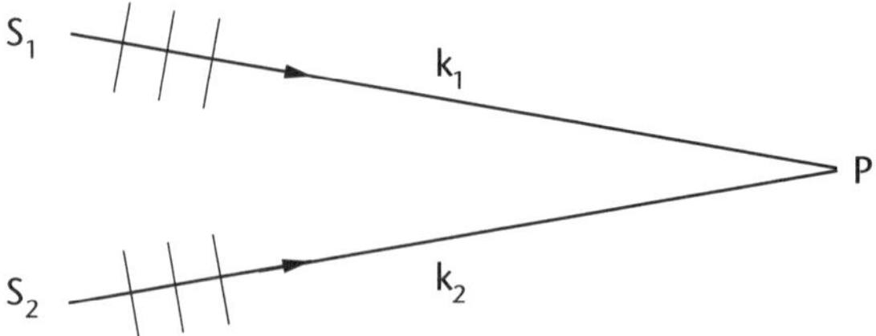

Figure 3.1 Interference of two monochromatic, linearly polarized plane waves.

Consider two monochromatic, linearly polarized plane waves of the same frequency propagating from sources S_1 and S_2 towards an observation point P, as shown in Figure 3.1. Much of the analysis is quite general and does not depend on the wavefront shape (see Exercise 3.1), but the assumption of plane waves considerably simplifies the algebra. Let these two waves have electric fields given by

$$\mathbf{E}_1(\mathbf{r},t) = \mathbf{E}_{01}\cos\left(\mathbf{k}_1\cdot\mathbf{r} - \omega_0 t + \varepsilon_1\right) \tag{3.1a}$$

and

$$\mathbf{E}_2(\mathbf{r},t) = \mathbf{E}_{02}\cos\left(\mathbf{k}_2\cdot\mathbf{r} - \omega_0 t + \varepsilon_2\right) \tag{3.1b}$$

where $\mathbf{E}_{0i}$, $i = 1,2$ is a vector whose amplitude and direction give the amplitude and direction of polarization of the electric field, respectively. $\mathbf{k}$ is the propagation vector with $|\mathbf{k}| = 2\pi/\lambda$, and $\varepsilon_{1,2}$ the initial phases of these waves.

The two waves meet at P where, according to the superposition principle, their electric fields add. What we can measure at P (with photodetectors, CCD cameras, or the human eye), however, is not the instantaneous electric field but the intensity I or more strictly, the irradiance. For a linear, homogeneous, isotropic dielectric, the irradiance is given by [2]:

$$I = \varepsilon v\left\langle E^2 \right\rangle \tag{3.2}$$

where $\langle,\rangle$ denotes time average and ε, v are the permittivity and speed in the medium, respectively. Further details may be found in Box 3.1. Since we are only ever interested in measuring relative intensities, we drop the constants and assume that the intensity $I = \langle \mathbf{E}^2 \rangle$.

BOX 3.1   Optical Intensity or Irradiance

The irradiance or intensity of an electromagnetic wave (such as light) is defined as *the average energy per unit area per unit time* and is equal to the time-averaged value of the magnitude of the Poynting vector [2]. For harmonic fields, this is given by

$$I = \langle S \rangle = c^2\varepsilon_0\left|\mathbf{E}_0\times\mathbf{B}_0\right|\left\langle\cos^2\left(\mathbf{k}\cdot\mathbf{r} - \omega_0 t\right)\right\rangle$$

where c is the speed of light and ε_0 is the permittivity of free space. By definition,

$$\left\langle \cos^2\left(\mathbf{k} \cdot \mathbf{r} - \omega_0 t\right) \right\rangle = \frac{1}{T} \int_t^{t+T} \cos^2\left(\mathbf{k} \cdot \mathbf{r} - \omega_0 t'\right) dt'$$

which, after some algebra, eventually yields

$$= \frac{1}{2} + \frac{1}{2}\frac{\sin\left(\omega_0 T\right)}{\omega_0 T}\cos\left(\mathbf{k} \cdot \mathbf{r} - 2\omega_0 t - \omega_0 T\right)$$

$$= \frac{1}{2} \quad \text{for } T \gg \tau.$$

The term $\sin(\omega_0 T)/\omega_0 T = \text{sinc}(\omega_0 T)$ and is shown plotted in Figure 3.2 with $u = \omega_0 T = 2\pi T/\tau$ where τ is the period of the electric field. For light in the visible region, $\tau \approx 10^{-15}$ s and so an integration period T of even a microsecond corresponds to $T \approx 10^9 \tau$, a value that is more than sufficient to drive the sinc function to a totally negligible value.

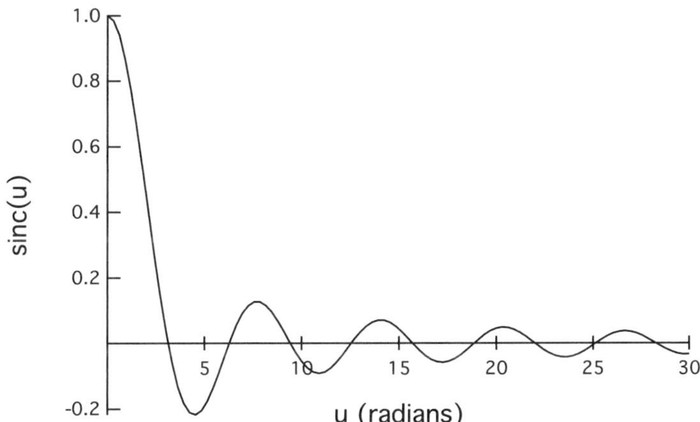

Figure 3.2   The sinc function; sinc(u) = sin(u)/u.

The intensity, therefore, is:

$$I \equiv \langle S \rangle = \frac{c^2 \varepsilon_0}{2}\left|\mathbf{E}_0 \times \mathbf{B}_0\right|$$

$$= \frac{c\varepsilon_0}{2} E_0^2$$

or, equivalently

$$I = c\varepsilon_0 \left\langle E^2 \right\rangle$$

in a vacuum.

With these points in mind, we can write the intensity at P as:

$$I = \left\langle \mathbf{E}^2 \right\rangle = \left\langle \left(\mathbf{E}_1 + \mathbf{E}_2\right) \cdot \left(\mathbf{E}_1 + \mathbf{E}_2\right) \right\rangle \tag{3.3a}$$

$$= \left\langle \mathbf{E}_1^2 \right\rangle + \left\langle \mathbf{E}_2^2 \right\rangle + 2\left\langle \mathbf{E}_1 \cdot \mathbf{E}_2 \right\rangle \tag{3.3b}$$

$$= I_1 + I_2 + 2\left\langle \mathbf{E}_1 \cdot \mathbf{E}_2 \right\rangle \tag{3.3c}$$

The interference term is $\langle \mathbf{E}_1 \cdot \mathbf{E}_2 \rangle$ which can be evaluated as follows:

$$\mathbf{E}_1 \cdot \mathbf{E}_2 = \mathbf{E}_{01} \cdot \mathbf{E}_{02} \cos\left(\mathbf{k}_1 \cdot \mathbf{r} - \omega_0 t + \varepsilon_1\right) \times \cos\left(\mathbf{k}_2 \cdot \mathbf{r} - \omega_0 t + \varepsilon_2\right) \tag{3.4a}$$

which, on expanding the cosine terms, yields

$$= \mathbf{E}_{01} \cdot \mathbf{E}_{02} \left\{ \cos\left(\omega_0 t\right)\cos\left(\mathbf{k}_1 \cdot \mathbf{r} + \varepsilon_1\right) + \sin\left(\omega_0 t\right)\sin\left(\mathbf{k}_1 \cdot \mathbf{r} + \varepsilon_1\right) \right\}$$
$$\times \left\{ \cos\left(\omega_0 t\right)\cos\left(\mathbf{k}_2 \cdot \mathbf{r} + \varepsilon_2\right) + \sin\left(\omega_0 t\right)\sin\left(\mathbf{k}_2 \cdot \mathbf{r} + \varepsilon_2\right) \right\} \tag{3.4b}$$

The evaluation of this expression is simplified if we make the following observations. Since $\sin(\omega_0 t)$ and $\cos(\omega_0 t)$ are orthogonal functions, the time-average of their product is zero and does not contribute to the interference. On the other hand, $\langle \cos^2(\omega_0 t)\rangle = \langle 1/2(1 - \cos(2\omega_0 t))\rangle = 1/2$ and similarly for the $\sin^2(\omega_0 t)$ term. Hence

$$\left\langle \mathbf{E}_1 \cdot \mathbf{E}_2 \right\rangle = \frac{1}{2}\mathbf{E}_{01} \cdot \mathbf{E}_{02} \left\{ \cos\left(\mathbf{k}_1 \cdot \mathbf{r} + \varepsilon_1\right)\cos\left(\mathbf{k}_2 \cdot \mathbf{r} + \varepsilon_2\right) \right.$$
$$\left. + \sin\left(\mathbf{k}_1 \cdot \mathbf{r} + \varepsilon_1\right)\sin\left(\mathbf{k}_2 \cdot \mathbf{r} + \varepsilon_2\right) \right\} \tag{3.4c}$$

which, when we recombine the trigonometric terms yields:

$$\left\langle \mathbf{E}_1 \cdot \mathbf{E}_2 \right\rangle = \frac{1}{2}\mathbf{E}_{01} \cdot \mathbf{E}_{02} \cos\left(\mathbf{k}_1 \cdot \mathbf{r} + \varepsilon_1 - \mathbf{k}_2 \cdot \mathbf{r} - \varepsilon_2\right) \tag{3.4d}$$

Substituting this back into (3.3c) yields

$$I = I_1 + I_2 + \mathbf{E}_{01} \cdot \mathbf{E}_{02} \cos\Delta\phi \tag{3.4e}$$

where $\Delta\phi = (\mathbf{k}_1 \cdot \mathbf{r} + \varepsilon_1 - \mathbf{k}_2 \cdot \mathbf{r} - \varepsilon_2)$ is the phase difference arising from the combination of path length and initial phase differences. Note that if $\mathbf{E}_{01}$ and $\mathbf{E}_{02}$ are orthogonally polarized, the interference term is zero, a finding that corresponds to one of the statements of the Fresnel-Arago laws (see Box 3.2 for details). Since it is advantageous experimentally to maximise the interference term, we usually ensure that $\mathbf{E}_1$ and $\mathbf{E}_2$ are polarized in the same plane. Finally, we note that since

$$I_1 = \left\langle \mathbf{E}_1^2 \right\rangle = \frac{E_{01}^2}{2} \tag{3.5}$$

and similarly for I_2, the interference term is $2\sqrt{I_1I_2}\cos\Delta\phi$, whereupon the interferometer equation is:

$$I = I_1 + I_2 + 2\sqrt{I_1I_2}\cos\Delta\phi \qquad (3.6a)$$

or

$$I = A + B\cos\Delta\phi \qquad (3.6b)$$

EXERCISE 3.1

Consider now the more general case where the electric fields are given by:

$$\mathbf{E}_1(\mathbf{r},t) = \mathbf{E}_1(\mathbf{r})\exp(-i\omega_0 t) \qquad (3.7a)$$

and

$$\mathbf{E}_2(\mathbf{r},t) = \mathbf{E}_2(\mathbf{r})\exp(-i\omega_0 t) \qquad (3.7b)$$

where the wavefront shapes are not specified and $\mathbf{E}_1$ and $\mathbf{E}_2$ are complex vectors depending on space and initial phase angle. Show that the interference term is now given by

$$\langle \mathbf{E}_1 \cdot \mathbf{E}_2 \rangle = \frac{1}{4}\left(\mathbf{E}_1 \cdot \mathbf{E}_2^* + \mathbf{E}_1^* \cdot \mathbf{E}_2\right) \qquad (3.8)$$

Show that (3.8) reduces to $\mathbf{E}_{01} \cdot \mathbf{E}_{02} \cos\Delta\phi$ for plane waves.

Solution

$$\mathbf{E}_1 \cdot \mathbf{E}_2 = \frac{1}{2}\left(\mathbf{E}_1\exp(-i\omega_0 t) + \mathbf{E}_1^*\exp(i\omega_0 t)\right) \cdot \frac{1}{2}\left(\mathbf{E}_2\exp(-i\omega_0 t) + \mathbf{E}_2^*\exp(i\omega_0 t)\right)$$

$$(3.9a)$$

Since $\Re(z) = 1/2(z + z^*)$ where $*$ denotes a complex conjugate.

$$\mathbf{E}_1 \cdot \mathbf{E}_2 = \frac{1}{4}\left(\mathbf{E}_1 \cdot \mathbf{E}_2\exp(-2i\omega_0 t) + \mathbf{E}_1^* \cdot \mathbf{E}_2^*\exp(2i\omega_0 t) + \mathbf{E}_1 \cdot \mathbf{E}_2^* + \mathbf{E}_1^* \cdot \mathbf{E}_2\right)$$

$$(3.9b)$$

The last two terms are time independent, while $\langle \mathbf{E}_1 \cdot \mathbf{E}_2 \exp(-2i\omega_0 t)\rangle \rightarrow 0$ and similarly for the second term above. Hence

$$\langle \mathbf{E}_1 \cdot \mathbf{E}_2 \rangle = \frac{1}{4}\left(\mathbf{E}_1 \cdot \mathbf{E}_2^* + \mathbf{E}_1^* \cdot \mathbf{E}_2\right) \qquad (3.9c)$$

If $\mathbf{E}_1(\mathbf{r},t)$ is a plane wave, then we may write it as

$$\mathbf{E}_1(\mathbf{r},t) = \mathbf{E}_{01}\exp\left(i\left[\mathbf{k}_1 \cdot \mathbf{r} + \varepsilon_1 - \omega_0 t\right]\right) \qquad (3.10)$$

and similarly for $\mathbf{E}_2(\mathbf{r},t)$ from which the final result is trivially derived.

BOX 3.2   The Fresnel-Arago Laws

A description of the Fresnel-Arago laws can be found in most optics textbooks, for example Hecht [2], while a complete treatment using Stokes vectors can be found in the paper by Collett [3]. The laws are as follows:

1. Two waves, linearly polarized in the same plane, can interfere.
2. Two waves, linearly polarized with perpendicular polarizations, cannot interfere.
3. Two waves, linearly polarized with perpendicular polarizations, if derived from perpendicular components of unpolarized light and subsequently brought into the same plane, cannot interfere.
4. Two waves, linearly polarized with perpendicular polarizations, if derived from the same linearly linearly polarized wave and subsequently brought into the same plane, can interfere.

The third law is essentially a statement that the two orthogonal components of unpolarized light (strictly, randomly polarized) are mutually incoherent.

Now that we have established the interferometer equation, we can state the key idea behind all phase stepping techniques, namely:

If we introduce *known* phase increments δ by altering the optical path length $\Delta\phi$, then we can acquire a number of images $I_0, \ldots, I_{N-1}$ from which it is possible to extract the three unknown quantities in (3.6b) and in particular, the initial phase $\Delta\phi$.

In the following sections we will describe in some detail a number of common phase stepping methods.

3.2   Three-Frame Method

Since there are three unknowns in the interferometer equation, namely A, B, and ϕ, we require at least three independent measurements to solve for each of these unknowns. Hence the simplest technique requires three frames (or images) and therefore two phase steps. As a concrete example, let's consider the standard Michelson interferometer shown in Figure 3.3. The light from a suitable laser, for example a HeNe, is first expanded and collimated until it is large enough to illuminate the test surface T. For the moment, suppose that a flat reference mirror M in the other arm of the interferometer is driven by a piezoelectric transducer (PZT) so that its position, and hence the optical path length, can be precisely controlled. Other methods for generating the necessary phase steps will be described later in this chapter. The two beams recombine at the beam splitter and a lens F forms an image of the test surface on a screen or, more commonly, the image plane of a CCD array.

The intensity of a given pixel in the image thus formed is:

$$I(x,y) = A(x,y) + B(x,y)\cos\left[\phi(x,y)\right] \tag{3.11}$$

where $\phi(x, y)$ is the two-dimensional phase distribution produced, in this instance, by variations in the height of the test surface T, and is the quantity we wish to measure. In general, the background level $A(x, y)$ and modulation depth $B(x, y)$ may be functions of the pixel position (x, y) primarily because the laser beam may not uniformly illuminate the test surface. A typical experimental procedure might be as follows:

1. Grab a fringe pattern with, for example, a frame grabber and personal computer. This image is $I_0(x, y)$.
2. Move the mirror with the PZT so that the phase changes by exactly $2\pi/3$. This is the *first* phase step. Grab another set of fringes: $I_1(x, y)$.
3. Move the mirror by another $2\pi/3$ rad; that is, total phase step is $4\pi/3$. Grab the fringes again: $I_2(x, y)$.
4. Finally, calculate $\phi(x, y)$ on a pixel-by-pixel basis.

For the three-frame method we would now have the following set of images:

$$I_0 = A + B \cos \phi \tag{3.12a}$$

$$I_1 = A + B \cos(\phi + 2\pi/3) \tag{3.12b}$$

$$I_2 = A + B \cos(\phi + 4\pi/3) \tag{3.12c}$$

Note that the explicit dependence of A, B, and ϕ on pixel position (x, y) has been dropped for clarity. Using standard trigonometrical identities to expand the cosine terms leads to the following equations for the three captured images:

$$I_0 = A + B \cos(\phi) \tag{3.13a}$$

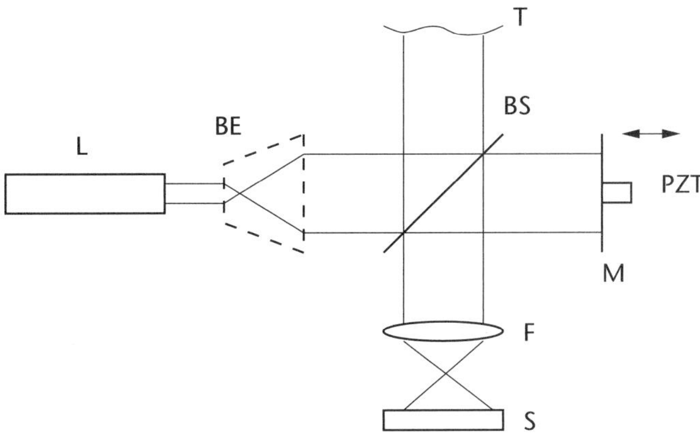

Figure 3.3 Michelson interferometer. L = laser; BE = beam expander; BS = beam splitter; M = reference mirror; PZT = piezoelectric transducer; T = test surface; F = focusing lens; S = screen or CCD camera.

$$I_1 = A - \frac{B}{2}\cos(\phi) - \frac{\sqrt{3}B}{2}\sin(\phi) \qquad (3.13b)$$

and

$$I_2 = A - \frac{B}{2}\cos(\phi) + \frac{\sqrt{3}B}{2}\sin(\phi) \qquad (3.13c)$$

Now it should be clear that if we take the following differences

$$\sqrt{3}\left(I_2 - I_1\right) = 3B\sin\phi \qquad (3.14a)$$

and

$$2I_0 - (I_1 + I_2) = 3B\cos\phi \qquad (3.14b)$$

then we are able to find the phase $\phi(x, y)$ on a pixel-by-pixel basis from

$$\phi(x,y) = \tan^{-1}\left\{\frac{\sqrt{3}\left(I_2 - I_1\right)}{2I_0 - \left(I_1 + I_2\right)}\right\} \qquad (3.15)$$

Note that the phase $\phi(x, y)$ will be returned modulo 2π and will need to be unwrapped. We have described this method with a phase step of $2\pi/3$, but a phase step of $\pi/2$ is also possible. While this method is the simplest to implement, it is also the most sensitive to errors in the phase step between images [4].

EXERCISE 3.2

Consider now the general case of equal phase steps α, that is

$$\delta_i = -\alpha,\ 0,\ \alpha, \qquad i = 1, 2, 3 \qquad (3.16a)$$

Find a general expression for the phase $\phi(x, y)$ and show that if $\alpha = \pi/2$ your expression reduces to:

$$\phi(x,y) = \tan^{-1}\left\{\frac{I_1 - I_3}{2I_2 - I_1 - I_3}\right\} \qquad (3.16b)$$

Now find an expression for $\phi(x, y)$ when $\alpha = 2\pi/3$. With a suitable reordering of the intensities, you should find that your expression agrees with (3.15).

Solution

The corresponding intensities are:

$$I_1 = A + B\cos(\psi - \alpha) = A + B\{\cos\psi\cos\alpha + \sin\psi\sin\alpha\} \qquad (3.17a)$$

$$I_2 = A + B\cos(\phi) \qquad (3.17b)$$

$$I_3 = A + B\cos(\phi + \alpha) = A + B\{\cos\phi\cos\alpha - \sin\phi\sin\alpha\} \qquad (3.17c)$$

from which it is easy to show that

$$I_1 - I_3 = 2B \sin \phi \sin \alpha \tag{3.17d}$$

$$2I_2 - I_1 - I_3 = 2B \cos \phi (1 - \cos \alpha) \tag{3.17e}$$

Hence

$$\frac{I_1 - I_3}{2I_2 - I_1 - I_3} = \tan \phi \frac{\sin \alpha}{1 - \cos \alpha} \tag{3.17f}$$

which yields the required result when $\alpha = \pi/2$.

When $\alpha = 2\pi/3$ the term $\sin \alpha/(1 - \cos \alpha)$ evaluates to $1/\sqrt{3}$ and if we make the mappings $I_1 \to I_2$, $I_2 \to I_0$ and $I_3 \to I_1$ then we recover (3.15) exactly.

3.3   Four-Frame Method

A much more common approach is the four-frame method since it leads to a particularly simple expression for the phase map. In this case, the phase steps required are $\pi/2$ and the corresponding images are:

$$I_0 = A + B \cos \phi \tag{3.18a}$$

$$I_1 = A + B \cos(\phi + \pi/2) = A - B \sin \phi \tag{3.18b}$$

$$I_2 = A + B \cos(\phi + \pi) = A - B \cos \phi \tag{3.18c}$$

$$I_3 = A + B \cos(\phi + 3\pi/2) = A + B \sin \phi \tag{3.18d}$$

from which we may readily derive the following expression for the phase:

$$\phi(x, y) = \tan^{-1} \left\{ \frac{I_3 - I_1}{I_0 - I_2} \right\}. \tag{3.19}$$

EXERCISE 3.3

Assuming perfect measurement of the fringe intensity distributions in (3.19), estimate the error in the final phase measurement caused by errors in the phase shifter. Assume that $A = B$ (that is, the fringe visibility is 1), and that typical errors in the phase shifter are of the order of $\lambda/10$.

Solution

We use partial differentiation and standard error propagation techniques to solve this problem. For notational convenience, let

$$\tan \phi = k = \frac{N}{D} \tag{3.20a}$$

where N and D represent the numerator and denominator, respectively, of (3.19). Then the error in ϕ is given by

$$\Delta\phi = \cos^2 \phi \Delta k \tag{3.20b}$$

where

$$\Delta k = \sqrt{\left(\frac{\Delta N}{D}\right)^2 + \left(\frac{N}{D^2}\Delta D\right)^2} \tag{3.20c}$$

Now:

$$\Delta N = \sqrt{\Delta I_3^2 + \Delta I_1^2} \tag{3.20d}$$

and

$$\Delta D = \sqrt{\Delta I_0^2 + \Delta I_2^2} \tag{3.20e}$$

If we let

$$I_i = A + B \cos(\phi + \theta_i), \qquad i = 0, \ldots, 3 \tag{3.20f}$$

where θ_i are the phase shifts, then we find that

$$\Delta I_i = -B \sin(\phi + \theta_i)\Delta\theta \tag{3.20g}$$

where $\Delta\theta$ is the phase shifter error. Substituting the (exact) phase steps θ_i yields expressions for each of the ΔI_i. Substituting those expressions into ΔN and ΔD and subsequently into Δk, yields, after some algebraic manipulation:

$$\Delta\phi = \frac{\Delta\theta}{\sqrt{2}}\sqrt{\cos^4 \phi + \sin^4 \phi}. \tag{3.20h}$$

The error, which is clearly periodic, is plotted in Figure Ex 3.1.

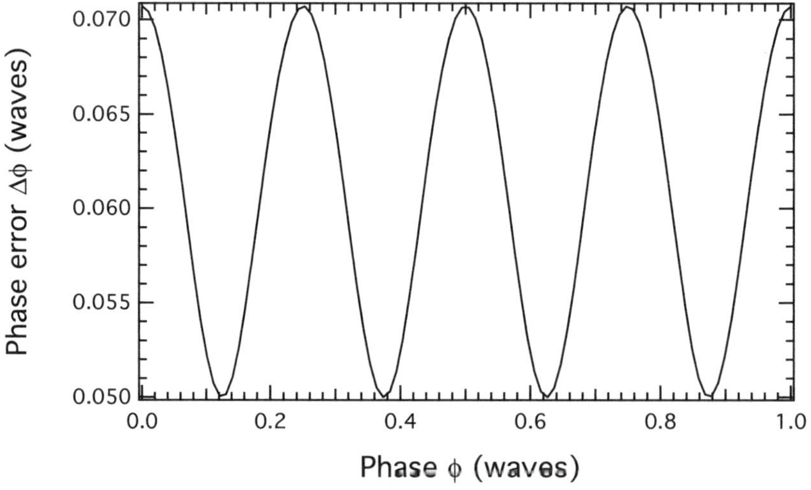

Figure Ex 3.1   Error in the phase ϕ for a phase shifter error of 0.1λ.

EXERCISE 3.4

Now estimate the effects of intensity measurement errors on the final phase measurement accuracy for the four-frame method. Assume that the phase shifter is perfect, the fringe visibility is 1, and that the fringes are measured using an 8-bit digitizer with random errors of ±4 least significant bits. The fringe intensity variations use the full dynamic range of the digitizer.

Solution

Since the fringe contrast is 1, we know that $A = B$ and given that the full dynamic range of the digitizer is used, $A + B = 256$; therefore $A = 128$ and $B = 128$. We have 4 bit error, so $\Delta I_i = 16$, $i = 1, \ldots, 4$ and thus $\Delta N = \Delta D = 16\sqrt{2}$. From the previous exercise and (3.20c), we have

$$\Delta k = \frac{1}{D}\sqrt{\Delta N^2 + \tan^2 \phi \Delta D} \tag{3.21a}$$

$$= \frac{16\sqrt{2}}{2B\cos\phi}\sqrt{1 + \tan^2 \phi} \tag{3.21b}$$

and substituting this into the equation for $\Delta\phi$ finally yields:

$$\Delta\phi = \frac{1}{8\sqrt{2}} \text{ rad.} \tag{3.21c}$$

We conclude this section with a small example that illustrates the essential steps of the algorithm. There are, of course, no restrictions on the shape of the test surface, but the effect of the phase stepping is perhaps best seen if we choose a simple surface such as a parabola. A MATLAB code fragment to generate the test surface and images is:

```
x = 1:256;
[X,Y] = meshgrid(x);
x0 = 128;
y0 = 128;
phase = 6*((X - x0).^2 + (Y - y0).^2)./128^2; I = 32*(1 +
cos(phase + step));
```

The image is 256×256 pixels grayscale and `step` is incremented appropriately for each image. The four frames thus generated are shown in Figure 3.4. The phase map, when calculated using (3.19) and subsequently unwrapped, yields the surface shown in Figure 3.5. Phase unwrapping is an important topic in its own right but well beyond the scope of this chapter. Interested readers are encouraged to consult the literature [5–8].

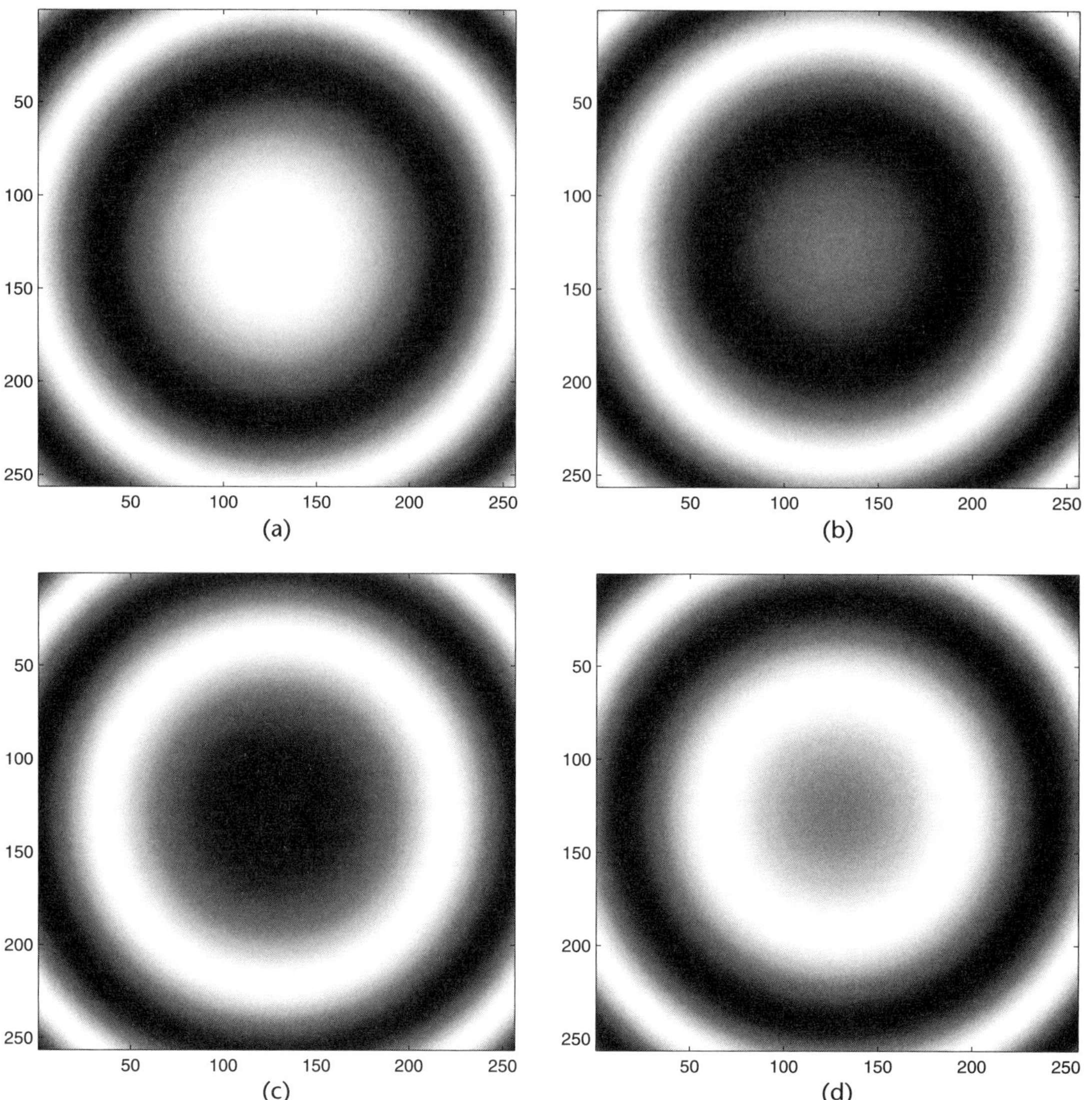

Figure 3.4 Fringe patterns for the parabolic surface $6((x - x_0)^2 + (y - y_0)^2)/128^2$ with phase steps of $\pi/2$. (a) I_1: no phase step. (b) I_2: phase step of $\pi/2$. (c) I_3: phase step of π. (d) I_4: phase step of $3\pi/2$

3.4   Least-Squares Algorithms

The measured intensity $I(x, y)$ at a given pixel position (x, y) depends on the three unknowns A, B, and ϕ and varies sinusoidally with the stepping phase, which has a known period. It is not surprising, therefore, that least-squares methods can be used to fit the measured intensities at each pixel position to a sinusoidal function. The general solution for N images, each shifted by δ_i, was first reported by Bruning et al. [9] and later by Morgan [10] and Greivenkamp [11]. To see how this method works, we first write (3.11) as

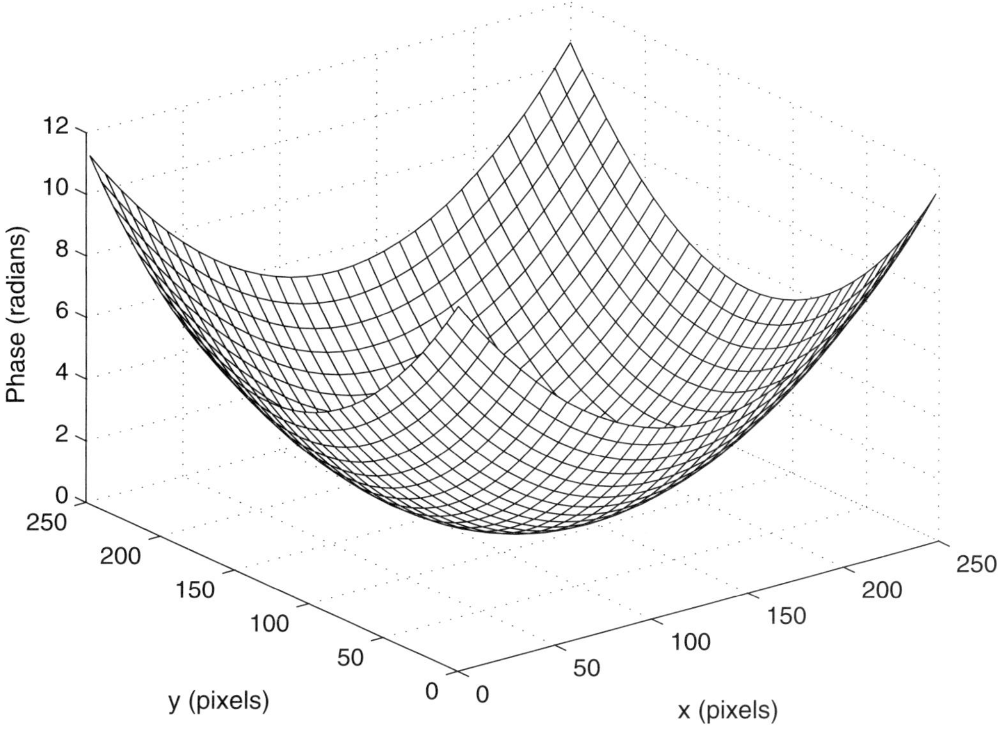

Figure 3.5   Recovered phase map from the images in Figure 3.4.

$$I_i(x, y) = A(x, y) + B(x, y) \cos(\phi(x, y) + \delta_i) \tag{3.22a}$$

$$= A + B \cos(\phi) \cos(\delta_i) - B \sin(\phi) \sin(\delta_i) \tag{3.22b}$$

$$= \alpha_0 + \alpha_1 \cos(\delta_i) + \alpha_2 \sin(\delta_i) \tag{3.22c}$$

where

$$\alpha_0 = A \tag{3.23a}$$

$$\alpha_1 = B \cos(\phi) \tag{3.23b}$$

$$\alpha_2 = -B \sin(\phi) \tag{3.23c}$$

The nature of the least-squares algorithm is most easily seen if we write this in matrix form:

$$\begin{bmatrix} I_1 \\ \vdots \\ I_N \end{bmatrix} = \begin{bmatrix} 1 & \cos\delta_1 & \sin\delta_1 \\ \vdots & \vdots & \vdots \\ 1 & \cos\delta_N & \sin\delta_N \end{bmatrix} \begin{bmatrix} \alpha_0 \\ \alpha_1 \\ \alpha_2 \end{bmatrix} \tag{3.24a}$$

or, more compactly as

$$\mathbf{I} = \mathbf{M}\alpha \tag{3.24b}$$

It is well known that the solution which minimizes the sum of the squared differences between the measured and predicted intensities is given by

$$\tilde{\alpha} = \left(M^T M\right)^{-1} M^T I \qquad (3.25)$$

where $\tilde{\alpha}$ is our best estimate of the parameters α_i and T denotes matrix transpose. It is then trivial to determine our best estimates of the three unknowns:

$$A = \tilde{\alpha}_0 \qquad (3.26a)$$

$$B = \sqrt{\tilde{\alpha}_1^2 + \tilde{\alpha}_2^2} \qquad (3.26b)$$

and

$$\phi = \tan^{-1}\left\{\frac{-\tilde{\alpha}_2}{\tilde{\alpha}_1}\right\} \qquad (3.26c)$$

It is worth noting that this algorithm enables us to use any combination of three or more values for the phase step δ_i, and moreover, that these values do not need to be evenly spaced.

However, for the special case when the δ_is are evenly spaced over one period; that is, when

$$\delta_i = \frac{2\pi i}{N} \quad i = 1, \ldots, N \qquad (3.27)$$

then the algorithm has a particularly simple form. We find that all the off-diagonal elements of

$$M^T M = \begin{bmatrix} N & \Sigma\cos(\delta_i) & \Sigma\sin(\delta_i) \\ \Sigma\cos(\delta_i) & \Sigma\cos^2(\delta_i) & \Sigma\cos(\delta_i)\sin(\delta_i) \\ \Sigma\sin(\delta_i) & \Sigma\cos(\delta_i)\sin(\delta_i) & \Sigma\sin^2(\delta_i) \end{bmatrix} \qquad (3.28a)$$

are zero and hence

$$\left(M^T M\right)^{-1} = \begin{bmatrix} 1/N & 0 & 0 \\ 0 & 1/\Sigma\cos^2\delta_i & 0 \\ 0 & 0 & 1/\Sigma\sin^2\delta_i \end{bmatrix} \qquad (3.28b)$$

Our best estimates for the parameters $\tilde{\alpha}$ are therefore:

$$\tilde{\alpha}_0 = \frac{\Sigma I_i}{N} \qquad (3.28c)$$

$$\tilde{\alpha}_1 = \frac{\Sigma I_i \cos\delta_i}{\Sigma\cos^2\delta_i} \qquad (3.28d)$$

and

$$\tilde{\alpha}_2 = \frac{\Sigma I_i \sin \delta_i}{\Sigma \sin^2 \delta_i} \tag{3.28e}$$

Since $\Sigma \sin^2 \delta_i = \Sigma \cos^2 \delta_i$, we readily find that

$$\phi = \tan^{-1}\left\{ \frac{-\Sigma I_i \sin(\delta_i)}{\Sigma I_i \cos(\delta_i)} \right\}. \tag{3.28f}$$

A justification of the two assertions in this derivation may be found in Box 3.3.

BOX 3.3   Additional Details in the Derivation of (3.28f)

First we need to show that the off-diagonal elements of $M^T M$ are zero. The key to this lies in the fact that the δ_i are *evenly spaced* over one period. For the sake of simplicity we choose $N = 5$ and show, plotted in Figure 3.6, the values of δ_i and $\sin(\delta_i)$.

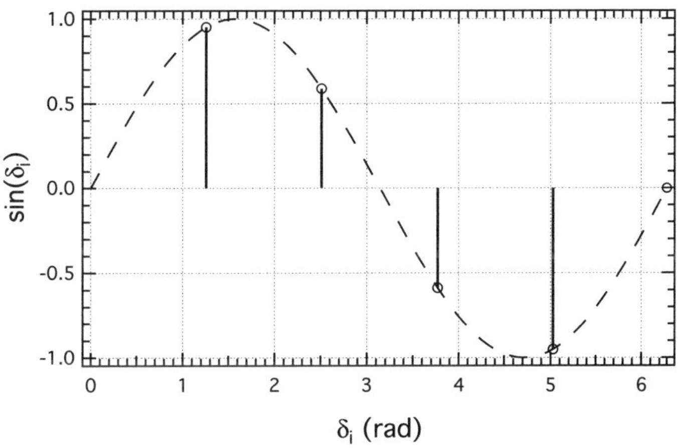

Figure 3.6 $\sin(\delta_i)$ for $\delta_i = 2\pi i/N$, $N = 1, \ldots, 5$.

It is clear that $\sin(\delta_2) = -\sin(\delta_3)$ and that $\sin(\delta_1) = -\sin(\delta_5)$ while the final value will always be zero since it corresponds to $\sin(2\pi)$. This pairwise canceling is true for any value of N and a plot for $\cos(\delta_i)$ would show that similar arguments also hold for this function. Therefore, $\Sigma \sin(\delta_i) = \Sigma \cos(\delta_i) = 0$.

Next, for each value of δ_i, $\cos(\delta_i) \sin(\delta_i) = 1/2 \sin(2\delta_i)$ which just corresponds to sampling one period at the coarser interval of $N/2$. Hence by the previous argument, $\Sigma \cos(\delta_i) \sin(\delta_i) = 0$ and all the off-diagonal elements of $M^T M$ are indeed zero.

Finally, we can write

$$\Sigma \sin^2 \delta_i = \frac{1}{2}\Sigma\left(1 - \cos(2\delta_i)\right) \tag{3.29a}$$

$$= \frac{1}{2}\Sigma 1 \tag{3.29b}$$

since $\cos(2\delta_i)$ represents a sampling at $N/2$ which we have already shown sums to zero. Similarly,

$$\Sigma \cos^2 \delta_i = \frac{1}{2}\Sigma\left(1 + \cos\left(2\delta_i\right)\right) \tag{3.29c}$$

$$= \frac{1}{2}\Sigma 1 \tag{3.29d}$$

and thus $\Sigma \sin^2 \delta_i = \Sigma \cos^2 \delta_i$. This then completes the missing steps in (3.28f).

The three- and four-frame algorithms described earlier that were derived analytically are identical to those that would be obtained from the least-squares solution of (3.28f). That is, the three- and four-frame algorithms are already optimal in the least-squares sense.

3.5   Carré's Algorithm

In the foregoing, we have assumed that it is possible to apply precise phase steps of known value between each frame. Clearly, if the actual phase shift differs from the assumed value, errors will be introduced into the reconstruction and we should therefore treat the phase step as one of the unknown quantities. Additionally, being able to treat the phase step as a variable is very usefully experimentally. For example, we might use a linear phase retarder such a Soleil-Babinet compensator to generate phase steps that we know are of equal increment but of unknown value. The best-known algorithm for these situations is Carré's algorithm [4, 12].

We begin by assuming that the phase step between each frame is 2α; that is

$$\delta_i = -3\alpha, -\alpha, \alpha, 3\alpha \quad i = 1, \ldots, 4 \tag{3.30}$$

and by writing the interferometer equation in symmetrical form:

$$I_0 = A + B \cos(\psi - 3\alpha) \tag{3.31a}$$

$$I_1 = A + B \cos(\phi - \alpha) \tag{3.31b}$$

$$I_2 = A + B \cos(\phi + \alpha) \tag{3.31c}$$

$$I_3 = A + B \cos(\phi + 3\alpha) \tag{3.31d}$$

Next, we expand the cosine terms as before and evaluate the following sums and differences:

$$3(I_1 - I_2) - (I_0 - I_3) = 6B \sin \phi \sin \alpha - 2B \sin \phi \sin 3\alpha \tag{3.32a}$$

which, on substituting $\sin 3\alpha = -4 \sin^3 \alpha + 3 \sin \alpha$, yields

$$= 8B \sin \phi \sin^3 \alpha \tag{3.32b}$$

Similarly,

$$(I_0 - I_3) + (I_1 - I_2) = 2B \sin \phi \sin 3\alpha + 2B \sin \phi \sin \alpha \tag{3.32c}$$

$$= -8B \sin \phi \sin^3 \alpha + 8B \sin \phi \sin \alpha \tag{3.32d}$$

which, after extracting a common factor of $8B \sin \phi \sin \alpha$, leaves

$$= 8B \sin \phi \sin \alpha \cos^2 \alpha \tag{3.32e}$$

Hence the unknown phase step may be estimated from

$$\alpha(x,y) = \tan^{-1} \left\{ \frac{3(I_1 - I_2) - (I_0 - I_3)}{(I_0 - I_3) + (I_1 - I_2)} \right\}^{1/2} \tag{3.33}$$

In order to recover the phase, we need the following sums and differences of the captured images:

$$(I_0 - I_3) + (I_1 - I_2) = 8B \sin \phi \sin \alpha \cos^2 \alpha \tag{3.34a}$$

and

$$(I_1 + I_2) - (I_0 + I_3) = -2B \cos \phi \,(-\cos \alpha + \cos 3\alpha) \tag{3.34b}$$

which, substituting $\cos 3\alpha = 4 \cos^3 \alpha - 3 \cos \alpha$ yields, after some simplification

$$= 8B \cos \phi \cos \alpha \sin^2 \alpha \tag{3.34c}$$

Since we have the phase step α from (3.33), we can recover the phase map from

$$\phi(x,y) = \tan^{-1} \left\{ \tan(\alpha[x,y]) \frac{(I_0 - I_3) + (I_1 - I_2)}{(I_1 + I_2) - (I_0 + I_3)} \right\} \tag{3.35a}$$

or, by inserting the result for $\tan \alpha$ from (3.33):

$$\phi(x,y) = \tan^{-1} \left\{ \frac{\left\{ [3(I_1 - I_2) - (I_0 - I_3)][(I_0 - I_3) + (I_1 - I_2)] \right\}^{1/2}}{(I_1 + I_2) - (I_0 + I_3)} \right\} \tag{3.35b}$$

Converting the result of the arctangent function in this expression to a wavefront phase $\phi(x, y)$ modulo 2π is not as straightforward as the other algorithms discussed thus far. The square root in the numerator of (3.35b) returns the absolute value of $\sin(\phi[x, y])$ and not the sine. And since the denominator can be either positive or negative, we must construct additional terms that are proportional to the sine and cosine of the phase [4, 13] in order to correctly return ϕ modulo 2π.

The most commonly used value for the phase step between frames is 90°; that is, $\alpha = 45°$, but the optimal value depends on the experimental conditions, for example, a phase step of 110° minimizes the influence of random intensity noise [14].

3.6 Hariharan's Method

A linear phase shift error produces a sinusoidal error at twice the fringe frequency in the reconstructed phase ϕ [15] and therefore there is much interest in phase-shifting algorithms that are relatively immune to this error. Hariharan's algorithm is one such method [16, 17].

We assume a phase shift α between frames:

$$\delta_i = -2\alpha, -\alpha, 0, \alpha, 2\alpha \quad i = 1, \ldots, 5. \tag{3.36}$$

We again write the intensities I_i in symmetrized form:

$$I_0 = A + B \cos(\phi - 2\alpha) \tag{3.37a}$$

$$I_1 = A + B \cos(\phi - \alpha) \tag{3.37b}$$

$$I_2 = A + B \cos(\phi) \tag{3.37c}$$

$$I_3 = A + B \cos(\phi + \alpha) \tag{3.37d}$$

$$I_4 = A + B \cos(\phi + 2\alpha) \tag{3.37e}$$

We expand the cosine terms as before and evaluate the following differences:

$$I_1 - I_3 = 2B \sin \phi \sin \alpha \tag{3.38a}$$

and

$$2I_2 - I_4 - I_0 = 2B \cos \phi (1 - \cos 2\alpha) \tag{3.38b}$$

$$= 4B \cos \phi \sin^2 \alpha \tag{3.38c}$$

We can combine these two expressions to produce:

$$\frac{\tan[\phi(x,y)]}{2\sin\alpha} = \frac{I_1 - I_3}{2I_2 - I_4 - I_0} \tag{3.39}$$

and since the phase shift α is a variable, we are free to choose it in such a way that it minimises variations in this expression to errors in the phase shift. Accordingly, differentiating with respect to α:

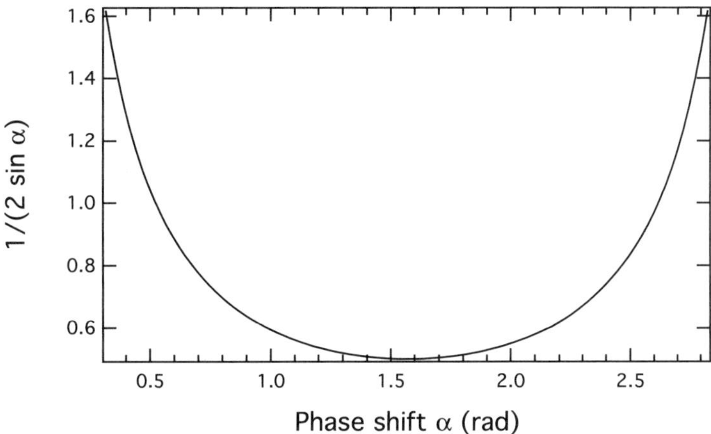

Figure 3.7   Plot of 1/2 sin α as a function of the phase step α.

$$\frac{d}{d\alpha}\left\{\frac{\tan[\phi(x,y)]}{2\sin\alpha}\right\} = \frac{-\cos\alpha\tan[\phi(x,y)]}{2\sin^2\alpha} \tag{3.40}$$

which goes to zero when $\alpha = \pi/2$. With this value for the phase shift, (3.39) reduces to:

$$\phi(x,y) = \tan^{-1}\left[\frac{2(I_1 - I_3)}{2I_2 - I_4 - I_0}\right] \tag{3.41}$$

If $\phi(x, y) = \pi/4$, then a plot of the left hand side of (3.39) has the form shown in Figure 3.7. What is notable is that the minimum, centered at $\pi/2$, is quite broad, which means that the Hariharan algorithm can tolerate quite large errors in the phase shift without causing signifiant errors in the recovered phase $\phi(x, y)$.

EXERCISE 3.5

Estimate the maximum error that would occur in the recovered phase $\phi(x, y)$ if there is a 2° error in the phase steps.

Solution

Suppose that the actual phase shift between measurements is $\pi/2 + \epsilon$ where ϵ is a small angle; 2° in this example. Let the true phase be $\phi(x, y)$ and the measured phase be $\phi'(x, y)$ so that

$$\phi'(x, y) = \phi(x, y) + \Delta\phi(x, y) \tag{3.42a}$$

Our aim is to find an estimate for $\Delta\phi(x, y)$. From (3.39), we have

$$\frac{\tan[\phi(x,y)]}{2\sin\alpha} = \frac{(I_1 - I_3)}{2I_2 - I_4 - I_0} = \frac{\tan[\phi'(x,y)]}{2} \tag{3.42b}$$

Since $\sin(\pi/2 + \varepsilon) = \cos\varepsilon \approx 1 - 1/2\,\varepsilon^2$, we have

$$\tan[\phi'(x,y)] = \frac{1}{1 - \dfrac{1}{2}\varepsilon^2}\tan[\phi(x,y)] \tag{3.42c}$$

$$\approx \left(1 + \frac{1}{2}\varepsilon^2\right)\tan[\phi(x,y)] \tag{3.42d}$$

Now, using standard trigonometric expansions, we can write

$$\tan\Delta\phi = \tan(\phi' - \phi) \tag{3.42e}$$

$$= \frac{\tan\phi' - \tan\phi}{1 + \tan\phi'\tan\phi} \tag{3.42f}$$

which, when we substitute in our approximate expression for ϕ' from (3.42d), yields:

$$\tan\Delta\phi \approx \frac{\dfrac{\varepsilon^2}{2}\tan\phi}{1 + \left(1 + \dfrac{\varepsilon^2}{2}\right)\tan^2\phi} \tag{3.42g}$$

$$\approx \frac{\varepsilon^2}{2}\frac{\tan\phi}{\left(1 + \tan^2\phi\right)} \tag{3.42h}$$

Finally,

$$\Delta\phi \approx \tan\Delta\phi \approx \frac{\varepsilon^2}{4}\sin[2\phi(x,y)] \tag{3.42i}$$

Since $\max|\sin(2\phi)| = 1$, we easily calculate that a 2° error in the phase step causes a maximum error $\Delta\phi$ of around 0.017°.

3.7   Design of Phase-Stepping Algorithms

At this point, we have examined a number of different phase-stepping algorithms, some of which are relatively robust to stepping errors and some of which are decidedly not. So it is only natural to ask: What causes an algorithm to have this desirable behavior? And can we design algorithms so that they have particular characteristics? A very elegant solution to these questions was proposed by Surrel [18, 4]. In essence, any phase-shifting algorithm can be associated with a polynomial, the roots of which determine the algorithm's sensitivity to harmonics and stepping errors.

If the intensity of a fringe pattern varies sinusoidally, then we may write it as:

$$I(\phi) = I_0(1 + \gamma\cos\phi) \tag{3.43a}$$

$$= I_0 + \frac{I_0\gamma}{2}\exp(i\phi) + \frac{I_0\gamma}{2}\exp(-i\phi) \tag{3.43b}$$

This notation is slightly more compact than that used in (3.5b). Here γ is the fringe visibility that is essentially a measure of the fringe contrast. See Box 3.4 for further details.

BOX 3.4　Fringe Visibility

A convenient and oft-used measure of the contrast of interference fringes is the visibility $\mathcal{V}$. From (3.8b), we easily see that the intensity has a maximum value $I_{max} = I_1 + I_2 + 2\sqrt{I_1I_2}$ when $\delta = 2m\pi$, m an integer. Similarly, it has a minimum $I_{min} = I_1 + I_2 - 2\sqrt{I_1I_2}$ when $\delta = (2m + 1)\pi$.

The visibility $\mathcal{V}$ is defined as:

$$\mathcal{V} = \frac{I_{max} - I_{min}}{I_{max} + I_{min}} \tag{3.44a}$$

$$= \frac{2\sqrt{I_1I_2}}{I_1 + I_2} \tag{3.44b}$$

$$= \frac{B}{A} = \gamma \tag{3.44c}$$

with $0 \le V \le 1$.

Since $I(\phi)$ is a periodic function, it can be expanded as a Fourier series:

$$I(\phi) = \sum_{m=-\infty}^{\infty} \alpha_m \exp(im\phi) \tag{3.45}$$

Evidently, from (3.43b)

$$\alpha_1 = \alpha_{-1} = \frac{I_0\gamma}{2} \tag{3.46}$$

and all other coefficients are identically zero. In phase stepping, we consider ϕ to be a given phase at a point (which we wish to measure), and we add phase steps δ. That is, we measure $I(\phi + \delta)$, which we can also express as a Fourier series expansion, thus:

$$I(\phi + \delta) = \sum_{m=-\infty}^{\infty} \left[\alpha_m \exp(im\phi)\right]\exp(im\delta) \tag{3.47a}$$

$$= \sum_{m=-\infty}^{\infty} \beta_m(\phi)\exp(im\delta) \tag{3.47b}$$

The main purpose of any phase-shifting algorithm is to evaluate the argument of the β_1 Fourier coefficient, corresponding to the fundamental harmonic of the intensity signal. Since

$$\beta_1 = \alpha_1(\cos\phi + i\sin\phi) \tag{3.48a}$$

it is clear that

$$\arg(\beta_1) = \frac{\mathfrak{I}(\beta_1)}{\mathfrak{R}(\beta_1)} = \tan\phi \tag{3.48b}$$

As we have seen from our earlier examples, M-step phase shifting algorithms are usually evaluated as the arctangent of the ratio of two linear combinations of values of $I(\delta)$. That is, the measured phase, denoted by $\tilde{\phi}$ to distinguish it from the actual phase ϕ, is given by

$$\tilde{\phi} = \tan^{-1}\left\{\frac{\sum_{k=0}^{M-1} b_k I(\phi + k\delta)}{\sum_{k=0}^{M-1} a_k I(\phi + k\delta)}\right\} \tag{3.49}$$

In order to see how this is related to polynomials, we consider the following complex linear combination:

$$S(\phi) = \sum_{k=0}^{M-1} c_k I(\phi + k\phi) \tag{3.50}$$

where $c_k = a_k + ib_k$ whence it is easy to see that

$$\tilde{\phi} = \arg[S(\phi)] \tag{3.51}$$

Using (3.47a), we can write the Fourier series of $S(\phi)$ as:

$$S(\phi) = \sum_{m=-\infty}^{\infty}\left\{\alpha_m \exp(im\phi)\sum_{k=0}^{M-1} c_k [\exp(im\delta)]^k\right\} \tag{3.52a}$$

$$= \sum_{m=-\infty}^{\infty} \alpha_m \exp(im\phi)P[\exp(im\delta)] \tag{3.52b}$$

where $P(z)$ is a polynomial of degree $M - 1$:

$$P(z) = \sum_{k=0}^{M-1} c_k z^k \tag{3.52c}$$

All the properties of any given phase shifting algorithm can be deduced from the roots of $P(z)$, which is called the *characteristic polynomial* of the algorithm. These roots lead to the following design rules [18]:

1. In order to detect the fundamental frequency, $\exp(-i\delta)$ must be a root of the characteristic polynomial, but not $\exp(i\delta)$.
2. If $\exp(im\delta)$ and $\exp(-im\delta)$ are roots of the characteristic polynomial ($m \neq 1$), then the algorithm is insensitive to the mth-harmonic of the intensity signal.
3. When only one of $\exp(-im\delta)$ or $\exp(im\delta)$ is a double root of the characteristic polynomial, but not the other, then that harmonic component can be detected, but the algorithm is insensitive to miscalibration of the phase shifter.
4. If both $\exp(-im\delta)$ and $\exp(im\delta)$ are double roots of the characteristic polynomial, then the algorithm is insensitive to the mth-harmonic as well as to miscalibration of the phase shifter.

In order to show how these rules work in practice, we will apply them to two well-known phase-shifting algorithms.

EXERCISE 3.6

Find the characteristic polynomial for Hariharan's method.

Solution

Reading the coefficients a_k and b_k directly from the denominator and numerator of (3.41), respectively, leads us to the characteristic polynomial for this method:

$$P(z) = -1 + 2iz + 2z^2 - 2iz^3 - z^4$$
$$= -(z - 1)(z + 1)(z + i)^2 \tag{3.53}$$

and which therefore has single roots at $z = \pm 1$ and a double root at $-i$. Surrel also described a very compact and convenient graphical method for displaying the features of the characteristic polynomial, an example of which is shown in Figure Ex 3.2 for this method. The phase steps are plotted as radial lines, with the angle of the line from the real axis equal to the phase step. For this method, the phase step is $\pi/2$, so we see that the unit circle is divided into four quadrants. Next, a solid dot is drawn at the intersection of a line and the unit circle if that particular harmonic has a root there. Double roots are indicated by a solid dot surrounded by an open circle. Finally, phase steps beyond 2π are indicated by a thick solid line along the real axis.

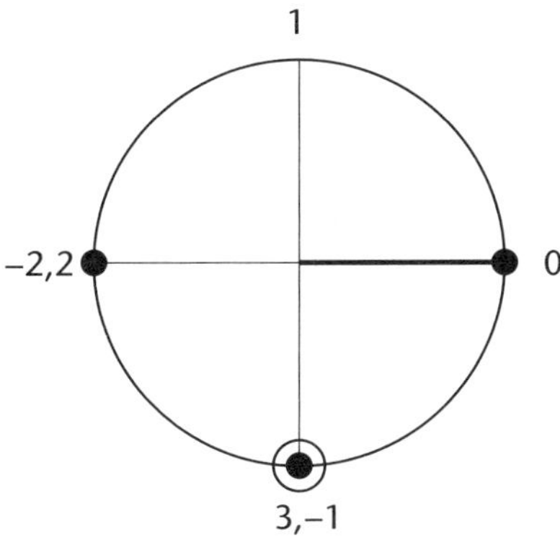

Figure Ex 3.2 Roots of the characteristic polynomial on the unit circle for Hariharan's method.

The diagram tells us that there is no root for $m = 1$, but there is for $m = -1$, consistent with Rule 1. Since the polynomial has a root for both $m = 2$ and -2, the algorithm is insensitive to the second harmonic by Rule 2. Finally, by Rule 3, the double root for $m = 3$ makes the algorithm insensitive to phase shift errors as we saw earlier.

EXERCISE 3.7

Find the characteristic polynomial for the three-frame method and comment on its sensitivity to phase-shift errors.

Solution

We asserted earlier that the three-frame method (with phase steps of $2\pi/3$) was the simplest possible algorithm, but also the most sensitive to stepping errors. This is now easily demonstrated using the characteristic polynomial. Reading the coefficients directly from (3.15), we have:

$$P(z) = 2 - \left(1 + \sqrt{3}i\right)z - \left(1 - \sqrt{3}i\right)z^2$$

$$= \frac{1}{2}\left(-1 + \sqrt{3}i\right)(z - 1)\left(2z + 1 + \sqrt{3}i\right) \tag{3.54}$$

Plotting these roots on the unit circle generates the diagram shown in Figure Ex 3.3. By Rule 1, we are able to detect the fundamental, but the lack of paired roots or a double root renders the method sensitive to all harmonics and to phase-shift errors.

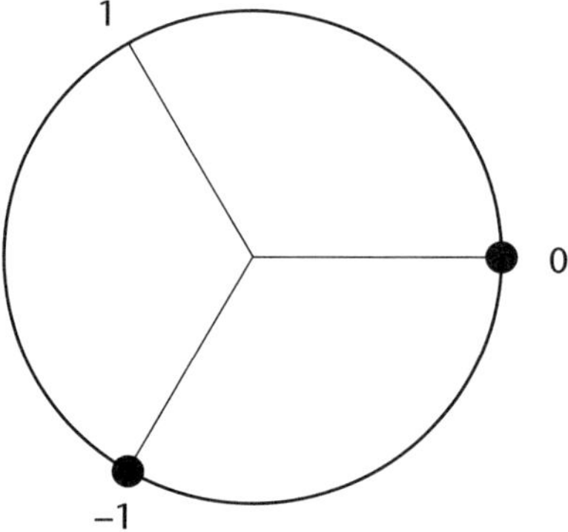

Figure Ex 3.3 Roots of the characteristic polynomial on the unit circle for the three-frame method.

3.8   Phase Stepping Techniques and Applications

3.8.1   Mechanical Stepping

One of the most common methods for producing the phase steps is to use a PZT to translate the reference mirror or some convenient optical surface. PZTs are ceramic materials, usually lead-zinc-titanate, that expand or contract on application of a modest dc voltage. For modern devices, around 100V is generally enough to cause a mechanical translation of a wavelength.

It is not surprising, therefore, to find that mechanical phase-stepping with PZTs is widely used in optical testing [4], usually with the interferometer arranged in the Tywman-Green configuration, as shown in Figure 3.8. The Michelson and Twyman-Green interferometers are ideally suited to mechanical phase-stepping because the reference mirror is moved parallel to the direction of propagation of the reference beam. The same technique can also be applied to a standard Mach-Zehnder interferometer, although the mirror movement is now at an oblique angle to the beam direction. Figure 3.9 shows the situation at the moving mirror. The phase shift acquired by the wavefront is [2]

$$\delta = \overline{AB} + \overline{BC} - \overline{AD} \tag{3.55a}$$

where

$$\overline{AB} = \overline{BC} = \frac{x}{\cos\theta} \tag{3.55b}$$

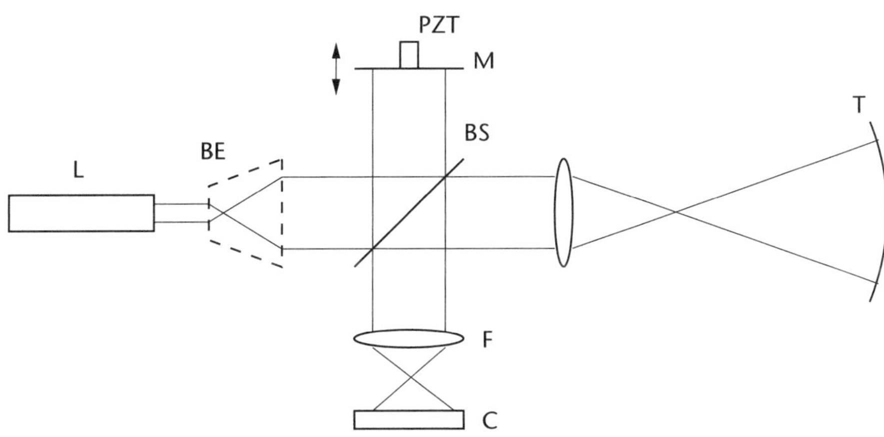

Figure 3.8 Twyman-Green interferometer. L = laser; BE = beam expander; BS = beam splitter; M = reference mirror; PZT = piezoelectric transducer; T = test surface; F = imaging lens; C = CCD camera.

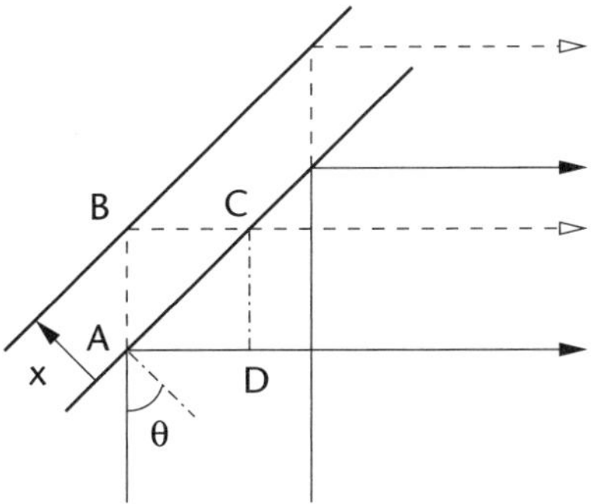

Figure 3.9 Phase step and beam displacement at one mirror of a Mach-Zehnder interferometer.

and

$$\overline{AD} = \overline{AC}\sin\theta \qquad (3.55c)$$

$$= 2x\tan\theta\sin\theta \qquad (3.55d)$$

which, on substituting back into (3.55a), yields:

$$\delta = 2x\cos\theta \qquad (3.55e)$$

where x is the mirror displacement and θ is the angle of incidence. Note that the beam has also been laterally displaced.

Mechanical phase-stepping is widely used in Fizeau interferometers, although some care must be taken if spherical surfaces are being tested (so-called spherical Fizeau cavities). The nature of the problem is illustrated in Figure 3.10, which shows a partial view of a spherical cavity Fizeau interferometer. The inner surface of the meniscus lens, with radius r_2 is the reference surface. The test surface, with radius r_1 is mounted so that these two radii have a common center. The exact analogue of the motion of the plane reference mirror in either the Michelson or Twyman-Green interferometers is to change either r_1 or r_2 without moving the center point, which clearly cannot be done. Moore and Slaymaker [19] consider the case in which the test surface is moved in a direction parallel to the interferometer axis and find that for numerical apertures $NA \leq 0.8$, the error is $< \lambda/100$.

A more convenient and commonly adopted approach is to step the reference surface [4], as shown in Figure 3.11. The errors associated with this approach have been carefully investigated by de Groot [20] who found that modern, multiframe algorithms could be used to extend the useable NA to almost 0.95.

More recently, an alternative solution to phase-stepping in Fizeau interferometers has been proposed [21, 22] that avoids the necessity to move either the reference or test surfaces. The essential idea behind this method is illustrated in Figure 3.12, in which two identical glass wedges are arranged so that they form a variable thickness parallel plate. Phase-shifting is achieved by laterally translating one of the wedges, say W_2. The optical path difference introduced by this lateral shift is given by

$$\delta = 2(n - 1)s\tan\theta \qquad (3.56)$$

where n and θ are the refractive index and wedge angle, respectively, of the wedged plates and s is the lateral displacement.

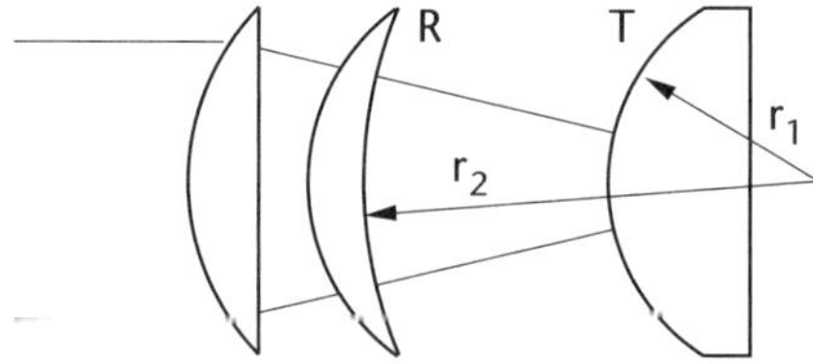

Figure 3.10 Partial schematic of a Fizeau interferometer; collimated laser light is incident from the left-hand side. T = test surface; R = reference surface.

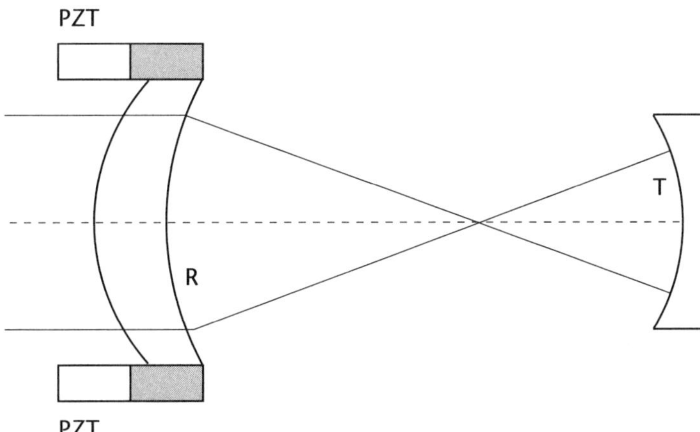

Figure 3.11   Fizeau interferometer with mechanically stepped reference surface. R = reference surface; T = test surface.

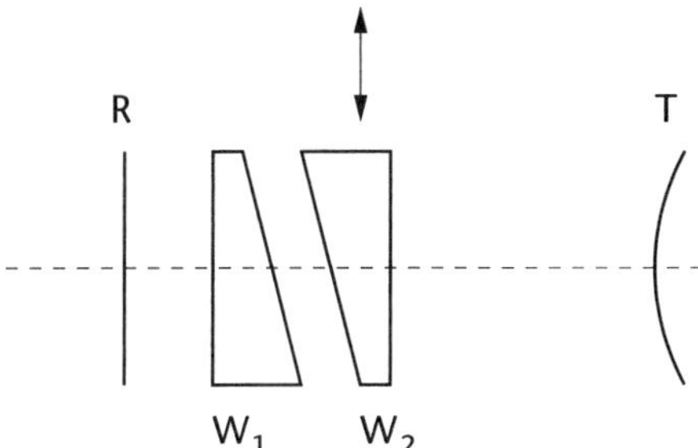

Figure 3.12   Fizeau cavity. R = reference surface; T = test surface; $W_{1,2}$ = identical glass wedges.

3.8.2   Grating Methods

So far we have considered phase-shifting in the spatial domain, which is typically achieved by physically moving an optical element in one arm of an interferometer. It is also possible to produce phase shifts in the temporal domain, usually by arranging for the reference and test beams to have slightly different optical frequencies. Strictly speaking, this constitutes heterodyne interferometry, but we include a brief discussion here as it provides an alternative method to phase step a Fizeau interferometer, as we shall show shortly.

First, consider two electric fields with slightly different optical frequencies. For convenience, we assume that both fields have the same polarization:

$$E_1 = E_{01} \exp\left(i\left[\phi_r(x,y) - 2\pi(\nu + \Delta\nu)t\right]\right) \tag{3.57a}$$

and

$$E_2 = E_{02} \exp\left(i\left[\phi_t(x,y) - 2\pi\nu t\right]\right) \tag{3.57b}$$

where $\phi_r = \mathbf{k}_1 \cdot \mathbf{r} + \epsilon_1$ is the phase in the reference arm and similarly for ϕ_t, the phase in the test arm. It is not difficult to show that the resulting intensity is given by:

$$I = I_1 + I_2 + 2\sqrt{I_1 I_2}\,\cos\bigl(\phi(x,y) + 2\pi\Delta\nu t\bigr) \qquad (3.57c)$$

where $I_{1,2} = |E_{1,2}|^2$, $\phi(x, y) = \phi^t - \phi^r$ and $\Delta\nu$ is the heterodyne or beat frequency. This frequency difference is equivalent to a linear (temporal) phase shift

$$\delta(t) = 2\pi\Delta\nu t \qquad (3.57d)$$

between the reference and test beams. It is clear, therefore, that sampling the sinusoidally varying intensity at each pixel will generate the phase steps required for any phase stepping algorithm.

There are many ways in which this frequency difference can be generated—moving mirrors, electro- and acousto-optic modulators being among the more popular—but we wish to focus here on grating methods. A diffraction grating that is moved perpendicular to the direction of propagation of the incident beam causes the diffracted orders to be Doppler-shifted in frequency. The frequency shift is proportional to the diffraction order and the velocity of the grating [4]. Beams that are diffracted in the same direction as the motion are shifted upwards in frequency and conversely for orders diffracted contrary to the motion. From a practical point of view, it is more convenient to rotate a radial grating, in which case the frequency offset $\Delta\nu$ is directly proportional to the grating angular velocity ω [23].

Fizeau interferometers are very stable, since the two interfering beams travel very nearly the same path, with the exception of the small region between the test plates. It would seem, therefore, that any optical frequency shifter would need to be placed in this region but such an approach is highly impractical. An elegant solution to this problem is described by Barnes [24], in which the heterodyning is performed *externally* to the interferometer with the aid of a rotating grating.

A schematic of the heterodyne Fizeau is shown in Figure 3.13. Light from the laser is diffracted by a rotating radial grating and the diffracted beams are focussed in the plane of spatial filter F_1 by lens L_1. This filter blocks all orders except the $+1$ and -1 orders. These two orders are collected and collimated by lens L_2 and L_3 so that two, nearly coaxial beams illuminate the test and reference surfaces. On reflection, the beams encounter a second spatial filter F_2, which is placed in a focal plane of the system. The surfaces are adjusted so that the focus of the -1 order from T is superimposed on the focus of the $+1$ order reflected from R. Spatial filter F_2 has a single central hole that only allows these two beams to pass through. Finally, lenses L_4 and L_5 form an image of the region between the test surfaces onto a screen or CCD camera. Note that each pixel in this image will exhibit a sinusoidal variation in intensity at the beat frequency.

One further application of diffraction gratings is as an achromatic phase shifter [25], shown schematically in Figure 3.14. The device is essentially a frequency domain optical delay line and operates on the well-known time-shifting property of the Fourier transform; namely, that a linear phase ramp in the frequency domain corresponds to a delay in the time domain. A broadband, collimated beam is incident on the transmission grating G and dispersed in such a way that the spectral component at the mean wavelength λ_0 travels along the optic axis z. The grating is placed in a focal plane of a lens L, with focal length f, which collimates the dispersed light and projects it onto a tiltable mirror M whose pivot point is located in

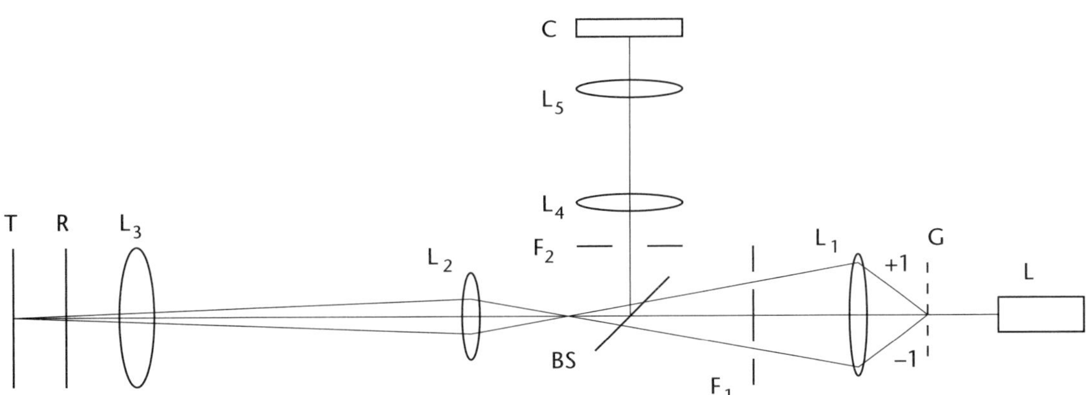

Figure 3.13 Heterodyne Fizeau interferometer. L = laser; G = radial grating; L_1 to L_5 = lenses; $F_{1,2}$ = spatial filters; BS = beam splitter; C = CCD camera; T,R = test and reference surfaces.

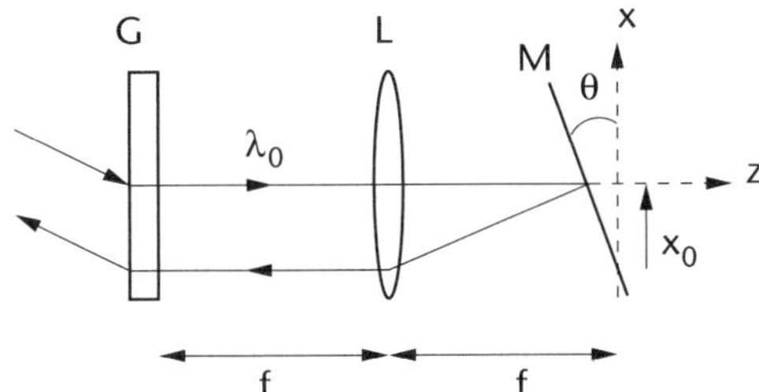

Figure 3.14 Achromatic phase shifter based on an optical delay line. G = grating; L = lens; M = tiltable mirror.

the other focal plane. The tilt angle θ of the mirror can be varied and the pivot point is offset from the optic axis by a distance x_0, as shown. The tilted mirror imparts a phase ramp on the spectral components of the beam that, for small angles, is linearly proportional to the tilt angle and a phase offset that is linearly proportional to the pivot offset. The reflected light is collected by the lens and projected back onto the grating, which recollimates the dispersed light into a beam. The phase $\delta(k)$ acquired by a spectral component with wavenumber k [25] is:

$$\delta(k) = 2\theta x_0 k - \frac{4\pi m\theta f\left(k - k_0\right)}{pk_0} \tag{3.58}$$

where θ is the tilt angle, m the order of diffraction, f is the focal length of the lens, and p is the grating period. If $x_0 = m\lambda_0 f/p$, then the group delay is zero and the phase shift is truly achromatic.

EXERCISE 3.8

If the drive current in a semiconductor laser diode is modulated, it leads to a linear change in the laser wavelength and this change can be exploited to generate phase steps in an interferometer with unequal arm lengths [26]. Let these arms lengths be L_1 and L_2. Derive an expression for the phase shift $\Delta\phi$ as a function of the path length difference $\Delta l = 2(L_1 - L_2)$ and the fractional wavelength change, $\Delta\lambda/\lambda_0$.

If $\lambda_0 = 635$ nm and $\Delta l = 100$ mm, find the wavelength shift $\Delta\lambda$ that will create a $\pi/2$ phase shift.

Solution

Assuming that we have a two-pass interferometer such as a Michelson, we can relate the optical path length of each arm to the wavelength as follows:

$$2L_1 = m_1\lambda_0 \tag{3.59a}$$

$$2L_2 = n_1\lambda_0 \tag{3.59b}$$

and when the wavelength increase by $\Delta\lambda$:

$$2L_1 = m_2(\lambda_0 + \Delta\lambda) \tag{3.59c}$$

$$2L_2 = n_2(\lambda_0 + \Delta\lambda) \tag{3.59d}$$

When $\Delta\lambda$ is such that $(m_1 - m_2) - (n_1 - n_2) = 1$, one fringe or 2π rad phase has occurred. Hence

$$1 \text{ fringe} = 2(L_1 - L_2)\left(\frac{1}{\lambda_0} - \frac{1}{\lambda_0 + \Delta\lambda}\right) \tag{3.59e}$$

But

$$\frac{1}{\lambda_0} - \frac{1}{\lambda_0 + \Delta\lambda} \approx \frac{\Delta\lambda}{\lambda_0^2} \tag{3.59f}$$

and

$$2(L_1 - L_2) = \Delta l \tag{3.59g}$$

is the path length difference. So the change in phase $\Delta\delta$ is

$$\Delta\delta = 2\pi\Delta l\frac{\Delta\lambda}{\lambda_0^2} \tag{3.59h}$$

Inserting the values given into the equation above yields a wavelength change $\Delta\lambda \approx$ 1 pm, a value that is easily achievable via a very modest change in the drive current.

3.8.3   Methods Based on Pancharatnam's Phase

The mechanical stepping that we have considered so far generates a phase step δ that is inversely proportional to the wavelength. For double-pass interferometers, such as the Michelson, Twyman-Green, and Fizeau, $\delta = 4\pi d/\lambda$ where d is the distance moved by the mirror or optical element. For broadband or white-light sources, this type of phase stepping is obviously incapable of generating the same phase step for

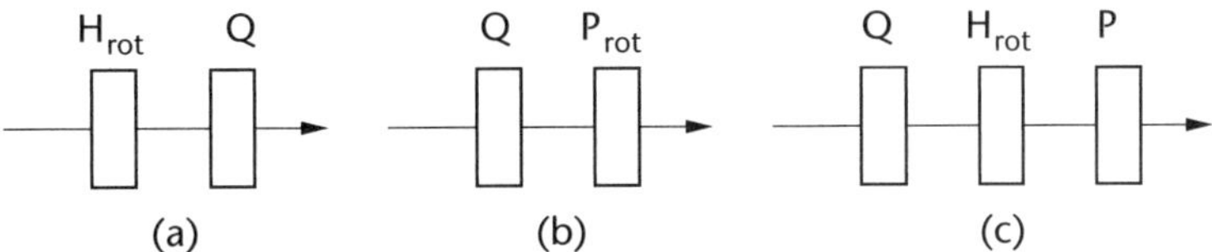

Figure 3.15 Three configurations for performing geometric phase shifts. Q = quarter-wave plate; H = halfwave plate; P = polarizer. Subscript `rot` denotes a rotating component. Configuration (a) is suitable for the input end of an interferometer, while (b) and (c) are suited for the output.

every wavelength. Fortunately, we can generate phase steps that are independent of the wavelength by exploiting Pancharatnam's phase (also called the geometric or Berry's phase).

Pancharatnam's phase is a geometric or topological phase associated with the continuous deformation of a polarisation state [27–30]. A polarization state that describes a closed loop on the Poincaré sphere acquires a geometric phase equal to half the solid angle subtended at the center of the Poincar´e sphere by the loop. See Box 3.5 for a brief description of the Poincar´e sphere. Since this phase shift is wavelength independent, it is particularly well suited to broadband sources. There are a number of ways in which polarisation phase shifting can be achieved [31, 32], three of which are reproduced in Figure 3.15. The configuration of optical components shown in (a) is designed to operate at the input of an interferometer and is in fact identical to the frequency shifter arrangement first published by Sommargren [33]. The other two configurations are designed for the output of an interferometer.

BOX 3.5   Jones Algebra and the Poincaré Sphere

We need a convenient system of mathematics in order to determine how a given polarization state evolves as a beam propagates through an arbitrary sequence of optical components. One such system is the Jones algebra, invented in 1941 by American physicist R. Clark Jones [34]. The x and y (or, equivalently, p- and s-) components of the electric field of a polarized beam are written as a column vector:

$$\mathbf{E} = \begin{bmatrix} E_{ox} \exp i\phi_x \\ E_{oy} \exp \phi_y \end{bmatrix} \tag{3.60}$$

where E_{ox}, E_{oy} are the amplitudes and ϕ_x, ϕ_y are the phases. Note that in general, the elements of this vector may be complex.

Every optical component, whether it operates in transmission or reflection, is represented by a 2×2 matrix of, in general, complex numbers. So for example, an ideal polarizer with its transmission axis oriented along the x-axis is represented by

$$J_P = \begin{bmatrix} 1 & 0 \\ 0 & 0 \end{bmatrix} \tag{3.61a}$$

whereas an (ideal) retarder or waveplate is represented by

$$J_R = \begin{bmatrix} 1 & 0 \\ 0 & \exp(-i\delta) \end{bmatrix} \tag{3.61b}$$

where δ is the retardation. Together with the standard rotation matrix $R(\theta)$, this formalism allows any arbitrary concatenation of optical components to be described by the multiplication of the corresponding Jones matrices. Note, however, that this algebra is *only* applicable to polarized waves; it cannot treat partially polarized waves. For that, one needs the alternative approach based on the Stokes parameters and Muller matrices. However, for laser interferometers of the type described in this chapter, the Jones algebra is more than adequate. Some further reading may be found in [2], with fuller descriptions in [35] and [36].

The Poincaré sphere is a sphere with unit diameter that is used to graphically represent polarization states as shown, for example, by Figure 3.16. The north and south poles of this sphere correspond to right- and left-hand circularly polarized light, respectively, while the equator contains all linearly polarized states. The northern hemisphere represents all right-handed elliptically polarized states and conversely for the southern hemisphere. A line of longitude contains all states with the same azimuthal angle, but with ellipticity varying from right-hand circular at the north pole, to linear at the equator and through elliptical states to left-hand circular at the south pole. Conversely, a line of latitude contains all states with the same ellipticity but with constantly varying azimuthal angle. The utility of the Poincaré sphere stems from the fact that there is a one-to-one mapping of the Stokes parameters onto points on the sphere. Further details may be found in [36, 35].

To see how the geometric phase is generated, we will use configuration (b) as an example and analyze the evolution of the input state of polarization. Suppose that the light incident on this phase shifter is arbitrarily elliptically polarised and is represented by

$$\mathbf{E}_{in} = \begin{bmatrix} E_{01}\exp(i\phi(x,y)) \\ E_{02} \end{bmatrix} \tag{3.62a}$$

$$= \begin{bmatrix} 1 \\ 0 \end{bmatrix} E_{01}\exp(i\phi(x,y)) + \begin{bmatrix} 0 \\ 1 \end{bmatrix} E_{02} \tag{3.62b}$$

where E_{01} and E_{02} are the electric field amplitudes and ϕ their relative phase. This elliptically polarized state is represented by the solid dot on the Poincaré sphere in Figure 3.16. The vectors $[1, 0]^T$ and $[0, 1]^T$ represent p- and s-polarized light, respectively, [35]. The p- and s-polarized components are incident on the quarter-wave plate, which is oriented at 45°, and consequently produces right-hand and

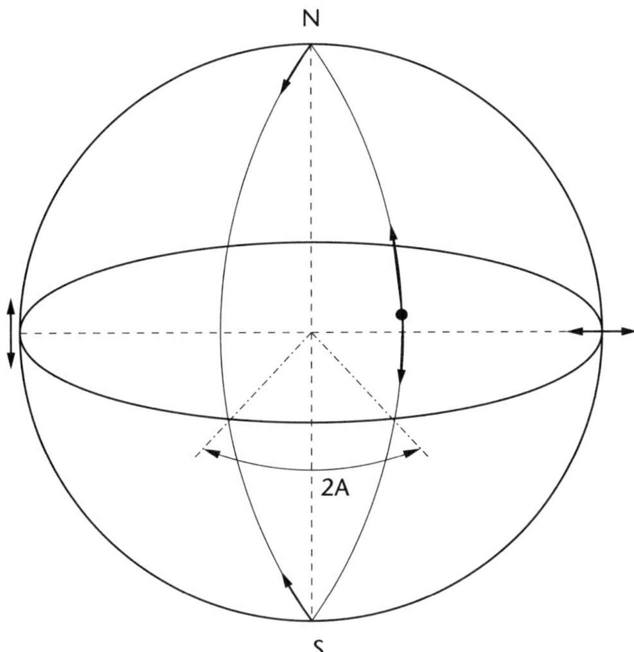

Figure 3.16 Poincaré sphere showing evolution of initial polarization state.

left-hand circularly polarized light, respectively, generating the initial trajectories shown on the Poincaré sphere. The analyser, oriented at an angle A, brings both of these circularly polarized states to the same point on the equator, completing the loop and thereby creating the geometric phase step [27].

We may arrive at the same result using Jones matrix algebra (see [35] and Box 3.5). The electric field after the analyzer is

$$\begin{bmatrix} E_x \\ E_y \end{bmatrix} = \begin{bmatrix} \cos^2 A & \cos A \sin A \\ \cos A \sin A & \sin^2 A \end{bmatrix} \begin{bmatrix} 1 & i \\ i & 1 \end{bmatrix} \begin{bmatrix} E_{01}\exp(i\phi) \\ E_{02} \end{bmatrix} \tag{3.63a}$$

where the first matrix represents the effect of a polarizer with its transmission axis oriented at an angle A to the horizontal, the second matrix represents an ideal quarter-wave plate oriented at $45°$, and the final vector is our input electric field. Multiplying out, we have:

$$= \begin{bmatrix} \cos^2 A\left(E_{01}\exp(i\phi) + iE_{02}\right) + \cos A \sin A\left(iE_{01}\exp(i\phi) + E_{02}\right) \\ \cos A \sin A\left(E_{01}\exp(i\phi) + iE_{02}\right) + \sin^2 A\left(iE_{01}\exp(i\phi) + E_{02}\right) \end{bmatrix} \tag{3.63b}$$

which may be factored as:

$$= \left(\cos A\left[E_{01}\exp(i\phi) + iE_{02}\right] + \sin A\left[iE_{01}\exp(i\phi) + E_{02}\right]\right)\begin{bmatrix} \cos A \\ \sin A \end{bmatrix} \tag{3.63c}$$

Now, if we take the common factor we can rewrite it as:

$$= \left(E_{01} \exp(i\phi)[\cos A + i\sin A] + iE_{02}[\cos A - i\sin A] \right) \begin{bmatrix} \cos A \\ \sin A \end{bmatrix} \qquad (3.63\text{d})$$

which, after applying Euler's equation to the terms in square brackets readily yields

$$= \left(E_{01} \exp(i[\phi + A]) + iE_{02} \exp(-iA) \right) \begin{bmatrix} \cos A \\ \sin A \end{bmatrix} \qquad (3.63\text{e})$$

Note how E_{01} and E_{02} have acquired opposite rotations. The intensity, at any wavelength, is then readily given by

$$I \propto E_{01}^2 + E_{02}^2 + 2E_{01}E_{02} \sin\left(2A + \phi(x,y)\right) \qquad (3.63\text{f})$$

where $2A$ is the phase step. We can therefore produce any number of phase-stepped images merely by setting the polarizer angle A to the appropriate value and since this can be done repeatably and accurately, without any of the hysteresis associated with PZTs, polarization phase-stepping is highly attractive.

EXERCISE 3.9

Use Jones algebra to derive the matrix corresponding to an ideal quarter-wave plate oriented with its fast axis at 45° with respect to the horizontal. Hence show that p- and s-polarized light incident on this plate will generate orthogonal, circularly polarized states.

Solution

The Jones matrix for a rotated component is $J_{\text{rot}} = R(-\theta)J_{\text{unrot}} R(\theta)$ [35] and since our quarter-wave plate is rotated through 45°, this gives

$$J_{\text{QWP}} = \frac{1}{\sqrt{2}} \begin{bmatrix} 1 & -1 \\ 1 & 1 \end{bmatrix} \begin{bmatrix} 1 & 0 \\ 0 & -i \end{bmatrix} \begin{bmatrix} 1 & 1 \\ -1 & 1 \end{bmatrix} \qquad (3.64\text{a})$$

$$= \frac{1}{2} \begin{bmatrix} 1-i & 1+i \\ 1+i & 1-i \end{bmatrix} \qquad (3.64\text{b})$$

$$= \frac{1-i}{2} \begin{bmatrix} 1 & i \\ i & 1 \end{bmatrix} = \frac{\exp(-i\pi/4)}{\sqrt{2}} \begin{bmatrix} 1 & i \\ i & 1 \end{bmatrix} \qquad (3.64\text{c})$$

as required. The common phase term is not important, so if p-polarized light is incident on this wave-plate we have

$$E = \frac{1}{\sqrt{2}} \begin{bmatrix} 1 & i \\ i & 1 \end{bmatrix} \begin{bmatrix} 1 \\ 0 \end{bmatrix} \qquad (3.64\text{d})$$

$$= \frac{1}{\sqrt{2}} \begin{bmatrix} 1 \\ i \end{bmatrix} \tag{3.64e}$$

which is the Jones vector for right-hand circularly polarized light. Conversely, if the incident light was s-polarized then left-hand circularly polarized light would be generated.

This method has been successfully applied in a number of application areas, for example: white-light ellipsometry [37], single-frequency imaging ellipsometry [38], and heterodyne interferometry [39] in which a beat frequency is created by continuously rotating the polarizer. Additionally, it has been used to implement single-snapshot interferometers, in which all four phase steps are acquired simultaneously.

To see why a single-snapshot interferometer is highly desirable, we need to consider the limitations to the accuracy of a typical phase-stepping instrument. Spatial variations in the mean intensity A and modulation B are perfectly acceptable. After all, the recovered phase $\phi(x, y)$ is independent of these two parameters. We do, however, require that the laser intensity and fringe contrast remain constant in the time it takes to acquire the fringes. Generally, this is not difficult to achieve. The biggest limitation of phase-shifting interferometry is sensitivity to the environment, from vibrations and air turbulence, both of which are often outside our control. Since each of the various phase-shifted frames are acquired at different times, vibration (or turbulence) causes the phase shifts between the frames to be different from what was desired. Bearing in mind that a 90° phase step in a two-pass interferometer requires a mirror displacement of $\lambda/8 \approx 79$ nm at the HeNe wavelength of 632.8 nm, it is easy to understand why unwanted vibrations can cause large phase shifting-errors.

Figure 3.17 shows schematically one way in which simultaneous phase stepping can be achieved [40, 41, 42]. A wire grid polarizer array is constructed in which each element of the array is the same size and is aligned with a detector pixel. To implement the four-frame algorithm, we take a group of four elements and arrange for the polarizers in each element to be oriented as shown in part (a) of the figure, repeating this pattern until the detector is covered. If the light illuminating this

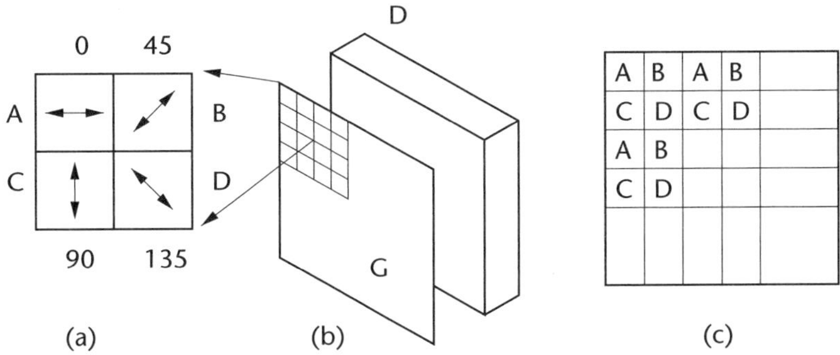

Figure 3.17 Snapshot phase stepping. (a) Orientation of wire grid polarizers. (b) Grid of polarizers G is aligned and sized to match the pixels in detector D. (c) Subimage containing all pixels labeled A constitute one of the phase-stepped images and similarly for subimages labeled B through D.

structure is a mixture of right-hand and left-hand circularly polarized light, then all the pixels labeled A correspond to a subimage $I_0(x, y)$ with a phase step of $0°$, B's to $I_1(x, y)$ with a step of $90°$, and so on. It is clear, then, that this arrangement enables us to acquire all four images required for the four-frame algorithm simultaneously and hence substantially reduces sensitivity to environmental disturbances.

3.8.4 Spatial Methods

In this final section, we will describe two methods that enable simultaneous acquisition of all four frames, but where these images are now laterally displaced. The simplest arrangement is shown in Figure 3.18 [40]. The laser is linearly polarized at $45°$ and only the light transmitted through the first beam splitter is used in the interferometer. An octet plate ($\lambda/8$) in the reference arm is oriented at $0°$. Since two passes through the octet plate is equivalent to a quarter-wave plate, the x and y components of the light reflected in the reference arm are $90°$ out of phase with each other. The reference and test beams recombine at the output beamsplitter and are spatially separated by a polarising beam splitter. Two fringe patterns are projected onto two CCD arrays, which allow two of the four frames to be captured simultaneously. The octet waveplate ensures that these two frames correspond to phase steps of $0°$ and $90°$.

Consider what happens to the reference and test beams when they are reflected from the dielectric beam splitter, as shown in Figure 3.19. One of these beams, the reference say, undergoes a reflection from an interface between a low index medium and a high index medium. Conversely, the test beam undergoes a reflection from a high index medium to a low index medium. There is therefore a $180°$ phase difference between these two beams, as can easily be verified from the Fresnel reflection coefficients [2]. The same conclusion can be arrived at by energy conservation arguments. If the path lengths in the test and reference arms are such that there is complete constructive interference at the interferometer output, then the beams propagating back towards the laser must experience compete destructive interference, otherwise energy is not conserved in the interferometer.

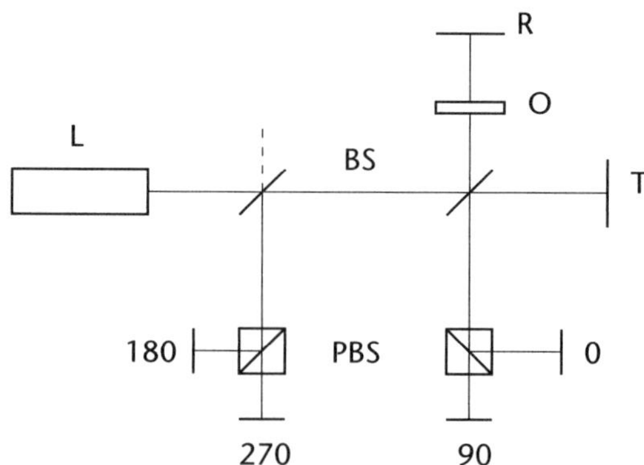

Figure 3.18 Simultaneous phase-stepping with spatially displaced images. L = laser; BS = dielectric beam splitters; R,T = reference and test surfaces, respectively; O = octet ($\lambda/8$) plate at $0°$; PBS = polarizing beam splitters.

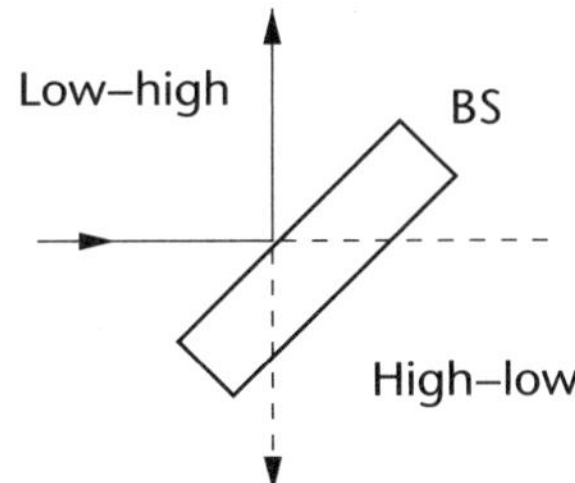

Figure 3.19   Phase shifts on reflection at a dielectric beam splitter.

The first beam splitter is used to pick off these backward-traveling waves and directs them to a second polarizing beam splitter that again spatially separates the *x* and *y* components. Two further CCD arrays acquire the frames corresponding to 180° and 270° phase steps. While this allows four frames to be acquired simultaneously, the use of four cameras is not without problems. In particular, the alignment (image registration) and calibration of the cameras is critical for good accuracy.

A far better approach is to have all four phase-shifted images fall onto a single CCD camera that can be achieved using the configuration shown in Figure 3.20 [40]. The polarizing beam splitter at the interferometer input causes the beams in the reference and test arms to have orthogonal, linear polarizations. The first pass through the quarter-wave plate converts these linear polarizations to circular (one left-handed and the other right-handed). Reflection from the mirrors reverses the handedness of the light and the second pass through the quarter-wave plates converts the beams back to linear polarization, but with p-polarized light now converted to s-polarisation and vice versa. Hence the beam that was initially reflected at the polarizing beam splitter is now transmitted and the beam that was reflected is now transmitted. The two orthogonally polarized beams exit the interferometer and are incident on a holographic element that splits the beam into four separate beams. These beams pass through a birefringent mask that introduces phase shifts between test and reference beams of 0, 90, 180, and 270 degrees. A polarizer oriented at 45° is placed after the phase mask and just in front of the CCD. Thus all

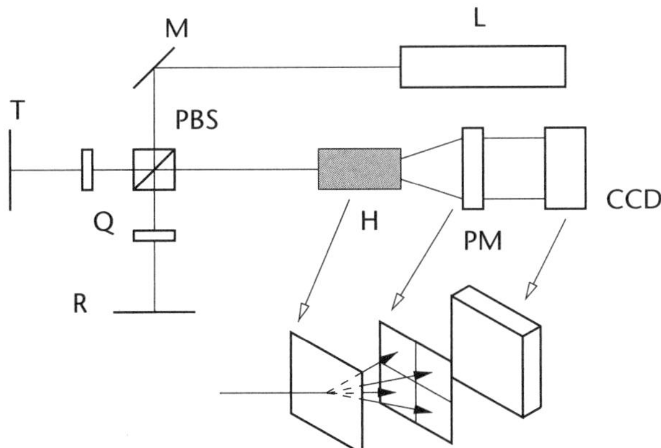

Figure 3.20   Simultaneous phase-shifting with a holographic element. L = laser; PBS = polarizing beam splitters; R,T = reference and test surfaces, respectively; Q = quarter-wave plates; H = holographic element; PM = phase mask and polarizers; CCD = CCD camera.

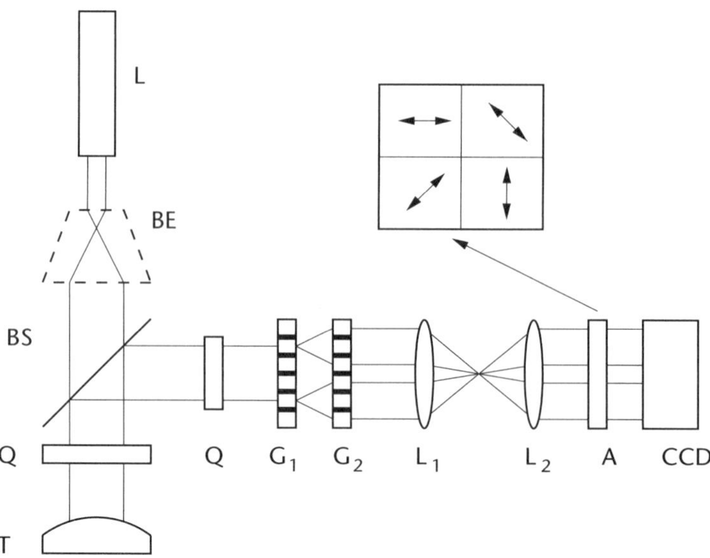

Figure 3.21 Fizeau interferometer with simultaneous phase shifting. L = laser; BE = beam expander; BS = beam splitter; Q = quarter-wave plates; T = test surface; G_1, G_2 = Ronchi gratings; PA = polarizer array; CCD = CCD camera.

four phase-shifted interferograms are detected simultaneously on a single CCD array. The use of a single array considerably simplifies alignment and calibration.

A method for phase-stepping a Fizeau interferometer has recently been suggested [43] that exploits the Pancharatnam phase for the stepping but uses spatially separated beams on a single detector to acquire all four frames at once. A sketch of the interferometer is shown in Figure 3.21. Vertically polarized light from a HeNe laser is expanded and collimated before entering the interferometer. The reference surface in this instance is a silica quartz quarter-wave plate with $\lambda/10$ flatness and ensures that the reflected reference and test waves have orthogonal polarizations. These orthogonally polarized beams encounter another quarter-wave plate when they exit the interferometer, oriented at 45°, so that two, orthogonal, circularly polarized waves are created. Two Ronchi phase gratings, G_1 and G_2, with their grating vectors perpendicular to each other diffract the reference and test waves into different orders. A polarizer array in front of the CCD array performs the phase shifting, as described earlier, and allows all four phase-stepped frames to be acquired simultaneously.

References

[1]   Kingslake, R., "The Analysis of an Interferogram," *Transactions of the Optical Society,* Vol. 28, No. 1, 1926, p.1.

[2]   Hecht, E., *Optics,* 4th Edition, San Francisco: Addison Wesley, 2002.

[3]   Collett, E., "Mathematical Formulation of Interference Laws of Fresnel and Arago," *American Journal of Physics,* Vol. 39, No. 12, 1971, pp. 1483–1495.

[4]   Malacara, D., *Optical Shop Testing,* Third Edition, Hoboken, NJ: Wiley, 2007.

[5]   Ghiglia, D. C., and M. D. Pritt, *Two-Dimensional Phase Unwrapping: Theory, Algorithms, and Software*, New York, Wiley, 1998.

[6]   Navarro, M. A., J. C. Estrada, M. Servin, J. A. Quiroga, and J. Vargas, "Fast Two-Dimensional Simultaneous Phase Unwrapping and Low-Pass Filtering," *Optics Express*, Vol. 20, No. 3, 2012, pp. 2556–2561.

[7]   Ghiglia, D. C., and L. A. Romero, "Robust Two-Dimensional Weighted and Unweighted Phase Unwrapping that Uses Fast Transforms and Iterative Methods," *Journal of the Optical Society of America A: Optics and Image Science, and Vision*, Vol. 11, No. 1, 1994, pp. 107–117.

[8]   Marroquin, J. L., and M. Rivera, "Quadratic Regularization Functional for Phase Uwrapping," *Journal of the Optical Society of America A*, Vol. 12, No. 11, 1995, pp. 2393–2400.

[9]   Bruning, J. H., D. R. Herriot, J. E. Gallagher, D. P. Rosenfeld, A. D. White, and D. J. Brangaccio, "Digital Wavefront Measuring Interferometer for Testing Optical Surfaces and Lenses," *Applied Optics*, Vol. 13, No. 11, 1974, pp. 2693–2703.

[10]   Morgan, C. J., "Least-Squares Estimation in Phase-Measurement Interferometry," *Optics Letters*, Vol. 7, No. 8, 1982, pp. 368–370.

[11]   Greivenkamp, J. E., "Generalized Data Reduction for Heterodyne Interferometry," *Optical Engineering*, Vol. 23, No. 4, 1984, pp. 350–352.

[12]   Carré, P., "Installation et Utilisation du Comparateur Photoéctrique et Interférentiel du Bureau International des Poids et Measures," *Metrologia*, Vol. 2, No. 1, 1966, pp. 13–23.

[13]   Creath, K., "Phase-Shifting Speckle Interferometry," *Applied Optics*, Vol. 24, No. 18, 1985, pp. 3053–3058.

[14]   Qian, K. M., F. J. Shu, and X. P. Wu, "Determination of the Best Phase Step of the Carre Algorithm in Phase Shifting Interferometry," *Measurement Science and Technology*, Vol. 11, No. 8, 2000, pp. 1220–1223.

[15]   Schwider, J., R. Burow, K. E. Elssner, J. Grzanna, R. Spolaczyk, and K. Merkel, "Digital Wave-Front Measuring Interferometry: Some Systematic Error Sources," *Applied Optics*, Vol. 22, No. 1, 1983, pp. 3421–3432.

[16]   Hariharan, P., "Digital Phase-Stepping Interferometry: Effects of Multiply Reflected Beams," *Applied Optics*, Vol. 26, No. 13, March 1987, pp. 2506–2507.

[17]   Hariharan, P., B. F. Oreb, and T. Eiju, "Digital Phase-Shifting Interferometry—A Simple Error-Compensating Phase Calculation Algorithm," *Applied Optics*, Vol. 26, No. 13, 1987, pp. 2504–2506.

[18]   Surrel, Y., "Design of Algorithms for Phase Measurements by the Use of Phase Stepping," *Applied Optics*, Vol. 35, No. 1, pp. 51–60.

[19]   Moore, R. C., and F. H. Slaymaker, "Direct Measurement of Phase in a Spherical-Wave Fizeau Interferometer," *Applied Optics*, Vol. 19, No. 13, 1980, pp. 2196–2200.

[20]   de Groot, P., "Phase-Shift Calibration Errors in Interferometers with Spherical Fizeau Cavities," *Applied Optics*, Vol. 34, No.16, 1995, pp. 2856–2862.

[21]   Chatterjee, S., "Measurement of Surface Figure of Plane Optical Surfaces Using Fizeau Interferometer with Wedge Phase-Shifter," *Optics & Laser Technology*, Vol. 37, 2004, pp. 43–49.

[22]   Chatterjee, S., and Y. P. Kumar, "Measurement of the surface form error of a spherical surface with a wedge phase shifting Fizeau interferometer," *Journal of Optics-Nouvelle Revue D Optique*, Vol. 42, No. 2, April 2013, pp. 122–127.

[23]   Stevenson, W. H., "Optical Frequency Shifting by Means of a Rotating Diffraction Grating," *Applied Optics*, Vol. 9, No. 3, 1970, pp. 649–652.

[24]   Barnes, T. H., "Heterodyne Fizeau Interferometer for Testing at Surfaces," *Applied Optics*, Vol. 26, No. 14, 1987, pp. 2804–2809.

[25]   Zvyagin A. V., and D. D. Sampson, "Achromatic Optical Phase Shifter–Modulator," *Optics Letters*, Vol. 26, No. 4, 2001, pp. 187–189.

[26] Tatsuno, K., and Y. Tsunoda, "Diode-Laser Direct Modulation Heterodyne Interferometer," *Applied Optics*, Vol. 26, No. 1, 1987, pp. 37–40.

[27] Frins, E. M., W. Dultz, and J. A. Ferrari, "Polarization-Shifting Method for Step Interferometry," *Pure and Applied Optics*, Vol. 7, 1998, pp. 53–60.

[28] Hariharan, P., and M. Roy, "A Geometric-Phase Interferometer," *Journal of Modern Optics*, Vol. 39, No. 9, 1992, pp. 1811–1815.

[29] Hariharan, P., H. Ramachandran, K. A. Suresh, and J. Samuel, "The Pancharatnam Phase as a Strictly Geometric Phase: A Demonstration Using Pure Projections," *Journal of Modern Optics*, Vol. 44, No. 4, 1997, pp.707–713.

[30] Kothiyal, M. P., and C. Delisle, "Shearing Interferometer for Phaseshifting Interferometry with Polarization Phase-Shifter," *Applied Optics*, Vol. 24, No. 24, 1985, pp. 4439–4442.

[31] Hariharan, P., "Geometric-Phase Interferometers: Possible Optical Configurations," *Journal of Modern Optics*, Vol. 40, No. 6, 1993, pp. 985–989.

[32] Sommargren, G. E., "Up-Down Frequency Shifter for Optical Heterodyne Interferometry," *Journal of the Optical Society of America*, Vol. 65, No. 8, pp. 960–961.

[33] Clark Jones, R., "A New Calculus for the Treatment of Optical Systems," *Journal of the Optical Society of America*, Vol. 31, No. 7, 1941, pp. 488–493.

[34] Azzam, R. M. A., and N. M. Bashara, *Ellipsometry and Polarized Light*, Amsterdam, Netherlands: North-Holland, 1987.

[35] Goldstein, D., *Polarized Light*, 2nd Edition, New York: Marcel Dekker, 2003.

[36] Watkins, L. R., and M. Derbois, "White-light Ellipsometer with Geometric Phase Shifter," *Applied Optics*, Vol. 51, No. 21, 2012, pp. 5060–5065.

[37] Chou, C., H.- K. Teng, C.- J. Yu, and H.- S. Huang, "Polarization Modulation Imaging Ellipsometry for Thin Film Thickness Measurement," *Optics Communications*, Vol. 273, No. 1, 2007, pp. 74–83.

[38] Hong, Z. H., "Polarization Heterodyne Interferometry Using A Simple Rotating Analyzer. 1. Theory and Error Analysis," *Applied Optics*, Vol. 22, No. 13, 1983, pp. 2052–2056.

[39] Millerd, J., N. Brock, J. Hayes, B. Kimbrough, M. Novak, M. North-Morris, and J. C. Wyant, "Modern Approaches in Phase Measuring Metrology," In *Optical Measurement Systems for Industrial Inspection IV*, Vol. Proc. SPIE 5856, 2005, Part I, pp. 14–22.

[40] Novak, M., J. Millerd, N. Brock, M. North-Morris, J. Hayes, and J. Wyant, "Analysis of a Micropolarizer Array-Based Simultaneous Phase-Shifting Interferometer," *Applied Optics*, Vol. 44, No. 32, 2005, pp. 6861–6868.

[41] Millerd, J., N. Brock, J. Hayes, M. North-Morris, M. Novak, and J. Wyant, "Pixelated Phase-Mask Dynamic Interferometer," in *Optical Science and Technology, SPIE 49th Annual Meeting*, pp. 304–314, International Society for Optics and Photonics, 2004.

[42] Adbelsalam, D. G., B. Yao, P. Gao, J. Min, and R. Guo, "Single-Shot Parallel Four-Step Phase Shifting Using On-Axis Fizeau Interferometry," *Applied Optics*, Vol. 51, No. 20, 2012, pp. 4891–4895.

CHAPTER 4

Systematic Approach to Digital Holographic Imaging and Its Applications in Interferometry

Rishikesh Kulkarni and Pramod Rastogi

4.1　Introduction

In 1948, Dennis Gabor, a Hungarian physicist, invented a new method of recording and reconstruction of optical waves scattered by an object [1], for which he was later awarded the Nobel prize. This method is called *holography*. Unlike photography wherein only the amplitude of the optical wave is recorded on a photosensitive medium, in holography, both the amplitude and phase informations are recorded. Holography involves the recording of the interference between the light waves scattered by an object, termed as an object wave, and the light unaffected by the object, termed as a reference wave. Both the light waves are derived from a single coherent laser source. Depending on the type of photosensitive medium used for hologram recording, holography can be classified into two groups: classical holography and digital holography.

In classical holography, the photosensitive medium is a two-dimensional photographic plate. The developed photographic plate is called a hologram. During the reconstruction step, the hologram is placed in its initial recording position and is illuminated with the reference wave that will now act as the reconstruction wave. The optically reconstructed virtual image of the object is visible to the observer looking through the holographic plate at the exact position where the object was placed during the hologram recording. If the object is present in the reconstruction step, the virtual image gets overlapped on the object. At this time, if the object undergoes any loading, its effect can be detected in the form of optical fringes that are generated on the object surface.

The use of wet chemical processing for the photographic plates is a key constraint in classical holography and has contributed to considerably slowing down its penetration in research and its applications in industrial environment. Consequently, an alternative to the optical reconstruction of a wavefield called numerical or digital reconstruction was developed [2]. In it, a hologram was electronically recorded by scanning a photosensitive surface with an electron gun in a video camera tube.

The recorded electronic signal was sampled and quantized to obtain the digitized hologram. Subsequently, the numerical reconstruction of the optical wavefield was performed using Fourier transform routines. This process is called digital holography. The numerical reconstruction is useful in the retrieval of phase information of the optical wavefield. A major breakthrough in digital holography came with the use of the CCD for direct digital recording of holograms [3]. CCD and CMOS digital sensor arrays do not involve the previously required scanning by electron gun and they operate with much faster image acquisition rates. The recent advances in computers and digital technology have further boosted their development and applications in digital holography, as detailed in [4–6].

4.2   Recording of a Digital Hologram

The optical configuration of digital holography is shown in Figure 4.1(a). A laser beam of wavelength λ is divided into two separate light beams by a beam splitter, one of which is used to illuminate the object. The other light beam acts as a reference beam and is allowed to incident directly on the CCD faceplate, which in our case represents the hologram recording plane. The hologram recording plane is located at $z = 0$. The coordinates of a point on the recording plane are represented by $(x_h, y_h, 0)$ as shown in Figure 4.1(b). The object is located at a distance z_o from the CCD.

The spherical object wave scattered by a point on the object located at $(x_o, y_o, -z_o)$ reaches the recording plane with a complex amplitude that can be represented as

$$O\left(x_h,y_h\right) = \frac{O_o}{\rho_o}\exp\left\{i\frac{2\pi}{\lambda}\rho_o\right\} \tag{4.1}$$

Similarly, the reference wave diverging from a point located at $(x_u, y_u, -z_u)$ reaches the recording plane with a complex amplitude that can be represented as

$$R\left(x_h,y_h\right) = \frac{R_u}{\rho_u}\exp\left\{i\frac{2\pi}{\lambda}\rho_u\right\} \tag{4.2}$$

where

$$\rho_o = \sqrt{\left(x_o - x_h\right)^2 + \left(y_o - y_h\right)^2 + z_o^2}$$

$$\rho_u = \sqrt{\left(x_u - x_h\right)^2 + \left(y_u - y_h\right)^2 + z_u^2}$$

Here, O_o and R_u represent the amplitudes of the object and reference waves, respectively. The object and reference waves meet each other at the recording plane and form an interference pattern, better known as a hologram. The intensity of the interference pattern can be expressed as

$$I\left(x_h,y_h\right) = \left|O + R\right|^2$$
$$= OO^* + RR^* + OR^* + O^*R \tag{4.3}$$

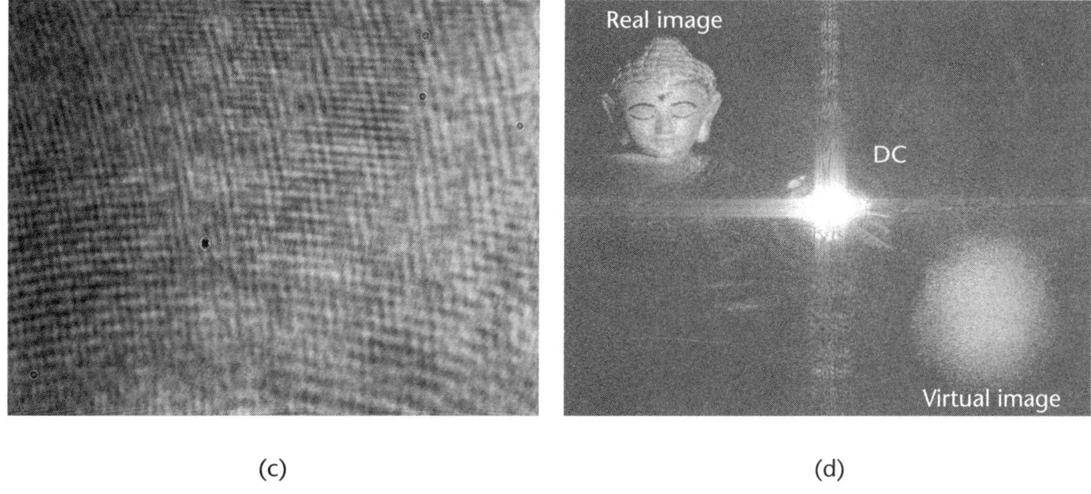

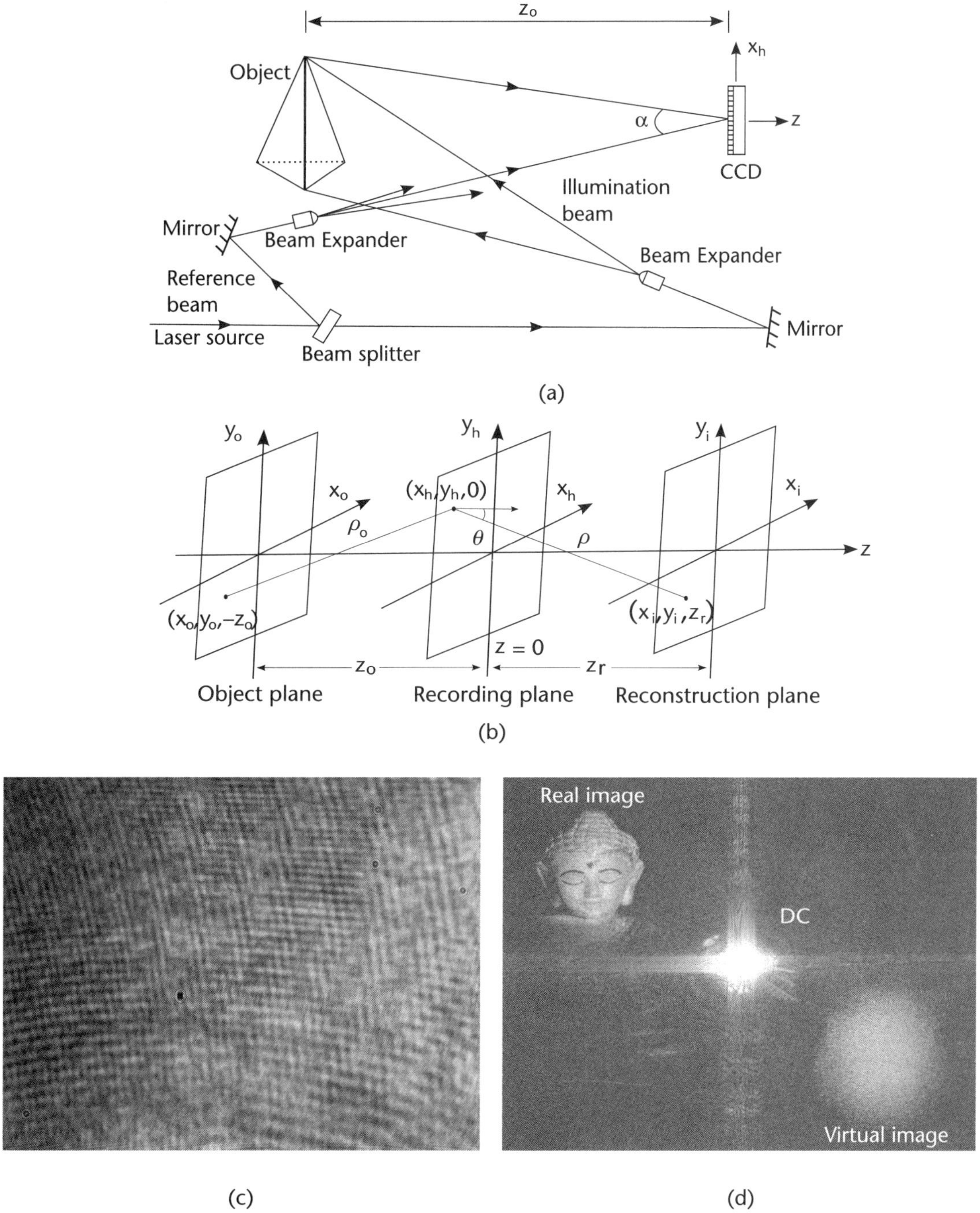

Figure 4.1   (a) Optical configuration of digital holography, (b) coordinates system, (c) experimentally recorded hologram, and (d) numerical reconstruction using Fresnel approximation.

Here, the functional dependence of complex amplitudes on the point coordinates (x_h, y_h) has been dropped for the sake of conciseness. The intensity distribution $H(x_h, y_h)$ recorded on the hologram is obtained by multiplying the intensity distribution given in (4.3) with the CCD factor

$$H(x_h, y_h) = [\text{CCD factor}]I(x_h, y_h) \tag{4.4}$$

where, the *CCD factor* depends on the physical properties of the CCD (exposure time, dynamic range, etc.). The hologram recording medium (i.e., the CCD) should be able to resolve the maximum spatial frequency f_{max} of the interference pattern generated by the object and reference beams. Considering α_{max} as the maximum angle between the object and reference beam at the recording plane, the dependence of f_{max} on α_{max} is given as

$$f_{max} = \frac{2}{\lambda} \sin\left(\frac{\alpha_{max}}{2}\right) \tag{4.5}$$

In classical holography, where photographic plates are used for hologram recording, the value of f_{max} goes over 3,000 linepairs per millimeter (Lp mm^{-1}). This allows α_{max} to take values over 100 degrees. However, in digital holography, the value of f_{max} is limited by the size of the CCD pixel. The maximum resolvable frequency based on Nyquist sampling theorem is given as

$$f_{max} = \frac{1}{2\max(\Delta x_h, \Delta y_h)} \tag{4.6}$$

where Δx_h and Δy_h are the sizes of the pixels in x_h and y_h directions, respectively. The typical pixel size of commercially available cameras vary from 2 to 10 μm. This limits the value of α_{max} to few degrees. From (4.5) and (4.6), the expression for α_{max} can be obtained as

$$\alpha_{max} = 2\sin^{-1}\left(\frac{\lambda}{4\max(\Delta x_h, \Delta y_h)}\right) \approx \frac{\lambda}{2\max(\Delta x_h, \Delta y_h)} \tag{4.7}$$

From the above equation, it can be seen that the value of α_{max} depends on the wavelength and the maximum pixel size. For an optimal recording of the hologram, the size of the object and the distance between the object and camera should also be taken into consideration. Considering a circular object of diameter L placed in the optical configuration shown in Figure 4.1(a), the maximum value of α_{max} between the corner of the object and CCD can be calculated as

$$\alpha_{max} \approx \frac{L + \max(N\Delta x_h, M\Delta y_h)}{z_o} \tag{4.8}$$

where N and M are the number of pixels along x_h and y_h directions, respectively. Comparing (4.7) and (4.8), we obtain the expression for the minimum required object to CCD distance as

$$z_{o_{min}} = 2\frac{\max(\Delta x_h, \Delta y_h)}{\lambda}\left[L + \max(N\Delta x_h, M\Delta y_h)\right] \tag{4.9}$$

EXERCISE 4.1

In a digital holographic interferometry setup shown in Figure 4.1(a), a green laser source of wavelength 532 nm is used. An object is placed at a distance of 1 m away from the CCD. The CCD imaging area is 1600×1200 pixels with each pixel $4.4\ \mu m$. What is the maximum size of the object that can be effectively imaged by the setup?

Solution

First, we obtain the maximum value of α (i.e., α_{max}) for the given setup. Here, $\lambda = 532$ nm, $z_0 = 1$ m, $\Delta x_h = \Delta y_h = 4.4\ \mu m$. Using (4.7), we obtain $\alpha_{max} = 0.0605$ rad. Substituting α_{max} in (4.8), we obtain $L = 5.35$ cm. Thus, a circular object of diameter up to 5.35 cm can be effectively imaged using the given setup.

4.3   Numerical Reconstruction in Digital Holography

In classical holography, the reconstruction of optical wavefield scattered by the object during hologram recording is performed by illuminating the hologram with the reference wave. In digital holography, the reconstruction is performed numerically, wherein the intensity $H(x_h, y_h)$ is multiplied with the numerical equivalent of the reference beam $R(x_h, y_h)$ to obtain the diffracted optical wavefield in the hologram recording plane. The reconstruction of the wavefield at a distance z_r is performed by computing the propagation of this diffracted light using Rayleigh-Sommerfeld diffraction integral given as [7],

$$\Gamma(x_i, y_i) = \frac{1}{i\lambda} \iint H(x_h, y_h) R(x_h, y_h) \frac{\exp\left(i\frac{2\pi}{\lambda}\rho\right)}{\rho} \cos(\theta)\, dx_h\, dy_h \tag{4.10}$$

with

$$\rho = \sqrt{(x_h - x_i)^2 + (y_h - y_i)^2 + z_r^2} \tag{4.11}$$

Here, $\cos(\theta)$ is the obliquity factor; (x_i, y_i, z_r) indicate the coordinates of a point in the reconstruction plane. The above integral is calculated numerically. Various numerical reconstruction methods have been developed for computing object images. Some of these methods are given below.

4.3.1   Reconstruction Based on Convolution Approach

From (4.10), it can be observed that the wavefield reconstruction is a linear process. Consequently, (4.10) can be expressed as

$$\Gamma(x_i, y_i) = \iint H(x_h, y_h) R(x_h, y_h) g(x_h - x_i, y_h - y_i) dx_h\, dy_h \tag{4.12}$$

where,

$$g(x_h, y_h) = \frac{1}{i\lambda} \frac{\exp\left(i\frac{2\pi}{\lambda}\sqrt{x_h^2 + y_h^2 + z_r^2}\right)}{\sqrt{x_h^2 + y_h^2 + z_r^2}} \tag{4.13}$$

Since for a sufficiently large reconstruction distance z_r, the value of $\cos\theta = z_r/\rho$ varies very little around 1, a valid substitution of $\cos\theta = 1$ has been made. Equations (4.12) and (4.13) indicate that the numerical reconstruction is a convolution between the product $H \cdot R$ and the impulse response function g. This allows us to represent the above integral as

$$\Gamma(x_i, y_i) = ((H \cdot R) * g)(x_i, y_i) \tag{4.14}$$

where, the symbol $*$ represents the convolution operator. For the numerical implementation, the impulse response function g can be represented as

$$g(n,m) = \frac{1}{i\lambda} \frac{\exp\left(i\frac{2\pi}{\lambda}\sqrt{(n - N/2)^2\Delta x_h^2 + (m - M/2)^2\Delta y_h^2 + z_r^2}\right)}{\sqrt{(n - N/2)^2\Delta x_h^2 + (m - M/2)^2\Delta y_h^2 + z_r^2}} \tag{4.15}$$

with $n \in [1, N]$ and $m \in [1, M]$. The coordinate shifts of $N/2$ and $M/2$ are provided on the account of symmetry. The convolution in (4.14) can be efficiently computed using the convolution property of the Fourier transform as

$$\Gamma(x_i, y_i) = \mathfrak{F}^{-1}\left[\mathfrak{F}(H \cdot R) \cdot \mathfrak{F}(g)\right] \tag{4.16}$$

where, $\mathfrak{F}$ represents the Fourier transform operator. Three FFT operations are involved in the above computation. The Fourier transform of the impulse response function g depends only on the optical setup parameters. Therefore, the expression $\mathfrak{F}(g)$ in (4.16) can be replaced by G, the Fourier spectrum of g. This saves one FFT calculation. Thus,

$$\Gamma(x_i, y_i) = \mathfrak{F}^{-1}\left[\mathfrak{F}(H \cdot R) \cdot G\right] \tag{4.17}$$

where,

$$G(k,l) = \exp\left\{i\frac{2\pi z_r}{\lambda}\sqrt{1 - \frac{\lambda^2\left(k + \frac{N^2\Delta x_h^2}{2z_r\lambda}\right)^2}{N^2\Delta x_h^2} - \frac{\lambda^2\left(l + \frac{M^2\Delta y_h^2}{2z_r\lambda}\right)^2}{M^2\Delta y_h^2}}\right\} \tag{4.18}$$

In convolution-based reconstruction, the pixel size of the image in the reconstruction plane is the same as that on the recording plane (i.e., $\Delta x_i = \Delta x_h$ and $\Delta y_i =$

Δy_h). This suggests that these methods are well suited for objects of sizes comparable to the CCD imaging area.

4.3.2   Reconstruction Based on Fresnel Approximation

The object size, however, could be much larger than the CCD imaging area. This is especially true in the case of interferometric applications in experimental mechanics and in nondestructive testing. Accordingly, the recording distance, and in turn the reconstruction distance will be much larger compared to the size of the object image and the CCD (i.e., x_i, y_i, x_h, $y_h \ll z_r$). This allows us to obtain the approximation of ρ using the binomial expansion of (4.11):

$$\rho = z_r \left(1 + \frac{\left(x_h - x_i\right)^2}{z_r^2} + \frac{\left(y_h - y_i\right)^2}{z_r^2} \right)^{1/2}$$

$$\approx z_r + \frac{\left(x_h - x_i\right)^2}{2z_r} + \frac{\left(y_h - y_i\right)^2}{2z_r} \tag{4.19}$$

Consequently, the Rayleigh-Sommerfeld diffraction integral in (4.10) can be represented as

$$\Gamma\left(x_i, y_i\right) = \frac{1}{i\lambda z_r} \exp\left(i\frac{2\pi}{\lambda} z_r\right) \iint H\left(x_h, y_h\right) R\left(x_h, y_h\right)$$

$$\times \exp\left[i\frac{\pi}{\lambda z_r}\left(\left(x_h - x_i\right)^2 + \left(y_h - y_i\right)^2\right)\right] dx_h dy_h \tag{4.20}$$

$$\Gamma\left(x_i, y_i\right) = \frac{1}{i\lambda z_r} \exp\left(i\frac{2\pi}{\lambda} z_r\right) \exp\left[i\frac{\pi}{\lambda z_r}\left(x_i^2 + y_i^2\right)\right] \iint H\left(x_h, y_h\right) R\left(x_h, y_h\right)$$

$$\times \exp\left[i\frac{\pi}{\lambda z_r}\left(x_h^2 + y_h^2\right)\right] \exp\left[-i\frac{2\pi}{\lambda z_r}\left(x_h x_i + y_h y_i\right)\right] dx_h dy_h \tag{4.21}$$

The above equation is called the *Fresnel approximation*. In (4.21) the terms preceding the integral sign are independent of the object and depend only on the optical setup parameters. Therefore, their computation can be dropped without any loss of generality. This is particularly useful in the case of interferometric applications where we will be concerned only with the changes in the phase of wavefields. For efficient computation of (4.21), the following substitution is introduced:

$$\nu = \frac{x_i}{\lambda z_r}; \mu = \frac{y_i}{\lambda z_r} \tag{4.22}$$

Consequently, (4.21) becomes

$$\Gamma(v,\mu) = \iint H\left(x_h,y_h\right)R\left(x_h,y_h\right)\exp\left[i\frac{\pi}{\lambda z_r}\left(x_h^2 + y_h^2\right)\right]$$
$$\times \exp\left[-i2\pi\left(x_h v + y_h \mu\right)\right]dx_h dy_h \tag{4.23}$$

From (4.23), it can be observed that the Fresnel approximation is nothing but the two-dimensional Fourier transform of the product $H(x_h, y_h)R(x_h, y_h)$ $\exp\left[i\left(\pi/\lambda z_r\right)\left(x_h^2 + y_h^2\right)\right]$. It provides the reconstruction of the wavefield at a distance z_r from the CCD. The above equation can be written as

$$\Gamma(v,\mu) = \mathfrak{F}\left\{H\left(x_h,y_h\right)R\left(x_h,y_h\right)\exp\left[i\frac{\pi}{\lambda z_r}\left(x_h^2 + y_h^2\right)\right]\right\}(v,\mu) \tag{4.24}$$

As mentioned previously, the CCD array consists of $N \times M$ number of pixels each of size Δx_h and Δy_h along x_h and y_h directions, respectively. The digitized form of $\Gamma(v, \mu)$ can thus be represented as

$$\Gamma(n,m) = \sum_{k=1}^{N}\sum_{l=1}^{M}H(k,l)R(k,l)\exp\left[i\frac{\pi}{\lambda z_r}\left(k^2\Delta x_h^2 + l^2\Delta y_h^2\right)\right]$$
$$\times \exp\left[-i2\pi\left(k\Delta x_h n\Delta v + l\Delta y_h m\Delta\mu\right)\right] \tag{4.25}$$

The maximum spatial frequency is determined by the sampling interval in the spatial domain, which corresponds to the pixel size of the CCD. Therefore, based on the Fourier transform theory, we have the following relationship between Δv, $\Delta\mu$ and Δx_h, Δy_h:

$$\Delta v = \frac{1}{N\Delta x_h}; \Delta\mu = \frac{1}{M\Delta y_h} \tag{4.26}$$

Using (4.22) and (4.26), we obtain

$$\Delta x_i = \frac{\lambda z_r}{N\Delta x_h}; \Delta y_i = \frac{\lambda z_r}{M\Delta y_h} \tag{4.27}$$

Substituting (4.26) into (4.25) gives

$$\Gamma(n,m) = \sum_{k=1}^{N}\sum_{l=1}^{M}H(k,l)R(k,l)\exp\left[i\frac{\pi}{\lambda z_r}\left(k^2\Delta x_h^2 + l^2\Delta y_h^2\right)\right]$$
$$\times \exp\left[-i2\pi\left(\frac{kn}{N} + \frac{lm}{M}\right)\right] \tag{4.28}$$

The above equation can be easily implemented using the 2D FFT routine.

According to (4.27), in the Fresnel approximation-based reconstruction, the pixel size of the image in the reconstruction plane is a function of the reconstruction distance. For illustration purpose, two holograms of the same object were recorded

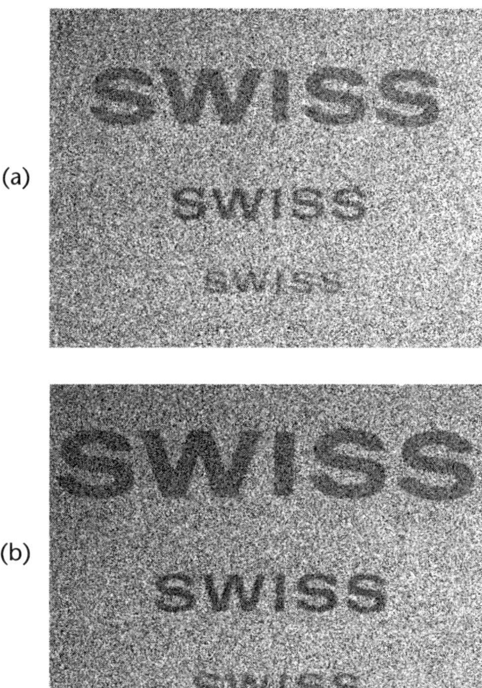

Figure 4.2 Numerical reconstruction of the hologram of the object kept at (a) z_o = 937 cm and (b) z_o = 750 cm.

for different values of object to CCD distance. The numerical reconstruction of these two holograms using Fresnel approximation is shown in Figures 4.2(a) and 4.2(b), respectively. It can be observed that the resolution of the image corresponding to larger object to CCD distance (Figure 4.2(a)) is lower than the resolution of the image corresponding to smaller object to CCD distance (Figure 4.2(b)).

When the number of pixels N and M are not the same along x_h and y_h directions, the zero padding operation is performed to match the size of the hologram in both the directions. The pixel sizes are, in general, same along x_h and y_h directions (i.e., $\Delta x_h = \Delta y_h$).

Since $x_o, y_o, x_h, y_h << z_o$, ρ_o can be written using the binomial expansion as

$$\rho_o \approx z_o + \frac{\left(x_o - x_h\right)^2}{2z_o} + \frac{\left(y_o - y_h\right)^2}{2z_o}$$

The object wave expression given in (4.1) can thus be written as

$$O\left(x_h, y_h\right) = O'_o \exp\left\{i\frac{\pi}{\lambda z_o}\left[\left(x_o - x_h\right)^2 + \left(y_o - y_h\right)^2\right]\right\}$$

where, O'_o represents the amplitude of the object wave. If a plane reference wave is used while recording the hologram (i.e., $z_u = \infty$), the expression for the reference wave at the recording plane becomes

$$R(x_h, y_h) = R_u \exp\left\{i\left(f_x x_h + f_y y_h\right)\right\} \tag{4.29}$$

where, $f_x = \vec{k} \cdot \hat{x}_h$ and $f_y = \vec{k} \cdot \hat{y}_h$ are the spatial frequencies associated with the reference wave whose values depend upon the value of angle α; $\hat{x}_h$ and $\hat{y}_h$ are the unit vectors along axes x_h ans y_h, respectively. Two often employed optical configurations used in digital holography depend upon the angle α: One is in-line holography and the other is off-axis holography.

BOX 4.1

In the case of small-sized objects that are placed close to the CCD, the approximation $\cos\theta = 1$ is not valid. Consequently, the convolution based or Fresnel approximation-based numerical reconstruction methods are not applicable. To overcome this limitation, angular spectrum based numerical reconstruction has been proposed [8]. In this method, the angular spectrum A_o (f_{x_h}, f_{x_y}, 0) (i.e., the Fourier transform of the complex amplitude of the object wave at the hologram plane) is computed. As a result, the object wave is divided into a number of plane waves traveling in different directions.

$$O(x_h, y_h, 0) = H(x_h, y_h) R(x_h, y_h) \tag{4.30}$$

$$A_o(f_{x_h}, f_{y_h}, 0) = \mathfrak{F}\left[O(x_h, y_h, 0)\right](f_{x_h}, f_{y_h}) \tag{4.31}$$

Here, fx_h and fy_h represent the spatial frequencies associated with the plane waves. The angular spectrum of the object wave at a distance $z = z_r$ is calculated as

$$A_o(f_{x_h}, f_{y_h}, z_r) = A_o(f_{x_h}, f_{y_h}, 0)\exp\left(i2\pi f_{z_r} z_r\right) \tag{4.32}$$

where $f_{z_r} = \frac{1}{\lambda}\left(1 - (\lambda f_{x_h})^2 - (\lambda f_{y_h})^2\right)^{1/2}$. Finally, the inverse Fourier transform of the angular spectrum provides the reconstructed object wave at the plane $z = z_r$

$$O(x_h, y_h, z_r) = \mathfrak{F}^{-1}\left[A_o(f_{x_h}, f_{y_h}, z_r)\right](x_h, y_h) \tag{4.33}$$

4.3.2.1 In-Line Holography

In this optical configuration, α is set equal to 0, which means that the plane reference wave is propagating parallel to the z-axis. Accordingly, the spatial frequencies f_x and f_y are set equal to zero. Thus, the reference wave expression given in (4.29) reduces to constant R_u. From (4.3), the intensity of the recorded hologram becomes

$$H\left(x_h, y_h\right) = \left|O_o'\right|^2 + \left|R_u\right|^2 + O_o' R_u \exp\left\{i\frac{\pi}{\lambda z_o}\left[\left(x_o - x_h\right)^2 + \left(y_o - y_h\right)^2\right]\right\}$$

$$+ O_o' R_u \exp\left\{-i\frac{\pi}{\lambda z_o}\left[\left(x_o - x_h\right)^2 + \left(y_o - y_h\right)^2\right]\right\} \tag{4.34}$$

Here, the *CCD factor* is considered to be unity for simplicity. In the numerical reconstruction step, substituting the above intensity expression into (4.24) and considering the complex amplitude of reconstruction wave equal to unity without any loss of generality, we obtain

$$\Gamma(v,\mu) = \mathfrak{F}\left(\left(\left|O_o'\right|^2 + \left|R_u\right|^2\right)\exp\left\{i\frac{\pi}{\lambda z_r}\left(x_h^2 + y_h^2\right)\right\}\right.$$

$$+ O_o' R_u \exp\left\{i\frac{\pi}{\lambda}\left[\frac{\left(x_o - x_h\right)^2}{z_o} + \frac{\left(y_o - y_h\right)^2}{z_o} + \frac{x_h^2 + y_h^2}{z_r}\right]\right\}$$

$$\left. + O_o' R_u \exp\left\{-i\frac{\pi}{\lambda}\left[\frac{\left(x_o - x_h\right)^2}{z_o} + \frac{\left(y_o - y_h\right)^2}{z_o} - \frac{x_h^2 + y_h^2}{z_r}\right]\right\}\right)(v,\mu) \tag{4.35}$$

In (4.35), the first term represents the undiffracted light (i.e., the DC term in the reconstructed image). Its presence in the image does not carry any useful information on the object and can be easily suppressed by DC suppression methods described in Section 4.4. The second and third terms represent the virtual and real images of the object, respectively. The virtual and real images correspond to the -1 and $+1$ order of diffraction of the reference beam by the hologram. Since we have considered a scattering point source on the object, it is expected that the numerical reconstruction generates the corresponding point image in the reconstruction plane. However, the presence of the quadratic phase term, $\exp\left[i\frac{\pi}{\lambda z_r}(x_h^2 + y_h^2)\right]$, broadens the image of the point source which in turn results in the severe broadening of the reconstructed image. From (4.35), it can be observed that the quadratic phase term vanishes for the virtual image if we set $z_r = -z_o$ and for the real image if we set $z_r = z_o$. Thus, numerical focusing of virtual or real image can be achieved by selecting the appropriate reconstruction distances. However, since f_x and f_y are set equal to 0 in the in-line holographic configuration, the DC and the virtual and real image terms spatially overlap each other. Therefore, filtering methods are required to remove the DC and the severely broadened image terms appearing during the reconstruction.

4.3.2.2   Off-Axis Holography

To avoid the overlapping of terms, corresponding to DC, virtual, and real images, an off-axis configuration in holography has been suggested [9]. In this configuration, an appropriate value of α is selected that results in an oblique illumination of the reference beam on the CCD surface. Effectively, we obtain

$$\Gamma(v,\mu) = \mathfrak{F}\left[\left(\left|O'_o\right|^2 + \left|R_u\right|^2\right)\exp\left\{i\frac{\pi}{\lambda z_r}\left(x_h^2 + y_h^2\right)\right\}\right.$$

$$+ O'_o R_u \exp\left\{i\left[\left(\frac{\pi}{\lambda}\left[\frac{\left(x_o - x_h\right)^2}{z_o} + \frac{\left(y_o - y_h\right)^2}{z_o} + \frac{x_h^2 + y_h^2}{z_r}\right]\right) - f_x x_h - f_y y_h\right]\right\}$$

$$\left. + O'_o R_u \exp\left\{i\left[-\left(\frac{\pi}{\lambda}\left[\frac{\left(x_o - x_h\right)^2}{z_o} + \frac{\left(y_o - y_h\right)^2}{z_o} - \frac{x_h^2 + y_h^2}{z_r}\right]\right) + f_x x_h + f_y y_h\right]\right\}\right](v,\mu)$$

$$(4.36)$$

From (4.36), the effect of the spatial frequencies on the spatial separation of reconstructed image components can be observed. Due to the presence of spatial frequencies f_x and f_y, the DC, virtual and real images are spatially separated. Thus, off-axis holography offers an advantage over the in-line configuration by providing a provision to separate the DC, the real and the virtual images from each other. However, spatial frequencies associated with the reference wave consumes a part of the maximum detectable spatial frequency of the CCD. This limits the available spatial frequency bandwidth for the object image.

An illustration of reconstruction based on off-axis configuration is provided in Figure 4.1. For the experimentally recorded hologram shown in Figure 4.1(c), numerical reconstruction is performed at $z_r = z_o$ using the Fresnel approximation. The separated virtual and real images and the DC can be observed in Figure 4.1(d).

4.3.2.3 Lensless Fourier Transform Holography

Lensless Fourier transform holography is an optical configuration wherein a spherical reference source is used during the hologram recording in the optical setup shown in Figure 4.3(a). The reference wave expression given in (4.1) can thus be written as

$$R\left(x_h, y_h\right) = R'_u \exp\left\{i\frac{\pi}{\lambda z_u}\left[\left(x_u - x_h\right)^2 + \left(y_u - y_h\right)^2\right]\right\} \tag{4.37}$$

where, R'_u represents the amplitude of the reference wave. From (4.3), the intensity of the recorded hologram can be written as

$$H\left(x_h, y_h\right) = \left|O'_o\right|^2 + \left|R'_u\right|^2 + O'_o R'_u \exp\left\{i\frac{\pi}{\lambda}\left[\frac{\left(x_o - x_h\right)^2 + \left(y_o - y_h\right)^2}{z_o} - \frac{\left(x_u - x_h\right)^2 + \left(y_u - y_h\right)^2}{z_u}\right]\right\}$$

$$+ O'_o R'_u \exp\left\{i\frac{\pi}{\lambda}\left[-\frac{\left(x_o - x_h\right)^2 + \left(y_o - y_h\right)^2}{z_o} + \frac{\left(x_u - x_h\right)^2 + \left(y_u - y_h\right)^2}{z_u}\right]\right\}$$

$$(4.38)$$

In the numerical reconstruction step, substituting the above intensity expression into (4.24) we obtain

$$\Gamma(\nu,\mu) = \mathfrak{F}\left(\left(\left|O'_o\right|^2 + \left|R'_u\right|^2\right)R\exp\left\{i\frac{\pi}{\lambda}\frac{\left(x_h^2 + y_h^2\right)}{z_r}\right\}\right.$$

$$+ O'_o R'_u R\exp\left\{i\frac{\pi}{\lambda}\left[\frac{\left(x_o - x_h\right)^2 + \left(y_o - y_h\right)^2}{z_o} - \frac{\left(x_u - x_h\right)^2 + \left(y_u - y_h\right)^2}{z_u} + \frac{x_h^2 + y_h^2}{z_r}\right]\right\}$$

$$\left. + O'_o R'_u R\exp\left\{i\frac{\pi}{\lambda}\left[-\frac{\left(x_o - x_h\right)^2 + \left(y_o - y_h\right)^2}{z_o} + \frac{\left(x_u - x_h\right)^2 + \left(y_u - y_h\right)^2}{z_u} + \frac{x_h^2 + y_h^2}{z_r}\right]\right\}\right)(\nu,\mu)$$

$$(4.39)$$

From the above equation, it can be observed that the quadratic term can be eliminated by substituting the reconstruction wave as a unit amplitude plane wave (i.e., $R = 1$), and the reconstruction distance z_r is given by

$$\frac{1}{z_r} = \pm\frac{1}{z_u} \mp \frac{1}{z_0} \tag{4.40}$$

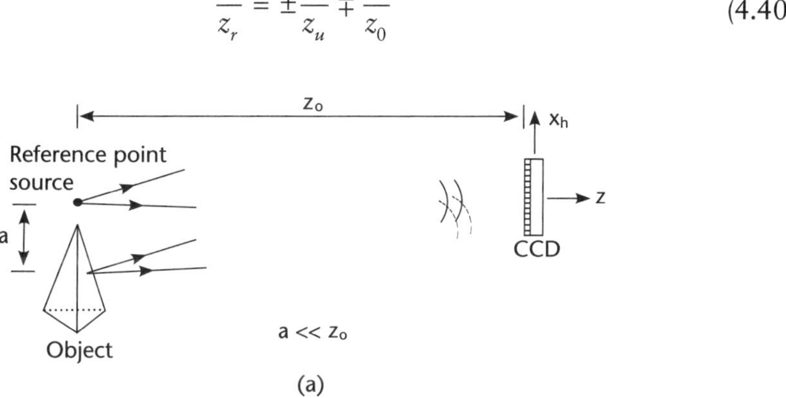

(b)

Figure 4.3 (a) Optical configuration of lensless Fourier transform holography, and (b) numerical reconstruction of the hologram.

In the special case wherein the spherical reference source is placed in the plane containing the object (i.e., $z_u = z_0$), the reconstruction is obtained at $z_r = \infty$. Accordingly, (4.39) reduces to

$$\Gamma(v,\mu) = \mathfrak{F}\left\{H\left(x_h,y_h\right)\right\} \tag{4.41}$$

The above equation indicates that with the lensless Fourier transform holography, the reconstruction algorithm consists of a single Fourier transform of the hologram intensity. An important thing to note in this case is that the quadratic term vanishes for both the virtual and real images. Consequently, in the numerical reconstruction we obtain focused virtual and real images of the object. A single hologram was recorded with the experimental set up shown in Figure 4.3(a). The Fourier transform of the hologram shown in Figure 4.3(b) shows the focused virtual and real images of the object.

EXERCISE 4.2

In the setup given in Exercise 4.1, if there are two points on the object located 60 μm apart, will they be resolved if the reconstruction is performed using the Fresnel approximation technique? If not, why? If we want them to be resolved in the reconstructed image, what could be the possible solutions?

Solution

Equation (4.27) provides the resolution in the image plane. Using the available data, we obtain the resolution in the image plane located at $z_r = z_o$ as $\Delta x_i = 75\ \mu$m and $\Delta y_i = 100\ \mu$m. It can be thus observed from the computed values that the two points on the object located 60 μm apart will not be resolved in the reconstructed image. In order to be able to resolve these two points on the object, the object to CCD distance should be decreased in keeping with the minimum object to CCD distance criteria mentioned in (4.9). Other option is to use a CCD camera with increased number of pixels or smaller pixel size.

4.3.3   Numerical Reconstruction with Adjustable Magnification

The ratio between the pixel size of the reconstructed image and that of the CCD is denoted as the *magnification* γ of the reconstruction method (i.e., $\gamma = \Delta x_i/\Delta x_h$). Since $\Delta x_i = \Delta x_h$ in the convolution-based reconstruction methods, the magnification is unity. On the other hand, in keeping with (4.27), the magnification in the Fresnel approximation-based reconstruction method is a function of wavelength λ, distance z_r, CCD pixel size, and the number of pixels. However, none of these methods provide the flexibility to adjust the magnification.

Various methods have been proposed for adjusting the magnification independently of the hologram recording parameters. Some of these methods based on the Fresnel approximation [10] are described below.

4.3.3.1 Fresnel-Bluestein Transform

A method for adjustable magnification of the reconstruction, called the Fresnel-Bluestein transform, has been proposed [11] by introducing the Bluestein substitution in the Fresnel kernel. Accordingly, the digitized form of (4.25) can be represented as

$$\Gamma(n,m) = \sum_{k=1}^{N}\sum_{l=1}^{M} H(k,l)R(k,l)\exp\left[i\frac{\pi}{\lambda z_r}\left(k^2\Delta x_h^2 + l^2\Delta y_h^2\right)\right]$$
$$\times \exp\left[-i\frac{2\pi}{\lambda z_r}\left(kn\Delta x_h\Delta x_i + lm\Delta y_h\Delta y_i\right)\right] \tag{4.42}$$

where we have undone the substitution of (4.22) in (4.25). Bluestein introduced a substitution wherein the products $2kn$ and $2lm$ in the exponential of (4.42) can be replaced by $2kn = k^2 + n^2 - (n-k)^2$ and $2lm = l^2 + m^2 - (m-l)^2$, respectively. This substitution results in the following equation:

$$\Gamma(n,m) = \sum_{k=1}^{N}\sum_{l=1}^{M} H(k,l)R(k,l)\exp\left[i\frac{\pi}{\lambda z_r}\left(k^2\Delta x_h\left(\Delta x_h - \Delta x_i\right) + l^2\Delta y_h\left(\Delta y_h - \Delta y_i\right)\right)\right]$$
$$\times \exp\left[i\frac{\pi}{\lambda z_r}\left((n-k)^2\Delta x_i\Delta x_h + (m-l)^2\Delta y_i\Delta y_h\right)\right] \tag{4.43}$$

Note that the terms corresponding to n^2 and m^2 preceding the summation sign have been dropped without any loss of generality. Considering $N = M$ and $\Delta x_h = \Delta y_h$, and substituting the value of magnification, $\gamma = \Delta x_i/\Delta x_h$, in the above equation, we obtain

$$\Gamma(n,m) = \sum_{k=1}^{N}\sum_{l=1}^{N} H(k,l)R(k,l)\exp\left[i\frac{\pi}{\lambda z_r}(1-\gamma)\left(k^2\Delta x_h^2 + l^2\Delta y_h^2\right)\right]$$
$$\times \exp\left[i\frac{\pi}{\lambda z_r}\gamma\left((n-k)^2\Delta x_h^2 + (m-l)^2\Delta y_h^2\right)\right] \tag{4.44}$$

The above formulation can be seen as nothing but the spatial convolution between the functions Γ_1 and Γ_2, which are defined as

$$\Gamma_1(k,l) = H(k;l)R(k;l)\exp\left[i\frac{\pi}{\lambda z_r}(1-\gamma)\left(k^2\Delta x_h^2 + l^2\Delta y_h^2\right)\right] \tag{4.45}$$

$$\Gamma_2(k,l) = \exp\left[i\frac{\pi}{\lambda z_r}\gamma\left(k^2\Delta x_h^2 + l^2\Delta y_h^2\right)\right] \tag{4.46}$$

The convolution can be efficiently implemented using the convolution property of Fourier transform

$$\Gamma(n,m) = \mathfrak{F}^{-1}\left[\mathfrak{F}\left(\Gamma_1(k,l)\right)\cdot\mathfrak{F}\left(\Gamma_2(k,l)\right)\right](n,m) \tag{4.47}$$

The magnification can thus be simply adjusted by changing the values of γ in the above equation. The intrinsic magnification $\gamma_0 = \lambda z_r / N \Delta x_h^2$ in the Fresnel approximation based reconstruction can be obtained by setting $\gamma = \gamma_0$.

4.3.3.2 Digital Quadratic Lens Method

Consider the following formulation of (4.20):

$$\Gamma(x_i,y_i) = \left(\left(H(x_h,y_h) \cdot R(x_h,y_h) \right) * \exp\left[i\frac{\pi}{\lambda z_r}\left(x_h^2 + y_h^2\right) \right] \right)(x_i,y_i) \quad (4.48)$$

It can be observed from the above equation that the numerical reconstruction can also be performed using convolution approach with the convolution kernel $g_f = \exp[i\frac{\pi}{\lambda z_r}(x_h^2 + y_h^2)]$ under Fresnel approximation. The presence of unit magnification in the case of convolution-based reconstruction method makes it unsuitable in the case of large size objects. The reason is that the spatial frequency bandwidths of the object and the convolution kernel g_f do not overlap with each other [12]. By adjusting the bandwidth of the convolution kernel such that it completely overlaps with the objects' spatial bandwidth, the convolution approach can be implemented to image large-sized objects. The spatial bandwidth of the convolution kernel in the reconstruction plane is given as

$$U \times V = \frac{1}{\Delta x_i} \times \frac{1}{\Delta y_i} = \frac{N \Delta x_h}{\lambda z_r} \times \frac{M \Delta y_h}{\lambda z_r} \quad (4.49)$$

One way of adjusting the spatial bandwidth of the kernel is to adjust the reconstruction distance. It is suggested that the reconstruction distance can be adjusted by using the numerical spherical reconstruction wave for reconstruction instead of a plane wave [12]. The virtual spherical wavefront has the effect of a quadratic lens. The spherical wavefront of radius R_c can be represented as

$$R(x_h,y_h) = \exp\left[-i\frac{\pi}{\lambda R_c}\left(x_h^2 + y_h^2\right) \right] \quad (4.50)$$

The reconstruction distance z_r, the radius of curvature of numerical spherical wave R_c and the object distance z_0 are related as

$$\frac{1}{z_r} = \frac{1}{R_c} - \frac{1}{z_0} \quad (4.51)$$

The adjustment of the reconstruction distance according to the above equation results in the change in magnification γ of the reconstructed image:

$$\gamma = -\frac{z_r}{z_0}. \quad (4.52)$$

From (4.51) and (4.52), the appropriate radius of curvature R_c can be computed as

$$R_c = \frac{\gamma z_0}{\gamma - 1} \tag{4.53}$$

Consequently, the convolution in (4.48) can be efficiently implemented using the FFT routine as follows:

$$\Gamma\left(x_i, y_i\right) = \mathfrak{F}^{-1}\left[\mathfrak{F}(H \cdot R) \cdot \mathfrak{F}\left(g_f\right)\right]\left(x_i, y_i\right) \tag{4.54}$$

Thus, the magnification of the reconstructed hologram can be adjusted independently to the recording parameters. Since the computation g_f depends only on the optical setup parameters, (4.54) can be further reduced by directly using the Fourier transform of g_f. Consequently, the hologram reconstruction is obtained as

$$\Gamma\left(x_i, y_i\right) = \mathfrak{F}^{-1}\left[\mathfrak{F}(H \cdot R) \cdot G_f\left(u, v\right)\right] \tag{4.55}$$

with

$$G_f(u, v) = \exp\left[-i\pi z_r \lambda\left(u^2 + v^2\right)\right] \tag{4.56}$$

where u and v indicate the spatial frequencies.

The experimental illustrations of the above-mentioned hologram reconstruction methods are given in Figure 4.4 for different values of magnification. The variable magnification capability of these methods allows us to either zoom in or zoom out the object images. However, the changes in the size of reconstructed images may either cause the replicas of the image to appear or cause aliasing to occur. In case where the size of the image is larger than the object size aliasing occurs whereas replicas appear in the opposite situation. To avoid such replicas and aliasing, additional operations are required to be performed. The replicas can be removed by considering only the $(\gamma_0/\gamma)N$ pixels in the image plane corresponding to the object image followed by zero padding operation. To deal with aliasing, a method for prefiltering of the hologram has been proposed [13].

4.4 Suppression of DC Term

The first two terms in (4.3), corresponding to zero-order diffraction, constitute the low frequency or in other words the DC part of the reconstructed image. The zero-order diffraction can be observed as a central bright spot in the reconstructed image shown in Figure 4.1(d). The spatial spread of the DC part depends on the spectrum of the object wave. In the in-line configuration, the DC part overlaps the in-focus image and in the off-axis configuration too, the object image is overlapped by the DC part in the case of a large-sized object. As the DC part does not carry any information on the phase of the object wave, it is desirable to remove its contribution from the reconstruction. A number of different techniques have been suggested to

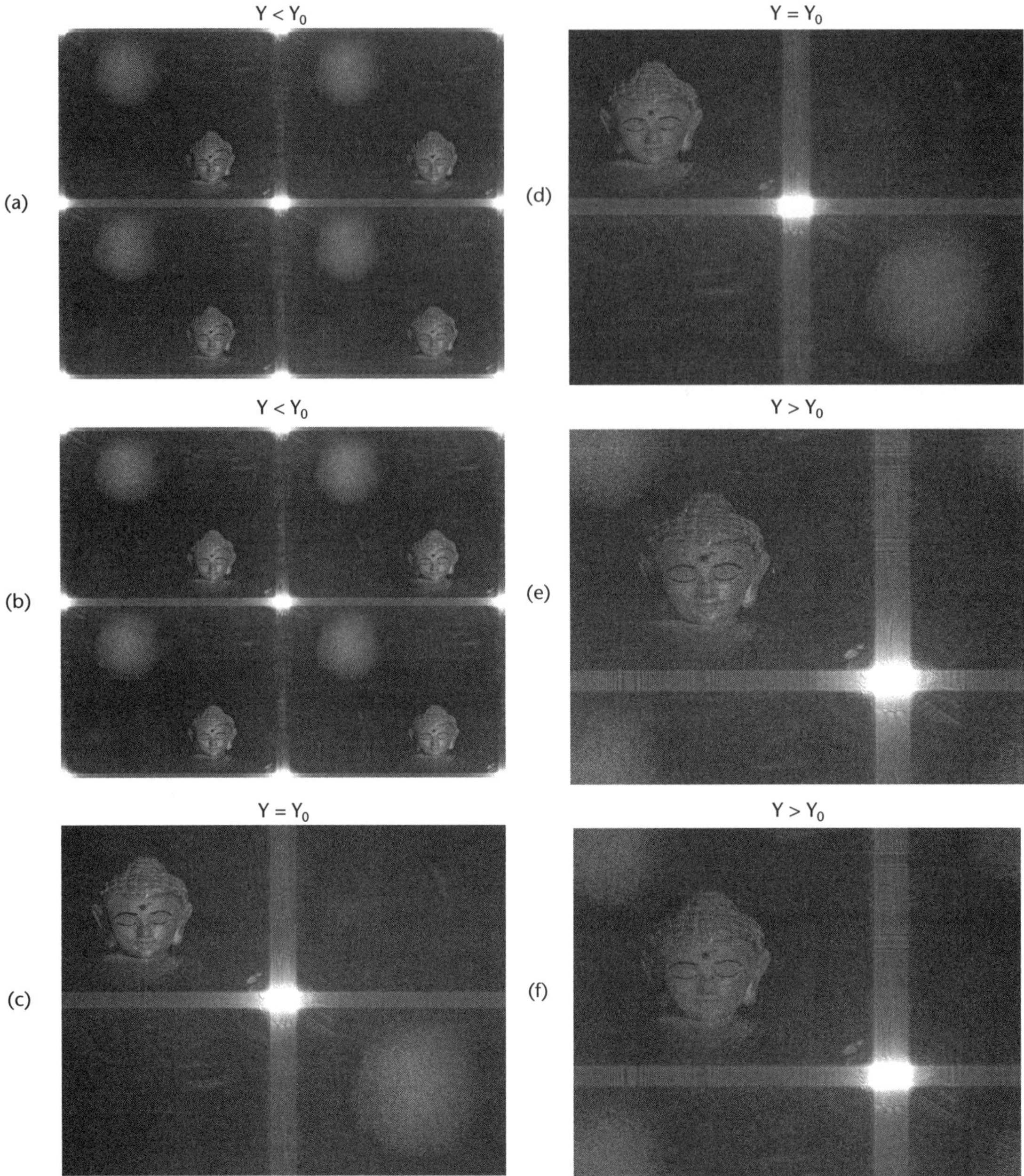

Figure 4.4 (a)(c)(e) Numerical reconstruction using the Fresnel-Bluestein transform, and (b)(d)(f) numerical reconstruction using the digital quadratic lens method.

suppress the DC term. One way is to subtract the average hologram intensity from each stored intensity before the reconstruction is performed. In this case, the new hologram function $H'(k, l)$ is calculated as

$$H'(k,l) = H(k,l) - \frac{1}{NM}\sum_{k=1}^{N}\sum_{l=1}^{M}H(k,l) \tag{4.57}$$

The DC suppression can also be achieved by subtracting the low-pass filtered hologram function from the original hologram function [14]. The low-pass filter can be designed as a 3 × 3 mean filter. Figure 4.5(a) shows an example of reconstruction that is similar to Figure 4.1(d) but with the suppressed DC term. Due to the global filtering effect involved in this method, the reconstructed object image also gets modified.

A nonlinear filtering method has been suggested to achieve the DC suppression [15]. In this method, the reference wave intensity is set up much higher than that of the object wave, and the off-axis holographic configuration is used in order to separate the image components. Consider the following rearrangement of (4.3)

$$I = |O + R|^2$$

$$\frac{I}{|R|^2} = \left(1 + \left(\frac{O}{R}\right)\right)\left(1 + \left(\frac{O}{R}\right)^*\right) \tag{4.58}$$

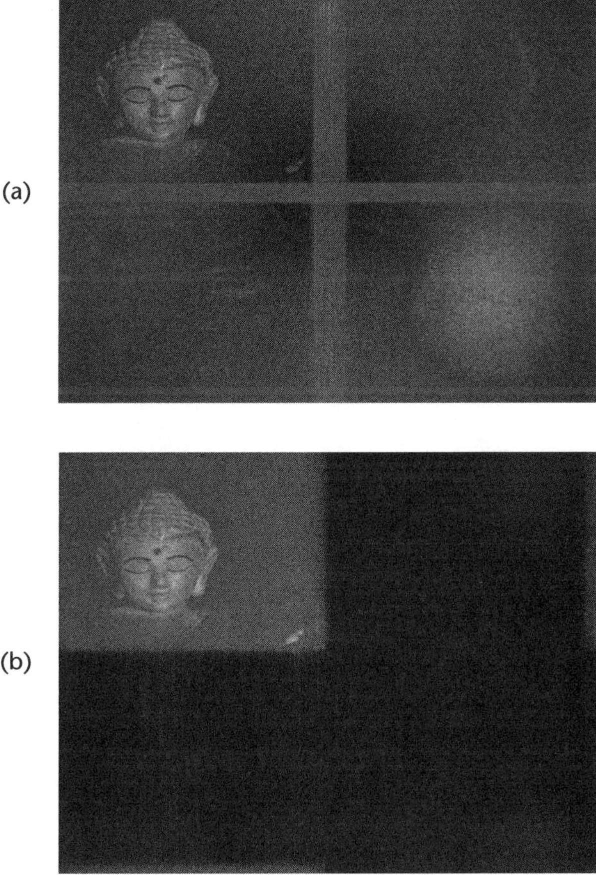

Figure 4.5   Suppression of DC term with (a) low-pass filtering and (b) nonlinear filtering.

The filtered hologram intensity $H'(x_h, y_h)$ is computed as

$$H'(x_h, y_h) = \log\left(1 + \left(\frac{O}{R}\right)\right)$$

$$= \mathfrak{F}^{-1}\left\{\mathfrak{F}\left\{\log\left(\frac{I}{|R|^2}\right)\right\} \times \mathbf{1}_{(N/2 \times M/2)}\right\} \tag{4.59}$$

The complex amplitude of the object wave at the hologram plane can then be computed as

$$O(x_h, y_h) = R(x_h, y_h)\left(\exp\left(H'(x_h, y_h)\right) - 1\right) \tag{4.60}$$

where $\mathbf{1}_{(N/2 \times M/2)}$ is an indicator function of the quadrant corresponding to the real image. The complex amplitude of the object wave at the object plane is next computed using the Fresnel approximation. This method essentially removes the DC term associated with the object wave and the unfocussed object image. To remove the DC part related to the reference wave, a simple calibration step is performed wherein the reference wave intensity is measured and is subtracted from the recorded hologram. Since the DC contribution of the object beam is suppressed, this method provides a large allowable maximum spatial bandwidth that the object can utilize. The DC and virtual image suppression is performed using nonlinear filtering method as shown in Figure 4.5(b).

BOX 4.2

A phase-shifting technique for digital holographic imaging has been proposed in [16]. Initially, the complex amplitude of the object wave $O(x_h, y_h)$ at the hologram recording plane is computed using the phase-shifting interferometry. The amplitude of the object wave $O_o(x_h, y_h)$ is obtained by recording the intensity corresponding only to the object wave scattered by the object. To compute the phase $\phi_o(x_h, y_h)$ of the object wave, four holograms $H(x_h, y_h)$ are recorded each for stepped phase differences of 0, $\pi/2$, π, and $3\pi/2$, respectively, imparted to the reference beam. The phase $\phi_o(x_h, y_h)$ is then obtained as

$$\phi_o(x_h, y_h) = \tan^{-1}\frac{H(x_h, y_h; 3\pi/2) - H(x_h, y_h; \pi/2)}{H(x_h, y_h; 0) - H(x_h, y_h; \pi)} \tag{4.61}$$

The complex amplitude of the object wave at the hologram recording plane is computed as

$$O(x_h, y_h) = O_o(x_h, y_h)\exp\left(i\phi_o(x_h, y_h)\right) \tag{4.62}$$

Note that this procedure essentially removes the contribution of the DC term in the computation of the object wavefield at the hologram recording plane. Thus, it can be considered as a technique for the suppression of the DC term. Finally, the complex amplitude of the object wave at the reconstruction plane, $O(x_i, y_i, z_r)$, is obtained by computing the propagation of $O(x_h, y_h)$ to the reconstruction plane using the Fresnel approximation.

Instead of recording four separate holograms corresponding to four phase shifts, a single hologram recording can be performed using a pixelated array and wave plates. In this technique, a micropolarizer array is placed in front of the CCD. The orthogonal circularly polarized object wave and the reference wave are allowed to interfere at the CCD plane. A unit cell in the micropolarizer is made up of four linear polarizers of different orientations. These polarizers introduce different phase shifts at the four neighboring pixels. Consequently, four sets of pixels with similar phase shifts are obtained. The intensity values at these pixels are interpolated to obtain the four phase-shifted holograms with their dimensions the same as that of the recorded hologram. The phase of the object wavefield is computed using these four holograms.

4.5   Reduction of Spatial Frequency Content Falling on CCD

As indicated in (4.7), the maximum spatial frequency that needs to be recorded depends on the maximum angle $\alpha_{\max}$ between the object and the reference beams. Since the maximum spatial frequency that can be recorded by the CCD is limited by its resolution, it imposes a strong constraint on $\alpha_{\max}$ and in turn, on the size of the object that can be faithfully imaged. Accordingly, (4.8) suggests that digital holography is suitable only for imaging small-sized objects. However, different methods have been proposed to overcome this limitation. One of these methods is to increase the distance z_o between the object and the CCD so that the criteria of $\alpha_{\max}$ given in (4.10) is satisfied. Some of the other methods for the reduction of spatial frequency content are explained below.

4.5.1   Reduction of the Object Angle on the CCD Array by the Use of a Divergent Lens

In this technique, the use of divergent lens is proposed to generate a demagnified virtual image of large-sized objects [17]. The divergent lens, placed in the imaging geometry between the object and the CCD, allows to achieve a substantial reduction in the object angle projected on the CCD surface. Effectively, a hologram corresponding to the virtual image is recorded by the CCD as shown in Figure 4.6. Given that the size of the virtual image is much smaller than the object size, the maximum spatial frequency that needs to be accommodated by the CCD is decreased significantly. A factor ς is defined as

$$\varsigma = \frac{\sin\varphi/2}{\sin\varphi'/2} \approx \frac{\tan\varphi/2}{\tan\varphi'/2} \tag{4.63}$$

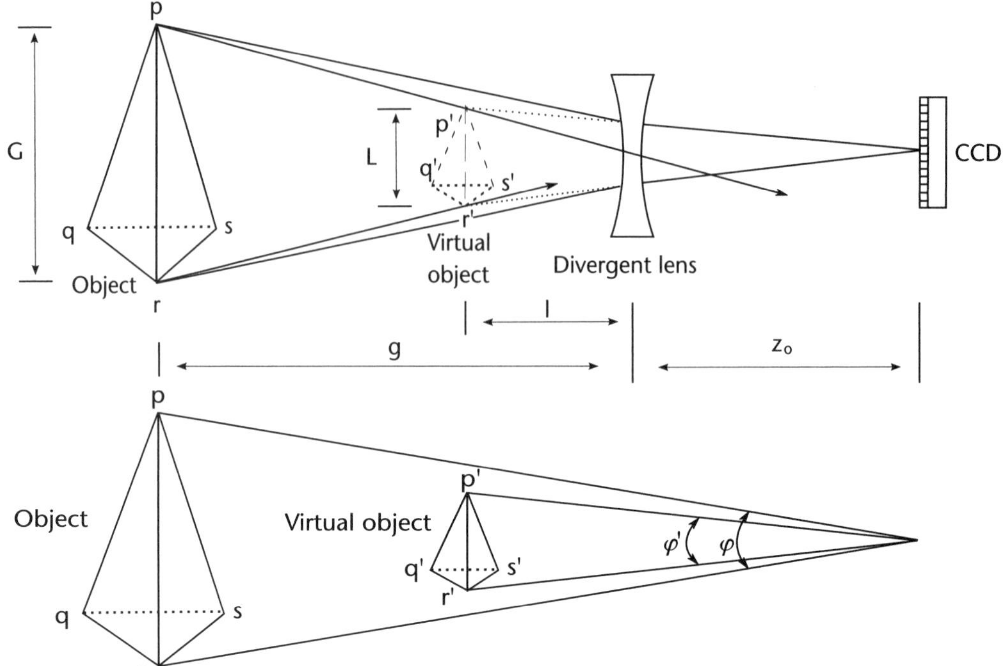

Figure 4.6   Schematic of the setup for the reduction of object angle on the CCD by the use of a divergent lens.

where φ and φ' are the angles projected by the object and virtual image on the CCD, respectively. From the optical configuration shown in Figure 4.6, ς can be rewritten as

$$\varsigma = \frac{G}{L} \times \frac{l + z_o}{g + z_o} = \frac{g}{l} \times \frac{l + z_o}{g + z_o} \tag{4.64}$$

where the significance of the quantities used in (4.64) are illustrated in Figure 4.6.

4.5.2   Reduction of the Object Angle on the CCD by the Use of an Aperture

An interesting method for recording holograms has been proposed [18] based on controlled selection of the spatial frequencies that are produced on the CCD. This is achieved by introducing an aperture of appropriate dimension between the object and the CCD, as shown in Figure 4.7. The aperture essentially limits the angles of the light rays scattered from the object surface and propagating in the direction of the CCD to lie within acceptable values. If these values are sufficient to satisfy the Nyquist sampling criteria, then the spatial frequencies would be recorded without aliasing.

It can be observed from Figure 4.7 that the aperture blocks the light rays scattered from the object surface which make relatively large angles, for example α_1 with the reference beam. This restricts the interference fringes between the reference and object beams to low spatial resolution frequency values and within the

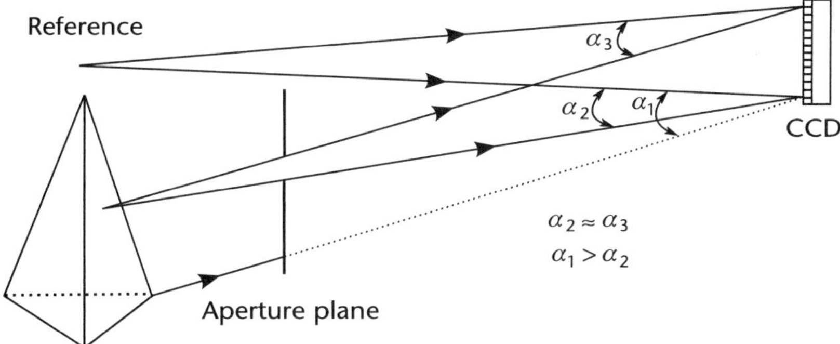

Figure 4.7   Schematic of the setup for the reduction of the object angle on the CCD by the use of an aperture.

range of spatial resolution values of the CCD. Thus with this technique, large-sized objects can be imaged but with a relatively large speckle. It is also to be noted that, since the aperture blocks the light coming from the object, the reduction in image brightness might occur if the aperture is not used properly.

4.6   Digital Holographic Interferometry

Interferometry is a well-established diagnostic tool in the field of physics and is widely used in engineering applications. Two coherent light waves derived from a single source are superimposed on the detector to generate an intensity fluctuation forming a pattern of bright and dark bands. This intensity pattern is termed as an interference fringe pattern. These interference fringes are formed due to the phase difference created by the different optical path lengths traveled by the two light waves. In an interferometric application, one light wave acts as a reference whose path length is generally maintained constant. The second wave travels such that the changes in its path length are proportional to the changes in the physical parameter whose measurement we are interested in. Subsequently, the resulting fringe pattern is used to provide the measurement of the changes in the aforementioned physical parameter.

Digital holography, on account of its capability to directly provide the complex amplitude of optical wavefield scattered by the object, lends itself well to applications requiring interferometric precision and measurement. In digital holographic interferometry, two holograms are recorded, each corresponding to a different object state. The change in object state causes a difference in the phases of wavefields. The phase associated with each object state can be calculated by numerically reconstructing the two holograms. The difference in phase values provides the information about the change in the object state at each point on the object. This whole-field and noncontact type measurement capability makes digital holographic interferometry an indispensable measurement tool in experimental mechanics and in nondestructive testing.

Consider that a hologram is recorded in the initial state of the object. The numerically reconstructed object wavefield can be represented as

$$\Gamma_b\left(x_o,y_o\right) = A_b\left(x_o,y_o\right)\exp\left[j\phi\left(x_o,y_o\right)\right] \tag{4.65}$$

where $\phi(x_o, y_o)$ is the phase and $A_b(x_o, y_o)$ is the amplitude of the wavefield. The object is next deformed and a separate hologram is recorded. The numerical reconstruction of this new hologram provides the complex amplitude of the wavefield associated with the new object state, and which can be represented as

$$\Gamma_a\left(x_o,y_o\right) = A_a\left(x_o,y_o\right)\exp\left[j\left(\phi\left(x_o,y_o\right) + \Delta\phi\left(x_o,y_o\right)\right)\right] \tag{4.66}$$

It can be observed that the phase has been altered to $\phi(x_o, y_o) + \Delta\phi(x_o, y_o)$. Due to surface roughness of the object, the phases of the wavefields corresponding to the two object states are random in nature. However, their difference $\Delta\phi(x_o, y_o)$ is deterministic and is proportional to the change in the object state. The interference field corresponding to the change in the object state is then computed by the conjugate multiplication of the complex amplitudes of the wavefields associated with the two object states. This is obtained as

$$\begin{aligned}\Gamma\left(x_o,y_o\right) &= \Gamma_a\left(x_o,y_o\right)\Gamma_b^*\left(x_o,y_o\right) \\ &= A\left(x_o,y_o\right)\exp\left(j\Delta\phi\left(x_o,y_o\right)\right)\end{aligned} \tag{4.67}$$

where $A(x_o, y_o) = A_a(x_o, y_o)A_b(x_o, y_o)$ is assumed to be constant or slowly varying, $\Gamma(x_o, y_o)$ is the interference field and $\Delta\phi(x_o, y_o)$ is termed as the *interference phase*. The fringe pattern can be observed as a real part of the interference field. Information on the change in the object state is encoded in the interference phase. This information can be derived by accurately extracting the interference phase from the interference field. The straightforward approach to extract the phase consists of using the arctan function (i.e., the inverse tangent operation). However, the principal values of the inverse tangent function lie in the interval $[-\pi, \pi]$, which essentially wraps the true phase value in this interval. This explains the need for the phase unwrapping operation to be performed by adding the integer multiples of 2π to the wrapped phase map.

4.6.1   Deformation Measurement

Measurement of object deformation plays a crucial role in the characterization of the object's behavior under loading. In deformation measurement using digital holographic interferometry, two holograms are recorded, one each before and after the deformation of the object. Consequently, the phase $\Delta\phi(x_o, y_o)$ of the reconstructed interference field carries the information on the object deformation. The relationship between the object surface displacement and the interference phase can be understood using the configuration shown in Figure 4.8(a).

Here, $\hat{i}$ is a unit vector along the object illumination direction; $\hat{o}$ is a unit vector along the observation direction. The unit sensitivity vector $\hat{s} = \hat{o} - \hat{i}$ is defined along the bisector between the illumination and observation directions. The relationship between the object displacement $\vec{d} = d_x\hat{x}_o + d_y\hat{y}_o + d_z\hat{z}$ and the interference phase $\Delta\phi(x_o, y_o)$ can be written as [19]

$$\begin{aligned}\Delta\phi\left(x_o, y_o\right) &= \frac{2\pi}{\lambda}\vec{d} \cdot \left(\hat{o} - \hat{i}\right) \\ &= \frac{2\pi}{\lambda}\vec{d} \cdot \hat{s}\end{aligned} \tag{4.68}$$

Thus, the fringe pattern carries information about the resolved part of the displacement vector along the sensitivity vector. If the interferometry setup shown in Figure 4.1 is arranged such that the sensitivity vector lies along the z direction, then the interference phase provides the out-of-plane displacement measurement of the object.

For illustration, an out-of-plane deformation measurement is performed. A circular membrane clamped along its edges is placed at 1 m away from the CCD. The object is illuminated with a laser source of wavelength $\lambda = 532$ nm. A hologram is recorded in the unloaded state of the object. The object is next deformed by applying a point load at its center, and a second hologram is recorded. The numerical reconstruction of the holograms is performed to obtain the complex amplitudes given in (4.65) and (4.66), respectively. The intensity map of the complex amplitude corresponding to the unloaded state of the object is shown in Figure 4.8(b). The object area selected for deformation analysis is also indicated in the figure. The phases corresponding to the two object states are shown in Figures 4.8(c) and 4.8(d), respectively. The phases are of random nature because of the micro-structure of the object surface. The difference in these phases is computed using (4.67). The interference phase in its wrapped form is shown in Figure 4.8(e). The fringe contours indicate the points on the object that have undergone equal displacements. Subsequently, a two-dimensional phase unwrapping algorithm is applied to obtain the true unwrapped interference phase map. Due to the noise associated with the measurement, filtering operation needs to be performed in order to obtain accurate 2D phase unwrapping. Different algorithms have been proposed in the literature to perform filtering and 2D-unwrapping [20]. Some of these algorithms are Goldstein, flood-fill, local histogram, regularized-phase tracking, and so forth. The unwrapped form of the interference phase is shown in Figure 4.8(f). The interference phase was computed using Goldstein algorithm that provides the quantitative measurement of surface displacement.

A different approach for phase estimation has been proposed [21, 22] by exploiting the continuous nature of the interference phase with respect to the spatial coordinates x_o and y_o. In this approach, the interference field in a given row y_o or column x_o is considered at a time. In accordance to Weierstrass approximation theorem, in this particular row/column, the interference phase can be approximated with a polynomial of an appropriate order. However, in the presence of rapid variations in phase, the order of the polynomial required for accurate representation of interference

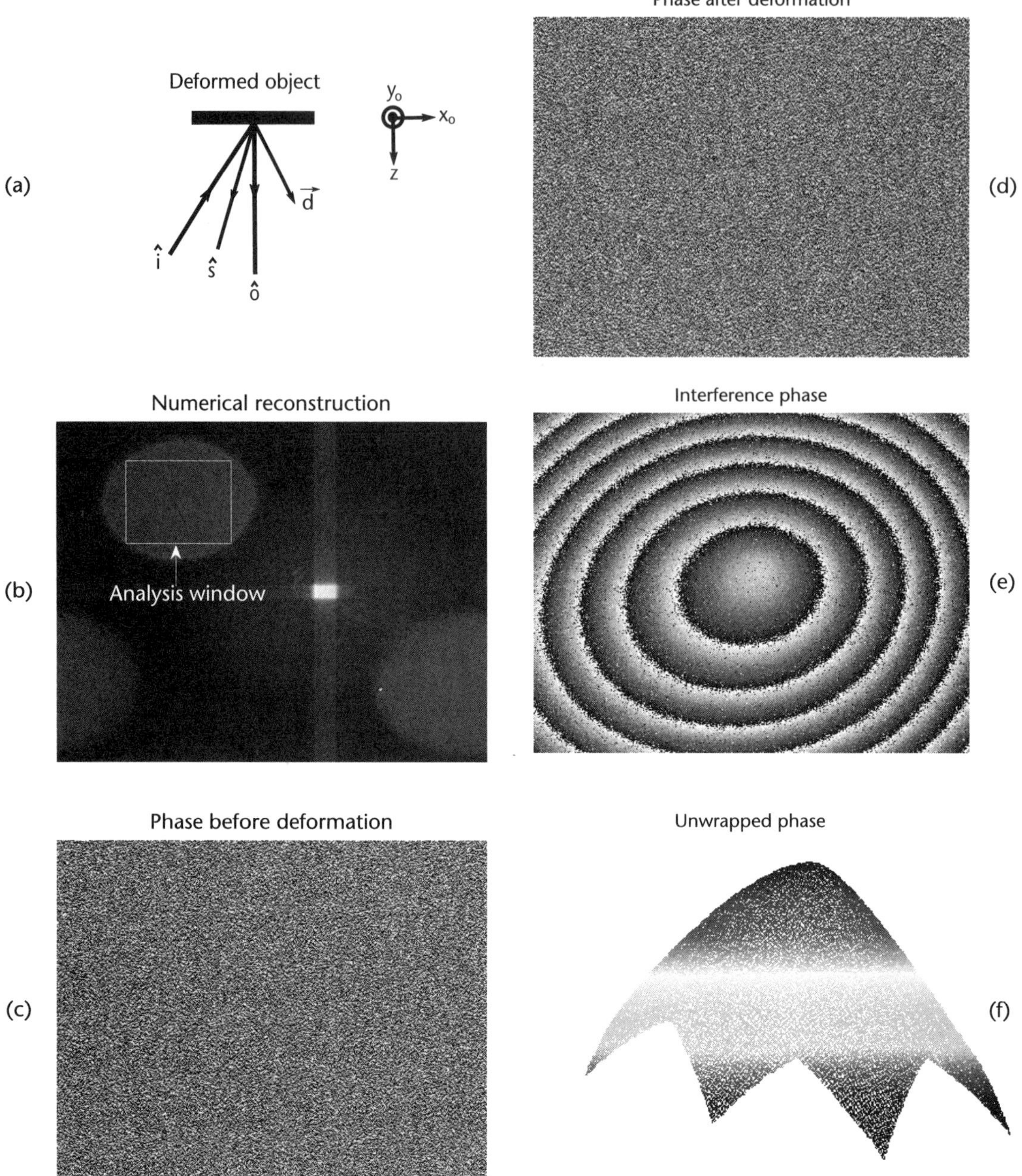

Figure 4.8 (a) Configuration indicating the relationship between the sensitivity and displacement vector and (b) numerical reconstruction of the hologram; phase of the complex amplitude corresponding to (c) the undeformed object state and (d) deformed object state; (e) interference phase in wrapped form and (f) interference phase after 2D phase-unwrapping.

phase will be high. Therefore, the interference field is divided into a number of segments and in each of these segments the interference phase is modeled as a lower order polynomial. Subsequently, an accurate estimation of polynomial coefficients is performed with which the estimate of interference phase is obtained. The local polynomial phase approximation provides quite accurate results. This method directly provides the unwrapped phase map without any requirement of filtering and unwrapping operation. We implemented this approach to directly obtain the unwrapped estimate of the interference phase for the measurement example shown in Figure 4.8. The estimated unwrapped phase in Figure 4.9(a). The wrapped form of the estimated phase is also shown in Figure 4.9(b) for the sake of comparison.

In nondestructive testing, direct estimation of displacement derivatives (i.e., strains), is highly desirable. Various methods based on digital shearing [23, 24], ridge detection in the space-frequency distribution of interference field [25, 26] and so forth, have been proposed for direct phase derivative estimation. While the digital shearing based methods are computationally efficient in estimating the wrapped form of phase derivative, the space-frequency distribution based methods provide the unwrapped phase derivative estimate with a better noise performance. We have thus implemented one of the space-frequency distributions, the pseudo-Wigner-Ville distribution (PWVD), to calculate the interference phase derivative for the measurement example cited above in this section. Figure 4.10(a) shows the phase derivative

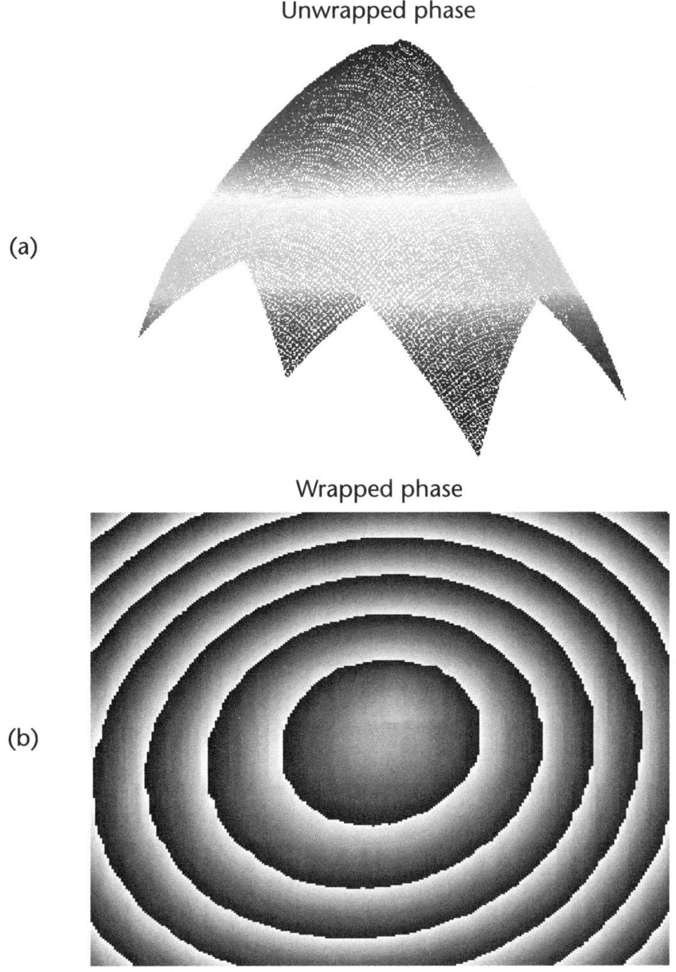

Figure 4.9   (a) Unwrapped estimate of interference phase using piecewise-polynomial approximation of phase and (b) corresponding wrapped form of the interference phase.

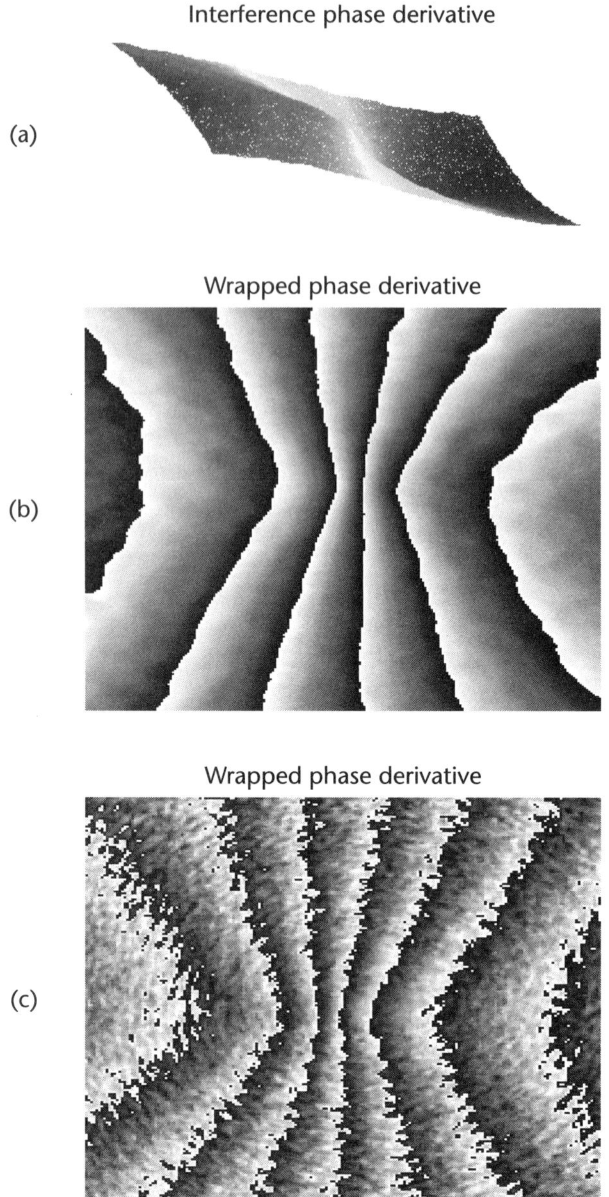

Figure 4.10   (a) Phase derivative estimate using pseudo-Wigner-Ville-distribution, (b) the corresponding wrapped form of the phase derivative, and (c) phase derivative estimate using the digital shearing method.

estimate that is directly related to the strain generated on the object surface due to loading. The wrapped forms of the appropriately scaled estimated phase derivative computed using the PWVD-based method and the digital shearing method are shown in Figures 4.10(b) and (c) for the sake of comparison.

For a sound characterization of the object deformation process, it might be advantageous to obtain the measurement of all the three or at least two components of deformation. For such measurements, a multibeam object illumination setup is generally used. The object is illuminated with two beams symmetrical to the observation direction pointing normal to the object surface. Each scattered object beam has a separate reference beam associated with it. An incoherent mixing of the scattered object beams corresponding to each of the two illumination beams can thus be achieved either using multiple color laser sources or using a single color laser source with orthogonally polarized reference beams [27–29]. These techniques allow the estimation of multiple interference phases from the multiple object-reference beam pair holograms recorded on the CCD. In another approach, a single reference beam has been used along with multiple object beams [30]. The holograms corresponding to each object-reference beam pair are recorded separately. This method does not allow simultaneous estimation of multiple interference phases. As usual, the algebraic relationship between the multidimensional deformation and the multiple interference phases can be obtained using the sensitivity vectors.

A digital holographic moiré based approach has been proposed [31] wherein an object is illuminated with two beams, each of which is placed symmetrical to the observation direction pointing normal to the object surface. The schematic of the object illumination and observation configuration is shown in Figure 4.11(a). A single reference beam is used to record the hologram. Two holograms are recorded, one each before and after the object deformation. This results in the formation of a moiré fringe pattern. Signal processing has been applied for the accurate estimation of the multiple interference phases present in the moiré fringe pattern [32, 33]. In one of these approaches, the interference field is divided into a number of non-overlapping segments in each row/column. In each segment, the interference field is represented as a multicomponent signal with second-order polynomial interference phases. A method based on product cubic phase function has been applied for the accurate estimation of polynomial phase coefficients. With the use of these polynomial coefficients, the interference phases are estimated in each segment. Further, these phase estimates are stitched together to obtain the complete 2D phase map. In this optical configuration, the sum and the difference of the estimated phases provide the measurement of the out-of-plane and in-plane displacement components, respectively.

An illustration is provided for multidimensional deformation measurement in a two-beam illumination setup. The object is subjected to out-of-plane deformation and in-plane rotation. The recorded moiré fringe pattern is shown in Figure 4.11(b). The estimated interference phases are shown in Figures 4.11(c) and 4.11(d). The sum and difference of estimated phases are shown in Figures 4.12(a)–(d), along with their wrapped forms.

4.6.2   Study of Refractive Index Distribution

Digital holographic interferometry has been a useful tool for investigating refractive index distributions in a transparent medium. When a light beam passes through a transparent medium, the change in the distribution of refractive index in the medium produces a change in the optical path traveled by the light. This change in the optical path length can be encoded in the interference phase. In a digital holographic interferometry setup, after passing through a transparent medium the laser beam is allowed to fall on the CCD. A hologram is recorded corresponding

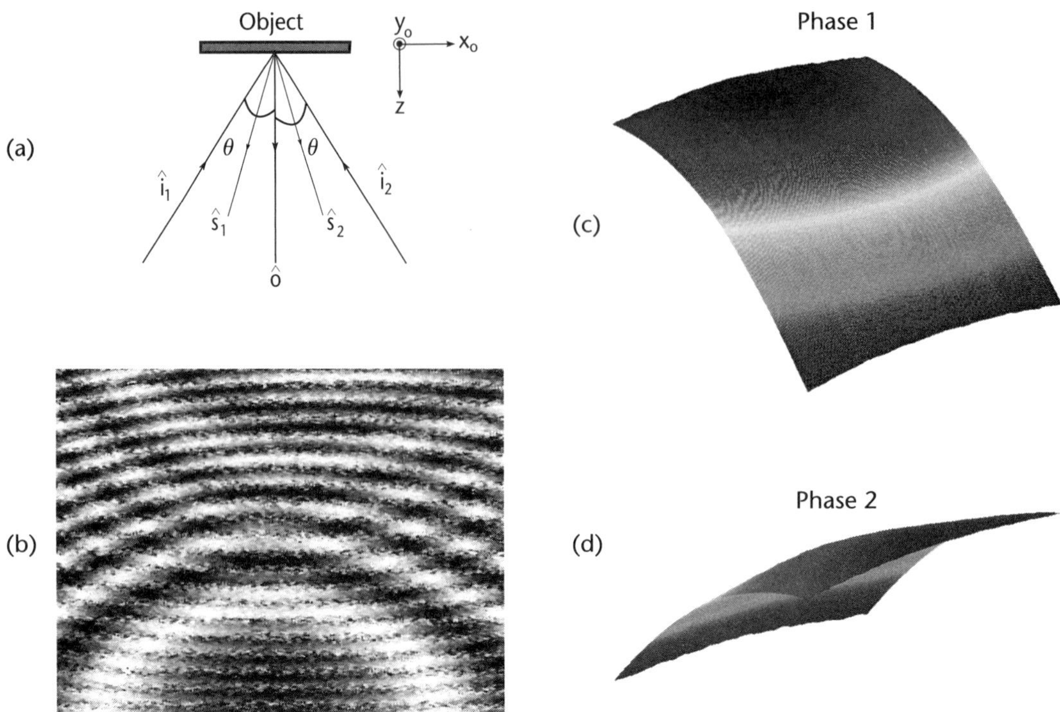

Figure 4.11   (a) Optical configuration for multidimensional deformation measurement indicating the illumination and observation direction, (b) experimentally recorded moiré fringe pattern, (c) estimated $\Delta\phi_1(x_o, y_o)$, and (d) estimated $\Delta\phi_2(x_o, y_o)$.

to the initial refractive index distribution state, $n(x_o, y_o, z)$, of the object. When the refractive index distribution changes to $n'(x_o, y_o, z)$, a second hologram is recorded. The interference phase of the reconstructed interference field due to the change in the refractive index distribution along the beam path can be expressed using the following equation as

$$\Delta\phi\left(x_o, y_o\right) = \frac{2\pi}{\lambda} \int_0^l \left(\Delta n\left(x_o, y_o, z\right)\right) dz \tag{4.69}$$

where $\Delta n(x_o, y_o, z) = n(x_o, y_o, z) - n'(x_o, y_o, z)$ and l is the length of the medium along z direction. The integration in the above equation indicates that the accumulation of the changes in the refractive index along the z direction produces the interference phase $\Delta\phi(x_o, y_o, z)$. If the change in the refractive index $\Delta n(x_o, y_o, z)$ remains constant along the z direction, the above relationship reduces to the following form:

$$\Delta\phi\left(x_o, y_o\right) = \frac{2\pi l}{\lambda} \left(\Delta n\left(x_o, y_o\right)\right) \tag{4.70}$$

The light beam passing through the transparent test medium can either be allowed to fall directly on the CCD or it can also be allowed to fall on a diffuse

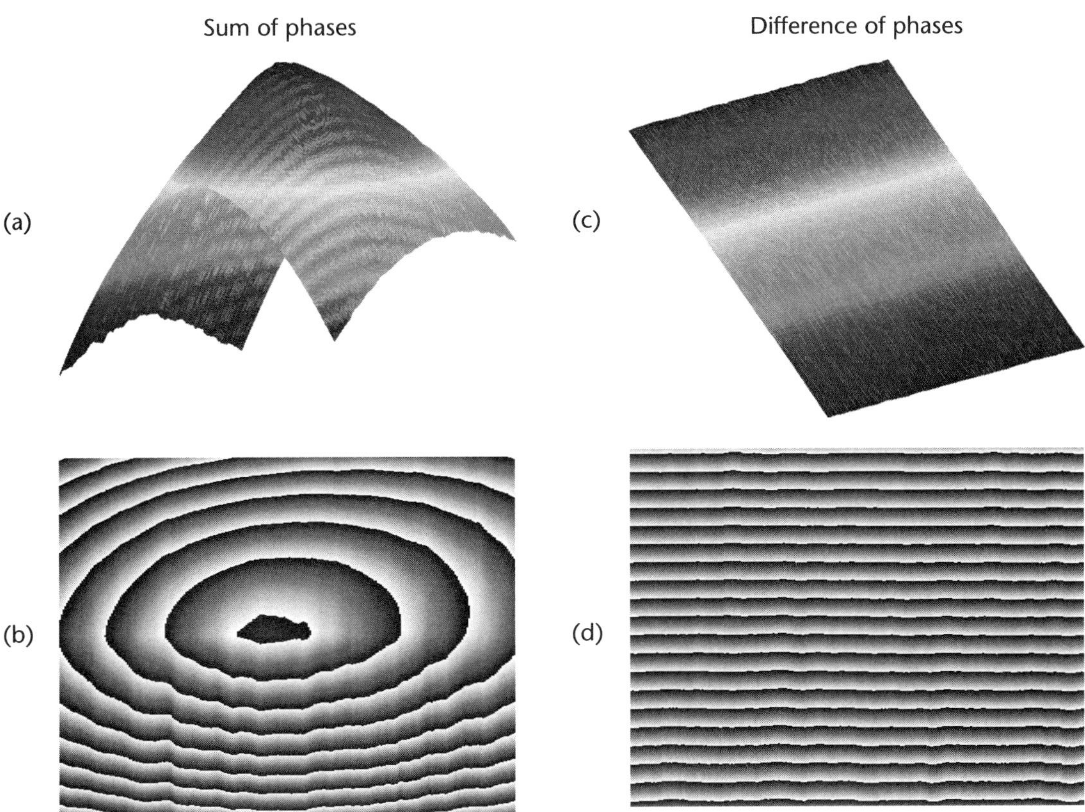

Sum of phases   Difference of phases

Figure 4.12   (a) Sum of phases, (b) wrapped form of sum of phases, (c) difference of phases, and (d) wrapped form of difference of phases.

surface. In the latter case, the light scattered off the diffuse surface falls on the CCD. To illustrate this fact, we placed a candle along the path of light reaching the CCD. A single hologram is recorded with the candle in its unlighted condition. A second hologram is recorded with the candle in its lighted state. The interference fringe pattern shown in the Figure 4.13 indicates the change in the refractive index of air induced by the change in air temperature caused by the lighted candle.

4.6.3   Vibration Measurement

Digital time-average holographic interferometry [34] has developed into an efficient and attractive tool for studying the resonant mode shapes and frequencies of vibrating objects. The method consists of exposing the CCD to the reference and the scattered object waves for a duration much larger than the time period with which the object is vibrating. Suppose that the object is vibrating with a frequency ω. The instantaneous wave scattered by the object for a sinusoidal vibration can be represented as $O(x_o, y_o, t) = O_0(x_o, y_o) \exp(i\phi(x_o, y_o, t))$. The instantaneous phase is $\phi(x_o, y_o, t) = \phi_0(x_o, y_o) + \phi_m(x_o, y_o) \sin(\omega t)$ where $\phi_0(x_o, y_o)$ is the initial random phase scattered by the object in its state of rest and ϕ_m is the maximum change in

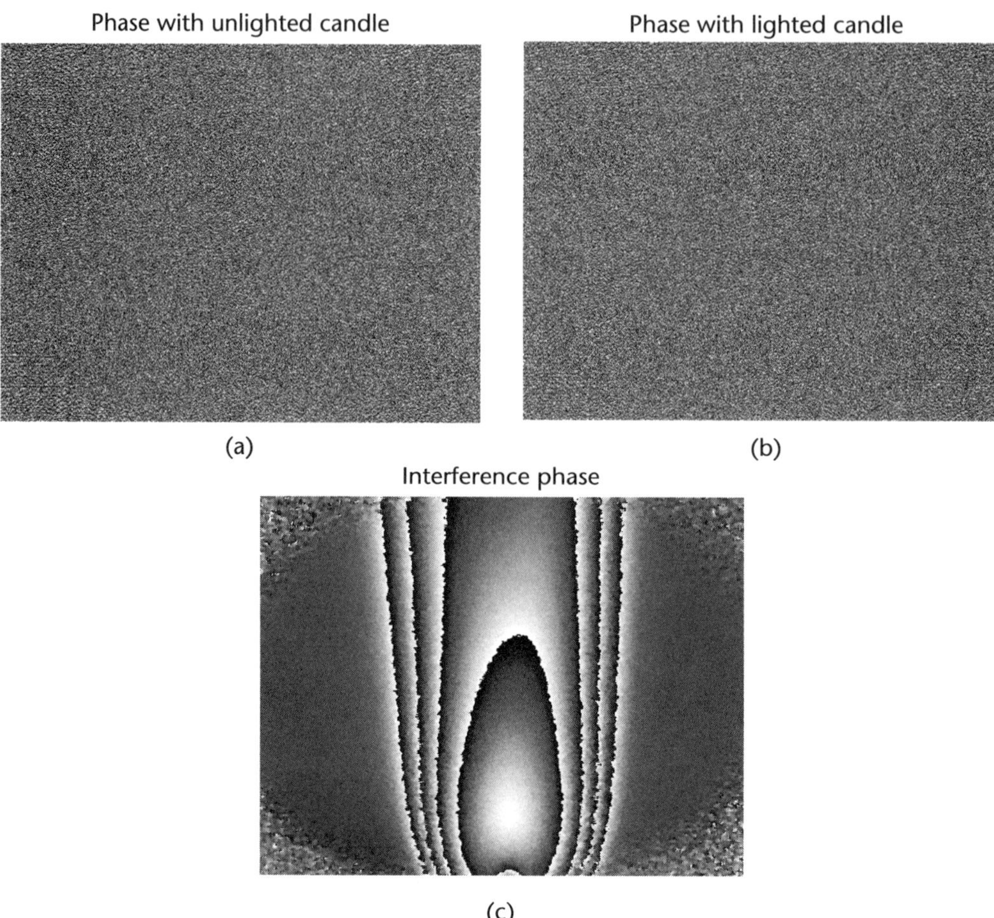

Figure 4.13   Phase corresponding to (a) unlighted candle, (b) lighted candle, and (c) change in the phase corresponding to the change in the refractive index of air.

the phase corresponding to the amplitude of vibration. The instantaneous wave at the hologram plane can be represented as

$$O\left(x_h, y_h, t\right) = \iint O\left(x_o, y_o, t\right) \exp\left[i\frac{\pi}{\lambda z_o}\left(x_o^2 + y_o^2\right)\right]$$
$$\times \exp\left[-\frac{i2\pi}{\lambda z_o}\left(x_o x_h + y_o y_h\right)\right] dx_o dy_o \tag{4.71}$$

To obtain a time-averaged hologram, the CCD exposure time T is adjusted such that $\omega T \gg 1$. Consequently, the intensity recorded by the CCD is given as

$$I\left(x_h, y_h\right) = \int_0^T \left| R\left(x_h, y_h\right) + O\left(x_h, y_h, t\right)\right|^2 dt \tag{4.72}$$

The term corresponding to the virtual image of the object can be given as

$$\int_0^T R^*(x_h,y_h)O(x_h,y_h,t) = R^*(x_h,y_h)\iint \tilde{A}(x_o,y_o)\exp\left[i\frac{\pi}{\lambda z_o}(x_o^2 + y_o^2)\right]$$

$$\times \exp\left[-\frac{i2\pi}{\lambda z_o}(x_o x_h + y_o y_h)\right]dx_o dy_o \tag{4.73}$$

where

$$\tilde{A}(x_o,y_o) = O_o(x_o,y_o)\exp\left(i\phi_0(x_o,y_o)\right)TJ_0\left[\phi_m(x_o,y_o)\right] \tag{4.74}$$

The numerical reconstruction of the time-averaged hologram shows the reconstructed object image to be modulated by the J_0^2 fringes, where J_0 is the zero-order Bessel function of the first kind. Typical fringe patterns obtained with a loudspeaker using digital time-average holographic interferometry are shown in Figure 4.14. The loudspeaker was driven at a sinusoidal frequency of 8.5 kHz and 9 kHz. The CCD camera was set with an exposure time of 250 ms. The first maximum corresponding to the nodes is visible in Figures 4.14(a) and 4.14(b). To show one of the numerous

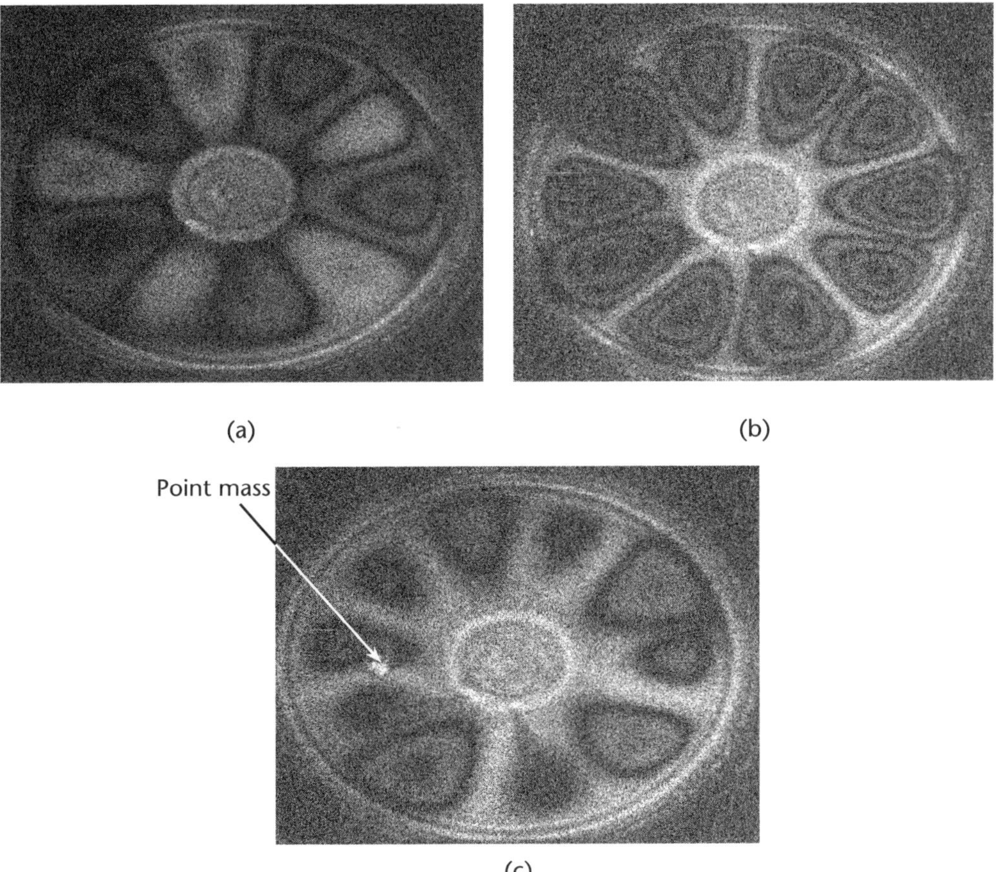

Figure 4.14 Vibration pattern corresponding to (a) 8.5 kHz and (b) 9 kHz. (c) Vibration pattern corresponding to 9 kHz with the loading of a point mass.

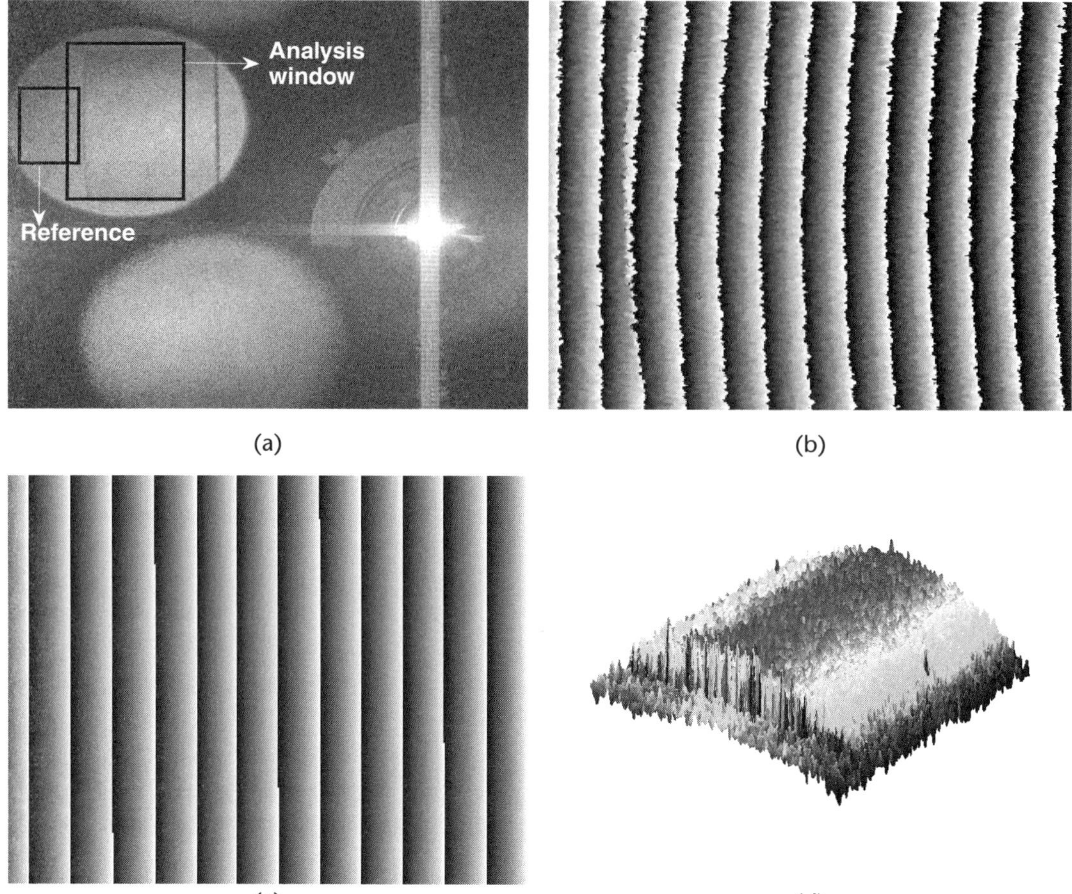

Figure 4.15   (a) Numerical reconstruction of the hologram with the analysis and reference window and (b) the wrapped form of the phase in the analysis window. (c) The wrapped form of the numerically computed carrier phase using the reference window and (d) the unwrapped phase in the analysis window after removal of the carrier.

advantages of the digital holography in vibration analysis, a tiny mass of 4 mm diameter and 0.33 mm thickness was glued at a point on the surface of the loudspeaker shown in 4.14(c). The loudspeaker was driven at a sinusoidal frequency of 9 kHz. The change in the nodal pattern with respect to the Figure 4.14(b) is indicative to the damping effect caused by the introduction of the tiny mass.

4.6.4   Object Contouring

Over the years, digital surface contouring has become an important tool in the arsenal of 3D shape measurement techniques. It has broad range of applications spanning from quality assurance in manufacturing, decreasing product development time, and so forth to 3D imaging of biological features. Of the various nondestructive optical techniques proposed for object contouring, digital holography belongs to a subset that can provide a high-resolution measurement capability. The basic principle behind holography-based contouring consists of introducing in its optical configuration, a phase change corresponding to the shape variations of the 3D object under study [35]. To achieve this change in the optical phase, different approaches based on two-illumination method, two-wavelength method, and two-refractive index method have been proposed.

In the setup used for object contouring, the object is illuminated by a collimated illumination beam making an angle β with respect to the surface normal. A single hologram is recorded with this configuration. A second hologram is recorded after changing the angle of illumination to $\beta + \Delta\beta$. The two holograms are reconstructed numerically, and a conjugate multiplication of the complex amplitudes gives rise to an interference pattern.

Note that since the object is illuminated at an angle β, there is an associated carrier present in the interference phase. This carrier phase is computed numerically and removed before obtaining the true unwrapped phase map, which provides the measurement of surface profile of the object.

A typical example illustrating object contouring using digital holographic interferometry is shown in Figure 4.15(a). The surface area of the object used for contouring is indicated as analysis window in the figure. The reference window, indicating a part of the reference plane, is used to numerically compute the carrier phase. Figure 4.15(b) shows the interference phase map. Figure 4.15(c) shows the phase map corresponding to the carrier phase term present in Figure 4.15(b). A phase-unwrapping algorithm is used after removal of the carrier phase term to obtain the 3D shape representation of the object. This is shown in Figure 4.15(d).

4.7 Conclusions

With the exponential increase in the power of computing and substantial improvements in the digital image processing and digital imaging technology, digital holography is emerging as a viable alternative to holography in as far as the image quality or its applications to real-world problems are concerned with the said advances only reined in by technological limitations. This chapter detailed a systematic approach to digital holography within the context of image formation that is inclusive of the suppression of the zero-order diffraction term and the reduction of spatial frequency content falling on the CCD, two parameters that are important contributions to the successful positioning of high-performance digital holography systems. The chapter also presented some of the existing applications of digital holography in optical metrology that have been developed over the past few years for determining the phase difference between two states of the same object directly from the digital holograms. These applications have been presented from the perspective of the measurement of object deformation, variations in refractive index, vibrations, and surface topography. Along with the rapid developments taking place in digital imaging, we strongly believe to be assisting in the opening of an era of new avenues in the development and applications of digital holography of which an immediate beneficiary without doubt will be optical metrology.

References

[1]   Gabor, D., "A New Microscopic Principle," *Nature*, Vol. 161, No. 4098, 1948, pp. 777–778.

[2]   Goodman, J. W. and R. W. Lawrence, "Digital Image Formation from Electronically Detected Holograms," *Applied Physics Letters*, Vol. 11, No. 3, 1967, pp. 77–79.

[3]   Schnars, U., and W. Juptner, "Digital Recording and Numerical Reconstruction of Holograms," *Measurement Science and Technology*, Vol. 13, 2002, pp. R85–R101.

[4]   Paturzo, M., P. Memmolo, A. Finizio, R. Näsänen, T. J. Naughton, and P. Ferraro, "Synthesis and Display of Dynamic Holographic 3d Scenes with Real-World Objects," *Optics Express*, Vol. 18, April 2010, pp. 8806–8815.

[5]   Locatelli, M., E. Pugliese, M. Paturzo, V. Bianco, A. Finizio, A. Pelagotti, P. Poggi, L. Miccio, R. Meucci, and P. Ferraro, "Imaging Live Humans Through Smoke and Flames Using Far-Infrared Digital Holography," *Optics Express*, Vol. 21, March 2013, pp. 5379–5390.

[6]   Matoba, O., E. Tajahuerce, and B. Javidi, "Real-Time Three-Dimensional Object Recognition with Multiple Perspectives Imaging," *Applied Optics*, Vol. 40, July 2001, pp. 3318–3325.

[7]   Goodman, J., *Introduction to Fourier Optics*, 2nd Edition, McGraw-Hill, 1996.

[8]   Yu, L., and M. K. Kim, "Wavelength-Scanning Digital Interference Holography for Tomographic Three-Dimensional Imaging by Use of the Angular Spectrum Method," *Optics Letters*, Vol. 30, No. 16, 2005, pp. 2092–2094.

[9]   Leith, E. N., and J. Upatniks, "Reconstructed Wavefronts and Communication Theory," *Journal of the Optical Society of America*, Vol. 52, October 1962, pp. 1123–1128.

[10]   Verrier, N., and M. Atlan, "Off-Axis Digital Hologram Reconstruction: Some Practical Considerations," *Applied Optics*, Vol. 50, 2011, pp. H136–H146.

[11]   Restrepo, J. F., and J. Garcia-Sucerquia, "Magnified Reconstruction of Digitally Recorded Holograms by Fresnel-Bluestein Transform," *Applied Optics*, Vol. 49, No. 33, 2010, pp. 6430–6435.

[12]   Picart, P., P. Tankam, D. Mounier, Z.-J. Peng, and J.-C. Li, "Spatial Bandwidth Extended Reconstruction for Digital Color Fresnel Holograms," *Optics Express*, Vol. 17, 2009, pp. 9145–9156.

[13]   Hennelly, B., D. Kelly, N. Pandey, and D. Monaghan, "Zooming Algorithms for Digital Holography," *Journal of Physics: Conference Series*, Vol. 206, No. 1, 2010, p. 012027.

[14]   Kreis, T. M., and W. P. O. Juptner, "Suppression of the DC Term in Digital Holography," *Optical Engineering*, Vol. 36, No. 8, 1997, pp. 2357–2360.

[15]   Pavillon, N., C. S. Seelamantula, J. Kuhn, M. Unser, and C. Depeursinge, "Suppression of the zero-order term in Off-Axis Digital Holography Through Nonlinear Filtering," *Applied Optics*, Vol. 48, No. 34, 2009, pp. H186–H195.

[16]   Yamaguchi, I., and T. Zhang, "Phase-Shifting Digital Holography," *Optics Letters*, Vol. 22, No. 16, 1997, pp. 1268–1270.

[17]   Schnars, U., T. M. Kreis, and W. P. O. Juptner, "Digital Recording and Numerical Reconstruction of Holograms: Reduction of the Spatial Frequency Spectrum," *Optical Engineering*, Vol. 35, No. 4, 1996, pp. 977–982.

[18]   Pedrini, G., S. Schedin, and H. J. Tiziani, "Lensless Digital-Holographic Interferometry for the Measurement of Large Objects," *Optics Communications*, Vol. 171, No. 1, 1999, pp. 29–36.

[19]   S. JE, "Holographic Interferometry Applied to Measurements of Small Static Displacements of Diffusely Reflecting Surfaces," *Applied Optics*, Vol. 8, No. 8, 1969, pp. 1587–1595.

[20]   Rajshekhar, G., and P. Rastogi, "Fringe analysis: Premise and perspectives," *Optics and Lasers in Engineering*, Vol. 50, No. 8, 2012, pp. iii–x.

[21]   Gorthi, S. S., and P. Rastogi, "Piecewise Polynomial Phase Approximation Approach for the Analysis of Reconstructed Interference Fields in Digital Holographic Interferometry," *Journal of Optics A: Pure and Applied Optics*, Vol. 11, No. 6, 2009.

[22]   Gorthi, S. S., and P. Rastogi, "Phase Estimation in Digital Holographic Interferometry Using Cubic-Phase-Function Based Method," *Journal of Modern Optics*, Vol. 57, No. 7, 2010.

[23]   Zou, Y., G. Pedrini, and H. Tiziani, "Derivatives Obtained Directly from Displacement Data," *Optics Communications*, Vol. 111, No. 5–6, 1994, pp. 427–432.

[24]   Liu, C., "Simultaneous Measurement of Displacement and Its Spatial Derivatives with a Digital Holographic Method," *Optical Engineering*, Vol. 42, No. 12, 2003, pp. 3443–3446.

[25]   Sciammarella, C. A., and T. Kim, "Determination of Strains from Fringe Patterns Using Space-Frequency Representations," *Optical Engineering*, Vol. 42, No. 11, 2003, pp. 3182–3193.

[26]   Rajshekhar, G., S. S. Gorthi, and P. Rastogi, "Strain, Curvature, and Twist Measurements in Digital Holographic Interferometry Using Pseudo-Wignerville Distribution Based Method," *Review of Scientific Instruments*, Vol. 80, No. 9, 2009.

[27]   Saucedo-Anaya, T., M. H. De La Torre-Ibarra, F. Mendoza Santoyo, and I. Moreno, "Digital Holographic Interferometer Using Simultaneously Three Lasers and a Single Monochrome Sensor for 3D Displacement Measurements," *Optics Express*, Vol. 18, No. 19, 2010, pp. 19867–19875.

[28]   Picart, P., E. Moisson, and D. Mounier, "Twin-Sensitivity Measurement by Spatial Multiplexing of Digitally Recorded Holograms," *Applied Optics*, Vol. 42, No. 11, 2003, pp. 1947–1957.

[29]   Rajshekhar, G., S. S. Gorthi, and P. Rastogi, "Simultaneous Measurement of In-Plane and Out-of-Plane Displacement Derivatives Using Dual Wavelength Digital Holographic Interferometry," *Applied Optics*, Vol. 50, No. 34, 2011, pp. H16–H21.

[30]   Morimoto, Y., T. Matui, M. Fujigaki, and A. Matsui, "Three-Dimensional Displacement Analysis by Windowed Phase-Shifting Digital Holographic Interferometry," *Strain*, Vol. 44, No. 1, 2008, pp. 49–56.

[31]   Rajshekhar, G., S. SivaGorthi, and P. Rastogi, "Simultaneous Multidimensional Deformation Measurements Using Digital Holographic Moire," *Applied Optics*, Vol. 50, No. 21, 2011, pp. 4189–4197.

[32]   Rajshekhar, G., S. S. Gorthi, and P. Rastogi, "Estimation of Multiple Phases from a Single Fringe Pattern in Digital Holographic Interferometry," *Optics Express*, Vol. 20, No. 2, 2012, pp. 1281–1291.

[33]   Kulkarni, R., and P. Rastogi, "Multiple Phase Estimation in Digital Holographic Interferometry Using Product Cubic Phase Function," *Optics and Lasers in Engineering*, Vol. 51, No. 10, 2013, pp. 1168–1172.

[34]   Picart, P., J. Leval, D. Mounier, and S. Gougeon, "Time-Averaged Digital Holography," *Optics Letters*, Vol. 28, No. 20, 2003, pp. 1900–1902.

[35]   Thalmann, R., and R. Dandliker, "Holographic Contouring Using Electronic Phase Measurement," *Optical Engineering*, Vol. 24, No. 6, 1985, pp. 930–935.

CHAPTER 5

Digital Speckle Pattern Interferometry

Matias R. Viotti

5.1 Introduction

The establishment of the basic laser theory concept by Albert Einstein, and the first experimental demonstration of an operating ruby laser by Theodore Maiman in May 1960, allowed the sudden availability of light sources with a high coherence degree. The growth of new kinds of lasers and laser applications has accelerated during the last half century. Nowadays, lasers are present in human life as never before. They fulfill an important role in consumer products, telecommunication, materials machining, medicine (usually for surgery), engineering, and scientific research. This latter topic will be described in this chapter, showing the application of laser light for metrological purposes.

Researchers working with laser light sources noticed that a high-contrast and fine-scale granular pattern was able to be seen by an observer when an optically rough surface was illuminated with laser light [1]. This effect was named speckle [2, 3], being characterized by a random distribution of scattered light which can deterministically be modified by displacements or rotations of a diffuser, where the light is reflected on or transmitted through. The speckle effect was considered a mere nuisance, especially in holographic techniques, during the first years following the advent of laser sources [4].

BOX 5.1

Speckle distributions are also modified by changes in the illumination and observation geometry, the wavelength of the laser light, and the refraction index of the medium in which the light travels. Thus, speckle distributions allow us to measure the following features: (a) out-of-plane and in-plane components of the surface deformation of a rough object, (b) three-dimensional shapes, and (c) derivatives of the surface displacements [5]. Taking in mind these possibilities, important research efforts started in the late 1960s and the early 1970s, focusing on the development of new methods for high sensitivity measurements on diffusely reflecting surfaces, all of which can be grouped by the global name of speckle interferometry [6–11].

In the beginning, photographic films were used to obtain images, and the recorded information was processed through optical methods. The lack of an optical table for processing and revealing films negative was the main drawback to overcome for robust measurements, mainly in industrial environments. As a consequence, research became oriented toward substituting the holographic film by television cameras and electronically processing the video signal. The new method was called electronic speckle pattern interferometry (ESPI). First results were a little discouraging because of the low detector resolution, low sensitivity, and low signal-to-noise ratio.

Constant advances in technology for high speed and resolution data acquisition systems made possible to connect a TV camera at first and nowadays a CCD or CMOS camera to a host computer in order to acquire a digital image of the surface illuminated with the laser light. Advances in data transmission enabled us to link cameras directly to the computer (IEEE-1394, USB, or GigE interfaces) and to transmit digital images without the need for additional elements digitizing these acquired images (such as frame grabbers). At this point, ESPI was called DSPI due to the use of digital images as well as digital processing techniques (the word "electronic" was replaced with "digital"). Thus, the technique is now extensively known as digital speckle pattern interferometry (DSPI).

Over the years, the technique has been thoroughly investigated, theoretically as well as experimentally, and has been improved continuously. For this reason, nowadays, speckle methods can be considered well-established experimental techniques and important tools to perform measurements in a laboratory as well as in industrial environments.

5.2   Speckle Principle

Huygens' principle states that "every point on a propagating wavefront works as a light source of spherical secondary wavelets whose envelope at some later time is the new propagated wavefront" [12].

When a surface is illuminated with coherent light from a laser, readers can consider that each point on the surface behaves like a small source that follows Huygens' principle, as shown in Figure 5.1. In addition, the magnitude of the optical field in each point of space will be determined by the complex coherent addition of all wavelets coming from each scattering point on the surface. If the laser light is incident on an optically rough surface (i.e., with height variations greater than the wavelength of the light λ), it will be scattered in all directions. The scattered waves will form an interference pattern consisting of dark and bright spots that are randomly distributed in space. This random light distribution called a speckle distribution can be seen in Figure 5.2. Readers can find the same effect when laser light is transmitted through an "optically rough" diffuser, where optically rough implies that the diffuser has thickness variations greater than λ.

The random amplitude value for each point of the speckle pattern is described by a set of vectors with a random phase that are superimposed, generating a resulting

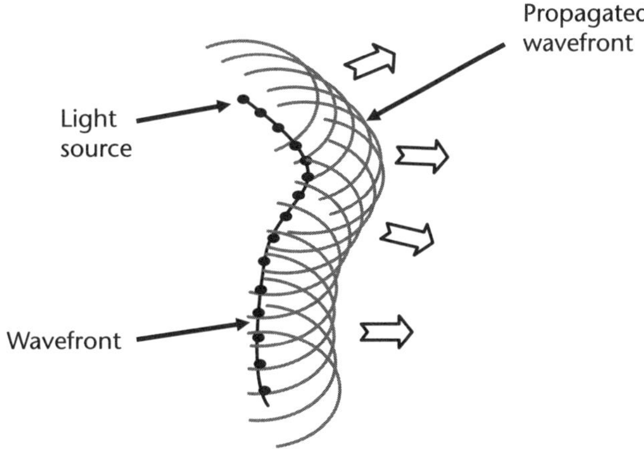

Figure 5.1   Huygens' principle: each point is considered a light source of the propagating wavelet.

vector with a random amplitude. This is a problem known in the literature as the random walk. The amplitude has a value that ranges from zero to the maximum value that is determined by the magnitude and the phase of the individual amplitudes. As the observation point is moved, the resulting amplitude and the intensity will have a different value.

Figure 5.2   Typical simulated speckle distribution.

> **BOX 5.2**
>
> Additional to the book written by Goodman [1], there are other very good books that readers can consult for further information about speckle statistics and complex addition of phasors, listed in [13–15]. Among them, the book edited by J. C. Dainty is the most widely cited, and the treatment found in the chapter written by M. H. Lehmann in the book edited by P. K. Rastogi [15] is an excellent reference for speckle properties applied to DSPI.

Speckle distributions can be classified depending on the optical configuration used to observe them. The first one is called objective speckle and corresponds to speckles in a freely propagating field that fills the three-dimensional space in front of the object surface. If an image is captured by using an image device such as a camera with an imaging lens, or the human eye, speckles will be presented in the image plane, and they will be called subjective speckles.

5.2.1 Objective Speckle Distributions

In order to find a representative value of the speckle size, a geometric computation can be performed with the help of Figure 5.3. This figure shows a rough surface being illuminated by laser light over an area with a cross-section of $l_0 \times l_0$. Points P_1 and P_2 belong to the boundaries of this surface. The optical path difference s from points P_1 and P_2 to $Q(x, y)$ can be expressed by

$$s = P_1Q - P_2Q \approx \frac{xl_0}{z} + \frac{1}{2}\frac{l_0^2}{z} \tag{5.1}$$

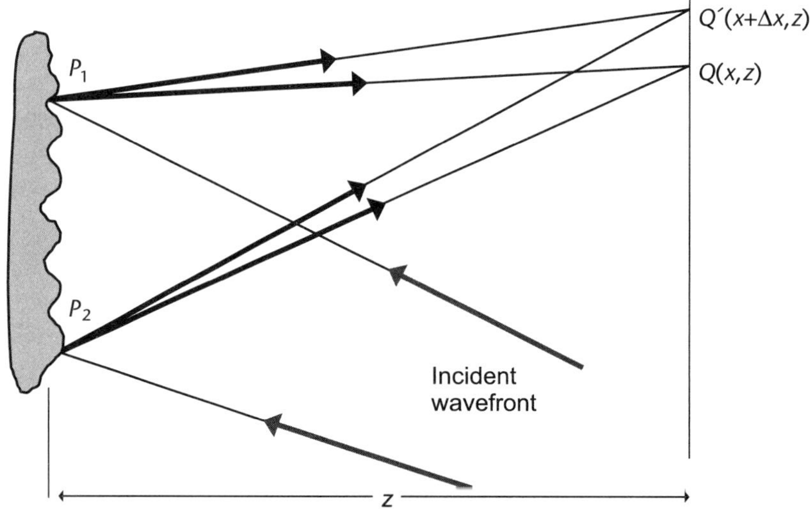

Figure 5.3 Optical setup for objective speckle distributions.

In the same way, the optical path difference for an adjacent point $Q'(x + \Delta x, y)$ will be described by

$$P_1Q' - P_2Q' \approx \frac{xl_0}{z} + \frac{1}{2}\frac{l_0^2}{z} + \frac{\Delta xl_0}{z} \tag{5.2}$$

Thus, the relative optical path difference between Q and Q' is expressed as

$$\Delta s = \frac{\Delta xl_0}{z} \tag{5.3}$$

For those points where $\Delta s < \lambda$, the relative phases of the components will approximately have the same value. On the other hand, if $\Delta s = \Delta xl_0/z \approx \lambda$, the relative phases will be completely different and the intensity in point Q' will not be correlated with the intensity in point Q. In consequence, the averaged speckle size d_{sp} will be expressed as

$$d_{sp} = \lambda z / l_0 \tag{5.4}$$

According to (5.4), it can be seen that the speckle size depends on the size of the illuminated area as well as on the distance between the screen where the scattered light is gathered and the object. In addition, the speckle size does not depend on the optical system used to observe it. For this reason, this kind of speckle is called as objective speckle.

5.2.2   Subjective Speckle Distributions

Figure 5.4 shows the case where the scattering surface is imaged by means of an optical system constituted by a lens and a circular aperture. Thus, a point P_1 that

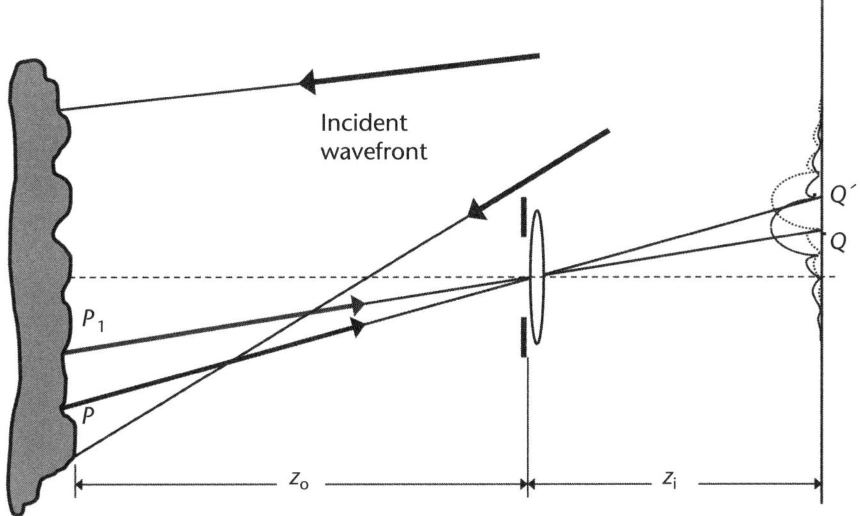

Figure 5.4   Optical configuration for subjective speckle generation.

belongs to the illuminated surface is imaged on the image plane as an intensity distribution centered at point Q (Airy disc). As it was shown for objective speckle, the light coming from point P_1 has a random phase related to the roughness of the scattering surface. The point Q receives light contributions from other points placed in the neighborhood of the point P_1. Thus, a set of Airy distributions with random phases will be superimposed in Q. In the same way, a new point P_2 can be now considered (see Figure 5.4), of which the diffraction pattern is centered at point Q'. Readers can see that, for this particular point, the first minimum intensity value of the Airy distribution will be coincident with point Q. Thus, point P_2 will not illuminate point Q. Additionally, points placed further away from point P_1 will also not contribute to illumination at point Q since the secondary maximum intensity values are significantly lower than the central maximum, thus producing a depreciated contribution.

As a result, the light intensity at point Q is given by contributions from a circular area of the illuminated object centered around point P_1, the radius of which equals the distance between points P_1 and P_2. The averaged speckle size d_{sp} corresponds to the radius of the Airy disc

$$d_{sp} = 1.22 \frac{\lambda z_i}{b} \tag{5.5}$$

where z_i is the distance between the aperture and the image plane, and b is the diameter of the aperture. The distance P_1P_2, being the radius of the scattering element placed on the illuminated surface that scatters the light in the point Q, is given by

$$\left(d_{sp} \right)_{obj} = 1.22 \frac{\lambda z_o}{b} \tag{5.6}$$

where z_o is the distance from the object to the aperture. This area is known as the resolution cell of the optical system and corresponds to the speckle grain on the illuminated object.

If M_g is the magnification of the optical system and F is the focal length, the following relations can be deduced

$$z_i = \left(1 + M_g \right) F \tag{5.7}$$

$$z_o = \frac{z_i}{M_g} = \frac{1 + M_g}{M_g} F \tag{5.8}$$

By considering $F_b = F/b$ as the numerical aperture of the optical system, the speckle size can be written as

$$d_{sp} = 1.22 \lambda F_b \left(1 + M_g \right) \tag{5.9}$$

Finally, the size of the speckle on the illuminated object can be expressed as

$$\left(d_{sp}\right)_{obj} = 1.22 \frac{\lambda F_b \left(1 + M_g\right)}{M_g} \tag{5.10}$$

Thus, small subjective speckles are related to large lens apertures. In other words, as the lens aperture is increased, smaller subjective speckles will be obtained. This can be easily verified by narrowing our eye aperture when looking at a speckle pattern.

BOX 5.3

A deep analysis of the first order statistics enables us to obtain expressions of probability distributions for the complex amplitude, phase, and intensity of the scattered wavefront, as well as for the speckle size. References [1] and [16] show expressions for these probability distributions. Additionally, second order statistics describe the variations of the intensity between a point and a neighboring point, enabling the estimation of the averaged speckle size as well as its distribution across space by means of a correlation function [1, 15,17,18]. As a result, (5.5) is obtained by the use of second order statistics.

Readers will see in the following paragraphs that speckle aliasing or undersampling makes measured speckle phase maps worse, thus decreasing the obtained signal-to-noise ratio. In order to avoid these effects, speckle size should be computed and the pixel size should be selected to be completely filled by a speckle grain. In other words, the speckle sizes should be the same as the pixel size. The following exercises show practical situations where a camera sensor is available for speckle interferometry, and a user should see if it is suitable for this application or select an adequate camera sensor.

EXERCISE 5.1

A CCD camera with a pixel size of 4.4 μm is used as a sensor device to obtain images of an illuminated surface. This camera is combined with a commercial imaging system with a focal length of 16 mm. The camera is 1.3 megapixels (1280 × 960 pixels) and the object area to be imaged is 25 mm in diameter. Is this sensor equipped to resolve a speckle grain generated with red laser light ($\lambda = 638.2$ nm)?

Solution

Since the CCD sensor has a pixel size of 4.4 mm, readers can see that the image size is 5.632 × 4.224 mm (1280 × 4.4 μm = 5.632 mm and 960 × 4.4 μm = 4.224 mm). As the object has a size of 25 mm, the small dimension of the CCD will be considered in order to image the whole object. Thus, the magnification will be 4.224 mm/25 mm = 0.1689. By using (5.9) and by considering a usual numerical aperture of 8, the speckle size will be

$$d_{sp} = 1.22 \times 0.6328 \, \mu m \times 8 \times (1 + 0.1689) = 7.22 \, \mu m \qquad (5.11)$$

Thus, readers can see that the speckle grain will be undersampled since it is larger than the pixel size. As a consequence, a different camera should be used with a larger pixel size in case the magnification is fixed.

EXERCISE 5.2

An object is illuminated with a laser source with a wavelength of 532 nm. Determine the distance between the object and the CCD of the camera if the following conditions are accomplished: (a) no lenses are used to image the object, (b) the illuminated area is 10 mm, (c) speckle will fit only one pixel, and (d) the pixel size is 3.8 μm.

Solution

The speckle obtained in this case is objective speckle since speckle is projected onto the sensor without lens. Additionally, the speckle should fit one pixel, having the following diameter $d_{sp} = 3.8 \, \mu m$. Thus, the wavelength of the light source will be

$$z = \frac{d_{sp} l_0}{\lambda} = \frac{3.8 \, \mu m \cdot 10 \, mm}{0.532 \, \mu m} = 71.42 \, mm \qquad (5.12)$$

EXERCISE 5.3

What would the speckle size be for the following conditions: (a) the object distance is 100 mm, (b) the distance between the aperture and the image plane is 50 mm, (c) the numerical aperture is 8, and (d) the laser light is green (532 nm). Define the pixel size in order to solve at least one speckle by pixel.

Solution

The magnification of the optical system will be $M_g = z_i/z_o = 50 \, mm/100 \, mm = 0.5$. Thus, the speckle size will be

$$d_{sp} = 1.22 \lambda F_b \left(1 + M_g\right) = 1.22 \cdot 0.532 \, \mu m \cdot 8 \cdot (1 + 0.5) = 7.788 \, \mu m \qquad (5.13)$$

The selected camera will have a speckle size of about 7.8 μm.

5.3   Speckle Interferometry

As it was indicated in the previous paragraphs, speckle distributions are deterministically modified by: (a) displacements produced on the scattering surface, (b) changes in the illumination and observation geometry, or (c) variations in the wavelength of the light source or in the refractive index of the medium where the light travels [1].

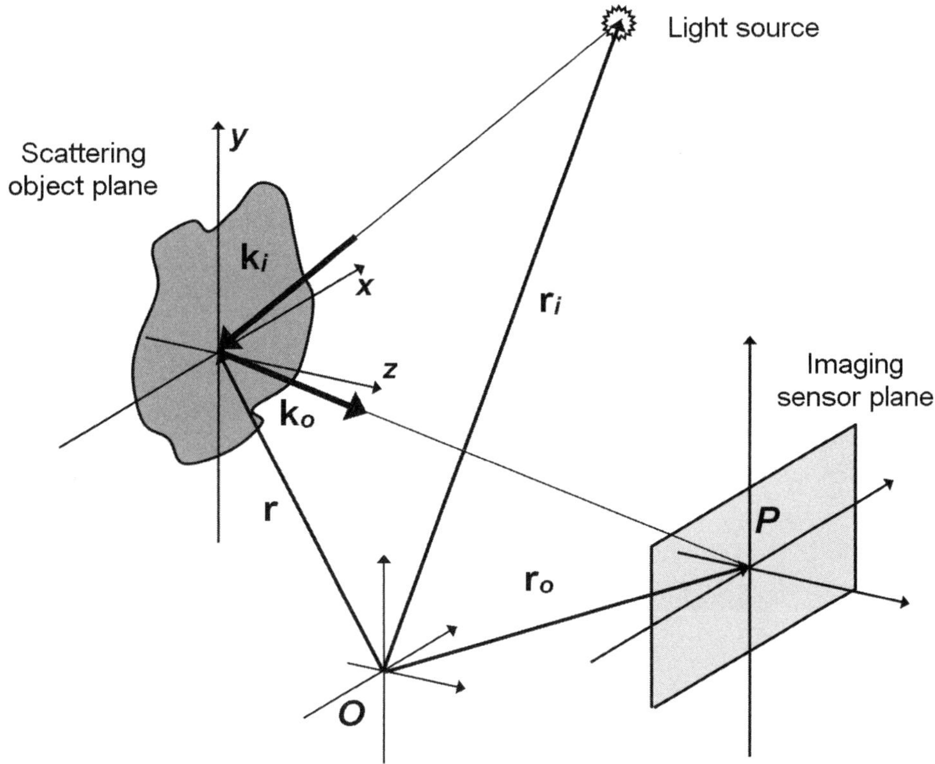

Figure 5.5 Geometry for the sensitivity of the phase of the speckle.

Applying the approximation of paraxial rays shows that the area on the diffuser surface (which contributes to the speckle distribution on the observation plane) forms a small angle with the observation point or with the axis of the optical system. In this case, the optical phase of the speckle distribution in each point P of the observation plane can be expressed as a function of the optical path traveled by the light from the light source in the following way (see Figure 5.5)

$$\varphi = \varphi_s + \psi = \varphi_s + \psi_i + \mathbf{k}_i \cdot (\mathbf{r} - \mathbf{r}_i) + \mathbf{k}_o \cdot (\mathbf{r}_o - \mathbf{r}) \tag{5.14}$$

The speckle's phase φ is composed of a random component φ_s and a deterministic one ψ. The magnitude of φ_s for each point of the speckle distribution depends on the roughness of the scattering surface along the area that contributes to the speckle in that point.

BOX 5.4

The deterministic part is made up of (a) the initial optical phase ψ_i (at the output of the light source), (b) changes in the phase produced along the path between the light source and the scattering surface ($\mathbf{k}_i \cdot (\mathbf{r} - \mathbf{r}_i)$), and (c) changes in the phase occurred along the path between the diffuser and the observation point

$(\mathbf{k}_o \cdot (\mathbf{r}_o - \mathbf{r}))$. For simplicity, the refractive index was omitted in (5.14), but it is important to highlight that variations in this index will also have an influence on the phase changes. The vector $\mathbf{r}$ is the position vector which is constant for all the points on the diffuser due to the paraxial assumption. Additionally, $\mathbf{r}_i$ indicates the curvature center of the incident wavefront, $\mathbf{r}_o$ is the position vector of the observation point, and $\mathbf{k}_i$ and $\mathbf{k}_o$ are the wave propagation vectors corresponding to the illumination and observation directions, respectively.

$$\mathbf{k}_i = \frac{2\pi}{\lambda}\hat{\mathbf{n}}_i, \quad \mathbf{k}_o = \frac{2\pi}{\lambda}\hat{\mathbf{n}}_o \tag{5.15}$$

where $\hat{\mathbf{n}}_o$ and $\hat{\mathbf{n}}_i$ are the unitary vectors.

The sensitivity vector is defined as (see Figure 5.6)

$$\mathbf{k} = \mathbf{k}_i - \mathbf{k}_o \tag{5.16}$$

This vector indicates the direction in which the diffuser produces the maximum variation of phase at the point P on the observation plane.

Equation (5.14) can be rewritten in the following way:

$$\varphi = \varphi_s + \psi_i + \mathbf{k}_o \cdot \mathbf{r}_o - \mathbf{k}_i \cdot \mathbf{r}_i + \left(\mathbf{k}_i - \mathbf{k}_o\right) \cdot \mathbf{r} \tag{5.17}$$

$$\varphi = \varphi_s + \psi' + \mathbf{k} \cdot \mathbf{r} \tag{5.18}$$

with

$$\psi' = \psi_i + \mathbf{k}_o \cdot \mathbf{r}_o - \mathbf{k}_i \cdot \mathbf{r}_i \tag{5.19}$$

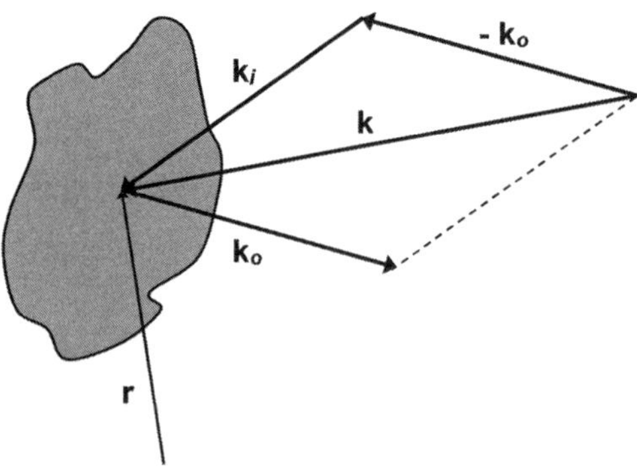

Figure 5.6   Determining the sensitivity vector.

When the diffuser is displaced or the illumination and observation directions are changed, the phase of the speckle distribution experiences a variation which can be expressed as

$$\Delta\varphi = \Delta\varphi_s + \Delta\psi' + \Delta(\mathbf{k} \cdot \mathbf{r}) \tag{5.20}$$

where the last term can be written as:

$$\Delta(\mathbf{k} \cdot \mathbf{r}) = \left[(\mathbf{k} + \Delta\mathbf{k}) \cdot (\mathbf{r} + \Delta\mathbf{r}) \right] - (\mathbf{k} \cdot \mathbf{r}) = \Delta\mathbf{k} \cdot \mathbf{r} + \mathbf{k} \cdot \Delta\mathbf{r} + \Delta\mathbf{k} \cdot \Delta\mathbf{r} \tag{5.21}$$

In this way

$$\Delta\varphi = \Delta\varphi_s + \Delta\psi' + \Delta\mathbf{k} \cdot \mathbf{r} + \mathbf{k} \cdot \Delta\mathbf{r} + \Delta\mathbf{k} \cdot \Delta\mathbf{r} \tag{5.22}$$

with

$$\Delta\psi' = \Delta\psi_i + \Delta(\mathbf{k}_o \cdot \mathbf{r}_o) - \Delta(\mathbf{k}_i \cdot \mathbf{r}_i) \tag{5.23}$$

The term $\mathbf{k} \cdot \Delta\mathbf{r}$ represents the increment in the optical phase generated by movement of the diffuser. On the other hand, $\Delta\psi' + \Delta\mathbf{k} \cdot \mathbf{r}$ is produced by changes in the geometry of the illumination and observation directions, in the wavelength, in the refractive index, and in the initial phase of the source. Finally, $\Delta\mathbf{k} \cdot \Delta\mathbf{r}$ denotes the combining effect and the variation of both parameters. If the microstructure related to the roughness of the scattering surface is kept relatively invariable, the random component φ_s will be constant. Consequently, its variation will be zero ($\Delta\varphi_s = 0$). Additionally, all changes in the phase distribution will be only produced by the deterministic component. Moreover, for most cases the diffuser moves without changes in the sensitivity vector. Thus, changes in $\Delta\mathbf{k} \cdot \Delta\mathbf{r}$ are negligible and have the following simplification:

$$\Delta\varphi \approx \Delta\psi = \Delta\psi' + \Delta\mathbf{k} \cdot \mathbf{r} + \mathbf{k} \cdot \Delta\mathbf{r} \tag{5.24}$$

Expression (5.24) can be used to describe all geometries and working conditions, and it enables us to model all interferometric techniques based on speckle distributions. For example, the experiment is designed such as to cancel all terms of (5.24), with exception of the term with $\Delta\mathbf{r}$, in order to only measure displacement fields. This is achieved when the illumination and observation directions are kept constant, as well as the refractive index. As a consequence, changes in the phase of the speckle will only reflect the diffuser displacements along the direction of the sensitivity vector.

$$\Delta\varphi \approx \Delta\psi = \mathbf{k} \cdot \Delta\mathbf{r} \tag{5.25}$$

The intensity distribution of the speckle does not change when small displacements are introduced in the diffuser. However, the phase distribution of the speckle experiences changes, which can be expressed by interfering the original speckle distribution with a reference beam coming from the same light source and with the same polarization direction. The reference beam can be a uniform or speckled beam.

Reference [12] shows that the intensity of the resultant beam after interfering two beams with intensities I_1 and I_2 can be written as

$$I = I_1 + I_2 + 2\sqrt{I_1 I_2}\,\cos\left(\varphi_1 - \varphi_2\right) \tag{5.26}$$

where φ_1 and φ_2 are the phases of the interfering wavefronts. Readers should note that I, I_1, I_2, φ_1, and φ_2 are functions of the position $\mathbf{r}(x, y, z)$ and the time t. Equation (5.26) can be expressed as [19]

$$I = I_0(1 + V\cos\phi) \tag{5.27}$$

where the averaged intensity I_0 (also called the background intensity) equals

$$I_0 = I_1 + I_2 \tag{5.28}$$

and the fringe visibility or contrast V is written as

$$V = \frac{2\sqrt{I_1 I_2}}{I_1 + I_2} \tag{5.29}$$

The relative phase between the interfering beams is computed as

$$\phi = \varphi_1 - \varphi_2 \tag{5.30}$$

Previous equations are valid for classic interferometry as well as for interferometry with speckle distributions. For the last case, (5.27) is a random distribution sensitive to changes in the optical phase, which is called a speckled interferogram or specklegram.

When the interfering optical fields are speckle distributions, the statistical properties of the interferogram are similar to the original distributions. A scaling factor should be included due to the increase of intensity generated by the superposition of two wavefronts [1, 20, 21]. On the other hand, when the speckle distribution interferes with a uniform optical field, the statistical properties are different from the original distribution. In particular, the contrast decreases and the average speckle size increases [1].

The phase in (5.27) carries information relative to the configuration of the interferometer. Regardless of the displacement of the diffuser, variations in the sensitivity vectors of each beam, changes in the wavelength of the light, or variation in the refractive index will produce changes in the relative phase ϕ.

In order to generalize techniques that use two speckle distributions, or one speckle distribution and a uniform wavefront as reference beam, the relative phase of the speckle can be expressed by using (5.18) and (5.30)

$$\phi = \varphi_s + \psi_r + \psi_o \tag{5.31}$$

where

$$\varphi_s = \left(\varphi_{s1} - \varphi_{s2}\right) \tag{5.32}$$

$$\psi_r = \left(\psi'_1 - \psi'_2\right) \tag{5.33}$$

$$\psi_o = \left(\mathbf{k}_1 \cdot \mathbf{r}_1 - \mathbf{k}_2 \cdot \mathbf{r}_2\right) \tag{5.34}$$

The phase labeled as φ_s belongs to the random components of the interfering phases and contributes to the speckle noise. For surfaces whose roughness exceeds the wavelength of the light, φ_s has a spatial distribution in the interval $(\pi, -\pi)$ [1]. This noise confers the granular aspect to the interferogram. The phase ψ_r is introduced deliberately by changing the initial phase of both beams, or by changing the beams' traveling distances or the direction of the illumination and observation paths (see (5.19)). The object phase ψ_o depends on the sensitivity vectors of the interfering beams and on the instantaneous configuration of the scattering surface (position in space). ψ_o is the term with the useful information in order to compute the deformation of the diffuser.

Usually, terms in (5.31) are functions of the coordinates $\mathbf{x}(x, y)$ in the observation plane and of the time t. However, experimental interferometers are designed in order to consider two simplifying hypotheses. The first one considers that the speckle distribution is neither decorrelated during the acquisition nor considerably displaced along a transverse direction during the measurement. Thus, φ_s, I_0, and V can be considered time independent. The second assumption takes into account that the illumination and observation directions are the same for the whole interferogram. In this way, $\mathbf{k}_1$, $\mathbf{k}_2$, and ψ_r are only time dependent parameters. In other words, beams incident on the diffuser, as well as the observation beams, are collimated [16]. In practice, collimated beams imply the use of big optical elements, which are typically very expensive. For this reason, diverging beams are often used with expander lenses placed far away from the object, having curvature radii as large as possible. By taking into account these considerations

$$\phi(\mathbf{x},t) = \varphi_s(\mathbf{x}) + \psi_r(t) + \psi_o(\mathbf{x},t) \tag{5.35}$$

with

$$\psi_o(\mathbf{x},t) = \mathbf{k}_1(t) \cdot \mathbf{r}_1(\mathbf{x},t) - \mathbf{k}_2(t) \cdot \mathbf{r}_2(\mathbf{x},t) \tag{5.36}$$

In this way, the intensity of the speckle interferogram corresponding to a specific state of the object can be written, based on (5.27), as follows:

$$I(\mathbf{x},t) = I_0(\mathbf{x})\left\{1 + V(\mathbf{x})\cos\left[\varphi_s(\mathbf{x}) + \psi_r(t) + \psi_o(\mathbf{x},t)\right]\right\} \tag{5.37}$$

For speckle interferometry, as well as for any other interferometric techniques, variations in the phase difference between the beams allow the measurement of displacements taking place in the diffuser. In this case, the change in the phase difference is

$$\Delta\phi = \Delta\varphi_s + \Delta\psi_r + \Delta\psi_o \tag{5.38}$$

If the speckle is not decorrelated between both acquisitions ($\Delta\varphi_s = 0$), and the deliberately introduced phase difference of the reference $\Delta\psi_r$ is known, the variation of the intensity in (5.37) will reveal only changes in the object phase $\Delta\psi_o$. From (5.24), it can be seen that changes in phase are only related to the displacements $\Delta\mathbf{r}$ of the scattering surface, if the sensitivity vectors are kept constant.

In order to separate the object phase ψ_o from the phase difference ϕ, two different states of the object need to be compared. Readers should note that a displacement field was introduced in the diffuser between both states without speckle decorrelation. As a consequence, $\Delta\varphi_s$ becomes equal to zero.

Readers will be guided through the following paragraphs in order to see the different methods used to encode the variation of the optical phase difference $\Delta\psi_o$ between two object stages. Equation (5.37) describes the intensity distribution of a speckle distribution for a particular state of the observed object. Consider

$$I_1(\mathbf{x},t_1) = I_0\left\{1 + V(\mathbf{x})\cos\left[\phi(\mathbf{x},t_1)\right]\right\} \tag{5.39}$$

and

$$I_2(\mathbf{x},t_2) = I_0\left\{1 + V(\mathbf{x})\cos\left[\phi(\mathbf{x},t_2)\right]\right\} \tag{5.40}$$

as the intensity distributions corresponding to the observed states before and after the deformation of the object, respectively, where

$$\phi(\mathbf{x},t_1) = \varphi_1 - \varphi_2 \quad \phi(\mathbf{x},t_2) = \varphi_1 + \Delta\varphi - \varphi_2 \tag{5.41}$$

and $\Delta\varphi$ is the variation of the phase of the speckle in the object beam produced by the displacements of the diffuser.

If $I_{12} = [I_1 - I_2]^2$ is evaluated, taking into account the following trigonometric identities

$$\cos(a) - \cos(b) = 2\sin\left(\frac{a+b}{2}\right)\sin\left(\frac{b-a}{2}\right) \tag{5.42}$$

$$\sin^2(a) = \frac{1}{2} - \frac{1}{2}\cos(2a) \tag{5.43}$$

I_{12} can thus be written as

$$I_{12} = I_{0f}\left[1 + V_f\cos(\Delta\phi)\right] \tag{5.44}$$

with

$$\Delta\phi = \phi(\mathbf{x},t_1) - \phi(\mathbf{x},t_2) \tag{5.45}$$

$$I_{0f} = I_0^2 V^2 \left\{1 - \cos\left[\phi(\mathbf{x},t_1) + \phi(\mathbf{x},t_2)\right]\right\}, \quad V_f = -1 \qquad (5.46)$$

where $\Delta\phi$ is the phase difference produced by the phase change between the speckle and object beam. By considering (5.29), (5.44) can be expressed as

$$I_{12} = I_{0f}\left[1 + V_f \cos\left(\Delta\psi_o - \Delta\psi_r\right)\right] \qquad (5.47)$$

where the random phase φ_s presented in the I_{0f} term causes the granular aspect of I_{12} and has components with high spatial frequencies. The cosine term produces low frequency modulation fringes and encodes phase differences between t_1 and t_2. If the phase of the reference beam is known, this term will only depend on changes of the object beam generated by the deformation. Thus, the subtraction of the specklegrams enables us to encode variations of intensities in the object phase. This modulation in the intensity is called correlation fringes. It is important to highlight the correspondence between (5.27) and (5.44), which respectively describe the intensity distribution of a specklegram and of correlation fringes. In the first case, the visibility and intensity are related to the speckle, whereas in the second case they are related to the correlation fringes.

BOX 5.5

Readers should see that correlation fringes can be interpreted as links between points with equal displacement along the sensitivity direction. Those points where $\Delta\psi_o = 2n\pi$ (with $n = 0, 1, 2, \ldots$) have identical speckle distributions, and the intensity I_{12} will be equal to zero, resulting in dark fringes. On the other hand, the points where $\Delta\psi_o = (2n = 1)\pi$ have bright fringes. In practice, it is easier to compute $|I_1 - I_2|$ than $[I_1 - I_2]^2$, and also a distribution is obtained with a slightly different form than represented by (5.47), however principally containing high order harmonics. Since the evaluation of the phase based on the correlation fringes involves a noise reduction step that eliminates these terms, the phase $\Delta\psi_o$ is equivalent.

To conclude this section, it should be mentioned that the direct visualization of correlation fringes is enough for the qualitative analysis of several problems. However, it is often desirable to obtain a quantitative value of $\Delta\phi$. This condition will be shown in the next section.

5.4   Phase Recovery for Speckle Interferometry

Two components can be separated in a system that uses DSPI: (a) the optical part composed by the interferometer, and (b) the electronic part that acquires the interferogram and generates the correlation fringes. In order to quantitatively evaluate

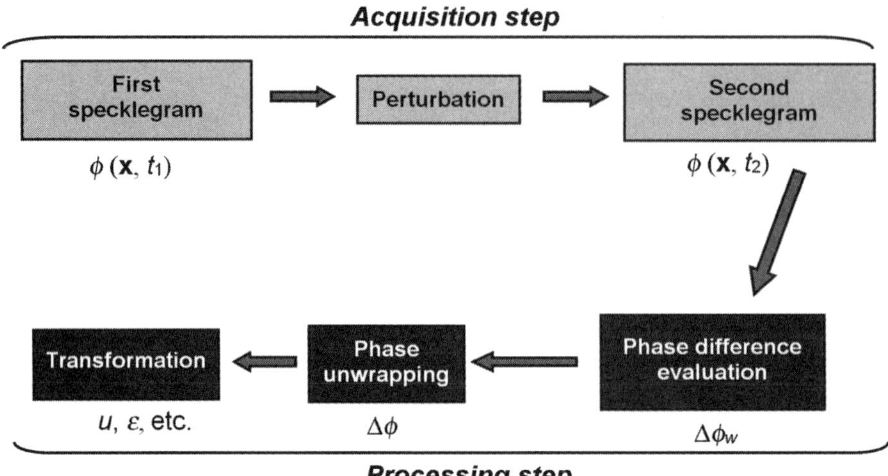

Figure 5.7   Steps required for the analysis of speckle interferograms.

the phase distribution, these fringes are numerically processed as presented in Figure 5.7. This figure shows the two steps involved in the generation and analysis of an interferogram. The first stage is called the acquisition step. Interferograms are acquired before and after the introduction of a perturbation to the object. As a consequence, two speckle distributions $I_1(\mathbf{x}, t_1)$ and $I_2(\mathbf{x}, t_2)$ are available that need to be processed in order to obtain the phase difference generated by the perturbation.

The next stage is called the processing step, which enables the analysis and processing of acquired interferograms. This stage can be split into two parts: (a) the evaluation of the phase difference $\Delta\phi$, and (b) the conversion of phase values in physical parameters that describe the perturbation (e.g., stresses, strains, displacements).

In the previous section, it was shown that correlation fringes have the following intensity distribution:

$$I_{12}(\mathbf{x},t) = I_{0f}(\mathbf{x},t)\left\{1 + V_f(\mathbf{x},t)\cos\left[\Delta\phi(\mathbf{x},t)\right]\right\}$$
$$= I_{0f}(\mathbf{x},t) + I_M(\mathbf{x},t)\cos\left[\Delta\phi(\mathbf{x},t)\right] \qquad (5.48)$$

where I_{0f} is the averaged intensity, V_f is fringe visibility, I_M is the modulation intensity, and $\Delta\phi$ is the variation in phase difference. In order to extract the value of $\Delta\phi$ from (5.48), its inversion is necessary. Since the cosine function is even with a period of 2π

$$\cos\Delta\phi = \cos(sig\Delta\phi + 2\pi q) \quad sig \in \{-1,1\} \quad q \in \mathbb{Z} \qquad (5.49)$$

the phase computed from only one I_{12} distribution will result undetermined in sign (sig) with an additive integer multiple q of 2π.

Nowadays, interferograms are acquired by using digital cameras. These cameras usually give eight-bit images (256 gray levels, where 0 is black and 255 is white). There are also cameras with 10 and 12 bits that allow us to work with low modulation signals. The digitization of N_t with a sensor of $N_n \times N_m$ pixels gives a discrete three-dimensional intensity distribution $I(n, m, t)$, where $n = 0, 1, 2, \ldots, N_n - 1$, $m = 0, 1, 2, \ldots, N_m - 1$, and $t = 0, 1, 2, \ldots, N_t - 1$. Thus, (5.48) can be written as

$$I(n,m,t) = I_0(n,m,t)\big[1 + V(n,m,t)\cos\phi(n,m,t)\big] \tag{5.50}$$

The nondimensional coordinates of the discrete image (n, m, t) are related to the dimensional and continuous coordinates $(\mathbf{x}, t)$ of the image plane, with $\mathbf{x} = x,y$ by using the relations $x = ndx \quad y = mdy \quad t = tdt$, where dx and dy are the distances between adjacent pixels in the x and y directions, and dt is the time interval between successive registrations.

5.4.1   Phase Evaluation by Fourier Transform

The Fourier transform evaluation method can be classified as a spatial method, since required data for phase computation are simultaneously acquired in only one interferogram [16, 22, 23]. In this method, a linear combination of harmonic functions is fitted to the measured intensity distribution $I_{12}(\mathbf{x}, t)$ corresponding to the correlation fringes. The spatial frequencies of the combined harmonic functions are defined in the Fourier space by through selection of a bandpass filter.

If a linear carrier function with spatial frequency $\mathbf{f}_c(f_{cn}, f_{cm})$ is introduced, (5.50) will have the following expression

$$I = I_0\left\{1 + V\cos\left[\phi + 2\pi\left(f_{c_n} n + f_{c_m} m\right)\right]\right\} \tag{5.51}$$

BOX 5.6

This method can be classified as a spatial phase shifting technique since continuous changes in the phase between adjacent pixels are introduced. In practice, the spatial carrier is introduced by tilting the reference or illumination beam between exposures. However, this method is equally valid for signals with temporal carriers [24], and several applications have been reported [25–28].

By defining the complex visibility V_c as

$$V_c(n,m) = \frac{1}{2}V(n,m)\exp\big[i\phi(n,m)\big] \tag{5.52}$$

Equation (5.51) can be reformulated in the following way

$$I = I_0\left\{1 + V_c\exp\left[2\pi\left(f_{c_n} n + f_{c_m} m\right)\right] + V_c^*\exp\left[2\pi\left(f_{c_n} n + f_{c_m} m\right)\right]\right\} \tag{5.53}$$

where $*$ denotes the complex conjugation.

By applying the discrete two-dimensional Fourier transform via a FFT algorithm [29], (5.53) yields

$$\tilde{I} = \tilde{I}_0 \otimes \left[\delta\left(f_n, f_m\right) + \tilde{V}_c\left(f_n - f_{c_n}, f_m - f_{c_m}\right) + \tilde{V}_c^*\left(f_n + f_{c_n}, f_m + f_{c_m}\right) \right]$$
$$= \tilde{I}_0 \otimes \delta\left(f_n, f_m\right) + \tilde{I}_0 \otimes \tilde{V}_c\left(f_n - f_{c_n}, f_m - f_{c_m}\right) + \tilde{I}_0 \otimes \tilde{V}_c^*\left(f_n + f_{c_n}, f_m + f_{c_m}\right)$$

$$(5.54)$$

where $\otimes$ represents the convolution, $\sim$ denotes the Fourier transform, δ is the Dirac delta function, and (f_n, f_m) is the spatial frequency. The explicit (f_n, f_m) dependence on the quantities $\tilde{I}$, $\tilde{I}_0$, and δ were omitted for clarity.

If the carrier frequency $\mathbf{f}_c$ is higher than the spatial variation of I_0, V, and ϕ, the three components of the spatial spectrum are well separated. Thus, the term $\tilde{I}_0 \otimes \tilde{V}_c$ can be separated by a bandpass filter. The resultant spectrum $\tilde{I}'$ will be

$$\tilde{I}' = \tilde{I}_0 \otimes \tilde{V}_c\left(f_n - f_{c_n}, f_m - f_{c_m}\right)$$

$$(5.55)$$

By applying the inverse Fourier transform

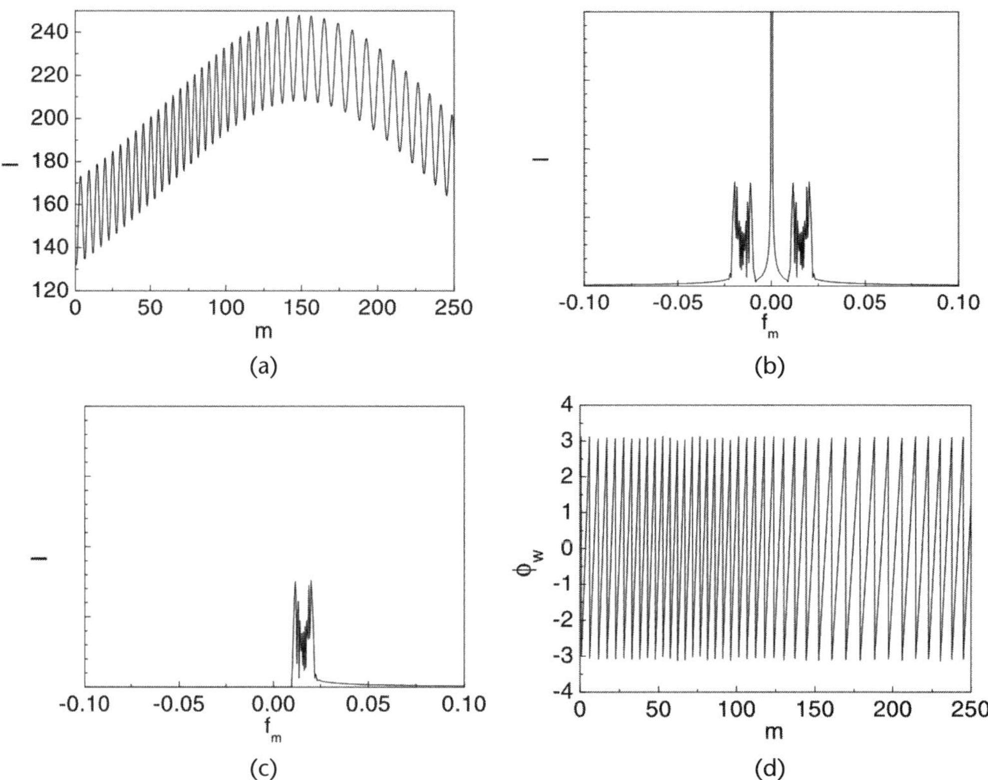

Figure 5.8 Phase evaluation using a one-dimensional Fourier transform: (a) intensity profile generated by Gaussian deformation combined with a linear spatial carrier; (b) signal spectrum computed from (a); (c) spectrum shown in (b) after a bandpass filter application; and (d) resultant phase (2π module). (Courtesy of G. E. Galizzi.)

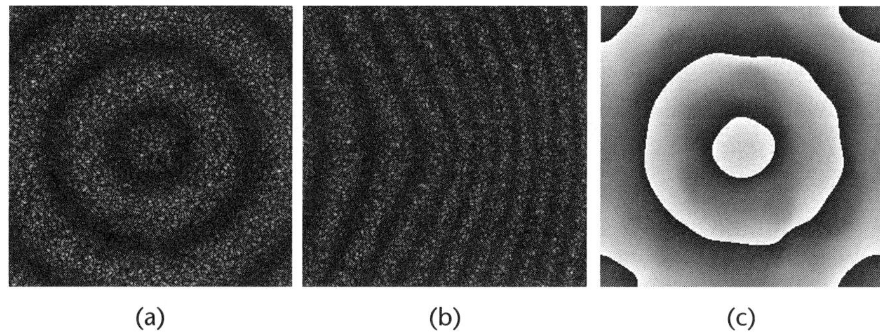

Figure 5.9 Phase evaluation with Fourier transform: (a) correlation fringes generated by a deformation, (b) the same fringes combined with a linear carrier, and (c) computed phase after removal of the carrier frequency. (Courtesy of G. E. Galizzi.)

$$
\begin{aligned}
I'(n,m) &= I_0(n,m)V_c(n,m)\exp\left[2\pi i\left(f_{c_n}n + f_{c_m}m\right)\right] \\
&= \frac{1}{2}I_0(n,m)V(n,m)\exp\left[i\phi(n,m) + 2\pi\left(f_{c_n}n + f_{c_m}m\right)\right]
\end{aligned}
\tag{5.56}
$$

The phase of $I'(n, m)$ is

$$
\phi(n,m) + 2\pi\left(f_{c_n}n + f_{c_m}m\right) = \tan^{-1}\left[\frac{\mathrm{Im}\{I'(n,m)\}}{\mathrm{Re}\{I'(n,m)\}}\right]
\tag{5.57}
$$

with values between $-\pi$ and $+\pi$. Figure 5.8 shows a one-dimensional example of the Fourier transform evaluation.

The linear phase ramp in (5.57) contributed by the carrier frequency can be removed in the spatial domain by using an estimated or known value of f_c [30]. An equivalent procedure can be performed in the frequency domain by shifting the filtered transformed peak I' to the origin before the computation of the inverse transform [22].

As an example, Figure 5.9(a) shows a set of circular correlation fringes, Figure 5.9(b) shows the combination of these fringes with a linear carrier, and Figure 5.9(c) shows the phase distribution after the application of the previous equations and subtraction of the carrier.

BOX 5.7

The Fourier method is sensitive to several error sources. Among them, the following can be highlighted:

- Aliasing generated by lower values of the sampling frequency;
- Leakage generated by physical edges of the object, which leads to discontinuous fringes;
- Insufficient filtering;
- Speckle noise.

5.4.2   Phase Shifting as a Quantitative Tool for Phase Evaluation

Readers can see that if the terms $I(n, m, t)$, $V(n, m, t)$, and $\phi(n, m, t)$ have fast variations in n and m, and slow changes along the time t, temporal phase shifting will be able to be used, usually being introduced in the object beam.

BOX 5.8

This condition implies that fractional changes $\delta I_o/I_o$, $\delta V/V$, and $\delta\phi/2\pi$ are less than unity during the time interval between successive interferograms.

This way, a finite number of interferograms N_t are acquired with a relative phase shift α_t. Thus, (5.50) can be rewritten as follows

$$I(n,m,t) = I_0(n,m)\left\{1 + V(n,m)\cos\left[\phi(n,m) + \alpha_t\right]\right\} \tag{5.58}$$

This method involves an independent data analysis for each pixel. For this reason, the dependence on pixel coordinates (n, m) will be omitted from the following equations.

Currently a wide variety of algorithms are available for phase computation, using three or more phase shifted images. For all cases, the phase value in a well-determined point is obtained by using the intensity at that point of the N_t interferograms. As a consequence, these methods are also referred to as local methods.

The more common types of these algorithms use known values of α_t that are equal for all points in each interferogram [31, 32]. In accordance with (5.58), there are three unknowns (I_0, V, ϕ), and at least three measurements are required in order to determine the phase values. One of the simplest cases, called the four-frame algorithm, is obtained by considering the case of four images shifted by $\alpha_t = \pi/2$. By substituting $t = 0, 1, 2$, and 3 into (5.58), the following set of equations are obtained [19, 33]:

$$I(0) = I_0 + I_M\cos\phi$$
$$I(1) = I_0 + I_M\cos(\phi + \pi/2) = I_0 - I_M\sin\phi$$
$$I(2) = I_0 + I_M\cos(\phi + \pi) = I_0 - I_M\cos\phi$$
$$I(3) = I_0 + I_M\cos(\phi + 3\pi/2) = I_0 + I_M\sin\phi \tag{5.59}$$

By rearranging this set of equations, the values of I_M and ϕ_w can be obtained as follows:

$$\phi_w = \tan^{-1}\left[\frac{I(3) - I(1)}{I(0) - I(2)}\right], I_M = \frac{\sqrt{\left[I(3) - I(1)\right]^2 + \left[I(0) - I(2)\right]^2}}{2} \tag{5.60}$$

The five-frame algorithm was proposed by Schwider et al. [34] as an improvement of the four-frame algorithm, being less susceptible to calibration errors. In this case, five interferograms with $\alpha_t = \pi/2$ are acquired, and the phase and modulation can be computed as follows:

$$\phi_w = \tan^{-1}\left\{\frac{2[I(3) - I(1)]}{I(4) + I(0) - 2I(2)}\right\}, I_M = \frac{\sqrt{4[I(3) - I(1)]^2 + [I(4) + I(0) - 2I(2)]^2}}{4}$$

$$(5.61)$$

An alternative method, in order to avoid calibration errors of the phase shifter, was introduced by Carré [35]. This method considers α_t as a further unknown. If the phase shifts between successive images are all equal, (5.58) will have four unknowns. If the phase shifts are selected as $\alpha_t = \alpha(2t - 3)/2$ $t = 0, 1, 2, 3$, the phase difference can be computed as [19]

$$\phi_w + \frac{3\alpha}{2} = \tan^{-1}\left[\frac{s\sqrt{[I(0) - I(3) + I(1) - I(2)]\{3[I(1) - I(2)] - I(0) + I(3)\}}}{I(1) + I(2) - I(0) - I(3)}\right] \quad (5.62)$$

where $s = \text{sign}[I(1) - I(2)]$. The *sign* function introduces the correct sign of the phase values for the four quadrants. The term $3\alpha/2$ disappears after calculation of the phase difference values, and it is produced by the choice of the carrier origin.

The wrapped phase, computed by (5.60), (5.61), and (5.62), has the initial random phase $\phi(n,m,0) = \varphi_s(n,m,0) - \psi_r(n,m,0) + \psi_o(n,m,0)$. The difference between the phase distributions, registered at t_1 and t_2, allows one to eliminate φ_s and Ψ_r. This condition is true if there is neither speckle decorrelation $\left[\varphi_s(n,m,t_1) = \varphi_s(n,m,t_2)\right]$, nor changes in the setup of the interferometer $\left[\psi_r(n,m,t_1) = \psi_r(n,m,t_2)\right]$. Thus, the phase distribution generated by the deformation of the diffuser can be expressed as

$$\Delta\phi_{ww}(n,m,t_1,t_2) = \phi_w(n,m,t_2) - \phi_w(n,m,t_1) \qquad (5.63)$$

Since ϕ_w values are between $-\pi$ and π, the phase $\Delta\phi_{ww}$ lies in the interval $[-2\pi, 2\pi]$. In order to correctly display the phase, it is wrapped back within the range $[-\pi, \pi]$ by using the wrapping operator $\Upsilon(\phi) = \phi - 2\pi\text{NINT}(\phi/2\pi)$, where NINT denotes rounding off toward the nearest integer. As consequence, the phase difference in the 2π module results

$$\Delta\phi_w(n,m,t_1,t_2) = \Upsilon\left[\Delta\phi_{ww}(n,m,t_1,t_2)\right] \qquad (5.64)$$

BOX 5.9

This method is called the difference of phases. There is another alternative method called the phase of differences, and it is based on the formation of correlation fringes. In this method one image is acquired before the deformation, and N_t

images are acquired afterward (or vice versa). The initial image is subtracted from the N_t images, and a set of N_t phase shifted correlation fringes is obtained. After this, they are filtered and can be analyzed by a standard phase shifting algorithm. This technique offers advantages for dynamic applications, since the unique image can be acquired during the transient event and the set of N_t images can be acquired with the object at rest. The main disadvantage of this method is the reduction in quality of the computed phase maps [19, 36].

EXERCISE 5.4

For the five-frame algorithm with $\alpha_t = \pi/2$, the phase is computed as follows

$$\phi_w = \tan^{-1}\left\{\frac{2[I(3) - I(1)]}{I(4) + I(0) - 2I(2)}\right\}$$ when the starting value is $\alpha_t = -\pi$. If the starting point was $\alpha_t = 0$, would the phase equation have the same expression than (6.59) or would

it have the following expression: $\phi_w = \tan^{-1}\left\{\frac{7[I(3) - I(1)]}{4I(0) - I(1) - 6I(2) - I(3) + 4I(4)}\right\}$?
Justify your answer.

Solution

A equation system similar to shown in (5.59) should be made for each case. For the case where $\alpha_t = -\pi$, the equation system should be

$$I(0) = I_0 + I_M \cos(\phi - \pi)$$
$$I(1) = I_0 + I_M \cos(\phi - \pi/2)$$
$$I(2) = I_0 + I_M \cos(\phi + 0)$$
$$I(3) = I_0 + I_M \cos(\phi + \pi)$$
$$I(4) = I_0 + I_M \cos(\phi + \pi/2) \tag{5.65}$$

For the other case where $\alpha_t = 0$, the equation system will be

$$I(0) = I_0 + I_M \cos(\phi + 0)$$
$$I(1) = I_0 + I_M \cos(\phi + \pi/2)$$
$$I(2) = I_0 + I_M \cos(\phi + \pi)$$
$$I(3) = I_0 + I_M \cos(\phi + 3\pi/2)$$
$$I(4) = I_0 + I_M \cos(\phi + 2\pi) \tag{5.66}$$

After solving, readers will see that both expressions are different and they are dependent on the starting point.

5.5   Phase Unwrapping Processing

The arc-tangent function used in (5.57), (5.60), (5.61), and (5.62) is multivalued and produces phase values between $-\pi$ and $+\pi$. For this reason, an additional computation is necessary to add the correct integral multiple of 2π to each pixel value, in order to solve jumps that have a continuous phase distribution. This extra procedure is shown in Figure 5.10 and is known as phase unwrapping. Reference [37] can be considered the most important book about phase unwrapping and is an excellent reference work for students and professors.

In order to unwrap a given phase distribution correctly, the original phase maps must have been correctly sampled in accordance with the Shannon sampling theorem [38]. If this requirement is satisfied for successive samples along the time axis, the phase at each pixel can be unwrapped as a function of time by adding multiples of 2π, as shown in Figure 5.10. This procedure is known as temporal phase unwrapping [39–42]. It is clear that for this case, the sampling theorem will also be satisfied for adjacent pixels in each image, allowing spatial phase unwrapping. For a phase difference $\Delta\phi_w(n,m,t_1,t_2)$, the sampling theorem will be violated for the time axis by changes occurring at times t_1 and t_2. However, this phase difference will satisfy the theorem for the spatial case, therefore allowing spatial phase unwrapping.

The simplest phase unwrapping method follows the principle shown in Figure 5.10, which integrates the phase by adding multiples of 2π, following a path that scans the whole image [37]. For a noise-free and correctly sampled phase, the unwrapped optical phase is independent on the integration path (see Figure 5.11(a)). By using Green's theorem, it can be demonstrated that this condition of path independence is valid, if and only if the integral of the gradient of the phase, in 2π module, along each and every closed loop in the whole image, equals zero [37]. Figure 5.11(b) shows a set of points (circles) that represent the position of four adjacent pixels. It is found that the sum of the gradient between each pixel is zero.

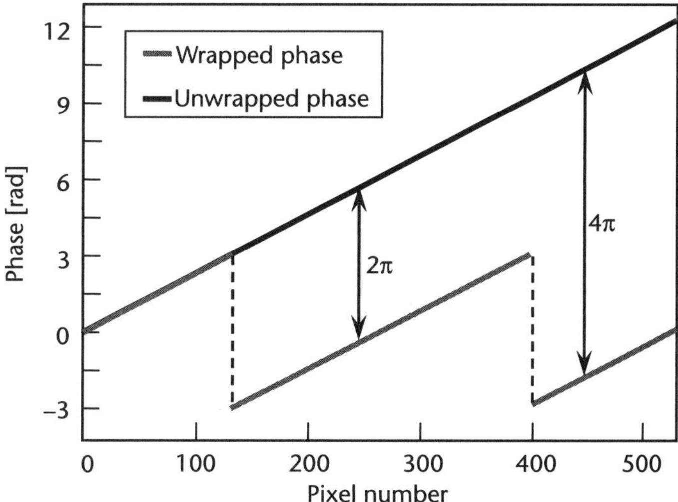

Figure 5.10   Unwrapped and wrapped phase, showing the phase unwrapping procedure.

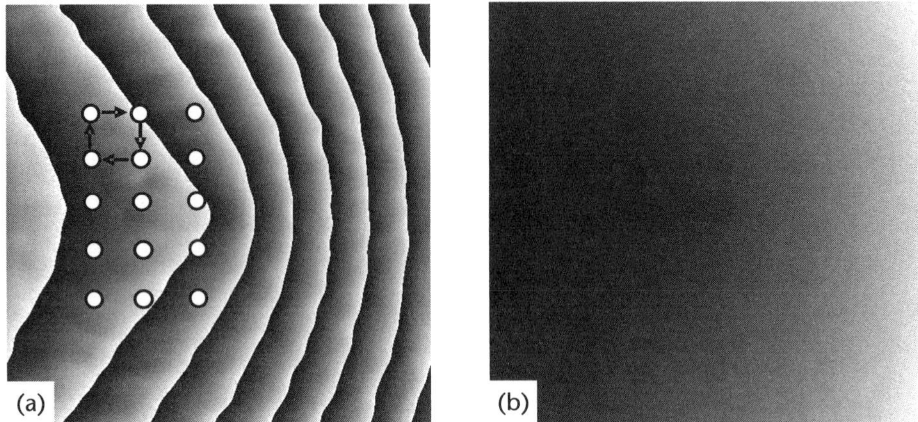

Figure 5.11   (a) Noise-free wrapped and (b) unwrapped phase. (Courtesy of G. E. Galizzi.)

On the other hand, phase maps measured in practical applications usually have phase discontinuities in localized parts of the image (see Figure 5.12 (a)). These discontinuities are produced by local speckle noise, local decorrelation of speckle, undersampling of global features such as edges of the specimen or holes, or cracks. For the case presented in Figure 5.12(a), it is found that the sum of gradients along a closed loop is zero at all pixels except for the two loops L_1 and L_2 around points 1 and 2, respectively. For these loops, the phase unwrapping procedure produces different phase values if different paths are followed. Thus, the unwrapped phase value will be different if the algorithm goes from point A to B along the P_1 or P_2 paths (see Figure 5.12 (b)). In other words, the unwrapping problem has path dependence. Points 1 and 2 are isolated points presented in the image, and they are called inconsistencies, residues, or discontinuity sources. Inconsistencies have positive and negative values, and they also appear together, in a configuration called dipoles. Reference [19] gives a more detailed explanation about dipoles and procedures that allow identifying discontinuities.

Nearly all of phase unwrapping algorithms, developed up to now, can be grouped in one of the following two categories: (a) those that select a single unwrapping

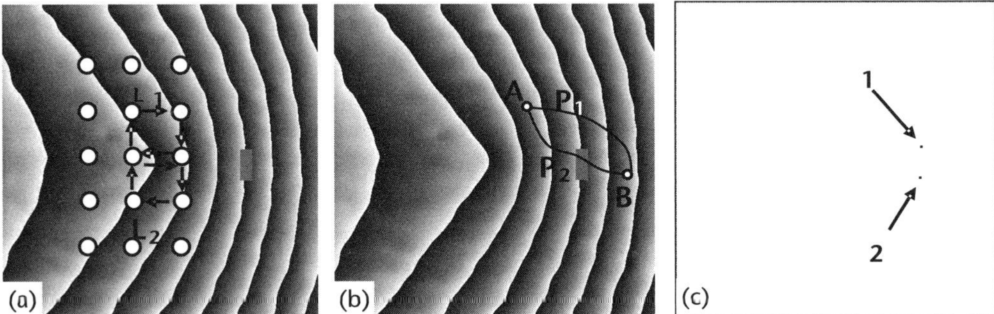

Figure 5.12   (a) Wrapped phase showing close loop around a phase fault, (b) unwrapping using two different paths, and (c) inconsistencies found in the wrapped phase in (a). (Courtesy of G. E. Galizzi.)

path, and (b) those that choose the computed unwrapped phase by means of global optimization [19]. Algorithms based on placing branch cuts that act as unwrapping barriers belong to the former group. For the latter group, the least square unwrapping is a well-known and appropriate approach.

BOX 5.10

When branch cut algorithms are considered, an interesting and peculiar observation can be made: branch cut algorithms were developed independently by researchers in satellite radar interferometry, experimental mechanics, and optical astronomy communities. For all cases, the method places barriers (called branch cuts) between residues, avoiding the unwrapping process across these points. Due to a judicious choice of branch cuts, it is always possible to make the phase maps into a single value. Additionally, [43] presents a statistical argument that shows that the most likely cut distribution is one that minimizes the sum of the squared cut lengths.

One of the simplest approaches is called the nearest-neighbor method. The idea in this method is to join two opposite residues (or one residue and the boundary of the image) that are separated by the shortest distance. In order to overcome some cases where the nearest-neighbor method fails, several optimization methods have been proposed in order to minimize the total cut length. Among these methods, the algorithm proposed in [44], based on a graph theory approach called the Hungarian algorithm, guarantees to find the minimum global cut length.

A family of global methods based on L^p norm minimization were developed by Ghiglia and Romero [37]. The main goal is to find $\Delta\phi(m, n)$ in order to minimize the following error term

$$
\begin{aligned}
\varepsilon^p = {} & \sum_{n=0}^{N_n-2} \sum_{m=0}^{N_m-1} \left| \Delta\phi(n+1,m) - \Delta\phi(n,m) - \Upsilon\left[\Delta\phi_w(n+1,m) - \Delta\phi_w(n,m)\right] \right|^p \\
& + \sum_{n=0}^{N_n-1} \sum_{m=0}^{N_m-2} \left| \Delta\phi(n,m+1) - \Delta\phi(n,m) - \Upsilon\left[\Delta\phi_w(n,m+1) - \Delta\phi_w(n,m)\right] \right|^p
\end{aligned}
\tag{5.67}
$$

where Υ is the wrapping operator defined in the previous section. Least-squares unwrapping is a special case where $p = 2$. For this case, the idea is to choose the $N_n \times N_m$ values of $\Delta\phi(m, n)$ so that the difference between neighboring pixels matches (in accordance with the least-squares approach) the corresponding phase difference computed from the measured wrapped phase. Reference [45] describes an efficient algorithm to compute this solution with a forward and inverse discrete cosine transform.

The introduction of weighting factors that take into account regions with poor data considerably improves this algorithm. However, the computational effort increases significantly since this algorithm is now iterative.

An interesting variation occurs when $p = 0$. Physically, this solution corresponds to the requirement that the solution phase gradients equal the input phase gradients in as many places as possible. It is therefore equivalent to unwrap with a set of branch cuts chosen to minimize the total branch cut length. The relationship between L^0 and branch cut algorithms are discussed in detail in [37].

Nowadays several other phase unwrapping algorithms have been developed as alternative methods to those described in the previous paragraphs. In order not to make this section too extensive, only the most transcendental ones, in the author's opinion, were presented.

5.6 Optical Configurations for Experimental Measurements

As was previously stated, changes produced in the optical phase of a speckle distribution are deterministically related to (a) the movement of the scattering surface, (b) changes in the illumination and observation geometry, or (c) variations in the wavelength of the light source or in the refractive index of the medium through which the light travels. If the wavelength of the laser light, refractive index, and geometry of the interferometer are kept constant during the measurement, the interferometric configuration will be able to measure only the displacement fields produced on the specimen under investigation. Several optical setups can be found in available literature, although they can be mainly grouped into two families: (a) interferometers with sensitivity to out-of-plane displacements [46–54], and (b) interferometers with sensitivity to in-plane displacements [55–59]. Additionally, there are optical arrangements that are also sensitive to the derivative of displacements; these are known as shearography interferometers [60–76].

5.6.1 Interferometers Sensitive to Out-of-Plane Displacements

Figure 5.13(a) shows a likely optical setup of a speckle interferometer with out-of plane sensitivity. The scattering surface is illuminated with a collimated beam that makes a small angle γ with the observation direction. In a practical context, the light source is placed as far as possible from the object in order to obtain an angle γ close to zero. By analyzing this figure, it is possible to observe that the speckle distribution scattered by the object interferes with a smooth reference beam ($\varphi_{s2} = 0$). Additionally, the interferogram's intensity is registered in the sensor of the camera located in the imaging plane.

The phase of the object beam is

$$\varphi_1 = \varphi_{s1} + \psi'_1 + \mathbf{k} \cdot \mathbf{r}$$

(5.68)

Since the phase of the reference beam is independent on the roughness of the object surface, it can be expressed as

$$\varphi_2 = \varphi_{s2} + \psi'_2$$

(5.69)

where φ_{s2} is equal to zero, since the reference beam is a smooth beam.

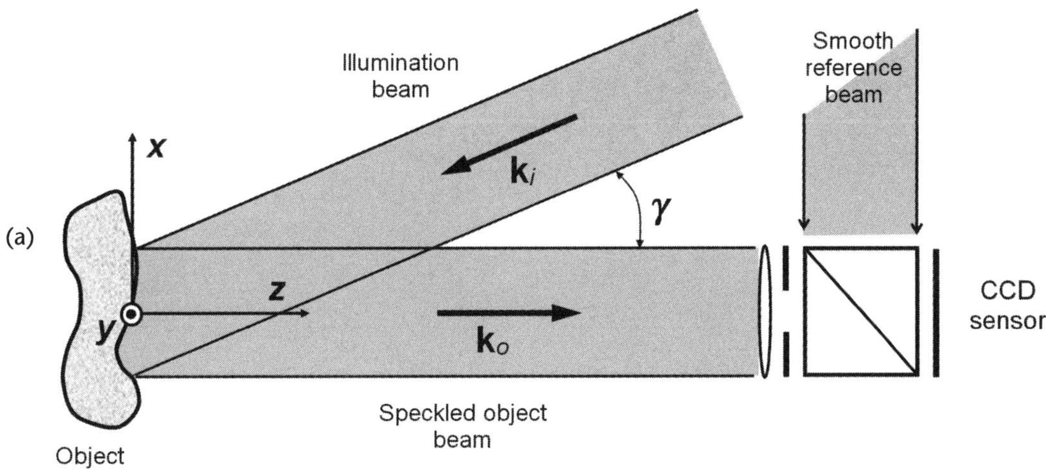

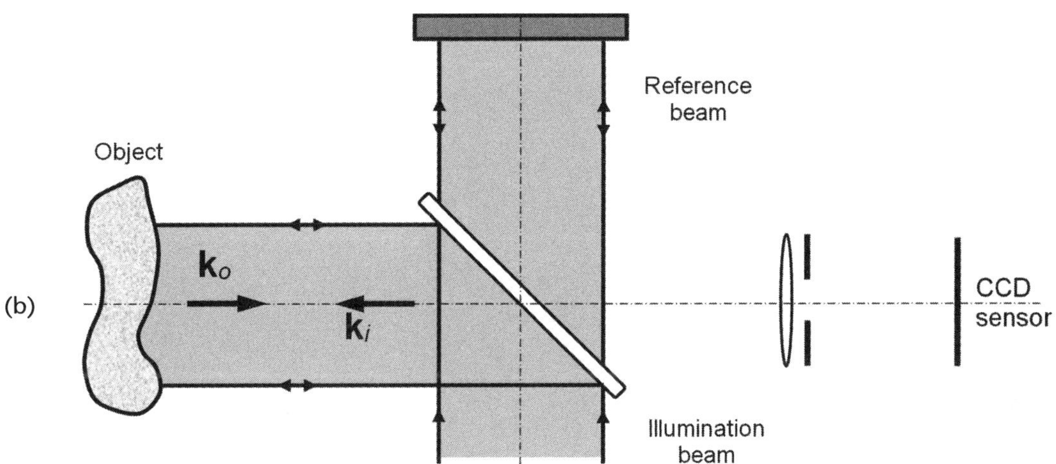

Figure 5.13 Optical setup for out-of-plane sensitivity: (a) with lateral illumination, and (b) with coincident illumination and observation directions.

By taking into account Figures 5.13, the sensitivity vector will be

$$\mathbf{k} = \mathbf{k}_i - \mathbf{k}_o = \frac{-2\pi}{\lambda}\left[\gamma\hat{\mathbf{i}} + (1 + \cos\gamma)\hat{\mathbf{k}}\right] \tag{5.70}$$

where $\hat{\mathbf{i}}$ and $\hat{\mathbf{k}}$ are the unitary vectors for the x and z directions, respectively, and λ is the wavelength of light source.

By using (5.36), the object phase can be expressed as

$$\psi_o(\mathbf{x},t) = \mathbf{k} \cdot \mathbf{r}(\mathbf{x},t) = \left\{\frac{-2\pi}{\lambda}\left[\gamma\hat{\mathbf{i}} + (1 + \cos\gamma)\hat{\mathbf{k}}\right]\right\} \cdot \mathbf{r}(\mathbf{x},t) \tag{5.71}$$

and the phase change becomes

$$\Delta\psi_o(\mathbf{x},t) = \mathbf{k} \cdot \Delta\mathbf{r}(\mathbf{x},t) = \left\{\frac{-2\pi}{\lambda}\left[\gamma\hat{\mathbf{i}} + (1 + \cos\gamma)\hat{\mathbf{k}}\right]\right\} \cdot \Delta\mathbf{r}(\mathbf{x},t) \qquad (5.72)$$

In accordance with the previous equations, readers can see that the variation of the phase is highly sensitive to displacements generated along the direction of observation when small values of γ are accomplished. Additionally, phase change is only slightly sensitive to displacements produced in the xy plane (see Figure 5.13(a)). When the directions of illumination and observation are parallel (Figure 5.13(b), being $\gamma = 0$), (5.71) and (5.72) can be rewritten as

$$\mathbf{k} = \frac{-4\pi}{\lambda}\hat{\mathbf{k}} \Rightarrow \Delta\psi_o(\mathbf{x},t) = \frac{-4\pi w}{\lambda} \qquad (5.73)$$

where w is the component of the object surface displacement along the z direction. The negative sign results from the selection of the reference system on the object. It should be highlighted that if the speckle is not decorrelated between both acquisitions ($\Delta\varphi_s = 0$), and the deliberately introduced phase difference of the reference $\Delta\psi_r$ is known, the variation of the intensity in (5.73) will reveal only changes in the object phase $\Delta\psi_o$.

For this type of interferometer, the maximum visibility of the correlation fringes will be reached when the optical aperture of the imaging system generates a subjective speckle distribution with a spatial frequency in accordance with the maximum frequency resolved by the camera sensor. Additionally, literature states that the intensity ratio between the object and reference beam must be 1.7 [2].

As an example, Figure 5.14 shows a practical configuration for a traditional DSPI interferometer with out-of-plane sensitivity. The laser light is divided by using a beamsplitter (*BS*). One of the beams is expanded with a lens (*L*) in order to illuminate the surface of the test specimen. The scattered light is captured by another lens (*L*) used as an optical imaging system. The second beam is guided by a set of mirrors (*M*), guided into the camera by means of a beamsplitter located between the *CCD* sensor and lens *L*. Mirrors are placed in the setup in order to achieve the same length between both interferometer arms. A piezoelectric actuator is attached to one of the mirrors in order to introduce a shift in the interferometer arm length, and compute the corresponding phase change by using phase shifting techniques.

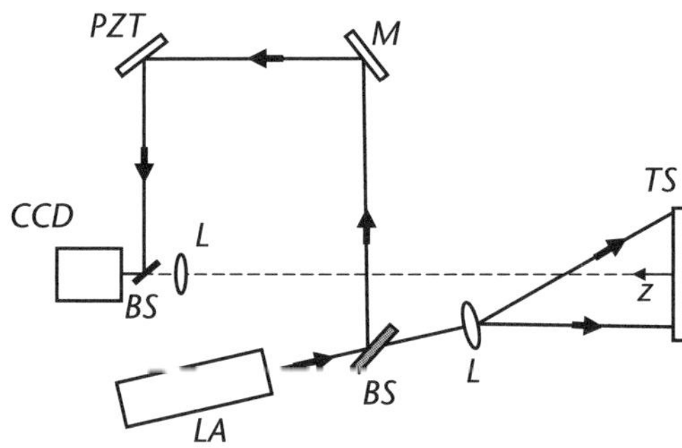

Figure 5.14 Out-of-plane interferometer: *LA*, laser source; *BS*, beam-splitters; *M*, mirrors; *PZT*, piezoelectric-driven mirror; *L*, lens; *CCD*, camera; *TS*, test specimen.

As an example, Figures 5.15(a)–(d) show the continuous optical phase and contour plot of the out-of-plane displacements measured with an optical layout similar to the one shown in Figure 5.14. The specimen (S) used in the tests was an aluminum circular plate of 100 mm in diameter and 2 mm in thickness, clamped along its edge.

Figure 5.15(a) shows the unwrapped phase generated by the out-of-plane displacement that was produced by the flawless specimen when it was continuously heated. Figure 5.15(b) shows the previously mentioned displacement field as a contour plot. As was expected, the contour plot is formed by a pattern of closed nearly circular curves.

On the other hand, Figures 5.15(c) and (d) show the unwrapped phase and contour plot of the displacement field obtained for a flawed plate subjected to continuous heating. It can be noticed that the displacement field was influenced by the presence of the flat-bottomed slots located on the back surface of the plate. The position of each of the three deepest slots is clearly depicted in this last image by the closed loop fringes. Also, a small perturbation in the vicinity of the slot with lowest depth can be seen, located at the bottom of the plate.

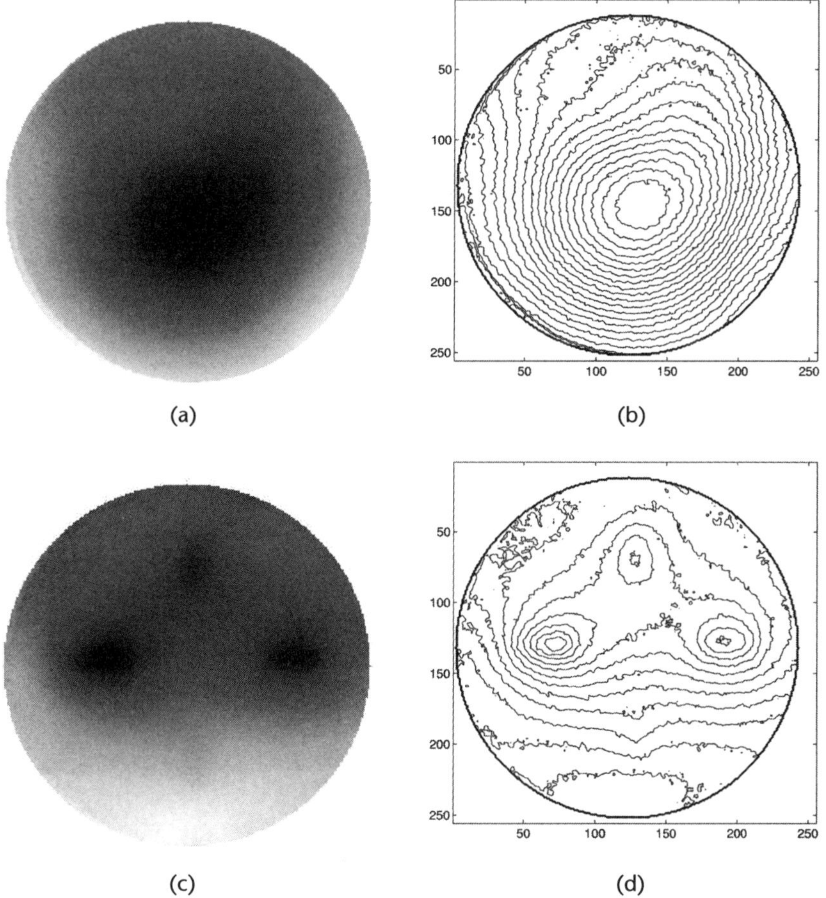

(a)

(b)

(c)

(d)

Figure 5.15 Unwrapped phase map (a) and contour plot (b) of the out-of-plane displacement obtained for the flawless specimen when it is heated continuously. Unwrapped phase map (c) and contour plot (d) of the out-of-plane displacement obtained for flawed plate continuously heated. (From [77].)

EXERCISE 5.5

What is the more suitable way to increase sensitivity in an out-of-plane interferometer? Justify your answer. How many fringes will be measured with a He-Ne laser light for a maximum displacement value of 5 μm?

Solution

In accordance with (5.73), the best and only way to change sensitivity in an out-of-plane interferometer is to modify the wavelength of the light source. Lesser wavelengths will have larger sensitivities.

A He-Ne laser has a wavelength of 632.8 nm,

$$\Delta\psi_o(\mathbf{x},t) = \frac{-4 \cdot \pi \cdot 5 \ \mu\text{m}}{0.6328 \ \mu\text{m}} = -99.29 \text{ rad} \tag{5.74}$$

Since a fringe order is related to 2π, the interferometer will approximately measure 16 fringes (15.8 fringes).

5.6.2   Interferometers Sensitive to In-Plane Displacements

Optical configurations for measuring in-plane displacements are usually based on the two-beam illumination arrangement, first described by Leendertz in 1970 [55]. These interferometers are generally capable of measuring the displacement component that is coincident with the in-plane direction.

Figure 5.16 shows the illumination and observation setup for this kind of interferometer. Two expanded and eventually collimated laser beams illuminate the object surface, forming two angles with the direction of illumination, namely γ_1 and γ_2. Thus, two speckle distributions coming from the object surface, with their respective sensitivity vectors $\mathbf{k}_{i1}$ and $\mathbf{k}_{i2}$, interfere in the imaging plane of the camera. The optical phase of these speckle distributions can be expressed as

$$\varphi_1 = \varphi_{s1} + \psi_1 + \mathbf{k}_1 \cdot \mathbf{r}_1 \tag{5.75}$$

$$\varphi_2 = \varphi_{s2} + \psi_2 + \mathbf{k}_2 \cdot \mathbf{r}_2 \tag{5.76}$$

By taking into account that $\mathbf{r}_1 = \mathbf{r}_2 = \mathbf{r}$ and (5.36), the object phase difference will be

$$\psi_o(\mathbf{x},t) = \left(\mathbf{k}_1 - \mathbf{k}_2\right) \cdot \mathbf{r}(\mathbf{x},t) = \mathbf{k} \cdot \mathbf{r}(\mathbf{x},t) \tag{5.77}$$

where $\mathbf{k}$ represents the sensitivity vector obtained from the subtraction between sensitivity vectors for each beam. It is perpendicular to the z direction of observation for the particular case where $\gamma_1 = \gamma_2 = \gamma_1$ and γ. In this case, if the illumination vectors are in the xy plane, the net sensitivity is

$$k_x \frac{-4\pi}{\lambda} \sin\gamma \tag{5.78}$$

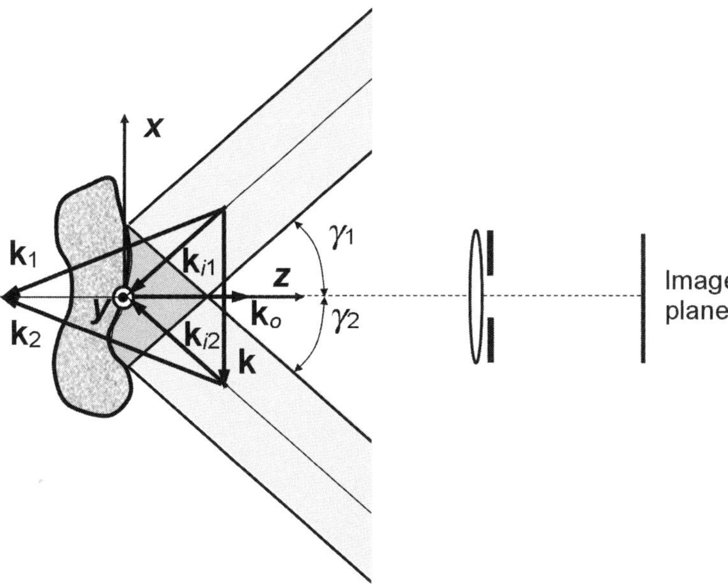

Figure 5.16   Optical configuration for in-plane sensitivity.

where k_x is the component of the sensitivity vector along the x direction. By analyzing this equation, readers can see that the magnitude of the angle γ can adjust the sensitivity of the interferometer from a null value (for the case of illumination perpendicular to the object surface) to a maximum limit value of $-4\pi/\lambda$ (for the case of illumination parallel to the object surface).

In order to obtain the phase difference for two object states (5.78) needs to be substituted into (5.77)

$$\Delta\psi_o(\mathbf{x},t) = \mathbf{k} \cdot \Delta\mathbf{r}(\mathbf{x},t) = k_x u(\mathbf{x},t) = \frac{-4\pi}{\lambda} u(\mathbf{x},t)\sin\gamma \qquad (5.79)$$

where u is the component of the displacement field along the x direction. As before, it should be highlighted that if the speckle is not decorrelated between both acquisitions ($\Delta\varphi_s = 0$), and the deliberately introduced phase difference of the reference $\Delta\psi_r$ is known, the variation of the intensity in (5.79) will reveal only changes in the object phase $\Delta\psi_o$.

For this kind of interferometer, the maximum visibility of subtraction fringes will be obtained when the optical system correctly resolves every speckle produced by the scattering surface, and the ratio between both illumination beam intensities is equal to 1 [2]. Figure 5.17 shows a sketch of a conventional in-plane digital speckle pattern interferometer with symmetrical dual-beam illumination. According to this figure, two expanders are used to illuminate the object. As the distance between the object and the expander lens is a hundred times larger than the measurement region, the variation of the sensitivity vector across the field of view can be considered negligible. As an example, Figures 5.18(a) and (b) show the wrapped and unwrapped phase maps measured with this kind of interferometer for an aluminum plate with

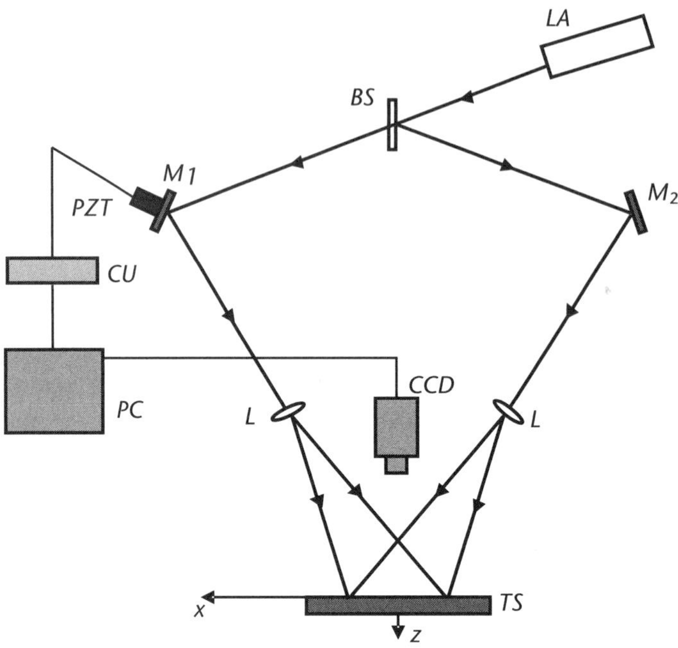

Figure 5.17 Dual-beam illumination interferometer: *LA*, laser source; *BS*, beam-splitter; M_1 and M_2, mirrors; *PZT*, piezoelectric-driven mirror; *L*, lens; *CCD*, camera; *CU*, control unit; *PC*, personal computer; *TS*, test specimen.

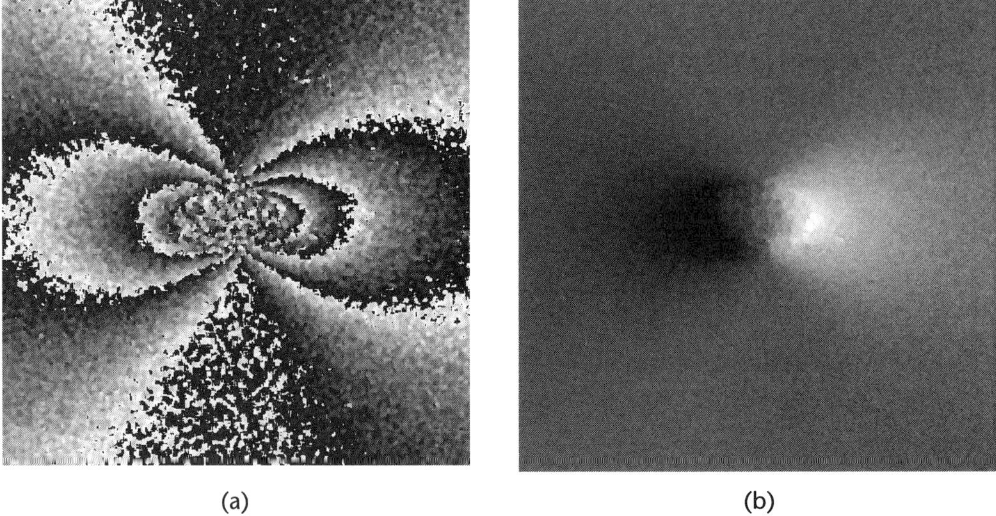

(a)   (b)

Figure 5.18 (a) Wrapped and (b) unwrapped phase map for a uniaxial residual stress along the horizontal direction.

an uniaxial residual stress along the horizontal direction. A small hole was introduced into the material in order to relieve the stresses, and the displacement field generated around the hole was measured with the interferometer.

BOX 5.11

In practical configurations, three-dimensional displacement fields are frequently separated into one component normal to the surface to be measured and two components along the tangential direction. For a plane or smooth surface, the former will be known as the out-of-plane displacement component and latter ones as the in-plane components. In-plane displacements are mainly more interesting for engineering applications where the main task is the determination of strain and stress fields applied in mechanical parts where their integrity has to be evaluated.

The determination of both in-plane components is impossible, using the interferometer shown in Figure 5.17, since it has sensitivity in only one direction (1D sensitivity). However, this kind of measurement can be achieved by using interferometers with 2D sensitivity. These systems consist of two interferometers sensitive to two orthogonal displacement directions and are based on polarization discrimination methods by using a polarizing beam splitter that splits the laser beam into two orthogonal linearly polarized beams [78, 79]. Additionally, other approaches have been applied, such as high-speed-switching-based methods (e.g., by using a liquid-crystal display that changes the polarization of the laser light [80, 81].) Two drawbacks can be found for all these approaches: (a) the test surface can appreciably depolarize the two dual-beam illumination sets, causing cross interference between them; and (b) the optical setup becomes more bulky and complex.

EXERCISE 5.6

An experimentalist would like to measure a displacement field generated by the elongation of a plane object (a metal bar). The maximum expected displacement magnitude will be 2.5 μm. What is the more adequate illumination angle to obtain at least four fringes with a wavelength of 632.8 nm? How many fringes will be obtained for a green laser?

Solution

Readers can see that main displacements presented into the object will be parallel with its surface; therefore, using an interferometer that is sensitive to in-plane displacements would be the best way to solve this problem. Of course, the sensitivity direction will be coincident with the elongation direction. By considering that the maximum displacement will be 2.5 μm ($u(\mathbf{x}, t) = 2.5$ μm) and for four fringes the object phase will be $\Delta\psi_o(\mathbf{x}, t) = 4 \times 2\pi = 8\pi$. By using (5.79), readers can see that

$$8\pi = \frac{4\pi}{0.6328 \ \mu m} \times 2.5 \ \mu m \times \sin \gamma \tag{5.80}$$

$$\sin \gamma = \frac{8\pi \times 0.6328 \ \mu m}{4\pi \times 2.5 \ \mu m} = 0.506 \Rightarrow \gamma \approx 30 \ \text{deg} \tag{5.81}$$

For an illumination angle of 30 degrees and for a green laser the maximum object phase will be

$$\Delta \psi_o = \frac{4\pi}{0.532 \ \mu m} \times 2.5 \ \mu m \times 0.5 = 9.39\pi \Rightarrow \Delta \psi_o \approx 4.7 \ \text{fringes} \tag{5.82}$$

Therefore, a short wavelength will increase the sensitivity of the interferometer in approximately 18%.

5.6.3 Interferometers Sensitive to Radial In-Plane Displacements

Figure 5.19(a) shows a cross section of a conical mirror used to obtain radial in-plane sensitivity [82–84]. The mirror is positioned near the specimen surface allowing double illumination of the specimen surface.

Figure 5.19(a) also displays two particular light rays chosen from a collimated illumination source. Each light ray is reflected by the conical mirror surface toward a point P on the specimen surface, reaching it with the same incidence angle. The illumination directions are indicated by the unitary vectors $\mathbf{n}_A$ and $\mathbf{n}_B$, and the sensitivity direction is given by the vector $\mathbf{k}$, obtained from the subtraction of both unitary vectors. As the angle is the same for both light rays, in-plane sensitivity is reached at point P. For the same cross section, and for any other point across the specimen surface, it can be verified that there is only one couple of light rays that merges at that point. Moreover, in the cross-section shown in Figure 5.19(a), the incidence angle is always the same for every point across the specimen surface and is symmetric with respect to the mirror axis. By taking into account unitary vectors, and by comparing Figures 5.16 and 5.19(a), readers can note similarities in both configurations. As a consequence, if the normal of the specimen surface and the axis of the conical mirror are parallel, then $\mathbf{n}_A$ and $\mathbf{n}_B$ will have the same angle. Therefore, the sensitivity vector $\mathbf{k}$ will be parallel to the specimen surface and in-plane sensitivity will be obtained. The previous description, presented for this particular cross section, can be extended toward any other cross section of the conical mirror. If the central point is kept out of this analysis, it can be demonstrated that each point of the specimen surface is illuminated by only one pair of light rays. As both rays are coplanar with the mirror axis, and they are also symmetrically oriented to it, full 360° radial in-plane sensitivity is obtained for a circular region across the specimen.

Finally, the radial in-plane displacement field $u_r(r, \theta)$ can be calculated from the optical phase distribution [85]

$$u_r(r,\theta) = \frac{\Delta\phi(r,\theta) \ \lambda}{4\pi \ \sin \gamma} \tag{5.83}$$

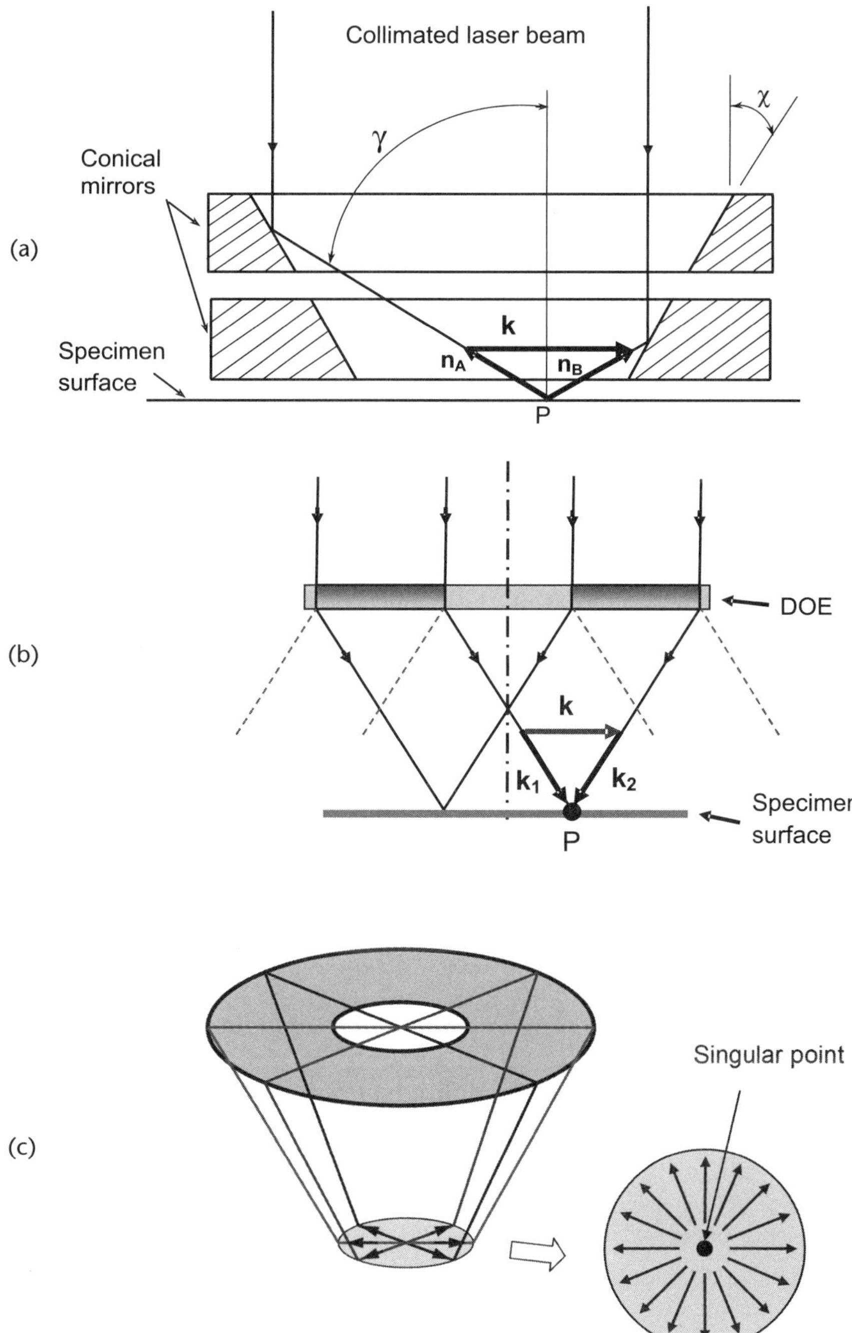

Figure 5.19   (a) Cross-section of the upper and lower parts of a conical mirror to show radial in–plane sensitivity, (b) cross-section of a diffractive optical element showing radial in-plane sensitivity, and (c) illumination principle.

As before, λ is the wavelength of the laser light, and γ is the angle between the illumination direction and the normal direction of the specimen surface.

The use of conical mirrors has two disadvantages when compared with diffractive elements: (a) it uses a relatively expensive, although high-quality, aluminum conical mirror, and (b) it requires wavelength stabilization of the laser used as light source, which cannot easily be achieved for a compact and cheap diode laser. As a consequence, applications outside the laboratory become complex or even unfeasible.

As is well known, diffractive structures are able to split white light into a spectrum of colors. On the other hand, the grating will generate an array of regularly spaced beams when the incident light is monochromatic. Thus, the diffractive structure will work as an optical element to split and shape the wavefront beam. The diffraction angle ξ of the spaced beams is given by the well-known grating equation [12, 86]

$$p_r \sin\xi = m\lambda \Rightarrow \sin\xi = \frac{m\lambda}{p_r} \tag{5.84}$$

where p_r is the period of the grating structure, and ξ is the diffraction angle for the m-order. From this equation it is clear that the -1 and $+1$ order form symmetrical angles with the incident rays.

Recent developments in microlithography manufacturing have allowed the production of diffractive optical elements (DOE) with a large variety of geometries and configurations. As consequence, a new and flexible family of optical elements with several optical functions has become available. Diffractive lenses, beam splitters, and diffractive shaping optics are just a few examples among the many possibilities. A special DOE can be designed to achieve radial in-plane sensitivity with DSPI. For this case, the DOE is composed of a circular diffraction grating with a binary profile, and a constant pitch p_r, as is shown in Figures 5.19(b) and 5.19(c). Thus, the symmetry of the -1 and $+1$ diffraction orders will produce double illumination with symmetrical angles, which generates radial in-plane sensitivity [87, 88]. Some advantages can be found for the use of DOE's compared to conical mirrors: (a) the manufacturing of diffractive structures has reached a certain maturity that makes it less expensive than specially fabricated conical mirrors, and (b) interferometer sensitivity is independent of the wavelength of the laser used as a light source [88].

By considering (5.84), the corresponding fringe equation becomes

$$u_r(r,\theta) = \frac{\lambda}{2\sin\gamma} \tag{5.85}$$

According to (5.85), the sensitivity of the method would be altered if angle γ or the wavelength of the light source were modified. For example, if the angle γ is increased, the sensitivity of the interferometer will also be increased. By observing Figure 5.19 (a), it is clear that diffraction angle ξ, as well as the angle between the direction of illumination and the normal to the specimen surface (γ), are the same; thus, $\sin\xi = \sin\gamma$. By substituting (5.85) into (5.84), only considering the first diffraction order ($m = 1$),

$$u_r(r,\theta) = \frac{\Delta\phi(r,\theta)\lambda}{4\pi\dfrac{\lambda}{p_r}} = \frac{\Delta\phi(r,\theta)p_r}{4\pi} \qquad (5.86)$$

In the same way, the corresponding fringe equation will become

$$u_r(r,\theta) = \frac{p_r}{2} \qquad (5.87)$$

Equations (5.86) and (5.87) show that the displacement field and the optical phase distribution are now related through the DOE grating's period. Additionally, they are independent of the wavelength of the laser. This particular and curious effect can be understood by considering that as the wavelength of the illumination source increases/decreases, the sine function of the diffraction angle decreases/increases with the same amount (5.85). As λ is divided by $\sin\gamma$ in (5.83), the ratio between them will remain constant.

EXERCISE 5.7

A diffractive optical element (see Figure 5.19(b)) with a grating period of 1.32 μm is illuminated with a collimated diode laser light with 660 nm in wavelength. For another case, the same optical setup uses a green laser with a wavelength of 532 nm. What is the sensitivity for both cases? Is it the same or not? What is the illumination angle?

Solution

In accordance with (5.84) the diffraction angle for the first diffraction order will be

$$\sin\xi_{red} = \frac{m\lambda}{p_r} = \frac{1 \times 0.660\ \mu\mathrm{m}}{1.32\ \mu\mathrm{m}} = 0.5 \Rightarrow \xi_{red} = 30\ \mathrm{deg}$$

$$\sin\xi_{green} = \frac{1 \times 0.532\ \mu\mathrm{m}}{1.32\ \mu\mathrm{m}} = 0.40 \Rightarrow \xi_{green} = 23.76\ \mathrm{deg} \qquad (5.88)$$

Therefore, the green laser will illuminate the specimen surface with a smaller angle. However, for the configuration shown in Figure 5.20(a) or (b) the sensitivity will be (by using (5.83))

$$u_{r(red)}(r,\theta,t) = \frac{\psi_o(r,\theta,t)\,\lambda_{red}}{4\pi\,\sin\gamma_{red}} = \frac{\psi_o(r,\theta,t) \times 0.660\ \mu\mathrm{m}}{4\pi\sin(30\ \mathrm{deg})} = 0.10504 \times \psi_o(r,\theta,t)$$

$$u_{r(green)}(r,\theta,t) = \frac{\psi_o(r,\theta,t)\,\lambda_{green}}{4\pi\,\sin\gamma_{green}} = \frac{\psi_o(r,\theta,t) \times 0.532\ \mu\mathrm{m}}{4\pi\sin(23.76\ \mathrm{deg})} = 0.10507 \times \psi_o(r,\theta,t)$$

$$(5.89)$$

In accordance with previous calculations, readers can see that sensitivity is the same for both interferometers, as previously expected.

The optical layout used to integrate the diffractive optical element can be seen in Figure 5.20(a). The light from a small diode laser (L) is expanded by a plane concave lens (E). Then, it passes through the elliptical hole of a 45° mirror M_1, illuminating mirrors M_2 and M_3, and then being reflected back to the mirror M_1. After that, the expanded laser light is directed toward the collimating lens (CL) in order to obtain a plane wavefront. Finally, the light is diffracted by the DOE, mainly in the 1st diffraction orders, toward the specimen surface. Residual nondiffracted light or light from higher diffraction orders are not a problem since they are not directed toward the central measuring area on the specimen surface.

BOX 5.12

One of the functions of the elliptical hole placed at M_1 is to allow that the light coming from the laser source illuminates mirrors M_2 and M_3. Additionally, this hole has further functions, namely, (a) to prevent that the laser light directly reaches the specimen surface (i.e., triple illumination), and (b) to provide a viewing window for the CCD camera.

Mirrors M_2 and M_3 are two special circular mirrors. The former is joined to a piezoelectric actuator (PZT), and the latter has a circular hole with a diameter slightly larger than the diameter of M_2. Mirror M_3 is fixed, while M_2 is mobile. The PZT actuator moves the mirror M_2 along its axial direction, thus generating a relative phase difference between the beam reflected by M_2 (central beam) and the one reflected by M_3 (external beam). The boundary between both beams is indicated in Figure 5.20(a) with dashed lines. In this figure, it is possible to see that every point across the illuminated area receives one ray from M_2 and another one from M_3. Thus, PZT enables the introduction of a phase shift in order to calculate the optical phase distribution by means of phase shifting algorithms.

The intensity of the light is not constant across the whole circular illuminated area of the specimen surface, and it is particularly higher at the central point because it receives light contributions from all cross sections. As a consequence, a very bright spot is visible at the central part of the circular measurement region, blurring the fringe quality. For this reason, the outlier diameter of mirror M_2 and the diameter of the central hole at M_3 are computed such as to obtain a gap of about 1.0 mm, thus enabling it to block light rays reflected to the center of the measurement area.

The optical layout of Figure 5.20(a) measures a circular area of approximately 10 mm in diameter. For some applications, smaller areas of illumination are more interesting and efficient for the measurement. As a consequence, the DOE should be as small as possible, and consequently the hole of mirror M_1 should also be smaller. For this reason, the manufacturing of this mirror becomes a complex task, since its diameter decreases considerably, and therefore an extra hole in a perpendicular axis (along the direction of observation) should be made. In order to overcome this practical problem, an off-axis configuration can be developed such as shown in Figure 5.20(b) [89].

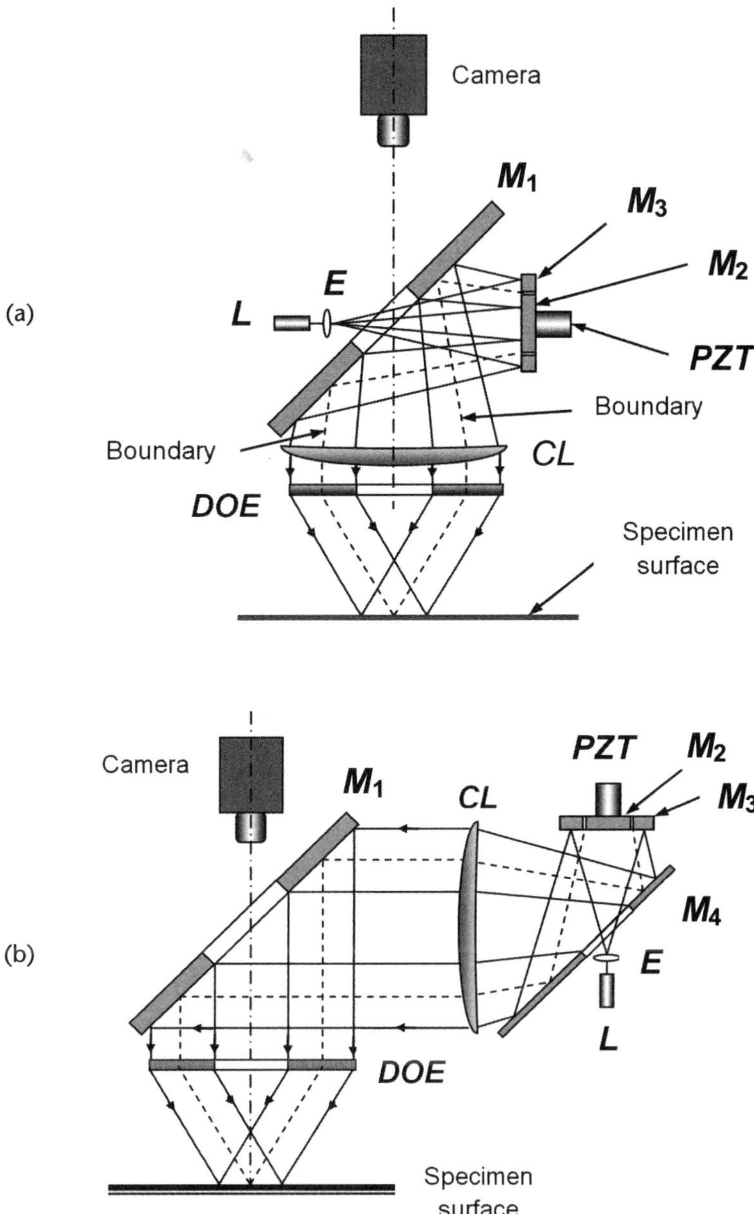

Figure 5.20 (a) Optical arrangement of a radial in-plane interferometer with DOE, and (b) alternative configuration for compactness.

Figures 5.21(a) and (b) show two photographs of radial in-plane interferometers built based on the optical layout of Figures 5.20 (a) and (b), respectively. The interferometer shown in Figure 5.21 (a) was developed to measure residual stresses by combining DSPI with the hole-drilling technique [90]. The black module on the right is the interferometer, and the smaller module on the left is the drilling module. They are interchangeable by rotating the black circular plate. On the other hand, Figure 5.21(b) shows an interferometer developed to measure mechanical stresses

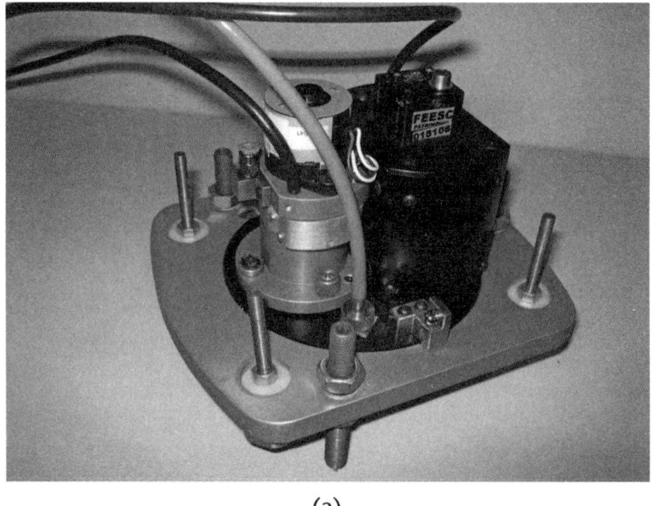

(a)

(b)

Figure 5.21   Radial in-plane interferometer for (a) residual stress measurement, and (b) mechanical stress measurement.

as an optical strain gage [89]. The measurement area is about 5 mm in diameter. For this reason, the off-axis configuration is more suitable for the materialization of the device. Finally, Figures 5.22(a) and (b) show two measured difference phase maps for residual and mechanical stress fields, respectively.

To conclude this section, Figure 5.23 shows an application outside the laboratory room for the radial in-plane interferometer. In this application, a previous modular version of the device shown in Fig 5.21(a) was used to measure residual stresses in a gas pipeline in working condition. The compact configuration provides an adequate robustness to allow the measurement in aggressive and noncooperative atmospheric conditions [91].

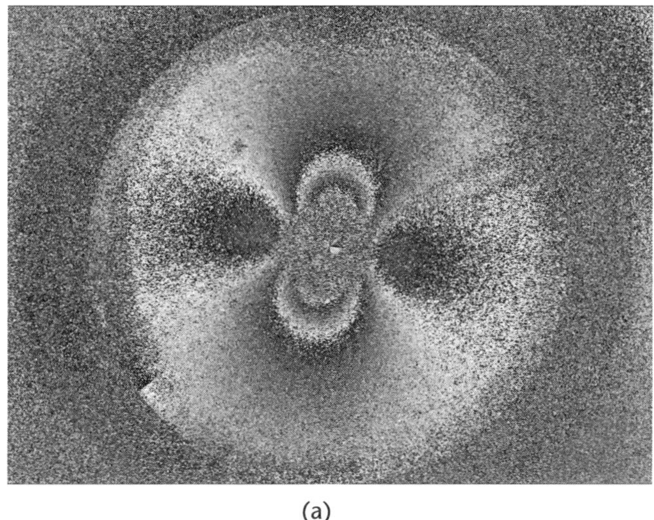

(a)

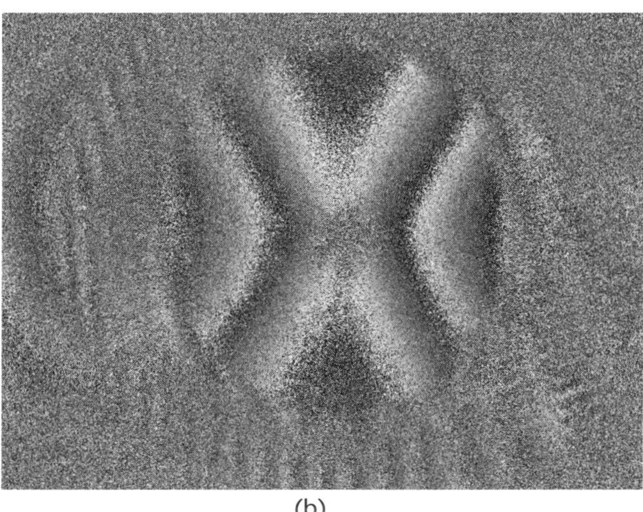

(b)

Figure 5.22   Difference phase map measured for (a) a residual stress field, and (b) mechanical stress field.

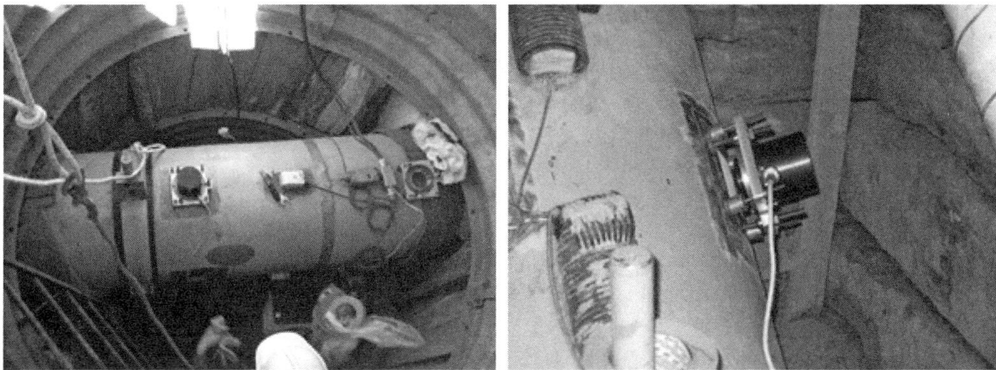

Figure 5.23   Residual stress measurement taking place outside laboratory.

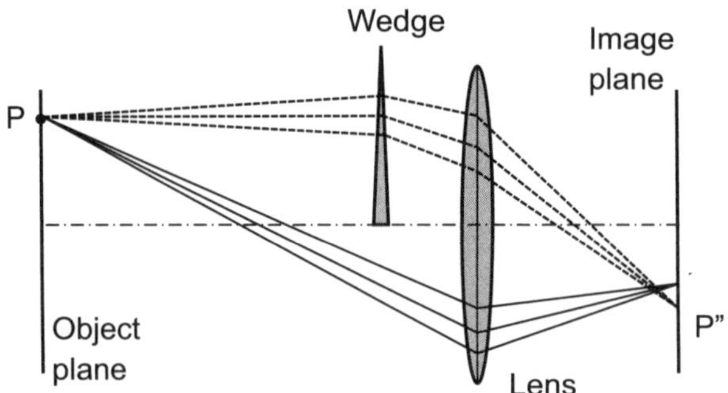

Figure 5.24   Shearing effect obtained by the introduction of a wedge.

5.6.4   Interferometers Sensitive to Derivatives of Displacements

Digital speckle shearing interferometry (also known as shearography) is used to measure derivatives of displacements that take place in the surface of an object under deformation. In order to accomplish this, a shearing device is introduced into the imaging system, producing two superimposed images of the object on the CCD camera. As a result, the object is duplicated or a point of the object is displaced by a determined lateral shift, which is called as shear.

The shearing effect can be produced in different ways. Figure 5.24 shows a possible solution using a wedge with a very small angle. Light rays coming from point P, drawn as continuous lines in the figure, are focused in point P' by the lower part of the lens, located on the image plane. Analysis of the upper part shows that the wedge slightly tilts the light rays coming from P (dashed lines) and focuses them on the image plane in a shifted position P'' from P'. Therefore, a double image of point P is visible on the image plane with a constant vertical shift.

In other words, each individual point in the image plane receives light that is scattered from two different points located on the object surface. Figure 5.25 shows this effect, where light from P_1 and P_2 merge into the same point $P' = P''$ on the image plane. If the rough surface of the object is illuminated by coherent light, speckles will be visible on the image plane. The superposition of two laterally shifted speckle patterns will produce an interference speckle pattern on the image plane, of which the phase difference will remain stable as long as the phase difference between P_1 and P_2 does.

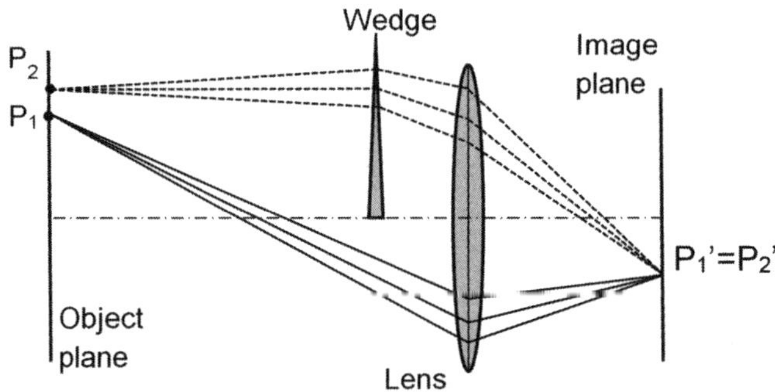

Figure 5.25   Interference between light rays coming from two points P_1 and P_2.

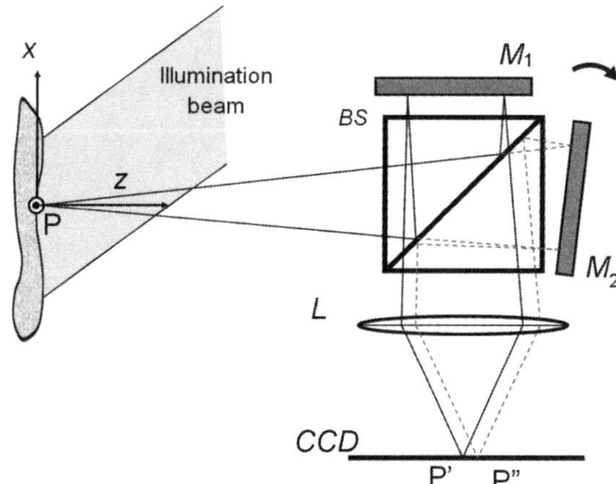

Figure 5.26   Production of a variable shearing effect by tilting a flat mirror.

The configuration shown in Figure 5.24 is very compact and robust since it involves nonmobile and adjustable parts. However, it produces a fixed shear amount, and in only one direction because the vertical shift is directly related to the angle of the wedge. A more flexible and functional configuration is shown in Figure 5.26. It comprises of two mirrors and a beamsplitter cube in a traditional Michelson configuration [60–62]. The object is illuminated by collimated light with an incidence angle γ. The scattered light is partially deflected by the beam splitter, illuminating a flat mirror M_1 (rays are represented by solid lines). Then, the light rays are reflected by this mirror toward the imaging system, focusing on the CCD sensor at point P. On the other hand, rays of light passing through the beamsplitter (represented by dotted lines) illuminate the flat mirror M_2. These rays are reflected back for this mirror and, after the deflection into the beamsplitter, they are imaged at point P'' on the CCD sensor. By tilting mirror M_2, point P' is laterally shifted from P'' by δx. The amount of tilting of mirror M_2 controls the magnitude of lateral shear of the image as well as the distance between points P' and P''. The tilting represented in Figure 5.26 produces a vertical shear. A horizontal shift of these points is achieved by rotating mirror M_2 around a vertical axis.

Additionally, the configuration shown in Figure 5.26 is very convenient in order to generate a phase shift between both interfering specklegrams. Mirror M_1 can be slightly moved up and down in a parallel way, changing the relative phase difference between the light rays that interfere by a small amount. That can easily be done by attaching the upper flat mirror to a micro translator, like a piezoelectric translator.

BOX 5.13

After a deep analysis of Figure 5.26, it is possible to see that the viewing angle of the optical setup is limited by the Michelson configuration to approximately an angle of 28°. In order to overcome this limitation, a modification in the Michelson interferometer, by introducing a $4f$ system, increases the viewing angle to approximately 57°, allowing the measurement of larger areas [64]. Another optical setup for shearography uses a diffractive grating instead of the beamsplitter cube that works as a diffractive optical beamsplitter. In this kind of interferometer, phase shifting is generated by the in-plane translation of the grating [36, 65]. Other authors have used other different shearing devices, namely, glass wedges, Fresnel's biprisms, Wollaston prism, and glass plates with parallel faces. A novel approach uses the birefringent properties of a liquid crystal spatial light modulator (SLM) as shearing element. As a consequence, this configuration has no moving mechanical parts in order to produce the shear. Thus, this configuration is robust and reliable since it features flexibility, high accuracy, and high reproducibility at the same time [92, 93].

BOX 5.14

Nowadays, several commercial interferometers based on DSPI are available for nondestructive applications. One of the worldwide leading suppliers of measurement systems is Dantec Dynamics. This company commercializes optical systems that allow a complete three-dimensional analysis for measuring strain and stress fields present into a mechanical part. The commercial name is Q-300 (called for the company as "3D ESPI system"). For the investigation of thermal expansions of MEMS and electronic components, the system named Q-300 TCT combines DSPI with a heating device in order to evaluate the thermal expansion coefficient of materials as well thermal stress of components. The company Optonor has the VibroMap 1000. This device is based on DSPI and allows the measurement of surface vibrations and static deflections.

On the other hand, it is important to highlight that shearography is the most widely used setup for nondestructive evaluation, especially in the aeronautical and aerospace industries. Several books and book chapters provide deep insight into the development of this technique during the past 20 years [94–97]. Nowadays, several companies have commercial shearographic systems and have the aeronautical, aerospace, and automotive industries as their main application areas. For example, Dantec Dynamics manufactures the Q-800 portable shearography system for nondestructive testing (NDT), as well as a portable version called Q-810. In the same way, Steinbichler Optotechnik GmbH also commercializes a portable shearographic device called ISISmobile 3000. According to the company's website, some of the main features of the measurement system are (a) flexible to use on various tasks, (b) easy adjustable, (c) one-man-operation, (d) customer-specific user interface, (e) real-time data display, and (f) data export in common file formats as well as network capability. The SNT 4045 from Optonor in Norway is a compact shearography system that is combined with dynamic excitation (using vibrational excitation) for NDT. The SNT 4040 is a system with very high resolution that uses vacuum and static excitation. Another very well-known company is ISI-sys, which also commercializes optical devices for NDT and strain measurements. Shearography systems developed by ISI-sys employ vibrational excitation for object loading through a piezoshaker, which is placed on the surface to be measured. Finally, Laser Technology Inc. (LTI) commercializes several systems (2100, 5000, 6000, and 9000 series) that can be used as portable devices as well as for production in-line inspection. Additionally, LTI offers inspection services for test or production application.

Finally, the technique called electronic holography introduced and developed by Karl Stetson can be considered as a pioneer in the application of optical techniques using laser sources as nondestructive evaluation tools. Today, Karl Stetson Associates, LLC, commercializes the K100/HOL optical head, which allows the identification of bonding flaws in real-time measurements. Furthermore, other important optical device suppliers use this principle in order to perform NDT evaluations. Among them, LTI manufactures the DH-8000 holographic inspection station used to identify and detect disbonds.

5.7   Concluding Remarks

The speckle effect is a consequence of the iteration between the coherent light of a laser source and the micro-geometry of rough surfaces. By using interferometric techniques, speckle properties can deterministically be related to displacement fields produced on the surface of the specimen under investigation.

Digital speckle pattern interferometry (DSPI) has common advantages compared to optical methods such as high inspection velocity, digital processing, automation of data analysis process, and noncontacting nature. Additionally, DSPI is adequate for the evaluation of real components without further preparation of the surface to be analyzed. Thus, DSPI is suitable for large-scale measurements such as can be found in industrial applications.

Traditional interferometers are influenced by atmospheric conditions and vibrations. For these reasons, they are suitable for laboratory rooms with temperature and humidity control as well as passive and active vibration isolation. On the other hand, interferometer setups can be built into portable devices in order to be applied outside the laboratory. In order to accomplish this, some requirements need to be fulfilled by the interferometer. The interferometer needs to be (a) robust, (b) flexible, (c) compact, (d) stable, (e) user friendly, and (f) cooperative. As an example, diffractive optical elements (DOE) bring new conceptual design features and properties for optical measurement systems conferring robustness and compactness. For interferometers sensitive to displacements, DOE can be used as beam splitters, providing independence on the wavelength of the laser source and allowing the use of cheap diode lasers without wavelength stabilization. For shearography, DOE allows the construction of compact devices with robustness against vibration.

For all these reasons, DSPI is currently more and more present in different areas outside research groups and laboratories, being mainly used as an inspection tool. Several commercial systems can be found around the world, including digital speckle interferometers and mainly shearographic interferometers that can solve several problems. There are new challenges in prospect, which will surely keep DSPI an up-to-date technique with a wide and comprehensive field of application.

5.8 Acknowledgments

The author is grateful for Prof. Gustavo E. Galizzi's help during the elaboration of some figures used along the text and for for several fruitful discussions with Dr. Gordon Craggs along the writing of the text and for his help with some special words.

References

[1] Goodman, J. W., *Speckle Phenomena in Optics: Theory and Applications*, Roberts and Company publishers, 2006.

[2] Jones, R., and K. Wykes, *Holographic and Speckle Interferometry*, 2nd ed. Cambridge University Press, 1989.

[3] Françon, M., *La Granularité Laser (Speckle)*, 2nd ed., Masson, 1978.

[4] Stetson, K. A., and W. R. Brohinsky, "Eletrooptic holography and its application to hologram interferometry," *Applied Optics*, Vol. 24, No. 21, 1985, pp. 3631–3637.

[5] Goodman, J. W., "Statistical properties of laser speckle patterns," In *Laser Speckle and Related Phenomena*, pp. 9–75, J. C. Dainty, (ed.),Springer Verlag, 1984.

[6] Ennos, A. E., "Speckle interferometry," In *Laser Speckle and Related Phenomena*, pp. 203–253, J. C. Dainty, (ed.),Springer Verlag, 1975.

[7] Butters, J. N., and J. A. Leendertz, "Holographic and video techniques applied to engineering measurement," *Journal of Measurement and Control*, Vol. 4, No. 12, 1971 pp.349–354.

[8] Macovski, A., D. Ramsey, and L. F. Schaefer, "Time lapse interferometry and contouring using television systems," *Applied Optics*, Vol. 10, No. 12, 1971, pp. 2722–2727.

[9] O. Schwomma. Austrian Patent No. 298830, 1972.

[10] Stetson, K. A., "A new design for the laser image-speckle interferometer," *Optics and Laser Technology*, Vol. 2, No. 4, 1970, pp. 179–181.

[11] Stetson, K. A., "Miscellaneus topics in speckle metrology," In *Speckle Metrology*, pp. 292–302, R. K. Erf (ed.), Academic, 1978

[12] Hetch, E., *Optics*, 4th ed., Addison Wesley, 2002.

[13] Dainty, J. C., *Laser Speckle and Related Phenomena*, 2nd ed. Springer Verlag, 1984.

[14] Zel'dovich, B. Y., A. V. Mamaev, and V. V. Shkunov, *Speckle-Wave Interaction in Application to Holography and Nonlinear Optics*, CRC Press, Inc., 1995.

[15] Lehmann, M., "Speckle statistics in the context of digital speckle interferometry." In *Digital Speckle Pattern Interferometry and Related Techniques*, pp. 60–139, P. K. Rastogi, (ed.), Wiley and Sons, 2001.

[16] Kreis, T., *Holographic Interferometry: Principles and Methods*, Wiley-VCH., 1996.

[17] Middleton, D., *An Introduction to Statistical Communication Theory*, New York: McGraw-Hill, 1960.

[18] Papoulis, A., *Probability, Random Variables and Stochastic Processes*, New York: McGraw-Hill, 1965.

[19] Huntley, J. M., "Automated analysis of speckle interferograms." In *Digital Speckle Pattern Interferometry and Related Techniques*, pp. 1–58, P. K. Rastogi, (ed.), Wiley and Sons, 2001.

[20] Løkberg, O. J., and G. A. Slettemoen. "Basic Electronic Speckle Pattern Interferometry." *Applied Optics and Optical Engineering*, pp 455–504. Shannon, R., and J. C. Wyant (eds.), Academic Press, 1987.

[21]   Sirohi, R. S., *Speckle Metrology*, Marcel Dekker, 1993.

[22]   Takeda, M., H. Ina, and S. Kobayashi, "Fourier-transform method of fringe-pattern analysis for computer-based topography and interferometry," *Journal of the Optical Society of America*, Vol. 72, No. 1, 1982, pp. 156–160.

[23]   Roddier, F., and C. Roddier, "Imaging with a multi-mirror telescope," *Proceeding of the ESO Conference on Optical Telescopes of the Future*, E. Pacini, W. Richter and R. N. Nilson, (eds.), pp 359–370, 1978.

[24]   Suetmatsu, M., and M. Takeda, "Wavelength-shift interferometry for distance measurements using the Fourier-transform technique for fringe analysis," *Applied Optics*, Vol. 30, No. 28, 1991, pp. 4046–4055.

[25]   Takeda, M., and H. Yamamoto, "Fourier-transform speckle profilometry—three dimensional shape measurements of diffuse objects with large height steps and/or spatially isolated surface," *Applied Optics*, Vol. 33, No. 34, 1994, pp. 7829–7837.

[26]   Joenathan, C., B. Franze, P. Haible, and H. J. Tiziani, "Large in-plane displacement measurement in dual-beam speckle interferometry using temporal phase measurement," *Journal of Modern Optics*, Vol. 45, No. 9, 1998, pp. 1975–1984.

[27]   Joenathan C., B. Franze, P. Haible, and H. J. Tiziani, "Shape measurement by use of temporal Fourier transformation in dual-beam illumination speckle interferometry," *Applied Optics*, Vol. 37, No. 16, 1998, pp. 3385–3390.

[28]   Joenathan C., B. Franze, P. Haible, and H. J. Tiziani, "Speckle interferometry with temporal phase evaluation for measuring large-object deformation," *Applied Optics*, Vol. 37, No. 13, 1998, pp. 2608–2614.

[29]   Brigham, E. O. *The Fast Fourier Transform*, Prentice Hall, 1974.

[30]   Preater, R. W. T., and R. C. Swain, "Fourier transform fringe analysis of electronic speckle pattern interferometry fringes from high-speed rotating components," *Optical Engineering*, Vol. 33, No. 4, 1994, pp. 1271–1279.

[31]   Creath, K., "Temporal Phase Measurement Methods." In *Interferogram Analysis: Digital Fringe Pattern Measurement Techniques*, pp. 94–140, Robinson, D. W., and G. T. Reid, (eds.), Institute of Physics Publishing, Bristol, 1993.

[32]   Wyant, J. C., "Interferometric optical metrology: basic principles and new systems," *Laser Focus*, Vol. 18, No. 5, 1982, pp. 65–71.

[33]   Kreis, T., *Handbook of Holographic Interferometry: Optical and Digital Methods*, Wiley-VCH, 2005.

[34]   Schwider, J., K. E. Burow, J. Grzanna, R. Spolaszyk, and K. Merkel, "Digital wave-front measuring interferometry-some systematic error sources," *Applied Optics*, Vol. 22, No.21, 1983, pp.3421–3432.

[35]   Carré, P., "Installation et utilisation du comparateur photoélectrique et interférentiel du Bureau International des Poids et Mesures," *Metrologia*, Vol. 2, 1966, pp.13–23.

[36]   Gundlach, A., J. M. Huntley, B. Manske, and J. Schwider, "Speckle shearing interferometry using a diffractive optical beamsplitter," *Optical Engineering*, Vol. 36, No. 5, 1997, pp. 1488–1493.

[37]   Ghiglia, D. C., and M. D. Pritt, *Two-Dimensional Phase Unwrapping*, John Wiley and Sons, 1998.

[38]   Shannon, C. E., "Communication in the Presence of Noise," *Proceedings of the IEEE*, Vol. 86, No. 2, 1998, pp. 447–457.

[39]   Huntley, J. M., G. H. Kaufmann, and D. Kerr, "Phase-shifted dynamic speckle pattern interferometry at 1 kHz," *Applied Optics*, Vol. 38, No. 31, 1999, pp. 6556–6563.

[40]   Huntley, J. M., and H. O Saldner, "Temporal phase-unwrapping algorithm for automated interferogram analysis," *Applied Optics*, Vol. 32, No. 17, 1993, pp. 3047–3052.

[41]   Saldner, H. O., and J. M. Huntley, "Profilometry using temporal phase unwrapping and a spatial light modulator-based fringe projector," *Optical Engineering*, Vol. 36, No. 2, 1997, pp. 610–615.

[42] Saldner, H. O., and J. M. Huntley, "Temporal phase unwrapping: application to surface profiling of discontinuous objects," *Applied Optics*, Vol. 36, No. 13, 1997, pp. 2770–2775.

[43] Huntley, J. M,. and J. R., Buckland, "Characterization of sources of 2π phase discontinuity in speckle interferograms," *Journal of the Optical Society of America*, Vol. 12, No. 9, 1995, pp. 1990–1996.

[44] Buckland, J. R., J. M. Huntley, and S. R. E. Turner, "Unwrapping noisy phase maps by use of a minimum-cost-matching algorithm," *Applied Optics*, Vol. 34, No. 23, 1995, pp. 5100–5108.

[45] Ghiglia, D. C., and L. A. Romero, "Robust two-dimensional weighted and unweighted phase unwrapping that uses fast transforms and iterative methods," *Journal of the Optical Society of America A*, Vol. 11, No. 1, 1994, pp. 107–117.

[46] Wykes, C., "Use of electronic speckle pattern interferometry (ESPI) in the measurement of static and dynamic surface displacements," *Optical Engineering*, Vol. 21, No. 3, 1982, pp. 400–406.

[47] Slettemoen, G. A., "Electronic speckle pattern interferometry based on a speckle reference beam," *Applied Optics*, Vol. 19, No. 4, 1980, pp. 616–613.

[48] Joenathan, C., and B. M. Khorana, "Quasi-equal path electronic speckle pattern interferometric system," *Applied Optics*, Vol. 32, No. 29, 1993, pp. 5724–5726.

[49] Virdee, M. S., D. C. Williams, J. E. Banyard, and N. S. Nassar, "A simplified system for digital speckle interferometry," *Optics and Lasers Technology*, Vol. 22, No. 5, 1990, pp. 311–315.

[50] Joenathan, C., and R. Torroba, "Modified electronic speckle pattern interferometry employing and off-axis reference beam," *Applied Optics*, Vol. 30, No. 10, 1991, pp. 1169–1172.

[51] Kato, J., I. Yamaguchi, and P. Qi, "Automatic deformation analysis by a TV speckle interferometer using a laser diode," *Applied Optics*, Vol. 32, No. 1, 1993, pp. 77–83.

[52] Wykes, C., and F. Flanagan, "The use of a diode laser in an ESPI system," *Optics and Lasers Technology*, Vol. 19, No. 1, 1987, pp. 37–39.

[53] Arizaga, R., H. Rabal, and M. Trivi, "Simultaneous multiple-viewpoint processing in digital speckle pattern interferometry," *Applied Optics*, Vol. 33, No. 20, 1994, pp. 4369–4372.

[54] Paoletti, D., G. Schirripa Spagnolo, M. Facchini, and P. Zanetta, "Artwork diagnostics with fiber-optic digital speckle pattern interferometry," *Applied Optics*, Vol. 32, No. 31, 1993, pp. 6236–6241.

[55] Leendertz, J. A., "Interferometric displacement measurement on scattering surfaces utilizing speckle effect," *Journal of Physics E: Scientific Instruments*. Vol. 3, No. 3, 1970, pp. 214–218.

[56] Viotti, M. R., and G. H. Kaufmann, "Accuracy and sensitivity of a hole drilling and digital speckle pattern interferometry combined technique to measure residual stresses," *Optics and Lasers in Engineering*, Vol. 41, No. 2, 2004, pp 297–305.

[57] Viotti, M. R., A. E. Dolinko, G. E. Galizzi, and G. H. Kaufmann, "A portable digital speckle pattern interferometry device to measure residual stresses using the hole drilling technique," *Optics and Lasers in Engineering*, Vol. 44, No.10, 2006, pp. 1052–1066.

[58] Dolinko, A. E., and G. H. Kaufmann, "Measurement of the local displacement field generated by a microindentation using digital speckle pattern interferometry and its application to investigate coating adhesion," *Optics and Lasers in Engineering*, Vol. 47, No. 5, 2009, pp. 527–531.

[59] Martinez, A., R. Rodriguez-Vera, J. A. Rayas, and H. J. Puga, "Error in the measurement due to the divergence of the object illumination wavefront for in-plane interferometers," *Optics Communications*, Vol. 223, No. 4–6, 2003, pp. 239–246.

[60] Leendertz, J. A., and J. N. Butters, "An image-shearing speckle pattern interferometer for measuring bending moments," *Journal of Physics E: Scientific Instruments*, Vol. 6 No. 11, 1973, pp. 1107–1110.

[61] Hung, Y. Y., "Shearography: a novel and practical approach to nondestructive testing," *Journal of Nondestructive Evaluation*, Vol. 8, No. 2, 1989, pp. 55–68.

[62] Hung, Y. Y., and C. E. Taylor, "Speckle-shearing interferometric camera—a tool for measurement of derivatives of surface displacements," *Proceeding of SPIE*, Vol. 41, 1973, pp. 169–175.

[63] Waldner, S., and S. Brem, "Compact shearography system for the measurement of 3D deformation," *Proceeding of SPIE*, Vol. 3745, 1999, pp. 141–148.

[64] Wu, S., X. He, and L. Yang, "Enlarging the angle of view in Michelson-interferometer-based shearography by embedding a 4*f* system," *Applied Optics*, Vol. 50, No. 21, 2011, pp. 3789–3794.

[65] Rabal, H., R. Henao, and R. Torroba, "Digital speckle pattern shearing interferometry using diffraction gratings," *Optics Communications*, Vol. 126, No. 4–6, 1996 pp. 191–196.

[66] Hung, Y. Y., and C. Y. Liang, "Image-shearing camera for direct measurement of surface strains," *Applied Optics*, Vol. 18, No. 7, 1979, pp. 1046–1051.

[67] Bhaduri, B., M. P. Kothiyal, and N. K. Mohan, "Curvature measurement using three-aperture digital shearography and fast Fourier transform," *Optics and Lasers in Engineering*, Vol. 45, No. 10, 2007, pp. 1001–1004.

[68] Ochoa, N. A., and A. A. Silva-Moreno, "Fringes demodulation in time-averaged digital shearography using genetic algorithms," *Optics Communications*, Vol. 260, No. 2, 2006, pp. 434–437.

[69] Hung, Y. Y., Y. H. Huang, L. Liu, S. P. Ng, and Y. S. Chen, "Computerized tomography technique for reconstruction of obstructed phase data in shearography," *Applied Optics*, Vol.47, No. 17, 2008, pp. 3158–3167.

[70] Mihaylova, E., M. Whelan, and V. Toal, "Simple phase-shifting lateral shearing interferometer," *Optics Letters*, Vol. 29, No. 11, 2004, pp. 1264–1266.

[71] Falldorf, C., E. Kolenovic, and W. Osten, "Speckle shearography using a multiband light source," *Optics and Lasers in Engineering*, Vol.40, No. 5–6, 2003, pp. 43–52.

[72] Hung, M. Y. Y, Y. S. Chen, S. P. Ng, S. M. Shepard, Y. Hou, and J. R. Lhota, "Review and comparison of shearography and pulsed thermography for adhesive bond evaluation," *Optical Engineering*, Vol. 46, No. 5, 2007, pp. 051007-1–051007-16.

[73] Hung, M. Y. Y, "Role of shearography in nondestructive testing," *Proceeding of SPIE*, Vol. 4537, 2002, pp. 1–8.

[74] Kalms, M. K., W. Osten, W. P. O. Jüptner, W. Bisle, D. Scherling, and G. Tober, "NDT on wide-scale aircraft structures with digital speckle shearography," *Proceeding of SPIE*, Vol. 3824, 1999, pp. 280–286.

[75] Osten, W., M. K. Kalms, W. P. O. Jüptner, G. Tober; W. Bisle, and D. Scherling, "A shearography system for the testing of large-scale aircraft components taking into account non-cooperative surfaces, " *Proceeding of SPIE*, Vol. 4101, 2000, pp. 432–438.

[76] Osten, W., M. K. Kalms, W. P. O. Jüptner, "Some ways to improve the recognition of imperfections in large scale components using shearography" *Proceeding of SPIE*, Vol. 3745, 1999, pp. 244–256.

[77] Kaufmann, G. H., M. R. Viotti, and G. E. Galizzi, "Flaw detection improvement in temporal speckle pattern interferometry using thermal waves," *Journal of Holography and Speckle*, Vol. 1, No. 2, 2004, pp. 80–84.

[78] Moore, A. J., and J. R. Tyrer, "An electronic speckle pattern interferometer for complete in-plane measurement," *Measurement Science and Technology*, Vol. 1, No. 10, 1990, pp. 1024–1030.

[79] Moore, A. J., and J. R. Tyrer, "Two-dimensional strain measurement with ESPI," *Optics and Lasers in Engineering*, Vol. 24, No. 5–6, 1996, pp. 381–402.

[80] Bowe, B., S. Martin, V. Toal, A. Langhoff, and M. Whelan, "Dual in-plane electronic speckle pattern interferometry system with electro-optical switching and phase shifting," *Applied Optics*, Vol. 38, No. 1, 1999, pp. 666–673.

[81] Facchini, M., and P. Zanetta, "An electronic speckle interferometry in-plane system applied to the evaluation of mechanical characteristics of masonry," *Measurement Science and Technology*, Vol. 6, No. 9, 1995, pp. 1260–1269.

[82] Albertazzi Jr., A., M. R. Borges, and C. Kanda. "A radial in-plane interferometer for residual stresses measurement using ESPI," In *Proceedings of SEM IX International Congress on Experimental Mechanics*, 2000, pp. 108–111.

[83] Viotti, M. R., R. Sutério, A. Albertazzi Jr., and G. H. Kaufmann, "Residual stress measurement using a radial in-plane speckle interferometer and laser annealing: preliminary results," *Optics and Lasers in Engineering*, Vol. 42, No. 1, 2004, pp. 71–84.

[84] Viotti, M. R., A. Albertazzi Jr., and G. H. Kaufmann, "Measurement of residual stresses using local heating and a radial in-plane speckle interferometer," *Optical Engineering*, Vol. 44 No.9, 2005, p. 093606.

[85] Rastogi, P. K., "Digital Speckle Pattern Interferometry and Related Techniques." In *Digital Speckle Pattern Interferometry and Related Techniques*, pp. 141–224, P. K. Rastogi, (ed.), Wiley and Sons, 2001.

[86] O'Shea, D. C., T. J. Suleski, A. D. Kathman, and D. W. Prather. "Diffractive optics: Design Fabrication and Test," *Tutorial Texts in OE*, TT62. 2003, SPIE.

[87] Viotti, M. R., A. Albertazzi Jr., and W. Kapp, "Experimental comparison between a portable DSPI device with diffractive optical element and a hole drilling strain gage combined system," *Optics and Lasers in Engineering*, Vol. 46, No. 11, 2008, pp. 835–841.

[88] Viotti, M. R., W. Kapp, and A. Albertazzi Jr., "Achromatic digital speckle pattern interferometer with constant radial in-plane sensitivity by using a diffractive optical element," *Applied Optics*, Vol. 48, No. 12, 2009, pp. 2275–2281.

[89] Viotti, M. R., A. Albertazzi Jr., and W. Kapp, "Mechanical stress measurement by an achromatic optical digital speckle pattern interferometry strain sensor with radial in-plane sensitivity: experimental comparison with electrical strain gauges," *Applied Optics*, Vol. 50, No. 7, 2011, pp. 1014–1022.

[90] Viotti, M. R., and A. Albertazzi Jr., "Compact sensor combining digital speckle pattern interferometry and the hole-drilling technique to measure nonuniform residual stress fields," *Optical Engineering*, Vol. 52, No. 10, 2013, p. 101905.

[91] Viotti, M. R., and A. Albertazzi Jr., "Industrial inspections by speckle interferometry: general requirements and a case study," In *Proceeding of SPIE*, Vol. 7389, 2009, p. 73890G.

[92] Falldorf, C., "Measuring the complex amplitude of wave fields by means of shear interferometry," *Journal of the Optical Society of America A*, Vol. 28, No. 8, 2011, pp 1636–1647.

[93] Falldorf, C., S. Osten, C. V. Kopylow, and W. Jüptner, "Shearing interferometer based on the birefringent properties of a spatial light modulator," *Optics Letters*, Vol. 34, No. 18, 2009, pp. 2727–2729.

[94] Steinchen, W., and L. Yang, *Digital Shearography: Theory and Application of Digital Speckle Pattern Shearing Interferometry*, SPIE Press, 2003.

[95] Hung, M. Y. Y., "Nondestructive Testing Using Shearography," In *Recent Advances in Experimental Mechanics. In Honor of Isaac M. Daniel*, pp. 397–408, E. E. Gdoutos (ed.), Kluwer Academic Publishers, 2002.

[96] Hung, M. Y. Y., "Digital shearography and applications" In *Trends in Optical Non-Destructive Testing and Inspection*, pp. 287–308, P. K. Rastogi and D. Inaudi (eds.), Elsevier, 2000.

[97] Bergmann, R. B., and P. Huke, "Advanced methods for optical nondestructive testing," In *Optical imaging and metrology: Advanced Technologies*. pp. 393–409, Osten, W., and N. Reingand (eds.), Wiley-VCH Verlag GmbH & Co. KGaA, 2012.

CHAPTER 6

Digital Image Correlation

Bing Pan

6.1 Introduction

Surface deformation measurement of materials and structural components subjected to various external loading (e.g., mechanical loading or thermal loading) is an important task of experimental solid mechanics. The measured displacements and strains directly reveal the deformation behavior or mechanism of the test material and structure. Also, combined with the knowledge of exerted mechanical or thermal loadings, various mechanical parameters of a material (e.g., Young's modulus, Poisson's ratio, coefficient of thermal expansion, and stress intensity factor) can be readily determined. Furthermore, these obtained kinematic fields help to verify the validity of theoretical predictions and FEM analyses.

Aside from the widely used pointwise strain gauge technique, which provides only the averaged strain readings within the gauge length, various noncontact full-field optical methods are available in the field of experimental mechanics. These methods include interferometric techniques, such as holography interferometry [1, 2], electronic speckle pattern interferometry [1] and moiré interferometry [3], and noninterferometric techniques, such as grid method [4] and digital image correlation (DIC) [5–7]. Interferometric metrology requires a coherent light source, and the measurements normally should be conducted on a vibration-isolated optical platform in the laboratory. Interferometric techniques measure the deformation by recording the phase difference of the scattered, reflected, or diffracted light wave from the test object surface before and after deformation. The measurement results are often presented in the form of fringe patterns; thus, further fringe processing and phase analysis techniques are required. In contrast, noninterferometric techniques determine the surface deformation by directly comparing the gray intensity changes of the images of the object surface before and after deformation using computer programs; thus, these techniques have less stringent requirements on experimental equipment and conditions.

As a representative noninterferometric optical technique, DIC has become widely accepted and commonly used as a powerful and flexible tool for surface deformation measurement in the field of experimental solid mechanics and in various other scientific research and engineering fields. In the past 30 years, a great amount of research and development activities have been dedicated to this easy-to-use and effective optical technique. Also, the applications of DIC have expanded from regular metal or polymeric materials to special composite and biological materials, from the macroscopic to the microscopic scale, from static or quasi-static loading to high-speed dynamic loading, and from common laboratory conditions to extreme high-temperature environments.

In principle, DIC is a noncontact, full-field, yet easy-to-implement optical technique based on digital image recording and digital image processing. It was first developed by a group of researchers at the University of South Carolina in the 1980s [8–10] when the digital image processing and numerical computing were still in their infancy. During the past three decades, the DIC technique has been extensively investigated and significantly improved for reducing computation complexity, achieving high accuracy deformation measurement and expanding application range. For example, the two-dimensional (2D) DIC method using a single fixed camera is limited to in-plane deformation measurement of nominal planar objects. To acquire accurate measurements, some requirements on the specimen deformation, loading device, and measuring system must be strictly met [11, 12]. If the test object has a curved surface, or three-dimensional (3D) deformation occurs after loading, the 2D-DIC method is no longer applicable. To overcome this deficiency of 2D-DIC, the 3D-DIC (or more exactly, stereo-DIC) based on the principle of binocular stereovision [13, 14] was developed. The 3D-DIC can be used for shape and deformation measurement of both planar and curved surfaces. Also, the digital volume correlation (DVC) method, as a direct 3D extension of the 2D-DIC method, has also been proposed by Bay and Smith et al. [15, 16]. DVC is capable of providing the internal deformation fields of solid objects by tracking the movement of volume units within digital image volumes of an object. The DVC method emerges as a novel and effective tool for quantifying the internal deformation of opaque cellular solid objects (e.g., trabecular bone, wood, rock, and cellular foam) and certain semi-transparent biological tissues (e.g., cells). Combining the projection of computer generated speckle pattern with a digital projector and the 2D-DIC technique, the speckle projection method [17, 18] was also developed for out-of-plane displacement and shape measurement. Table 6.1 lists the three main correlation-based DIC techniques and their applicability.

Compared with the aforementioned high-sensitivity interferometric optical techniques, DIC techniques offer the following special and attractive advantages:

1. Simple experimental setup and specimen preparation. Only one fixed CCD camera (2D-DIC) or two cameras (3D-DIC) are needed to record the digital images of the test specimen surface before and after deformation. Specimen preparation is unnecessary (if the natural texture on specimen surface has a random gray intensity distribution) or can be simply made by spraying paints onto the specimen surface.

2. Low requirements on measurement environment. DIC does not require a laser source. A white light source, natural light, or monochromatic light can be used for illumination during loading. Thus, it is especially suitable for both laboratory and field applications.

Table 6.1   Various Correlation-Based DIC Techniques and Their Applicability

DIC Technique	Imaging Devices Used	Applicability
2D-DIC	Optical imaging systems, SEM, AFM, and so on	In-plane deformation measurement of nominal planar objects' surface
3D-DIC	Stereovision system, light stereo microscope, and so on	3D deformation or shape measurement of planar or curve objects' surface
DVC	x-CT, LSCM, micro-MRT, and so on	3D internal deformation measurement of cellular solid object or biological tissues

3. Wide range of measurement sensitivity and resolution. Since the DIC method deals with digital images, the digital images recorded by various high-spatial-resolution digital image acquisition devices can be directly processed by DIC. For example, 2D-DIC can be coupled with various microscopes [19–29], such as optical microscopy, laser scanning confocal microscope (LSCM), scanning electron microscopy (SEM), and atomic force microscopy (AFM) to realize microscale to nanoscale deformation measurements. The 3D-DIC can be used with light stereo microscopes. Similarly, the instantaneous 2D (or 3D) deformation measurement can be realized by analyzing a dynamic sequence of digital images recorded with high-speed digital image recording equipment using the 2D- (or 3D-) DIC method [30–32].

More importantly, with the constant emergence of high-spatial-resolution and high-time-resolution image acquisition equipment, and continuous improvements that are being made to correlation algorithms, it is expected that the DIC technique can be easily applied to new areas and solve new mechanics problems. Hence it is fair to say that DIC is one of the most active optical measurement technologies currently used in the field of experimental mechanics and that it demonstrates increasingly broad application prospects.

6.2   Two-Dimensional Digital Image Correlation

The 2D-DIC method using a fixed camera is limited to in-plane deformation measurement of nominal planar objects. In general, the implementation of the 2D-DIC method comprises of the following three consecutive steps, namely, (1) specimen and experimental preparation, (2) the recording of digital images of the planar specimen surface before and after loading, and (3) the processing of the acquired images using a computer program to obtain the desired displacement and strain information. In this section, issues on specimen preparation and image capture are introduced first. Then, the basic principles and concepts of 2D-DIC are described. Next, the most commonly used displacement and strain estimation algorithms are outlined. Finally, a simple but typical application of 2D-DIC for the determination of tensile strains and elastic properties of an aluminum specimen is demonstrated.

6.2.1   Specimen Preparation and Image Capture

Figure 6.1 is the schematic illustration of a typical experimental setup using optical imaging device for 2D-DIC method. The test specimen surface must have a random gray intensity distribution, also known as random speckle pattern, which deforms together with the specimen surface as a carrier of deformation information. The speckle pattern can be a natural texture of the specimen surface or be artificially made by spraying black and/or white paints or by other techniques. The camera is placed with its optical axis perpendicular to the nominal planar specimen surface, imaging the planar specimen surface at different loading states onto its sensor plane.

In general, a simple ideal pinhole camera model, as shown in Figure 6.1, is implicitly used to describe the mathematical relationship between the coordinates of an object point and its corresponding image point on sensor plane [12]. That means the measured sensor plane displacements (u, v) are linearly proportional to the object plane displacements (U, V) according to a simple linear relationship of $u = MU$, $v = MV$, where M denotes the time-invariant constant magnification factor across the image plane. However, in practical applications of 2D-DIC, the magnification factor M cannot be seen as a constant across the entire image plane, unless the following requirements are met:

1. The imaging system is perfect and does not suffer from any geometric distortion. Hence the magnification factor M can be assumed as a constant across the image plane.
2. The specimen surface must be flat and should be placed to be parallel to the camera sensor target. Both the object surface and image surface should remain in the same plane without any motion or rotation during experiment. This assumption implies that the object distance Z and image distance L will not change after loading. Thus, the magnification factor M can be regarded as a time-invariant constant.

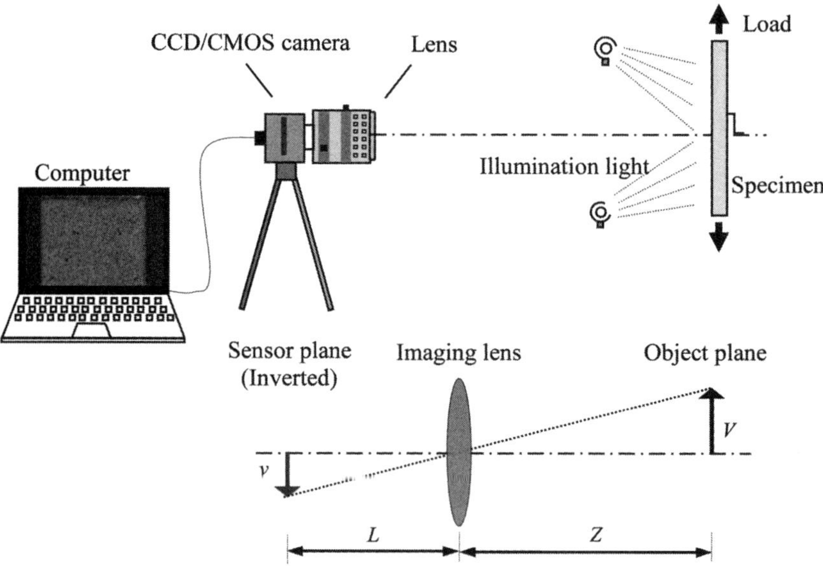

Figure 6.1 Typical optical image acquisition system for the 2D-DIC method and the corresponding pinhole camera model.

Regrettably, it has been convincingly shown [12] that the ideal pinhole imaging model assumed for an ordinary 2D-DIC system using a common imaging lens is neither perfect nor stable. That means the magnification factor M varies at different pixel locations and also changes slightly at different times. The reasons are as follows:

1. The pinhole imaging model is not perfect, because any real camera lens, especially a low-cost lens, unavoidably contains a certain amount of lens distortion due to the inherent aberration, machining errors, and assembly errors of the lens [33, 34]. In other high-resolution imaging systems including SEM, LSCM, and AFM, geometric distortion is more or less present, which also impairs the ideal linear correspondence between the physical point and imaged point and produces additional displacements. Due to the nonuniform characteristics of lens distortion, small in-plane motion of an object surface will cause different amounts of image motion, as schematically show in Figure 6.2(c). Recently, theoretical analysis and experimental

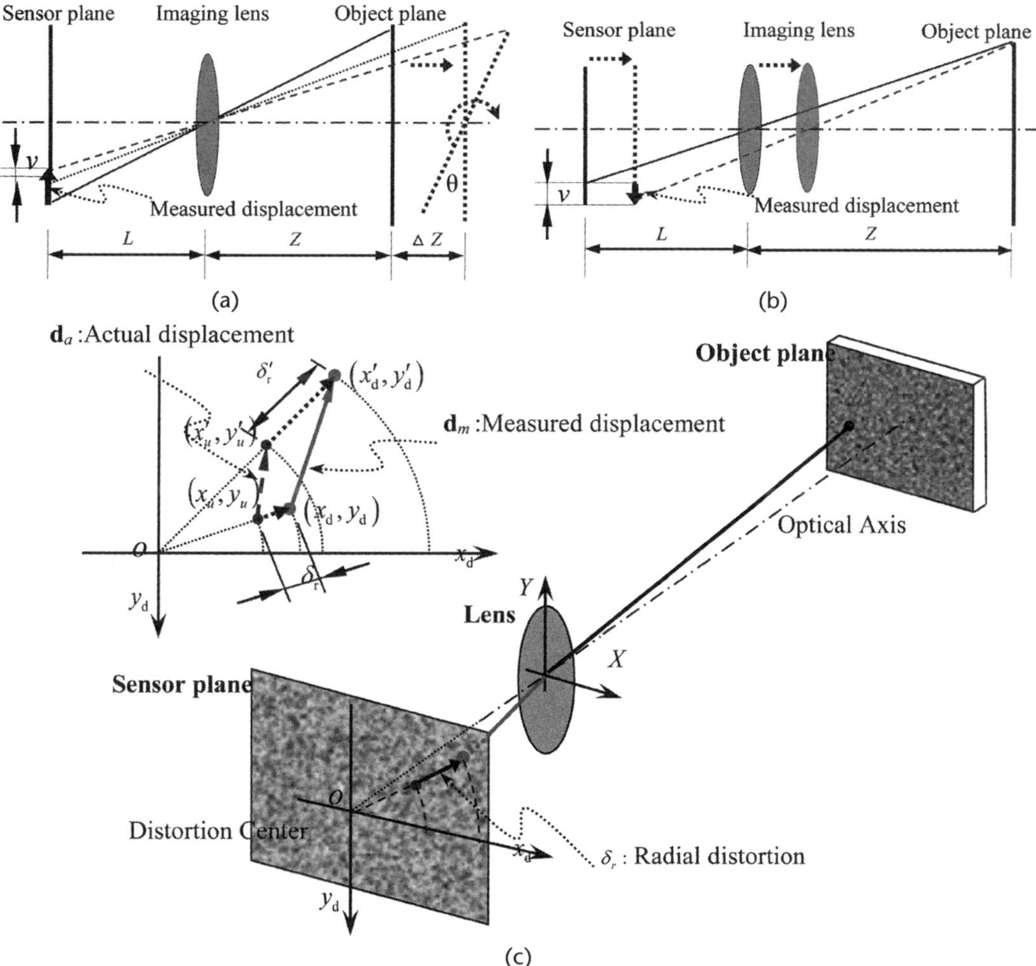

Figure 6.2 (a) Out-of-plane motion of test object alters the object distance Z of the imaging system; (b) self-heating of a camera changes the imaging distance L and/or the object distance Z of the imaging system; (c) schematic figure of the pinhole camera model considering the influence of lens distortion. The inserted figure shows the influence of radial lens distortion on the displacement vector measured by 2D-DIC.

    results carried out in [12] convincingly verified that the magnification factor M depends on several factors and is not a constant across the image.

2. The pinhole imaging model is not stable, because after a test sample is loaded, the object distance of the 2D-DIC system may change due to out-of-plane motion of the sample surface [11], as shown in Figure 6.2(a). Also, the image and/or object distances also change due to self-heating and/or temperature variation of a camera [35], as indicated in Figure 6.2(b). The changes in object distance and/or image distance alter the magnification factor M of the imaging system. In certain cases, the imperfection and instability of a low-cost 2D-DIC system may give rise to hundreds or even thousands of microstrain errors.

Consequently, in order to obtain high-fidelity 2D-DIC measurements, a near perfect (with negligible lens distortion) and very stable (insensitive to out-of-plane motions in object surface and imaging plane) optical system using a well-design bilateral telecentric lens is highly recommended for practical use [12]. Alternatively, an effective error compensation technique using an unstrained compensating specimen can be used to eliminate these unwanted measurement errors [36].

6.2.2   Basic Principles and Concepts

After recording the digital images of the specimen surface before and after deformation, DIC computes the motion of each point of interest by comparing the digital images of the test object surface at different states. The image displacements (in pixels) can be further differentiated using an appropriate numerical algorithm to estimate full-field strains. In the following, the basic principles and concepts involved in 2D-DIC are introduced.

6.2.2.1   Basic Principles

In the routine implementation for 2D-DIC, a region of interest (ROI) in the reference image should be specified at first. The ROI is further divided into an evenly spaced virtual grids as shown in the left part of Figure 6.3. The displacements are computed at each point of the virtual grid to obtain the full-field deformation. The basic principle of 2D-DIC is the tracking (or matching) of the same pixel points between the two images recorded before and after deformation as schematically illustrated in Figure 6.3. In order to compute the displacements of point P, a square reference subset of $(2M + 1) \times (2M + 1)$ pixels centered at a point $P(x_0, y_0)$ is chosen in the reference image and is used to track the corresponding location in the deformed image. The reason why a square subset, rather than an individual pixel, is selected for matching is that the subset, comprising a wider variation in gray levels, will distinguish itself from other subsets and can therefore be more uniquely and accurately identified in the deformed image.

    To quantitatively evaluate the degree of similarity between the reference subset and the deformed subset, a cross-correlation (CC) criterion or sum-squared difference (SSD) correlation criterion must be predefined as an objective function. The matching procedure is completed by seeking the peak position of the distribution of the correlation coefficient. Once the correlation coefficient extremum is detected, the position of the deformed subset is determined. The differences of the positions

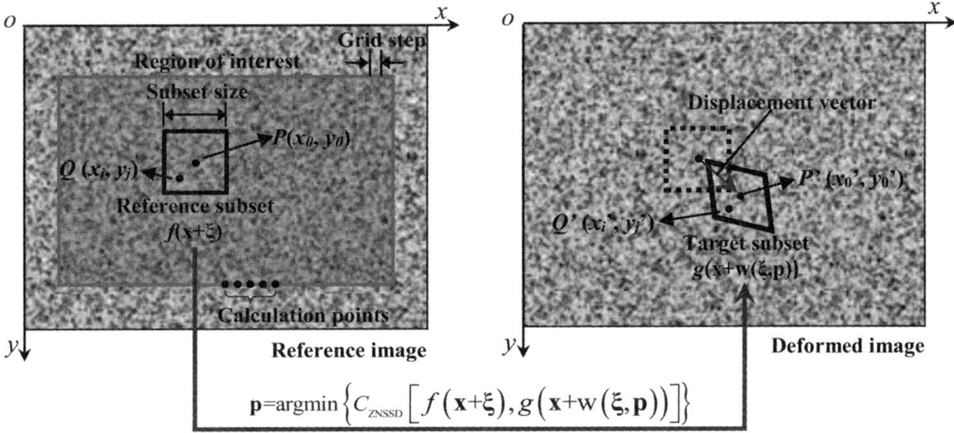

Figure 6.3 Left: reference image; the imposed black square is the subset used for tracking the motion of its center point, and the intersection points of red grid denotes the points to be calculated; right: the displacement vector of the considered point is computed by optimizing a correlation criterion that quantifies the similarity degree between reference square subset and target (or deformed) subset.

of the reference subset center and the target subset center give the in-plane displacement vector at point $P(x_0, y_0)$, as illustrated in Figure 6.3.

6.2.2.2 Shape Function/Displacement Mapping Function

Without loss of generality, one can assume that both the position and shape of the reference subset are changed in the deformed image as shown in Figure 6.3. However, based on the assumption of deformation continuity of a deformed solid object, a set of neighboring points in a reference subset should remain as neighboring points in the target subset. Thus, the coordinates of the point $Q(x_i, y_j)$ around the subset center $P(x_0, y_0)$ in the reference subset can be mapped to the point $Q'(x_i', y_j')$ in the target subset according to the so-called shape function [37] or displacement mapping function [38]:

$$x_i' = x_i + \xi\left(x_i, y_j\right)$$
$$y_j' = y_j + \eta\left(x_i, y_j\right) \qquad (i, j = -M : M) \qquad (6.1)$$

If only rigid body translation exists in the reference subset and deformed subset (in other words, the displacements of all points in the subset are the same (see Figure 6.4 (a)), the zero-order shape function can be used:

$$\xi_0\left(x_i, y_j\right) = u$$
$$\eta_0\left(x_i, y_j\right) = v \qquad (6.2)$$

Obviously, the zero-order shape function is not sufficient to depict the shape change of the deformed subset. Thus, the first-order shape function shown in Figure 6.4(b), which allows translation, rotation, shear, normal strains, and their combinations in the subset, is most commonly used:

$$\xi_1\left(x_i, y_j\right) = u + u_x \Delta x + u_y \Delta y$$
$$\eta_1\left(x_i, y_j\right) = v + v_x \Delta x + v_y \Delta y \tag{6.3}$$

Furthermore, the second-order shape functions proposed by Lu et al. [38] can be used to depict more complicated deformation states of the deformed subset:

$$\xi_2\left(x_i, y_j\right) = u + u_x \Delta x + u_y \Delta y + \frac{1}{2}u_{xx}\Delta x^2 + \frac{1}{2}u_{yy}\Delta y^2 + u_{xy}\Delta x \Delta y$$
$$\eta_2\left(x_i, y_j\right) = v + v_x \Delta x + v_y \Delta y + \frac{1}{2}v_{xx}\Delta x^2 + \frac{1}{2}v_{yy}\Delta y^2 + v_{xy}\Delta x \Delta y \tag{6.4}$$

In (6.2), (6.3), and (6.4), $\Delta x = x_i - x_0$, $\Delta y = y_j - y_0$, u, v are the x and y directional displacement components of the reference subset center, $P(x_0, y_0)$, u_x, u_y, v_x, v_y are the first-order displacement gradients of the reference subset, and u_{xx}, u_{xy}, u_{yy}, v_{xx}, v_{xy}, v_{yy} are the second-order displacement gradients of the reference subset. The influences of various deformation parameters on the final shape of the target subset are illustrated in Figure 6.4.

6.2.2.3 Correlation Criterion

As noted earlier, to evaluate the similarity (or difference) degree between the reference subset and its deformed counterpart, a correlation criterion must be defined, which is of fundamental importance in DIC. Basically, the commonly used correlation criteria can be classified into three categories according to their mathematical definitions [39] (i.e., cross-correlation (CC), sum of squared difference (SSD), and parametric sum of squared difference (PSSD)), as listed in Table 6.2.

$$\bar{f} = \frac{1}{n}\sum_{i=1}^{n} f_i, \quad \bar{g} = \frac{1}{n}\sum_{i=1}^{n} g_i, \quad \tilde{f}_i = f_i - \bar{f}, \quad \tilde{g}_i = g_i - \bar{g}$$

with n denoting the total number of pixels in a subset.

In a practical measurement, the local intensity within the target image may differ significantly from that of the reference image. The intensity changes between images recorded at different configurations may be induced for various reasons. A robust correlation criterion should be used to accommodate the variations in the intensity of the deformed image; otherwise, significant displacement measurement errors may occur due to the mismatch of the intensity change model [39–41]. Based on this consideration, the three robust correlation criteria listed at the bottom row of Table 6.2, say, a ZNCC, a ZNSSD criterion, and a PSSD$_{ab}$ criterion with two additional unknown parameters, have been highly recommended for practical use, as they are invariant against linear intensity changes of the target subset.

At first glance, the mathematical expressions of these three robust correlation criteria seem fully different. Yet these three correlation criteria have the same performance against the offset and scale changes in the deformed subset intensity. Recently, rigorous theoretical analysis performed by Pan [40] convincingly proved that these three correlation criteria are mathematically equivalent. Specifically, the ZNCC criteria are related to the ZNSSD criteria according to

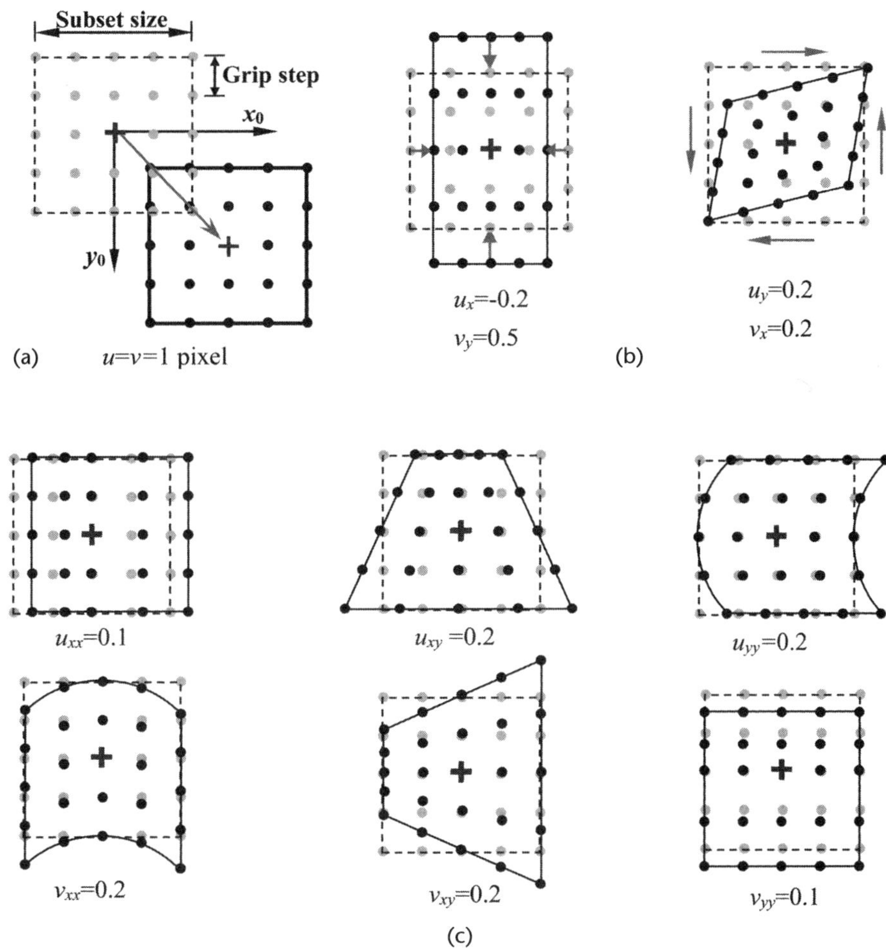

Figure 6.4 Examples showing the effects of different displacement parameters on the shape of the target subset: (a) zero-order, (b) first-order, and (c) second-order shape function.

Table 6.2 Three Types of Correlation Criteria Commonly Used in DIC for Evaluating the Similarity Degree between the Reference Subset and the Deformed Subset

Performance	CC Criteria	SSD Criteria	PSSD Criteria
Sensitive to all changes of the deformed subset intensity	$C_{CC} = \sum f_i g_i$	$C_{SSD} = \sum (f_i - g_i)^2$	
Insensitive to offset changes of the deformed subset intensity	$C_{ZCC} = \sum (\overline{f_i}\,\overline{g_i})$	$C_{ZSSD} = \sum \left[\overline{f_i} - \overline{g_i}\right]^2$	$C_{PSSDb} = \sum (f_i + b - g_i)^2$
Insensitive to scale changes of the deformed subset intensity	$C_{NCC} = \dfrac{\sum f_i g_i}{\sqrt{\sum f_i^2 \sum g_i^2}}$	$C_{NSSD} = \sum \left(\dfrac{f_i}{\sqrt{\sum f_i^2}} - \dfrac{g_i}{\sum g_i^2} \right)^2$	$C_{PSSDa} = \sum (a f_i - g_i)^2$
Insensitive to the scale and offset changes of the deformed subset intensity	$C_{ZNCC} = \dfrac{\sum \overline{f_i}\,\overline{g_i}}{\sqrt{\sum \overline{f_i}^2 \sum \overline{g_i}^2}}$	$C_{ZNSSD} = \sum \left(\dfrac{\overline{f_i}}{\sqrt{\sum \overline{f_i}^2}} - \dfrac{\overline{g_i}}{\sum \overline{g_i}^2} \right)^2$	$C_{PSSDab} = \sum (a f_i + b - g_i)^2$

$$C_{ZNSSD} = \sum \left(\frac{\bar{f_i}}{\sqrt{\sum \bar{f_i}^2}} - \frac{\bar{g_i}}{\sqrt{\sum \bar{g_i}^2}} \right)^2 = \sum \left(\frac{\bar{f_i}^2}{\sum \bar{f_i}^2} - 2\frac{\bar{f_i}^2 \bar{g_i}^2}{\sqrt{\sum \bar{f_i}^2} \sqrt{\sum \bar{g_i}^2}} + \frac{\bar{g_i}^2}{\sum \bar{g_i}^2} \right)$$

$$= 2 - 2\frac{\sum \bar{f_i}^2 \bar{g_i}^2}{\sqrt{\sum \bar{f_i}^2} \sqrt{\sum \bar{g_i}^2}} = 2\left[1 - C_{ZNCC} \right] \tag{6.5}$$

Similarly, it was also proven that the $PSSD_{ab}$ criterion is also equivalent to the ZNNC criterion [39], and the mathematical relationship between them can be written as:

$$C_{PSSDab} = \sum \bar{g_i}^2 \left(1 - C_{ZNCC}^2 \right) \tag{6.6}$$

However, the ZNSSD criterion is highly recommended in practical application, because it can be efficiently computed using the sum-table approach [42] and is easily optimized using nonlinear numerical algorithms.

EXERCISE 6.1

Assume that the subset size defined for correlation analysis is 5×5 pixels, and the local gray intensities of the reference subset $f(x_i, y_j)$ and target subset $g(x_i, y_j)$ are listed in the following tables. Calculate the ZNCC coefficient between the two subsets.

$$f(x_i, y_j) =$$

106	112	104	98	100
91	120	96	68	66
69	102	69	39	39
59	92	57	32	28
61	105	76	47	35

$$g(x_i, y_j) =$$

107	112	104	99	101
91	120	96	68	66
68	104	69	38	40
57	97	57	31	25
62	103	75	46	35

Solution

First, calculate the mean intensity values of the two subsets

$$\bar{f} = \frac{1}{25} \sum_{i=-2}^{2} \sum_{j=-2}^{2} f(x_i, y_j) = \frac{1}{25}(106 + 112 + \cdots + 47 + 35) = 74.84$$

$$\bar{g} = \frac{1}{25} \sum_{i=-2}^{2} \sum_{j=-2}^{2} g(x_i, y_j) = \frac{1}{25}(107 + 115 + \cdots + 49 + 38) = 76.44$$

Then, after subtracting the mean value, the intensity of the two subsets now can be written as

$$\bar{f}(x_i, y_j) = f(x_i, y_j) - \bar{f} =$$

30.5600	38.5600	27.5600	22.5600	25.5600
18.5600	45.5600	22.5600	−7.4400	−5.4400
−8.4400	27.5600	−4.4400	−38.4400	−33.4400
−19.440	20.5600	−17.4400	−41.4400	−51.4400
−14.440	29.5600	−1.4400	−27.4400	−38.4400

$$\bar{g}(x_i, y_j) = g(x_i, y_j) - \bar{g} =$$

31.1600	37.1600	29.1600	23.1600	25.1600
16.1600	45.1600	21.1600	−6.8400	−8.8400
−5.8400	27.1600	−5.8400	−35.8400	−35.8400
−15.8400	17.1600	−17.8400	−42.8400	−46.8400
−13.8400	30.1600	1.1600	−27.8400	−39.8400

Thus, the two terms in the denominator of ZNCC coefficient can be calculated as

$$\sqrt{\sum \bar{f}^2} = \frac{1}{25} \sum_{i=-2}^{2} \sum_{j=-2}^{2} f(x_i, y_j)^2 =$$

$$\frac{1}{25}\left[31.16^2 + 37.16^2 + \cdots + (-27.84)^2 + (-39.84)^2 \right] = 137.7583$$

$$\sqrt{\sum \bar{g}^2} = \frac{1}{25} \sum_{i=-2}^{2} \sum_{j=-2}^{2} \bar{g}(x_i, y_j)^2 =$$

$$\frac{1}{25}\left[30.56^2 + 38.56^2 + \cdots + (-27.44)^2 + (-38.44)^2 \right] = 139.9506$$

Finally, we can compute the ZNCC coefficient according to

$$C_{ZNCC} = \frac{\sum \bar{f}_i \bar{g}_i}{\sqrt{\sum \bar{f}_i^2 \sum \bar{g}_i^2}} =$$

$$\frac{\displaystyle\sum_{i=-2}^{2} \sum_{j=-2}^{2} \left[\begin{array}{l} 31.16 \times 30.56 + 37.16 \times 38.56 + \cdots \\ +(-27.84 \times -27.44) + (-39.84 \times -38.44) \end{array} \right]}{137.7583 \times 139.9506} = 0.9974$$

6.2.2.4   Interpolation Scheme

As can be seen from (6.1), the coordinates of point (x_i', y_j') in the deformed sub-set may be located between pixels (i.e., subpixel location). Before evaluating the similarity between reference and deformed subsets using the correlation criterion defined in Table. 6.2, the intensity of these points with subpixel locations must be provided. Thus, a certain subpixel interpolation scheme should be utilized. In the literature, various subpixel interpolation schemes including bilinear interpolation, bicubic interpolation, bicubic B-spline interpolation, biquintic B-spline interpolation, and bicubic spline interpolation schemes have been used. The detailed algorithms

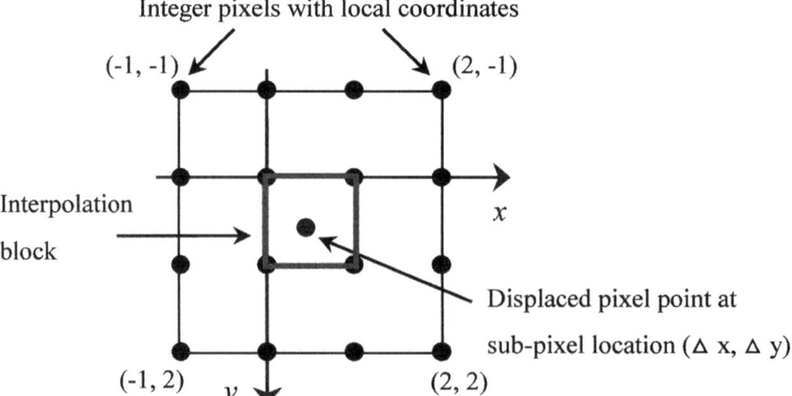

Figure 6.5 A schematic showing bicubic interpolation at subpixel location using the surrounding 16 integer pixels.

of these interpolation schemes can be found in numerical computing books [43]. However, high-order interpolation schemes (e.g., bicubuic spline interpolation or biquintic spline interpolation) are highly recommended [44] since they provide higher registration accuracy and better convergence character to the algorithm than the simpler interpolation schemes do.

Nevertheless, bicubic interpolation has been regularly used due to its efficiency and accuracy. In using the generalized bicubic interpolation scheme [45], the gray values and first-order gray gradients at subpixel locations can be estimated as:

$$g(\Delta x, \Delta y) = \sum_{m=0}^{3}\sum_{n=0}^{3}\alpha_{mn}(\Delta x)^m(\Delta y)^n$$

$$g_x(\Delta x, \Delta y) = \sum_{m=1}^{3}\sum_{n=0}^{3}\alpha_{mn}m(\Delta x)^{m-1}(\Delta y)^n \qquad (6.7)$$

$$g_y(\Delta x, \Delta y) = \sum_{m=0}^{3}\sum_{n=1}^{3}\alpha_{mn}n(\Delta x)^m(\Delta y)^{n-1}$$

In bicubic interpolation, the 16 unknown coefficients (i.e., a_{00}, a_{01}, ... , a_{33}, of (6.7)) can be determined by the gray intensity of the neighboring 4×4 pixels centered at the subpixel location, as illustrated in Figure 6.5. It should be emphasized here that the 16 coefficients remain the same for each interpolation block, regardless of the specific subpixel locations within it. For this reason, an efficient interpolation coefficient look-up table approach can be used to significantly speed the computational efficiency of DIC.

EXERCISE 6.2

A pixel is displaced to a subpixel location of $\Delta x = 0.4$ pixels, $\Delta y = 0.6$ pixels. The local gray intensities of the adjacent 4×4 pixels are listed in the table. Determine the intensity and two intensity gradients at the subpixel location using bicubic interpolatoin.

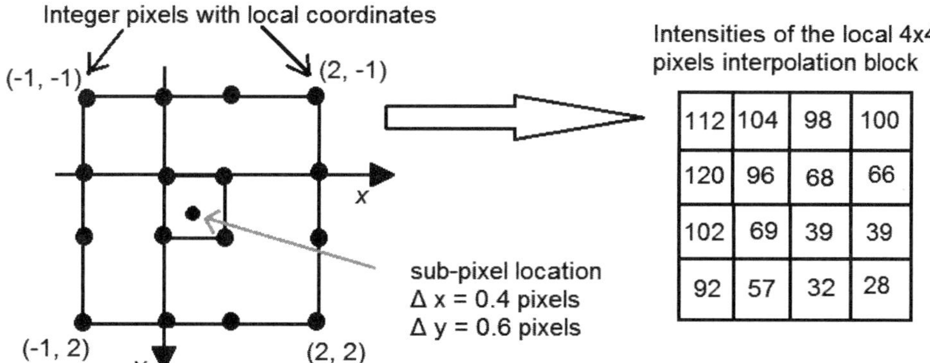

Solution

For each integer pixel in the local interpolation block, the relationship between its grayscale and its local coordinate can be written as

$$g(\Delta x, \Delta y) = a_{00} + a_{10}\Delta x + a_{20}\Delta x^2 + a_{30}\Delta x^3 + a_{01}\Delta y + a_{11}\Delta x\Delta y$$
$$+ a_{21}\Delta x^2\Delta y + a_{31}\Delta x^3\Delta y + a_{02}\Delta y^2 + a_{12}\Delta x\Delta y^2 + a_{22}\Delta x^2\Delta y^2$$
$$+ a_{32}\Delta x^3\Delta y^2 + a_{03}\Delta y^3 + a_{13}\Delta x\Delta y^3 + a_{23}\Delta x^2\Delta y^3 + a_{33}\Delta x^3\Delta y^3$$

As the local interpolation block contains 16 pixels with known local coordinates, these 16 equations can be rewritten in matrix form $\mathbf{G} = \mathbf{X}\mathbf{A_{mn}}$. That is,

$$
\begin{pmatrix}
g(-1-1) \\
g(-1,0) \\
g(-1,1) \\
g(-1,2) \\
g(0,-1) \\
g(0,0) \\
g(0,1) \\
g(0,2) \\
g(1,-1) \\
g(1,0) \\
g(1,1) \\
g(1,2) \\
g(2,-1) \\
g(2,0) \\
g(2,1) \\
g(2,2)
\end{pmatrix}
=
\begin{bmatrix}
1 & -1 & 1 & -1 & -1 & 1 & -1 & 1 & 1 & -1 & 1 & -1 & -1 & 1 & -1 & 1 \\
1 & -1 & 1 & -1 & 0 & 0 & 0 & 0 & 0 & 0 & 0 & 0 & 0 & 0 & 0 & 0 \\
1 & -1 & 1 & -1 & 1 & -1 & 1 & -1 & 1 & -1 & 1 & -1 & 1 & -1 & 1 & -1 \\
1 & -1 & 1 & -1 & 2 & -2 & 2 & -2 & 4 & -4 & 4 & -4 & 8 & -8 & 8 & -8 \\
1 & 0 & 0 & 0 & -1 & 0 & 0 & 0 & 1 & 0 & 0 & 0 & -1 & 0 & 0 & 0 \\
1 & 0 & 0 & 0 & 0 & 0 & 0 & 0 & 0 & 0 & 0 & 0 & 0 & 0 & 0 & 0 \\
1 & 0 & 0 & 0 & 1 & 0 & 0 & 0 & 1 & 0 & 0 & 0 & 1 & 0 & 0 & 0 \\
1 & 0 & 0 & 0 & 2 & 0 & 0 & 0 & 4 & 0 & 0 & 0 & 8 & 0 & 0 & 0 \\
1 & 1 & 1 & 1 & -1 & -1 & -1 & -1 & 1 & 1 & 1 & 1 & -1 & -1 & -1 & -1 \\
1 & 1 & 1 & 1 & 0 & 0 & 0 & 0 & 0 & 0 & 0 & 0 & 0 & 0 & 0 & 0 \\
1 & 1 & 1 & 1 & 1 & 1 & 1 & 1 & 1 & 1 & 1 & 1 & 1 & 1 & 1 & 1 \\
1 & 1 & 1 & 1 & 2 & 2 & 2 & 2 & 4 & 4 & 4 & 4 & 8 & 8 & 8 & 8 \\
1 & 2 & 4 & 8 & -1 & -2 & -4 & -8 & 1 & 2 & 4 & 8 & -1 & -2 & -4 & -8 \\
1 & 2 & 4 & 8 & 0 & 0 & 0 & 0 & 0 & 0 & 0 & 0 & 0 & 0 & 0 & 0 \\
1 & 2 & 4 & 8 & 1 & 2 & 4 & 8 & 1 & 2 & 4 & 8 & 1 & 2 & 4 & 8 \\
1 & 2 & 4 & 8 & 2 & 4 & 8 & 16 & 4 & 8 & 16 & 32 & 8 & 16 & 32 & 64
\end{bmatrix}
\begin{pmatrix}
a_{00} \\
a_{01} \\
a_{02} \\
a_{03} \\
a_{10} \\
a_{11} \\
a_{12} \\
a_{13} \\
a_{20} \\
a_{21} \\
a_{22} \\
a_{23} \\
a_{30} \\
a_{31} \\
a_{32} \\
a_{33}
\end{pmatrix}
$$

Substituting the intensity of each pixel into this equation, the unknown 16 coefficients (i.e., $a_{00}, a_{01}, \ldots, a_{33}$) can be determined as $\mathbf{X^{-1}G} = \mathbf{A_{mn}}$. That gives

$$A_{mn} = \begin{pmatrix} 96.0000000000000 \\ -31.0000000000000 \\ -2 \\ 5 \\ -23.1666666666667 \\ -12.5277777777778 \\ 1.33333333333333 \\ 1.36111111111111 \\ -9.50000000000000 \\ 9 \\ 3.25000000000000 \\ -2.25000000000000 \\ 5.66666666666667 \\ -1.47222222222222 \\ -1.08333333333333 \\ 0.388888888888889 \end{pmatrix}$$

Finally, by substituting $\Delta x = 0.4$ pixels, $\Delta y = 0.6$ pixels and the already determined interpolation coefficients a_{mn} into (6.7), the subpixel intensity $g(\Delta x, \Delta y)$ and intensity gradients $g_x(\Delta x, \Delta y)$　$g_y(\Delta x, \Delta y)$ are estimated as $g = 65.9497$, $g_x = -33.3623$, $g_y = -23.4825$, respectively.

6.2.3　Displacement Field Measurement Using DIC

Because strains are normally estimated by differentiating the obtained displacement fields, displacement field measurement is generally considered a key step in DIC. In the literature, various displacement tracking algorithms, including the subset-based local approach [46–48] and the element-based global approach [49–51] have been developed. However, the Newton-Raphson algorithm, originally proposed by Bruck et al. [46] and subsequently refined by other researchers [47, 45], has been considered as a classic and standard algorithm. A detailed comparison between NR algorithm and other subpixel registration algorithms (e.g., the peak-finding algorithm and the gradient-based algorithm) reveals that NR algorithm offers the most accurate and stable measurement [48]. In this section, the basic principle of NR algorithm is described in detail.

6.2.3.1　Subpixel Displacement Measurement Using Forward Additive Gauss-Newton Algorithm

If we take the relative deformation (i.e., shape change and rotation) between the reference and target subsets into account, the correlation function becomes a non-linear function with respect to the desired mapping parameters vector. For example, if the first-order shape function is used, the desired mapping parameters vector is

$\mathbf{p} = (u, u_x, u_y, v, v_x, v_y)^T$. These desired deformation parameters can be found using certain optimization algorithms. In this subsection, a forward additive matching strategy using the widely accepted NR algorithm is described. In the NR algorithm, the target subset is deformed by $\mathbf{W}(\boldsymbol{\xi};\mathbf{p})$ with $\mathbf{p}$ denoting an initial guess of deformation. Then an incremental parameter vector $\Delta\mathbf{p}$ is further applied to the deformed subset and is then compared with the original reference subset to solve for $\Delta\mathbf{p}$, as schematically shown in Figure 6.6.

Specifically, to quantitatively evaluate the similarity between the reference and target subsets and determine the desired deformation parameters, the robust ZNSSD criterion, combined with an affine transform warping function, is used.

$$C_{ZNSSD}(\Delta\mathbf{p}) = \sum_{\xi}\left\{\frac{[f(\mathbf{x}+\boldsymbol{\xi})-\overline{f}]}{\Delta f} - \frac{[g(\mathbf{x}+\mathbf{W}(\boldsymbol{\xi};\mathbf{p}+\Delta\mathbf{p}))-\overline{g}]}{\Delta g}\right\}^2 \tag{6.8}$$

where $f(\mathbf{x})$ and $g(\mathbf{x})$ denote the grayscale levels at $\mathbf{x} = (x, y, 1)^T$ of reference image and deformed image, and $\overline{f} = 1/N\sum_{\xi}f(\mathbf{x}+\boldsymbol{\xi})$, $\overline{g} = 1/N\sum_{\xi}g(\mathbf{x}+\mathbf{W}(\boldsymbol{\xi};\mathbf{p}+\Delta\mathbf{p}))$ are the mean intensity values of the two subsets with N denoting the total number of points within the subset, $\Delta f = \sqrt{\sum_{\xi}\left[f(\mathbf{x}+\boldsymbol{\xi})-\overline{f}\right]^2}$ and $\Delta g = \sqrt{\sum_{\xi}\left[g(\mathbf{x}+\mathbf{W}(\boldsymbol{\xi};\mathbf{p}+\Delta\mathbf{p}))-\overline{g}\right]^2}$.

$\boldsymbol{\xi} = (\Delta x, \Delta y, 1)^T$ are the local coordinates of the pixel in each subset. (Note that all image coordinates are expressed in a form of homogenous coordinates for notation brevity.) $\mathbf{W}(\boldsymbol{\xi};\mathbf{p})$ is the warp function, also known as displacement mapping function, depicting the position and shape of the target subset relative to the reference subset; $\Delta\mathbf{p}$ is the desired incremental deformation vector applied to the target subset when it is deformed by a initial guess of deformation $\mathbf{p}$.

In general, a practical affine warp function (i.e., linear displacement mapping function) allowing rigid body translation, rotation, normal strain, shear strain, and their combinations of the target subsets, is used.

$$\mathbf{W}(\boldsymbol{\xi};\mathbf{p}) = \begin{bmatrix} 1+u_x & u_y & u \\ v_x & 1+v_y & v \\ 0 & 0 & 1 \end{bmatrix}\begin{pmatrix} \Delta x \\ \Delta y \\ 1 \end{pmatrix} \tag{6.9}$$

where $\mathbf{p} = (u, u_x, u_y, v, v_x, v_y)^T$ is the precomputed deformation parameter vector exerted on the target subset. It should be noted that although only first-order displacement mapping is used, the following derivation can be easily extended to a second-order shape function.

To solve for $\Delta\mathbf{p}$, we first perform a first-order Taylor expansion of (6.8) with respect to $\Delta\mathbf{p}$. That yields

$$C_{ZNSSD}(\Delta\mathbf{p}) = \sum_{\xi}\left[\frac{f(\mathbf{x}+\boldsymbol{\xi})-\overline{f}}{\Delta f} - \frac{g(\mathbf{x}+\mathbf{W}(\boldsymbol{\xi};\mathbf{p}))+\nabla g\frac{\partial\mathbf{W}}{\partial\mathbf{p}}\Delta\mathbf{p}-\overline{g}}{\Delta g}\right]^2 \tag{6.10}$$

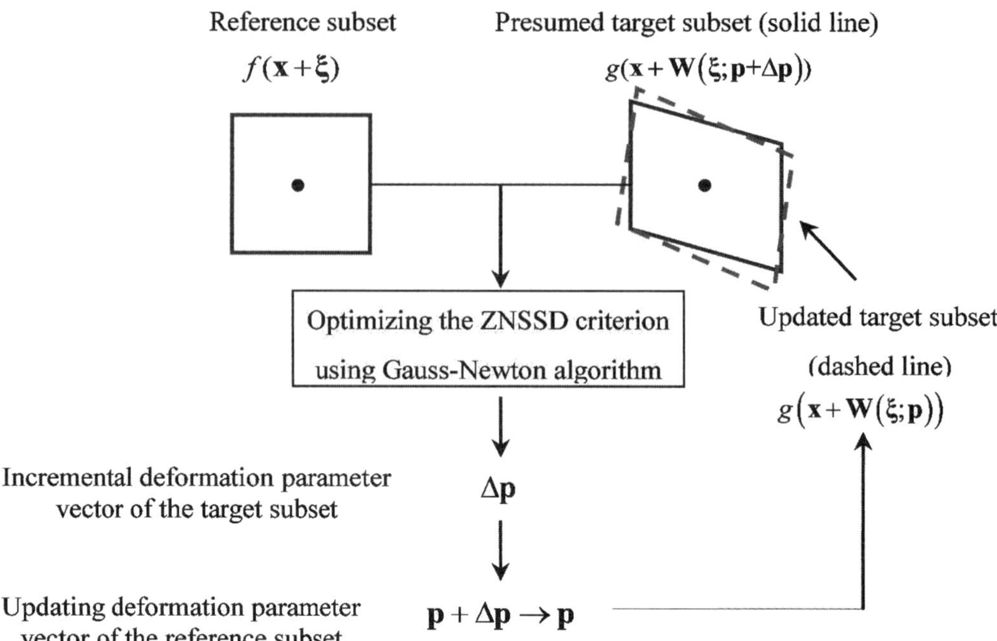

Figure 6.6 Schematic drawing of the forward matching strategy of the classic NR algorithm.

where $g(\mathbf{x} + \mathbf{W}(\boldsymbol{\xi};\mathbf{p}))$, $\nabla g = \big(\partial g(\mathbf{x} + \mathbf{W}(\boldsymbol{\xi};\mathbf{p}))/\partial x, \partial g(\mathbf{x} + \mathbf{W}(\boldsymbol{\xi};\mathbf{p}))/\partial y\big)$ are the intensities and intensity gradients of the target subset. Note that these values are located in subpixel positions; thus, a subpixel intensity interpolation algorithm (e.g., the bicubic interpolation scheme described in preceding section) should be used.

$\dfrac{\partial \mathbf{W}}{\partial \mathbf{p}} = \begin{pmatrix} 1 & \Delta x & \Delta y & 0 & 0 & 0 \\ 0 & 0 & 0 & 1 & \Delta x & \Delta y \end{pmatrix}$ is the Jacobian of the warp function.

Minimization of the $C_{ZNSSD}(\Delta\mathbf{p})$ with respect to $\Delta\mathbf{p}$ (i.e., $\partial C_{ZNSSD}(\Delta\mathbf{p})/\partial(\Delta\mathbf{p}) = 0$, gives the least-squares solution of $\Delta\mathbf{p}$).

$$\Delta\mathbf{p} = -\mathbf{H}_{6\times6}^{-1} \times \sum_{\xi}\left\{\left(\nabla g\frac{\partial\mathbf{W}}{\partial\mathbf{p}}\right)_{6\times1}^{T} \times \left[\frac{\Delta g}{\Delta f}\big(f(\mathbf{x}+\boldsymbol{\xi})-\bar{f}\big)-\big(g(\mathbf{x}+\mathbf{W}(\boldsymbol{\xi};\mathbf{p}))-\bar{g}\big)\right]\right\}$$

$$(6.11)$$

where $\mathbf{H}$ is the 6×6 Hessian Matrix given as

$$\mathbf{H} = \sum_{\xi}\left[\left(\nabla f\frac{\partial\mathbf{W}}{\partial\mathbf{p}}\right)^{T} \times \left(\nabla f\frac{\partial\mathbf{W}}{\partial\mathbf{p}}\right)\right] \qquad (6.12)$$

Based on the obtained incremental parameter vector $\Delta\mathbf{p}$, the updated deformation parameter vector of the target subset can be computed as

$$\mathbf{p} \leftarrow \mathbf{p} + \Delta\mathbf{p} \qquad (6.13)$$

Equation (6.11) is repeated until a predefined convergence condition is satisfied. In our recent results [52], the convergence conditions are set to ensure that variations in the norm of the incremental deformation parameter is equal to or less than 0.001 (i.e., $\|\Delta\mathbf{p}\| = \sqrt{(\Delta u)^2 + (\Delta v)^2} \leq 0.001$ pixels, or a maximum iteration has been achieved).

6.2.3.2   Initialization of NR Algorithm and Calculation Path

Recall that the regularly spaced pixel points within ROI are considered points to be correlated, and the displacements of all these calculation points should be computed through the NR algorithm. However, as a local optimization algorithm, the NR algorithm requires an initial guess of deformation sufficiently close to the true value to converge rapidly and correctly. Otherwise, the algorithm may converge to incorrect values or doesn't converge at all. Although we can estimate the initial guess separately for each point, this approach is either impractical because it is extremely time consuming or impossible if large deformation or rotation are present in the deformed image.

BOX 6.1   Reliability-guided displacement tracking strategy

More recently, a robust and universally applicable reliability-guided displacement tracking algorithm is proposed [53, 54] for reliable and efficient image deformation measurement. In the method, ZNCC coefficient is used to identify the reliability of a computed point. The correlation calculation begins with a seed point and is then guided by the calculated ZNCC coefficient. The neighbors of the point with highest ZNCC coefficient in a queue for computed points will be processed using NR algorithm first. Meanwhile, the computed deformation parameters of this point will be transferred to its neighbors according to deformation continuity assumption. The advantages of the reliability-guided displacement tracking algorithm are triple. First, the calculation path is always along the most reliable direction and possible error propagation of conventional DIC method can be entirely avoided. Second, because a very accurate initial guess can be predicted for the points to be processed, the NR algorithm performed on these points generally converge rapidly after minimal iterations. Third, it is universally applicable to the deformation measurement of images with area and/ or deformation discontinuities

---

6.2.3.2.1   Initial Guess for the Seed Point

As for the seed point, if the relative deformation or rotation between the reference subset and the deformed subset is quite small, it is easy to get an accurate estimation of the initial displacements of seed points by a simple searching scheme implemented either in the spatial domain or in the frequency domain. In the spatial domain, the accurate locations of the target subset can be determined by a fine search routine, pixel by pixel, performed within the specified range in the deformed image. Some schemes, such as the coarse-to-fine, the nested searching scheme [55], and sum-table approach [52], can be used to speed up the calculation. These search schemes yield 1 pixel resolution. Alternatively, the correlation between the reference subset and the deformed subset can also be implemented in the Fourier domain as conducted and advocated by Chen et al. [56]. In Fourier domain, the correlation between two subsets is calculated as the complex multiplication of the first subset's Fourier spectrum by the complex conjugate of the second subset's spectrum. Since the FFT can be implemented with very high speed, the Fourier domain correlation method is also extremely fast. However, the in-plane translation is implicitly assumed in Fourier domain method, and small strains and/or rotations that occur between the two subsets will lead to significant errors.

The previously mentioned technique performs well for most cases. Difficulties occur in certain cases when large rotation and/or large deformation is present between the reference and target subsets. In these cases, some pixels of the reference subset run out of the area of the assumed subset within the deformed image; consequently, the similarity between the reference subset and the assumed deformed subset will decrease substantially. As can be seen from Figure 6.7(a), if only rigid body translation exists between the reference and deformed subsets, a single peak can be found in the correlation coefficient distribution. In contrast, while a 20 degree relative rotation occurs, there will not be even a single sharp peak in the correlation distribution map as indicated in Figure 6.7(b); thus, this will result in a failure in integer pixel displacement searching.

In these cases, other techniques are therefore required to achieve a reliable initial guess of deformation. For example, a technique was presented by Pan et al. [57] to achieve a reliable initial guess for the NR method for these cases. The initial guess of deformation is determined by manually selecting three or more points $(x_i, y_i)(i = 1, 2, \ldots, n,$ and $n \geq 3)$ with distinct features around the reference subset center, and their corresponding locations $(x_i', y_i')(i = 1, 2, L, \ n,$ and $n \geq 3)$ in the deformed image. Thus, the desired initial guess can be resolved from their coordinate correspondence (depicted by the first-order shape function) using the least squares method. Alternatively, the genetic algorithm [58] and the robust feature matching algorithm called SIFT [59] can also be used as a fully automated technique for determining the initial guess of the first calculation point.

6.2.3.2.2 Initial Guess for Rest Calculation Points Using Reliability-Guided Transfer Scheme

To provide an accurate initial guess for the rest calculation points, a simple but effective initial guess transfer scheme, guided by the ZNCC correlation coefficients of computed points, is used in this work. Specifically, remove the first point, which has the maximum ZNCC coefficient, from the top of the queue. Then, analyze each of its four (or eight) neighboring points. If it is a valid point and has not been computed, it will then be computed using the NR algorithm to refine its correlation coefficient and deformation parameters. Note that (as shown in Figure 6.8), the initial guess of the neighboring points (denoted by $\mathbf{P}_{NP}^0 = (u_{NP}^0, u_{x\_NP}^0, u_{y\_NP}^0, v_{NP}^0, v_{x\_NP}^0, v_{y\_NP}^0)^T$) is computed from the computed deformation parameters of the removed point (denoted as $\mathbf{P}_{RP} = (u_{RP}, u_{x\_RP}, u_{y\_RP}, u_y, v_{RP}, v_{x\_RP}, v_{y\_RP})^T$) through the following formula

$$u^0_{NP} = u_{RP} + u_{x\_RP} \times \Delta x + u_{y\_RP} \times \Delta y$$

$$u^0_{x\_NP} = u_{x\_RP}$$

$$u^0_{y\_NP} = u_{y\_RP}$$

$$v^0_{NP} = v_{RP} + v_{x\_RP} \times \Delta x + v_{y\_RP} \times \Delta y \qquad (6.14)$$

$$v^0_{x\_NP} = v_{x\_RP}$$

$$v^0_{y\_NP} = v_{y\_RP}$$

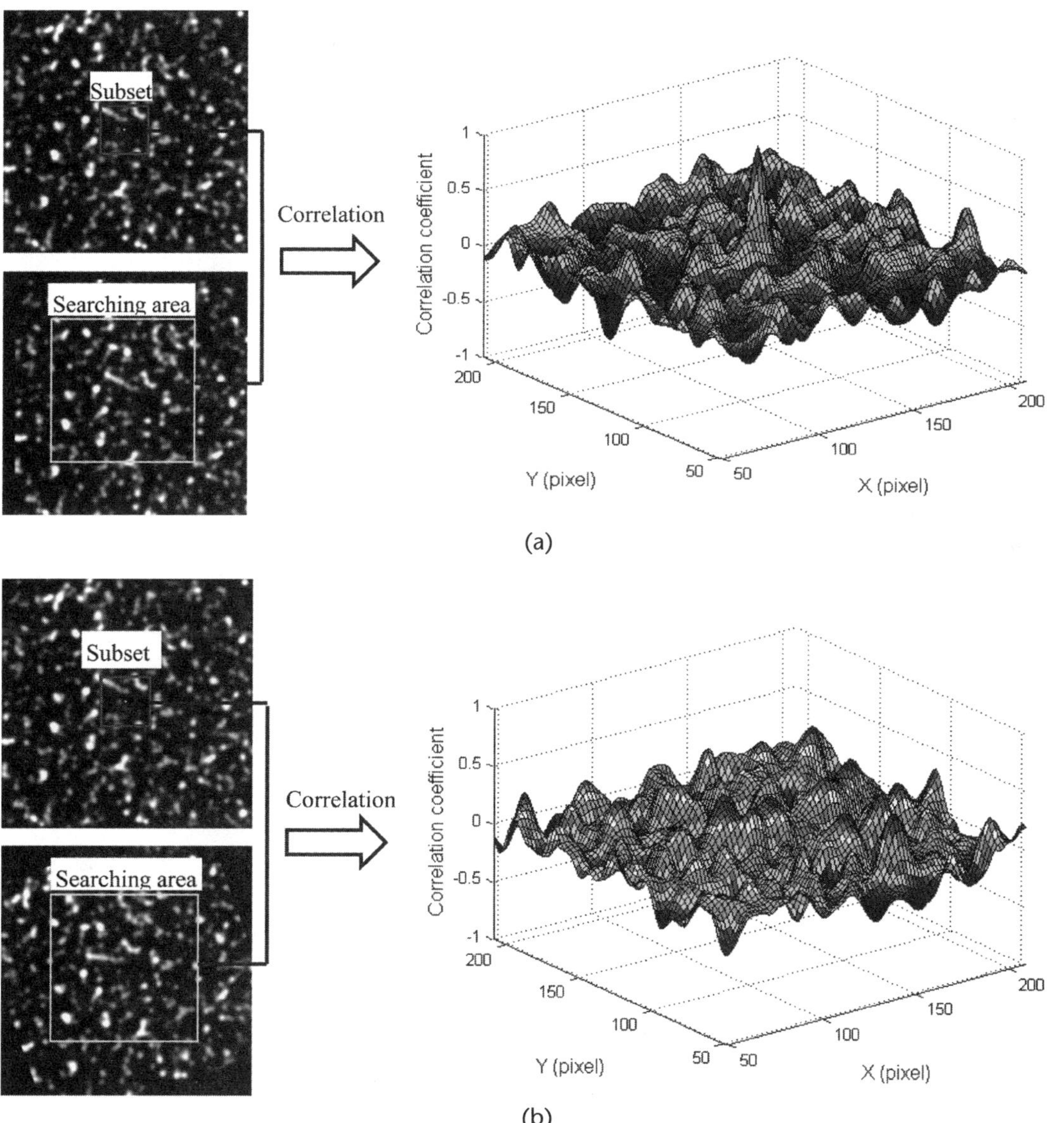

Figure 6.7 Computed whole-field cross-correlation coefficient distribution when the deformed image is subjected to (a) rigid body translation, and (b) 20 degree relative rotation.

where Δx, Δy are the distances from the neighboring point to the removed point (as indicated in Figure 6.8). We should note that (6.14) will provide very accurate initial guess estimations by taking the local displacement gradients into consideration. When the deformed images involve large relative deformation or rotation, the NR algorithm, whose initial guess is provided by (6.14), generally exhibits better convergence characteristics, resulting in a higher average ZNCC coefficient (indicating higher registration accuracy) and lower average iteration number (indicating higher computational efficiency) [60].

Afterward, each of the just calculated points is labeled as calculated points and inserted into the queue according to its ZNCC coefficient value. The initial guess procedure described here is repeated until the queue is empty, which means that all the valid points defined in the ROI have been computed and the DIC analysis is completed.

6.2.3.3　An Example of RG-NR Algorithm

To demonstrate the effectiveness and robustness of the reliability-guided displacement tracking, the experimental images of a dog-bone shaped specimen made of aluminum with dimensions 2mm × 8mm × 0.13mm is used. Prior to the experiment, an artificial speckle pattern was made on the specimen by spraying white and black paint onto its surface. The specimen was clamped tightly at both ends and the corresponding image was recorded as the reference image. Afterward, horizontal uniaxial tensile loading was exerted on the right side and a deformed image was recorded. The image pair shown in Figure 6.9 will be processed with the proposed DIC technique to extract the full-field displacement fields.

Since the specimen is of relatively complex shape with two semicircular cut-outs, the ROI is approximated with a polygon as shown in Figure 6.9(a). Figure 6.9(c) intuitively shows all the valid points (highlighted in white) defined in the reference image, and all the pixels in the ROI are labeled as 1 in the corresponding binary mask M_v. For the points within an irregular ROI, the conventional DIC method

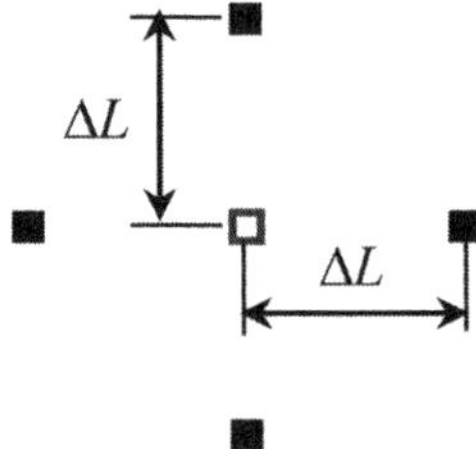

□　Calculated point removed from the top of the queue with an estimated displacement vector of $\mathbf{P}_{\mathrm{RP(removed\ point)}} = \left(u_{\mathrm{RP}}, u_{x\_\mathrm{RP}}, u_{y\_\mathrm{RP}}, u_y, v_{\mathrm{RP}}, v_{x\_\mathrm{RP}}, v_{y\_\mathrm{RP}} \right)^T$

■　Neighboring points to be processed, its initial guess of deformation $\mathbf{P}_{\mathrm{NP(Neighboring\ point)}} = \left(u^0_{\mathrm{NP}}, u^0_{x\_\mathrm{NP}}, u^0_{y\_\mathrm{NP}}, v^0_{\mathrm{NP}}, v^0_{x\_\mathrm{NP}}, v^0_{y\_\mathrm{NP}} \right)$ can be estimated using Eq.(6.14)

Figure 6.8　Schematic drawing showing the improved initial guess transfer scheme considering the distance and displacement gradients between adjacent calculation points.

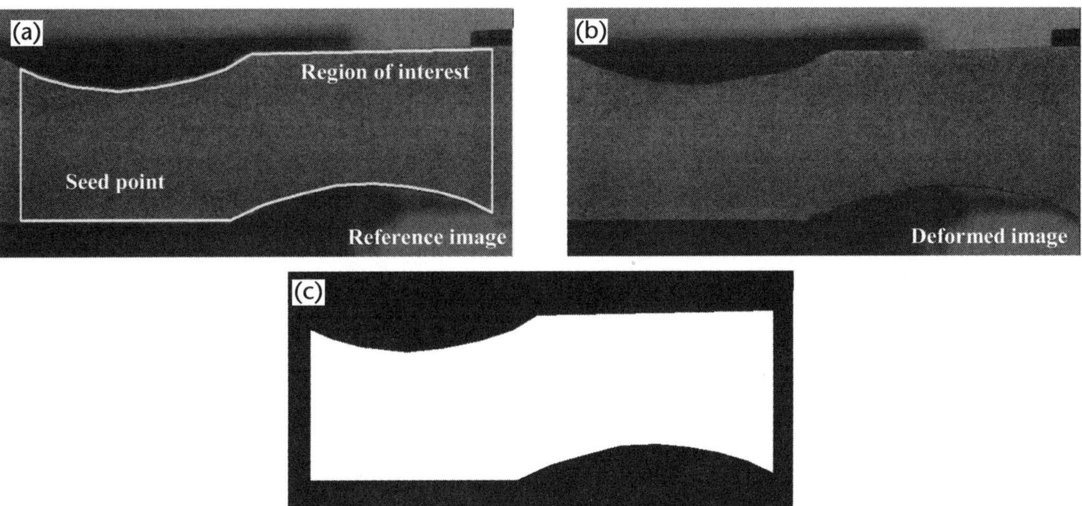

Figure 6.9 Experimental images of a dog-bone specimen subjected to uniaxial tensile loading: (a) reference image, (b) deformed image, (c) a binary image shows all the valid points within the specified ROI.

is not easy to process and is prone to yielding incorrect results at some locations. Using the reliability-guided displacement tracking approach, the calculation starts from the seed point (the seed point and its surrounding subset are plotted in red in Figure 6.9(a)) and is then guided by the ZNCC coefficients of computed points. Figure 6.10 gives two intermediate stages of the process along with the final results of the computed u-, v-displacement fields, and ZNCC coefficient map. Note that the void areas within the contour plots shown in Figure 6.10 denote the points that will be computed later. By observing the ZNCC coefficient distributions shown in bottom row of Figure 6.10, it can be clearly seen that the displacements were determined in an order from the points with higher ZNCC coefficients to the points with lower ZNCC coefficients. The displacements of the irregular ROI containing all boundary points were all correctly computed.

6.2.4   Strain Field Estimation with Pointwise Least Squares Algorithm

Using the displacement mapping algorithm, we can get the full-field displacements to subpixel accuracy. However, in many tasks of experimental solid mechanics, such as mechanical testing of materials and structural stress analysis, full-field strain distributions are more important and desirable. This is because strains are closely related to the many mechanical and physical quantities (such as stress state, elastic modulus, Poisson's ratio, coefficient of thermal expansion) of the test sample. Unfortunately, less work has been devoted on the reliable estimation of strain fields. One reason for this can be attributed to the fact that the displacement gradients (i.e., strains) can be directly calculated using the displacement estimation algorithms (e.g., the NR algorithm).

However, it should be mentioned first that the error of the estimated displacement gradients using the NR algorithm normally limits its use only to local strains greater than approximately 0.010 [46]. Additionally, although the relationship between the

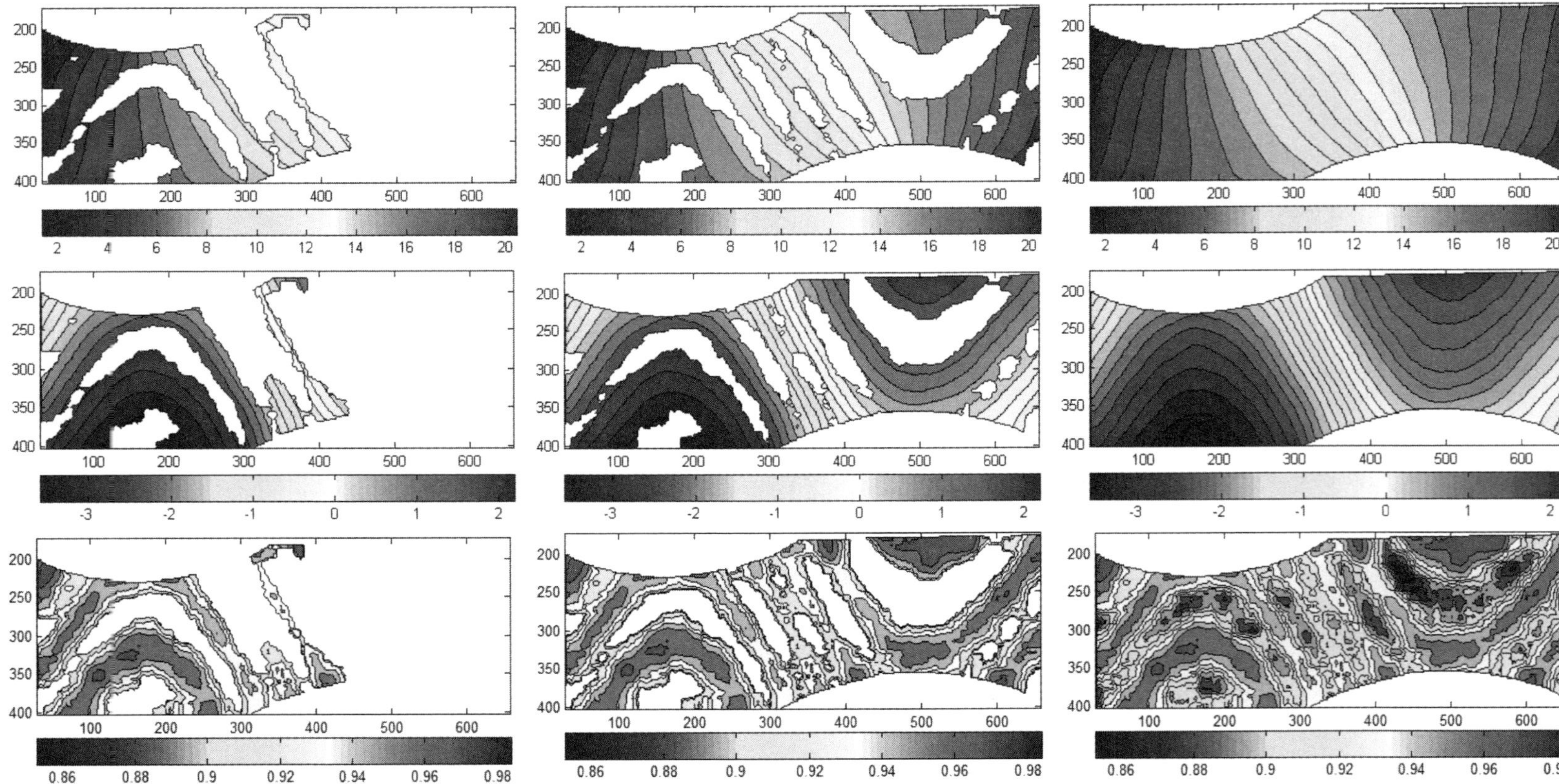

Figure 6.10 Two intermediate results and the final results of the computed u-displacement (top), v-displacement (middle), and ZNCC coefficient map (bottom) using the proposed method. The void areas within the contour plots denote the points with relatively lower ZNCC coefficients and will be processed later.

strain and displacement can be described as a numerical differentiation process in mathematical theory, unfortunately, the numerical differentiation is considered as an unstable and risky operation, because it can amplify the noise contained in the computed displacement. Therefore, the resultant strains are untrustworthy if they are calculated by directly differentiating the estimated noisy displacements. For example, if the error of the displacement estimation is estimated as ± 0.02 pixels and the grid step is 5 pixels, the error of resultant strain calculated by forward difference is $\Delta\varepsilon = (\,|\pm 0.02| + |\pm 0.02|\,)/5 = 8000\ \mu\varepsilon$ and by central difference is $\Delta\varepsilon = (\,|\pm 0.02| + |\pm 0.02|\,)/10 = 4000\ \mu\varepsilon$ [61]. An error to that extent will probably hide the underlying strain information of the tested specimen and is unbearable in most cases.

It is therefore believed that the accuracy of strain estimation will be improved by smoothing the computed displacement fields first and subsequently differentiating them to calculate strains. Based on these considerations, several researchers [62–64] proposed a technique involving first smoothing the computed displacement fields with the finite element method and subsequently differentiating them to calculate strains. In addition, the thin plate spline smoothing technique and three other smoothing algorithms were also introduced [65–68], respectively, to remove the noise contained in displacement fields. Because the noise level contained in the displacement field is significantly decreased after smoothing operations, these techniques substantially increase the precision in the resulting strain estimation. However, these smoothing techniques come with two shortcomings. First, it is very difficult to conduct an optimal smoothing. The resulting displacement field tends to be over-smoothed or not sufficiently smoothed. Second, the small fluctuations in the smoothed displacement fields can also be amplified by the numerical differentiation.

At present, the more practical technique for strain field estimation is the pointwise local least squares fitting technique used and advocated by Wattrisse et al. [69] and Pan et al. [41]. In order to analyze the strain localization phenomena that occurs during the tension of thin flat steel samples, Wattrisse et al. [69] implemented a local least squares method to estimate the strains from the discrete and noisy displacement fields computed by DIC method. To obtain the strains of the points located at calculation boundaries, a continuity extension of the displacement field is performed at the image boundary. A similar technique was also used by Pan et al. [41] with a simpler and more effective data processing procedure for the calculation of strains for the points located at image boundary, holes, cracks, and other discontinuity areas.

The implementation of local least squares fitting technique for strain estimation can be explained as follows. As shown in Figure 6.11, suppose we want to compute the strains of a calculation point (x, y). We first select a square window (also known as strain calculation window) containing $(2N + 1) \times (2N + 1)$ discrete points around it. Note that the real size of strain calculation window is $(2N \times \Delta L)^2$ pixels, with ΔL standing for the grid step between adjacent points. If the strain calculation window is small enough, the local displacement distributions in it can be approximated using a bilinear function; thus, we have

$$
\begin{aligned}
u(x + i, y + j) &\cong a_0 + a_1 i + a_2 j + a_3 ij \\
v(x + i, y + j) &\cong b_0 + b_1 i + b_2 j + b_3 ij
\end{aligned}
\tag{6.15}
$$

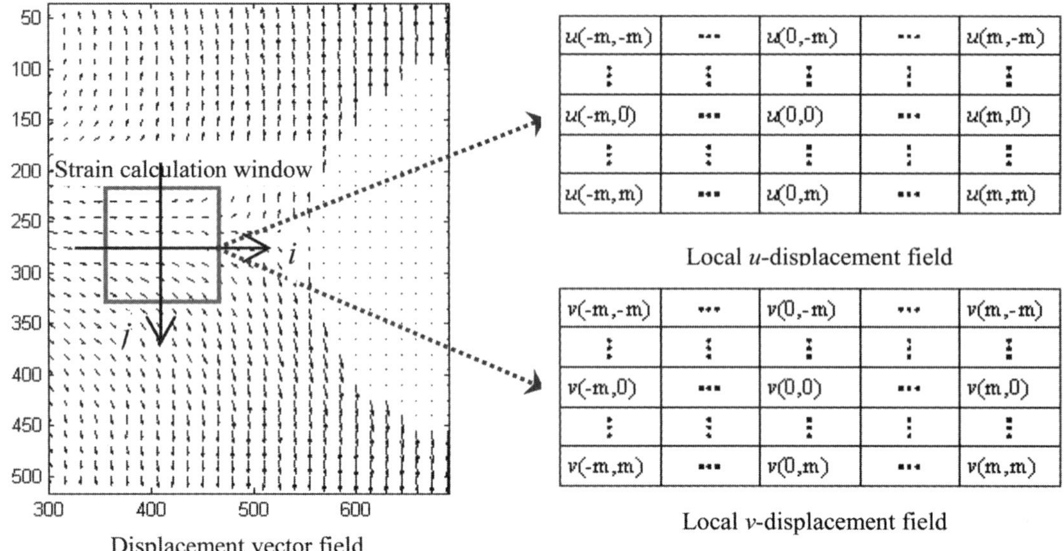

Figure 6.11 Local strain calculation window containing $(2m + 1) \times (2m + 1)$ discrete displacement data used for strain estimation.

where $i, j = -N \times \Delta L : N \times \Delta L$ are the local coordinates within the strain calculation window, $u(x + i, y + j)$ and $v(x + i, y + j)$ are the original displacements obtain by DIC as indicated in Figure 6.11, and $a_{i=0,1,2,3}$, $b_{i=0,1,2,3}$ are the unknown polynomial coefficients to be determined. These unknown parameters can be estimated by minimizing the following two objective functions:

$$\chi^2\left(a_0, a_1, a_2, a_3\right) = \sum_{i,j \in W_s} \left[u(x + i, y + j) - \left(a_0 + a_1 i + a_2 j + a_3 ij\right)\right]^2$$

$$\chi^2\left(b_0, b_1, b_2, b_3\right) = \sum_{i,j \in W_s} \left[v(x + i, y + j) - \left(b_0 + b_1 i + b_2 j + b_3 ij\right)\right]^2 \tag{6.16}$$

It is important to note that, for the points located at the image boundary or in the vicinity of discontinuity area, the strain calculation window may contain less than $(2m + 1) \times (2m + 1)$ points. However, we can still compute strain components using least squares fitting by simply neglecting these invalid points within the local strain calculation window from (6.16) [41].

Once these coefficients are obtained, the desired Cauchy strains or Green strains at the center point of a local subregion can be computed. For example, the Cauchy strains can be calculated as:

$$\varepsilon_x = \frac{\partial u}{\partial x} = a_1, \quad \varepsilon_y = \frac{\partial v}{\partial y} = b_2, \quad \gamma_{xy} = \frac{\partial u}{\partial y} + \frac{\partial v}{\partial x} = a_2 + b_1 \tag{6.17}$$

BOX 6.2   Proper strain window selection for heterogeneous deformation measurement

Since the random noise can be largely removed in the process of local fitting, the accuracy of calculated strains is greatly improved. However, it is quite necessary to note that, to obtain reasonable and accurate full-field strain estimation, the following two aspects are important and should be considered. One is the accuracy of displacement fields obtained by DIC, and the other is the size of the local strain calculation window used for fitting. For homogeneous deformation, a large strain calculation window is preferred. However, for inhomogeneous deformation, a proper size of strain calculation window should be selected with care to get a balance between strain accuracy and smoothness. Because a small strain calculation window cannot suppress the noise of the displacements, while a large strain calculation window may lead to the unreasonable linear approximation of deformation within the strain calculation window.

6.2.5   Application of 2D-DIC for Strain Measurement

As an easy-to-implement yet effective optical technique, 2D-DIC has been successfully used in various fields for in-plane surface deformation measurement of nominal planar objects. In this section, as a simple but representative application example, 2D-DIC is used to measure the in-plane strains of an aluminum specimen subjected to uniaxial tensile loading. To show the accuracy and precision of 2D-DIC, the strains measured by 2D-DIC are carefully compared with those measured by strain gauge rosettes. Figure 6.12 illustrates the established high-accuracy 2D-DIC

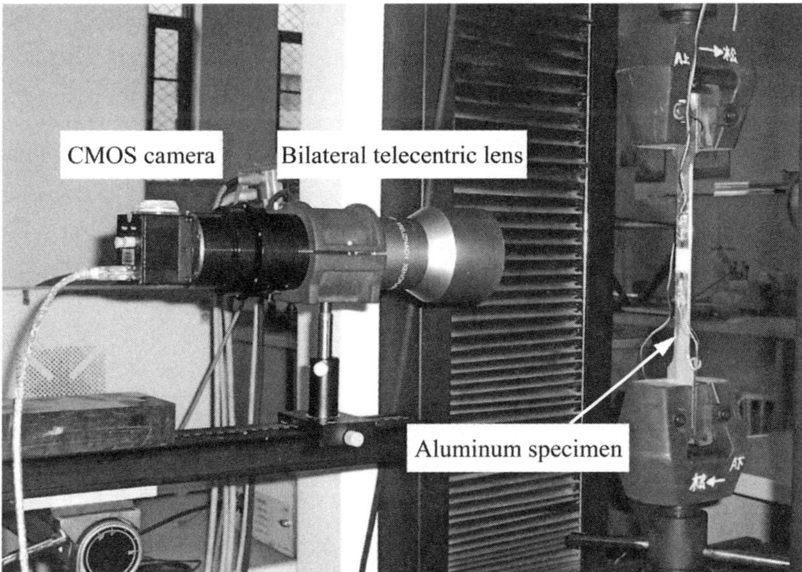

Figure 6.12   Aluminum tension specimen installed in the material testing machine to compare the strains measured by 2D-DIC and strain rosettes.

system using an 8-bit digital CMOS camera with a resolution of 1280×1024 pixels (DH-HV1351UM, Daheng Image Co., Ltd., Beijing, China) and a well-designed commercial bilateral telecentric lens (Xenoplan 1:5, Schneider Optics, Germany). The reason a high-quality telecentric lens is used in the 2D-DIC system instead of common industrial zoom lens is that the high-quality bilateral telecentric lens can offer two obvious merits over the common industrial lens: (1) it can maintain constant magnification even though the object surface and image plane undergo serious out-of-plane motion, and (2) the lens distortion is small enough to be neglected.

The geometry of the tension specimen is shown in Figure 6.13. The test sample is a dog-bone aluminum alloy measuring 20 mm wide by 4 mm thick. The overall length of the specimen is 360 mm, and the distance between two grips is approximately 280 mm. Before testing, white and black paints were sprayed onto the examined area of the specimen to form random speckles suitable for DIC analysis. For comparison, two strain gauge rosettes were attached above and under the area of the random speckle, respectively.

The tensile experiment is performed in a common universal testing machine (WDW-100A, Jinan Shidai Shijing Yiqi, Co., Ltd., Shandong Province, China). The specimen is tightly clamped at both ends on the test machine. During the test, the

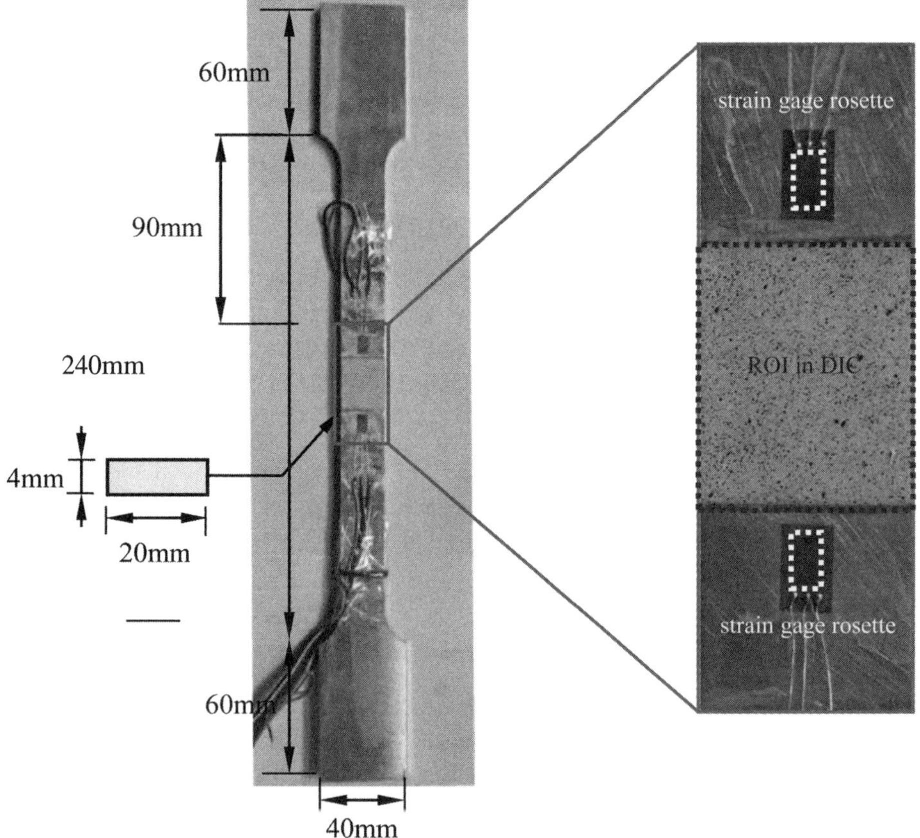

Figure 6.13 Photograph of the aluminum specimen. Random speckle patterns were made in the middle of the specimen; two rectangular strain rosettes were attached just above and below the ROI of DIC measurement.

movable upper clamper stretches the specimen with the constant loading rate value of 50 N/s. Note that the image acquired at the load stage of 1 KN is regarded as the reference image in consideration of the assembling error of the test machine, and the strain gauge reading was cleared. Subsequently, the image recorded every load step of 0.5 KN served as a deformed image, and simultaneously the data of strain gauge rosettes was recorded. When the load of the test machine reached 8.5 KN, the test was suspended, and the load was reduced to 1 KN. As the maximum load during the test is much less than yield load of the specimen, the specimen can be reused for testing.

For deformation analysis using DIC, a calculation area, with the size of 680 × 530 pixels, as indicated in Figure 6.14, is first selected. Then the calculation area is divided into a regularly distributed 68 × 53 virtual grid, and the subset size is chosen as 41 × 41 pixels for correlation matching in each calculation pointwise of the grid. The image series were first analyzed using fast and robust NR algorithm and then the computed displacements were locally fitted using the point least squares

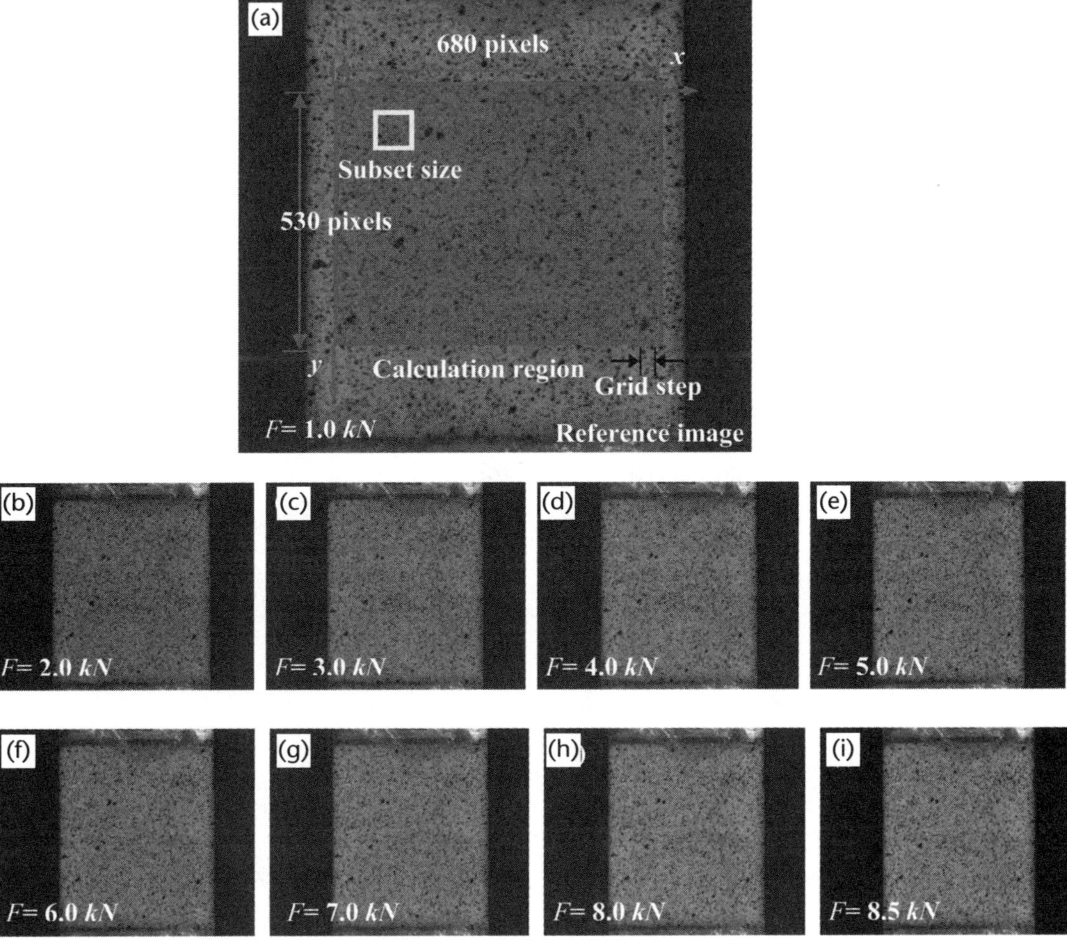

Figure 6.14   (a) Reference image: the red rectangle is the calculation area, of which each intersection point represents a calculation point. The yellow square stands for the subset size. (b)–(i) Recorded deformed images at various load stages.

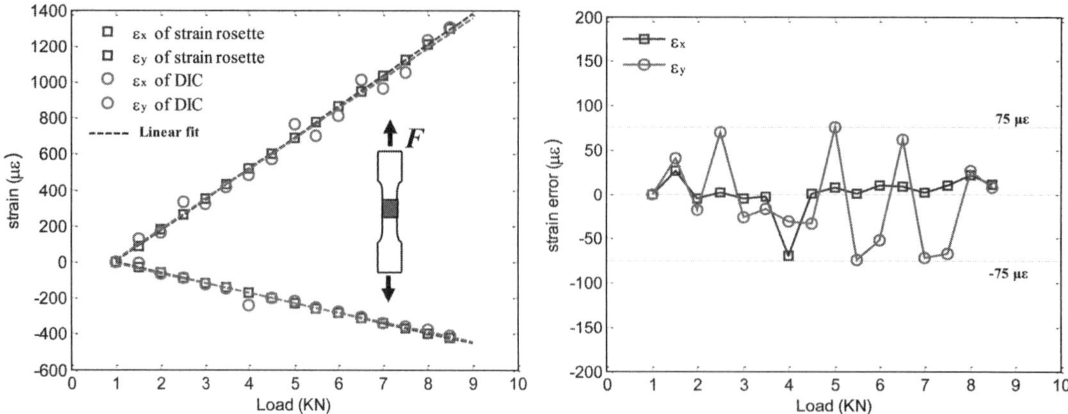

Figure 6.15 (a) Measured axial and transversal strains as a function of the applied loads using strain gauge rosettes and the established 2D-DIC system; (b) difference in ε_x and ε_y values measured by the two techniques.

strain estimation method with a strain window of 21×21 points (corresponding to a region of 200×200 pixels). Finally, the average value of the calculated full-field strain is regarded as the resulting strain and is analyzed in comparison to the average strain measured by the two strain gauge rosettes.

Figure 6.15(a) plots the x- and y-directional average strains measured by the proposed high-accuracy 2D-DIC system and strain gauge rosettes at various load stages. It is noted that the strain is zero at the load stage of 1 KN because the state is regarded as the initial state and the strain gauge reading was cleared. As the potential strain errors resulting from various negative factors were minimized using the Schneider bilateral telectentric lens, the 2D-DIC results agree well with the strain gauge rosettes results. It is also seen that the strains measured by two different techniques are in linear proportional to the applied loads. However, as 2D-DIC is susceptible to ambient illumination variations, the linearity of DIC measurement is a little inferior to that of strain rosette measurement. Taking the strain rosette measurement as a reference, the strain errors of the proposed high-accuracy 2D-DIC system were plotted in Figure 6.15(b). The average error in the axial strain ε_x was estimated as -7 ± 50 $\mu\varepsilon$, whereas the mean error for the transversal strain ε_y was observed to be 1 ± 21 $\mu\varepsilon$.

In addition, the measurements of elastic modulus and Possion's ratio, which objectively represent mechanical properties, were also performed. Table 6.3 summarizes

Table 6.3 The Elastic Modulus and Poisson's Ratio Measured by the Established 2D-DIC System and Strain Gauge Rossettes

Testing No.	Elastic Modulus (GPa)		Poisson's Ratio	
	Strain Gauge	2D-DIC	Strain Gauge	2D-DIC
No.1	72.6	73.4	0.326	0.322
No.2	72.7	74.6	0.327	0.362
No.3	72.6	74.3	0.326	0.360
Average	72.6±0.06	74.1±0.62	0.326±0.01	0.348±0.02

the elastic modulus and Possion's ratio measured by the established 2D-DIC system and strain gauge rosettes. The elastic modulus and Poisson's ratio measured by 2D-DIC system were found as 74.1±0.62 GPa and 0.348±0.02, respectively. The two values measured by strain rosettes were estimated as 72.6±0.06 GPa and 0.326±0.01, respectively. Though the average values of the two measurements are very close, the standard deviation errors of DIC measurements are slightly larger.

6.3   Three-Dimensional Digital Image Correlation

To overcome the limitation of 2D-DIC, the more practical 3D-DIC technique using two cameras was developed for shape and deformation measurements of both planar and curved objects. Though the subset-based matching algorithm can be directly used in 3D-DIC for determining disparities and image motions, stereovision calibration techniques are needed to estimate the intrinsic and extrinsic parameters of the two cameras, which are critical to the final measurements. In this section, the basic principles and implementation procedures of 3D-DIC are briefly described. Two typical applications of 3D-DIC for shape and deformation measurement are demonstrated.

Recall that 2D-DIC described in the preceding section can only be used for in-plane deformation of nominal planar objects. So, for deformation measurement of macroscopic objects such as structural component, industrial products with a curved surface, the advanced 3D-DIC is more practical and effective because it can be used for 3D profile and deformation measurement of both planar and curved surfaces, and is capable of detecting rather than being affected by out-of-plane displacements. Also, to meet the requirement of measuring shape and deformation of small-scale objects, 3D-DIC has been implemented on light stereo microscopes [70] as well as fluorescent stereo microscopy systems [71], which enables shape and deformation measurement of small objects with size varying from several millimeters to several centimeters. It is also interesting to note that 3D-DIC measurements can be realized by combining a single camera with mirrors [72], a biprism lens [73], as well as a diffraction grating [74, 75]. The main advantages of a singe-camera 3D-DIC system lie in its relatively low cost and convenience for time-resolved measurements, because the need for hardware or software synchronization of two expensive high-speed cameras can be avoided.

6.3.1   Basic Principles and Concepts

The 3D-DIC method is based on a combination of DIC and binocular stereo vision. It retains the advantages of the noncontact, full-field measurement of 2D-DIC but overcomes its inherent disadvantages. Compared with 2D-DIC, the advantages of 3D-DIC are embodied in the following two aspects: (1) 3D-DIC is suitable for the shape and deformation measurements of both planar and curved objects; (2) 3D-DIC can measure all three components of displacement of the test object. Thus, small amounts of out-of-plane motion of an object occurring during loading can be accurately detected without affecting the accuracy of the other two in-plane displacement components [11].

Figure 6.16 schematically illustrates the basic principle of the 3D-DIC method. In Figure 6.16, O_L and O_R are optical centers of the left and right cameras, respectively. It can be observed that a physical point P is imaged as point P_L in the image plane of the left camera and as point P_R in the image plane of the right camera. The 3D-DIC technique aims to accurately recover the 3D coordinates of point P with respect to a world coordinate system from image points P_1 and P_2. The world coordinate system can be erected by camera calibration techniques widely used in computer vision. Generally, a planar calibration target using evenly spaced features (circular dots or corners) is used. The two cameras first simultaneously capture images of the calibration target from different views. Then, the positions of these feature points on the calibration target images are detected using an image processing technique and input to the calibration model to optimize the desired intrinsic parameters (including effective focal length, principle point coordinates, and lens distortion coefficients) and extrinsic parameters (including the 3D position and orientation of the camera relative to a world coordinate system) of each camera. These calibration parameters together with the disparity data of each calculation point will be used to calculate the 3D position according to the triangulation principle. Obviously, the differences between the 3D coordinates reconstructed before and after deformation give the 3D displacement vector of a measurement point. From this procedure, it can be concluded that there are two crucial steps in the implementation of the 3D-DIC technique: (1) camera calibration (i.e., stereovision sensor calibration), and (2) stereo matching [14].

6.3.2   Binocular Stereovision System Calibration

6.3.2.1   Imaging Model in Stereovision System

Camera calibration is the procedure to determine the intrinsic parameters (e.g., effective focal length, principle point, and lens distortion coefficient) and extrinsic parameters (including the 3D position and orientation of the camera relative to a

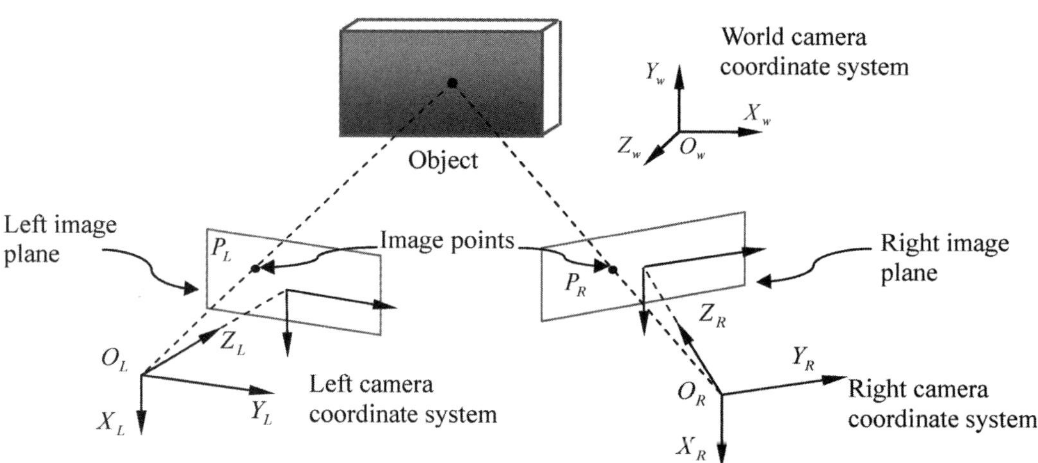

Figure 6.16   Schematic drawing of principle of binocular stereovision used in a 3D-DIC system.

world coordinate system) of a camera. For 3D-DIC, the calibration also involves determining the relative 3D position and orientation between the two cameras. As the accuracy of the final measurement heavily depends on the obtained calibration parameters, camera calibration plays a crucial role in the 3D-DIC measurement.

As shown in Figure 6.16, the projection of a 3D point $P(x_w, y_w, z_w)$ in units of millimeters onto the image plane (u, v) in units of pixels can be represented by a ideal pinhole model. The relationship between a 3D point $\mathbf{P}$ and its image points $\mathbf{P}_l$ and $\mathbf{P}_r$ are

$$s\begin{bmatrix} u \\ v \\ 1 \end{bmatrix} = \mathbf{A}\begin{bmatrix} \mathbf{R} & \mathbf{t} \\ \mathbf{0} & 1 \end{bmatrix}\begin{pmatrix} x_w \\ y_w \\ z_w \\ 1 \end{pmatrix}, \quad \mathbf{A} = \begin{bmatrix} \alpha & \gamma & u_0 & 0 \\ 0 & \beta & v_0 & 0 \\ 0 & 0 & 1 & 0 \end{bmatrix} \tag{6.18}$$

where s is an arbitrary scale factor, $\mathbf{A}$ are intrinsic parameters, and $\mathbf{R}$, $\mathbf{t}$ are extrinsic parameters (three for rotation and three for translation of each camera). In the camera intrinsic matrix $\mathbf{A}$, $(u_0; v_0)$ are the coordinates of the principal point, α and β are the scale factors in image u and v axes, and γ describes the skewness of the two image axes [76].

Due to nonlinear optical distortion, this equation is not sufficient for accurate camera calibration. Based on the reports in the literature, it is likely that the distortion function is totally dominated by the radial components, and especially dominated by the first term. It has also been found that any more elaborate modeling not only would not help to provide higher accuracy, but would also cause numerical instability. For this reason, radial distortion only considering first two terms is generally used.

Let (u, v) be the distortion-free pixel image coordinates, and $(\tilde{u}, \tilde{v})$ the corresponding real observed image coordinates. If radial distortion is considered, we have

$$\begin{bmatrix} \tilde{u} \\ \tilde{v} \end{bmatrix} = \begin{bmatrix} u \\ v \end{bmatrix} + \begin{bmatrix} (u - u_0)\left[k_1 r^2 + k_2 r^4 \right] \\ (v - v_0)\left[k_1 r^2 + k_2 r^4 \right] \end{bmatrix} \tag{6.19}$$

where k_1 and k_2 are coefficients of the radial distortion, and r is the distance between an image point (u, v) and the principle point.

Though calibration of a single camera can be conducted using widely accepted techniques (e,g., Zhang's method [76]), the calibration of binocular stereovision system needs more additional work [77]. In a stereovision system, aside from the intrinsic parameters of each camera, the relative position and orientation of the two cameras should be determined precisely. A spatial point P is assumed to be the 3D world coordinate and its image coordinate in left and right image are denoted by $\mathbf{x}_w$, $\mathbf{x}_l$ and $\mathbf{x}_r$, respectively. We have

$$\begin{aligned} x_l &= R_l x_w + t_l \\ x_r &= R_r x_w + t_r \end{aligned} \tag{6.20}$$

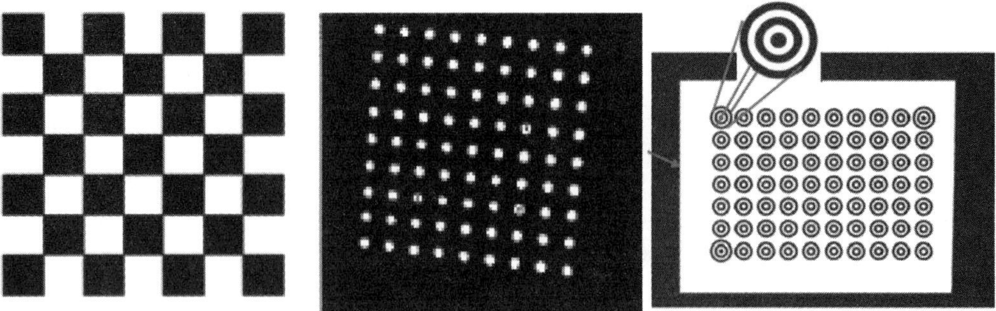

Figure 6.17 (from left to right) a chessboard calibration target, a circular dot calibration target, and ring template

Elimination of $\mathbf{x}_w$ from the these two formulas gives

$$x_l = R_l R_r^{-1} x_r + t_l - R_l R_r^{-1} t_r \qquad (6.21)$$

Let

$$\begin{aligned} R_{r2l} &= R_l R_r^{-1} \\ t_{r2l} &= t_1 - R_l R_r^{-1} t_r \end{aligned} \qquad (6.22)$$

Thus, we get

$$x_l = R_{r2l} x_r + t_{r2l} \qquad (6.23)$$

In summary, A_l, A_r, R_{r21}, t_{r21}, $k_{1,l}$, $k_{2,l}$, $k_{1,r}$, $k_{2,r}$ are parameters that need to be calibrated for a binocular stereovision system.

6.3.2.2 Stereovision System Calibration

Camera calibration needs a 3D stereo or 2D plane calibration target [76–79], in which the characteristic points with known 3D coordinates can be accurately and automatically extracted using a digital image processing algorithm. Figure 6.17 shows several typical calibration templates used for the 3D-DIC technique.

BOX 6.3 Determination of Calibration Parameters Using a Non-Linear Optimization Algorithm

Using Zhang's technique [76], the intrinsic and extrinsic parameters of the right and the left camera (i.e., A_l, $R_{l,i}$, $T_{l,i}$ and A_r, $R_{r,i}$, $T_{r,i}$) can be determined for each position of the calibration target. During capturing the images of the target, the relative position and orientation of the camera is fixed. So, all these parameters should be the same in theory. However, due to the presence of various noise and calculation errors, these parameters obtained using each calibration image are not exactly equal. Therefore, a nonlinear least squares optimization of the intrinsic parameters of the two cameras and the relative positions and orientations can be optimized globally using the following objective function [79]:

$$
\sum_{i=1}^{n}\sum_{j=1}^{m_{l,i}}\left\|m_{l,ij} - \tilde{m}\left(A_l, k_{1,l}, k_{2,l}, R_{1,i}, t_{1,i}, X_{1j}\right)\right\|^2
$$
$$
+ \sum_{i=1}^{n}\sum_{j=1}^{m_{r,i}}\left\|m_{r,ij} - \tilde{m}\left(A_r, k_{1,r}, k_{2,r}, R_{1,i}, t_{1,i}, R_{r21}, t_{r21}, X_{1j}\right)\right\|^2
$$

(6.24)

where $\tilde{m}(A_l, k_{1,l}, k_{2,l}, R_{1,i}, t_{1,i}, X_{1j})$ and $\tilde{m}(A_r, k_{1,r}, k_{2,r}, R_{1,i}, t_{1,i}, R_{r21}, t_{r21}, X_{1j})$ are the image coordinate of jth feature point in the left image and right image predicted using the non-linear imaging model; $m_{l,ij}$ and $m_{r,ij}$ are the corresponding real image coordinates obtained using a feature extraction technique. The cost function can be optimized using Levenberg-Marquart algorithm to determine the calibration parameters.

6.3.3   Stereo Matching

The goal of the stereo matching is to precisely match the same physical point in the two images captured by the left and right cameras. This task, commonly considered as the most challenging issue in stereovision measurement, can now be readily accomplished by the well-established subset-based matching algorithm adopted in 2D-DIC.

To reconstruct the profile of a sample surface of an object, a region of interest (ROI) should be first specified, within which all the calculation points are defined, in the left image. Then, these calculation points with known image coordinates are searched in the right image using a subset-based correlation algorithm to determine their corresponding image coordinates, as schematically shown in the top part of Figure 6.18. Afterward, these mapped image coordinates can be used for reconstruction of the profile of the ROI. Figure 6.18 also indicates the detailed procedure for 3D displacement measurements. After defining a ROI in the left image of the reference image pair, each measurement point within ROI is then searched in three other different diffraction images (i.e., right image in the reference state, left and right images in the deformed state) using the same subset-based matching method. Based on the disparity information and the calibration parameters of the two cameras, the world coordinate of the interrogated point P before and after deformation can be reconstructed as $P(x_{w0}, y_{w0}, z_{w0})$ and $P'(x_{w1}, y_{w1}, z_{w1})$, respectively. Clearly, the differences in the two coordinates yield the three displacement components caused by external loading.

The detailed matching procedures are summarized as follows:

1. The robust ZNSSD criterion combined with a twelve-parameter displacement mapping function can be defined as the objective function. Note that

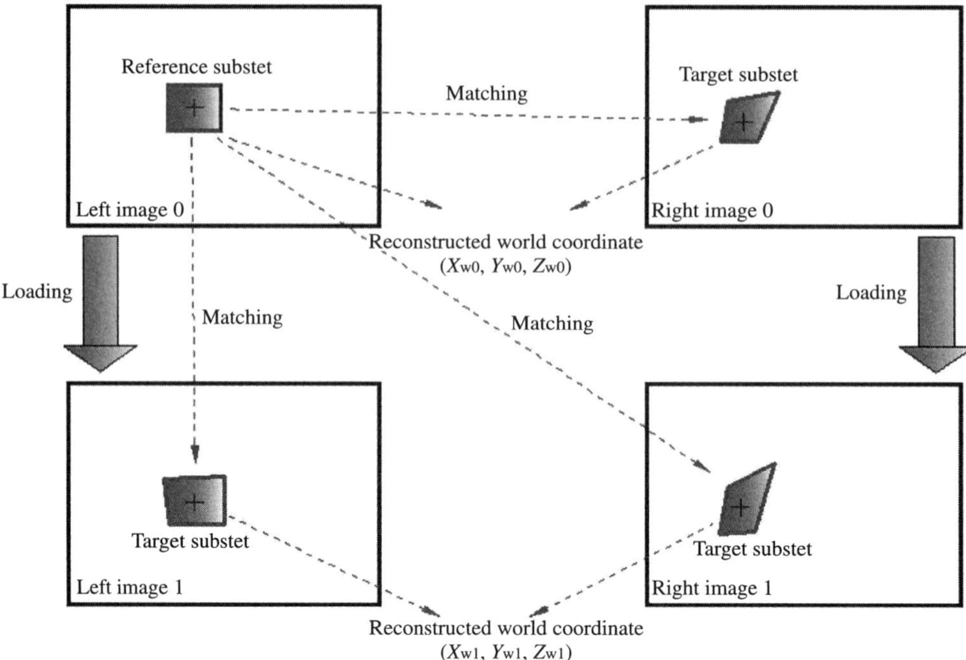

Figure 6.18 Subset-based 2D-DIC method is used to track the positions of each calculation point defined in left image 0 in the rest of images.

in stereovision matching, the second-order shape function generally offers better results than the simple first-order shape function, because it is capable of realistically approximating more complicated deformations of a target subset imaged by different cameras.

2. The nonlinear ZNSSD criterion is subsequently optimized using the classic (NR) algorithm and a bicubic interpolation scheme to determine the desired parameters. It is also worth noting that a simple but robust reliability-guided displacement tracking scheme [53] is used to guide the calculation path of the NR algorithm, which can greatly improve the computational efficiency of the NR algorithm by ensuring accurate initial guess transfer between consecutive measurement points. The same procedure is repeated to track the corresponding positions of the considered points of interest in the two images recorded after deformation.

6.3.4   3D Profile Reconstruction and Deformation Calculation

6.3.4.1   3D Profile Reconstruction

Based on the obtained calibration parameters of each camera and the calculated disparities of points in the images, the 3D world coordinates of the points in the regions of interest on the specimen surface can be easily determined using the classical triangulation method.

$$z_{c1}\begin{bmatrix} u_1 \\ v_1 \\ 1 \end{bmatrix} = \begin{bmatrix} m_{11}^1 & m_{12}^1 & m_{13}^1 & m_{14}^1 \\ m_{21}^1 & m_{22}^1 & m_{23}^1 & m_{24}^1 \\ m_{31}^1 & m_{32}^1 & m_{33}^1 & m_{34}^1 \end{bmatrix}\begin{pmatrix} x_w \\ y_w \\ z_w \\ 1 \end{pmatrix} \tag{6.25}$$

$$z_{c2}\begin{bmatrix} u_2 \\ v_2 \\ 1 \end{bmatrix} = \begin{bmatrix} m_{11}^2 & m_{12}^2 & m_{13}^2 & m_{14}^2 \\ m_{21}^2 & m_{22}^2 & m_{23}^2 & m_{24}^2 \\ m_{31}^2 & m_{32}^2 & m_{33}^2 & m_{34}^2 \end{bmatrix}\begin{pmatrix} x_w \\ y_w \\ z_w \\ 1 \end{pmatrix} \tag{6.26}$$

By eliminating the z_{c1}, z_{c2} from these equations, four linear equations as a function of the desired world coordinates (x_w, y_w, z_w) can be obtained.

$$\begin{aligned} \left(u_1 m_{31}^1 - m_{11}^1\right)x_w + \left(u_1 m_{32}^1 - m_{12}^1\right)y_w + \left(u_1 m_{33}^1 - m_{13}^1\right)z_w &= m_{14}^1 - u_1 m_{34}^1 \\ \left(v_1 m_{31}^1 - m_{21}^1\right)x_w + \left(v_1 m_{32}^1 - m_{22}^1\right)y_w + \left(v_1 m_{33}^1 - m_{23}^1\right)z_w &= m_{14}^1 - v_1 m_{34}^1 \\ \left(u_2 m_{31}^2 - m_{11}^2\right)x_w + \left(u_2 m_{32}^2 - m_{12}^2\right)y_w + \left(u_2 m_{33}^2 - m_{13}^2\right)z_w &= m_{14}^2 - u_2 m_{34}^2 \\ \left(v_2 m_{31}^2 - m_{21}^2\right)x_w + \left(v_2 m_{32}^2 - m_{22}^2\right)y_w + \left(v_2 m_{33}^2 - m_{23}^2\right)z_w &= m_{14}^2 - v_1 m_{34}^2 \end{aligned} \tag{6.27}$$

The geometric meanings of (6.25) and (6.26) are the spatial lines passing thorough $o_{c1}p$ and $o_{c2}p$; thus, the intersection point of these two lines gives the exact world coordinates of point P. However, it is clear that (6.27) is an over-determined function, and hence the world coordinates (x_w, y_w, z_w) are generally solved using linear least squares. By repeating the same procedure at other points of interest, the 3D shape of the ROI can be reconstructed.

6.3.4.2 Displacement and Strain Estimation

By tracking the coordinates of the same physical points before and after deformation, the 3D motions of the each point can be determined. Similarly, the same technique proposed in the previous section can be utilized for the strain estimation based on the obtained displacement fields. Matching the left reference image and a right deformed image can increase computational accuracy, as opposed to matching the left reference image and the corresponding right image. Subtracting world coordinates after and before deformation, three displacement components caused by external loading are computed.

Strain estimation uses point-wise local least squares to deal with the three noisy and discrete displacement components $u(x_w, y_w, z_w)$, $v(x_w, y_w, z_w)$, and $w(x_w, y_w, z_w)$, where x_w, y_w and z_w are the coordinates. The method is written by

$$\begin{aligned} u\left(x_w, y_w, z_w\right) &= a_0 + a_1 x_w + a_2 y_w + a_3 x_w y_w \\ v\left(x_w, y_w, z_w\right) &= b_0 + b_1 x_w + b_2 y_w + b_3 x_w y_w \\ w\left(x_w, y_w, z_w\right) &= c_0 + c_1 x_w + c_2 y_w + c_3 x_w y_w \end{aligned} \tag{6.28}$$

The linear least squares can be used to calculate these fitting coefficients, based on which the six strain components $\partial u/\partial x, \partial u/\partial y, \partial v/\partial x, \partial v/\partial y, \partial w/\partial x, \partial w/\partial y$ can be estimated accordingly.

6.3.5 Applications of 3D-DIC for Shape and Deformation Measurement

6.3.5.1 Profile Measurement of Satellite Antenna Surface Using 3D-DIC [80]

A carbon fiber composite satellite antenna supplied by Beijing Institute of Space Craft Environment Engineering (BISCEE) was made from wound carbon fiber (T-700) with epoxy resin as the matrix. Figure 6.19(a) shows a schematic drawing of the experimental antenna sample with geometry information included, and Figure 6.20(b) is an image of the antenna surface recorded by the left camera. It should be noted that watercolor paint had been applied onto the antenna surface to generate random speckle-like patterns before testing. The watercolor paint was used here because it can be easily and completely removed by flushing the antenna with water after testing. The sample was tightly fixed to a base to prevent undesired rigid body motions during image capturing.

Figure 6.19 shows the schematic diagram of the 3D-DIC experimental setup. Two CCD cameras (1302UM, Daheng Image Co., Ltd, Beijing, China) with a resolution of 1280×1024 pixels and dynamic range of 256 gray levels were used to simultaneously record the digital images of the specimen surface from two directions. Since the diameter of the antenna is rather large, two imaging lenses with a focal length of 8 mm were employed for proper imaging. To ensure high image resolution and high image quality, the cameras and lenses were adjusted to make the specimen fill the entire width of the image (as shown in Figure 6.19), and the lens aperture was set to its minimum value to achieve a maximum depth of focus. In addition, an optical fiber cold light source was used to illuminate the antenna specimen surface for better imaging.

During the experiment, six pairs of images of the chessboard calibration pattern at different orientations were first taken by the two cameras. Then, two images of the antenna specimen surface were simultaneously captured, followed by a digital image processing and analysis using the 3D-DIC software developed in Visual Basic.

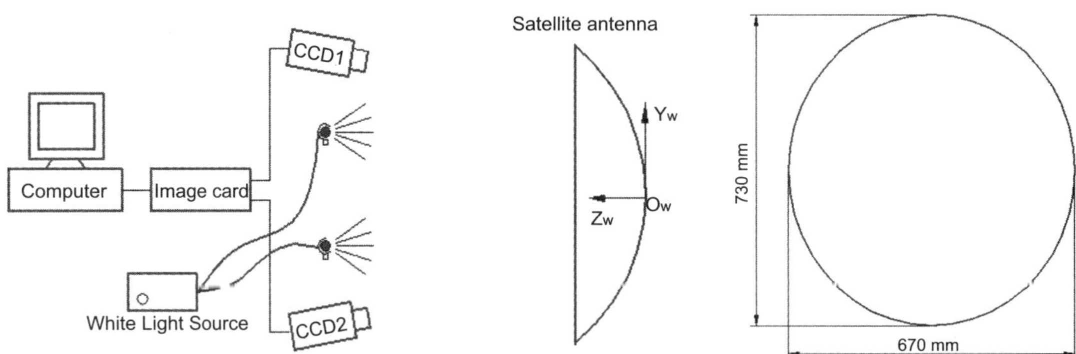

Figure 6.19 Schematic of the 3D-DIC experimental setup and the test satellite.

Table 6.4 Intrinsic Parameters of the Left and Right Cameras

Intrinsic Parameters	(u_0, v_0)	f_x *(pixel)*	f_y *(pixel)*	k_1	k_2
Left camera	(599.56, 478.13)	1600.65	1597.49	−0.08866	0.176398
Right camera	(675.44, 451.81)	1586.03	1587.11	−0.10054	0.165968

Table 6.5 Relative Position and Orientation Between the Left and Right Cameras

$\alpha(°)$	$\beta(°)$	$\gamma(°)$	t_x *(mm)*	t_y *(mm)*	t_z *(mm)*
−1.266	−22.944	−1.123	344.399	−9.759	61.513

Using the aforementioned camera calibration technique, the intrinsic parameters of each camera can be obtained, as shown in Table 6.4. It is observed that the standard deviation of reprojective errors for left camera and right camera are 0.0936 and 0.1028, respectively, using Zhang's technique. After global optimization using Levenberg-Marquardt algorithm, the standard deviation of reprojective error for all views is decreased to 0.0924.

In Table 6.4, (u_0, v_0) are the coordinates of the principle point; f_x and f_y are the effective focal lengths in x and y axes, respectively; k_1 and k_2 are the first- and second-order radial distortion coefficients of the imaging lens, respectively. The relative orientation and position of the left camera to the right camera can also be obtained based on their extrinsic parameters, which are listed in Table 6.5.

Figure 6.20 shows the images of the antenna surface captured by the left and right cameras. In order to acquire full-field measurement results, a polygon was plotted to approximate the antenna rim, and regularly distributed points (separated by the step size) inside the polygon were selected as calculation points. To guarantee a reliable stereo matching, all the pixels within the reference subset, which is centered at the current calculation point, should be located within the polygon as well. During the process of correlation calculation, the subset size was set to 41×41pixels,

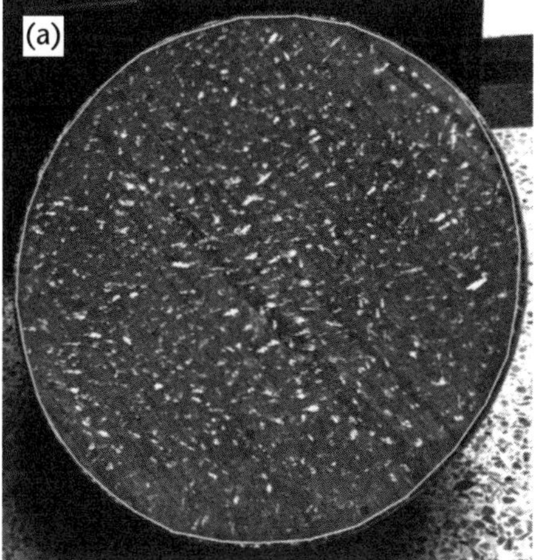

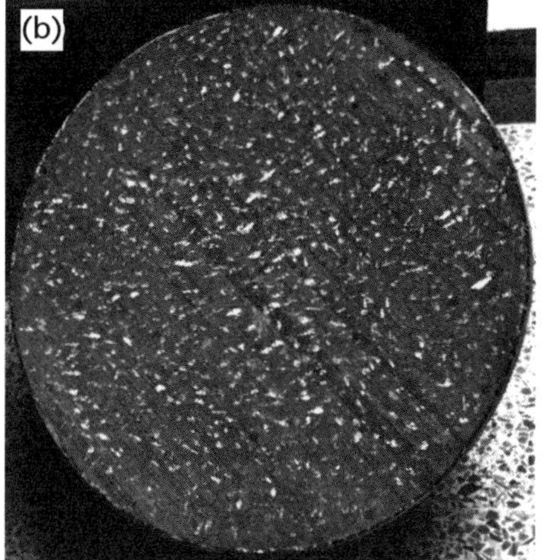

Figure 6.20 Images of the satellite antenna surface captured by (a) the left camera and (b) the right camera.

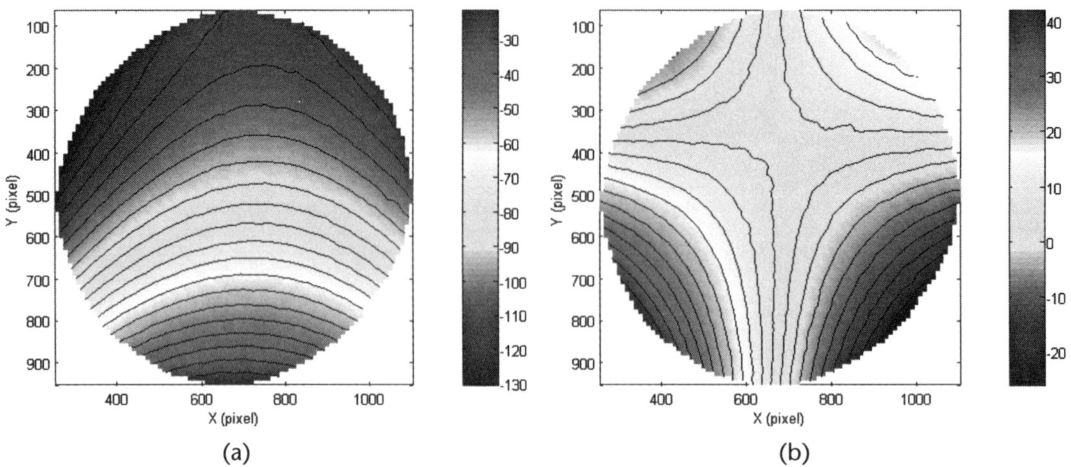

Figure 6.21   Map of the disparity between the left and right images: (a) in x direction, and (b) in y direction (unit: pixel).

and the step size, which is the distance between two adjacent processing points, was set to 5 pixels. The disparities in x and y directions obtained by using the stereo matching algorithm previously described are shown in Figure 6.21.

Using the calibration data shown in Tables 6.4 and 6.5 and the disparity data shown in Figure 6.21, the surface profile of the satellite antenna can be determined and reconstructed, and the results are shown in Figure 6.22. Figure 6.22(a) is a 2D contour plot of the satellite antenna surface in which the small fluctuations of contour lines indicate the influences from inevitable noises. Figure 6.22(b) shows a 3D topographic plot of the satellite antenna surface, which provides a more intuitive look. The measurement results indicate that the surface of the carbon fiber composite satellite antenna follows a parabolic function, which is in good accordance with the original design.

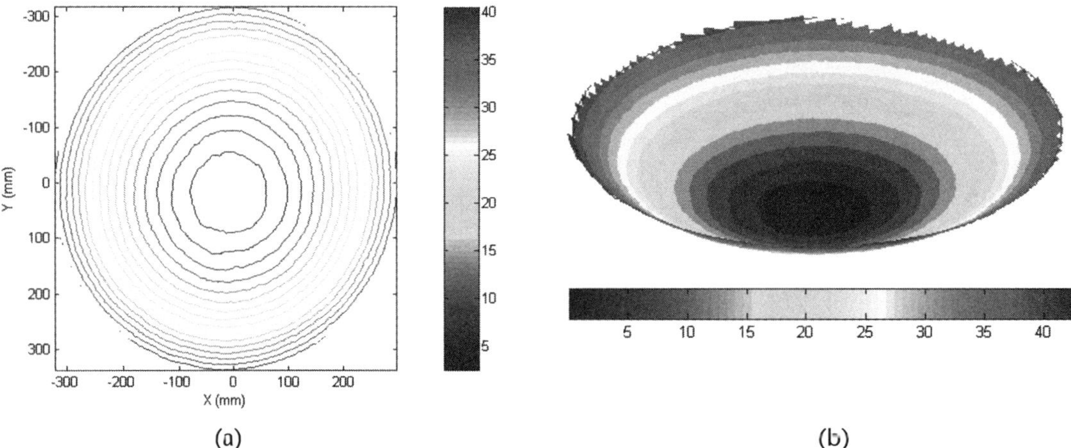

Figure 6.22   Measured surface profile of the satellite antenna: (a) 2D contour plot, (b) 3D topographic plot (unit: mm).

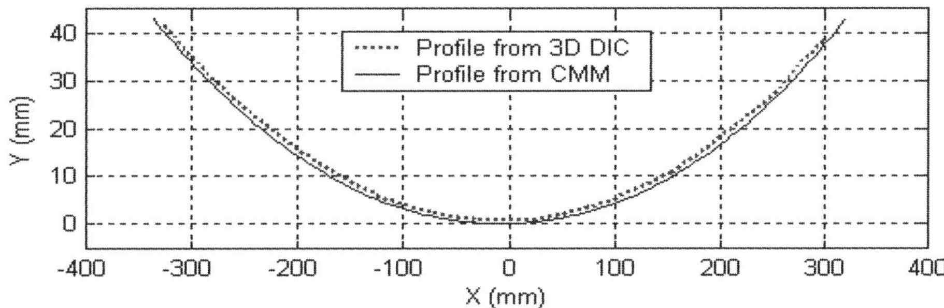

Figure 6.23 Comparison of the profiles obtained from 3D-DIC technique and 3D coordinate measuring machine.

To verify the measurement accuracy, the obtained surface profile has been compared with the one measured by a commercial 3D coordinate measuring machine (CMM) (IOTA 1203, made by China Precision Engineering Institute for Aircraft Industry, Beijing, China). The 3D CMM uses a contact probe to detect 3D coordinates of object surface points and the measurement accuracy can reach 5 μm. The result of comparison is shown in Figure 6.23. For a clear illustration, it is noted that only the surface profile along the horizontal centerline is plotted in the figure. It can be seen from Figure 6.8 that the two profiles are in good agreement, and the maximum discrepancy is less than 0.8 mm. It is also noticed from the figure that the profile measured by the 3D CMM is a little lower than that measured by the 3D-DIC. This can be explained by the fact that the touching probe of the CMM contacts the specimen surface with a certain amount of force (0.1N in this experiment), which leads to a slightly lower profile detection.

6.3.5.2   High-Temperature Strain Measurement Using Active Imaging 3D-DIC [81]

Existing 3D-DIC systems generally use a white light source or natural light to illuminate the test sample surface, and then the reflected light is passively collected by two CCD or CMOS cameras outfitted with ordinary imaging lenses. This regular optical imaging system can receive all light within the sensitive wavelength range of the camera. Being an image-based optical measuring technique, the contrast of the recorded speckle pattern has an important influence on 3D-DIC measurements. Therefore, to guarantee an accurate measurement, the illumination intensity and ambient light should remain stable during the experiment. Though linear changes in the intensity of illumination lighting can be well compensated by use of a robust correlation criterion(e.g., the ZNSSD criterion), large nonuniform variations in the illumination source and/or ambient light will lead to corresponding nonlinear intensity changes or even saturation in the recorded image, resulting in large errors in or a failure of 3D-DIC measurements.

In the laboratory, the requirement of stable illumination and ambient light can be readily satisfied. However, in certain uncommon measurement environments such as nonlaboratory measurements or high-temperature environments, this requirement is difficult or impossible to meet. The primary challenge of applying 3D-DIC in the previously mentioned harsh or extreme environments lies in how to acquire high-quality digital images with constant image contrast and prevent the presence of the speckle pattern decorrelation effect. To acquire high-quality images with a constant image contrast over a long period of time or a broad temperature range, the influences of ambient light variations and thermal radiation of a hot object on the captured image must be minimized. Based on this consideration, a robust active imaging 3D-DIC system is established by combining monochromatic light illumination with bandpass filter imaging. Figure 6.24(a) is a picture of the established active imaging 3D-DIC system, which consists of a self-developed monochromatic source, two optical bandpass filters tightly mounted just before the two fixed-focus lens, and two digital CCD cameras (MV-VS141FM, Weishi Image Co., Ltd, Xian, China) with a spatial resolution of 1392×1040 pixels at 256 gray levels.

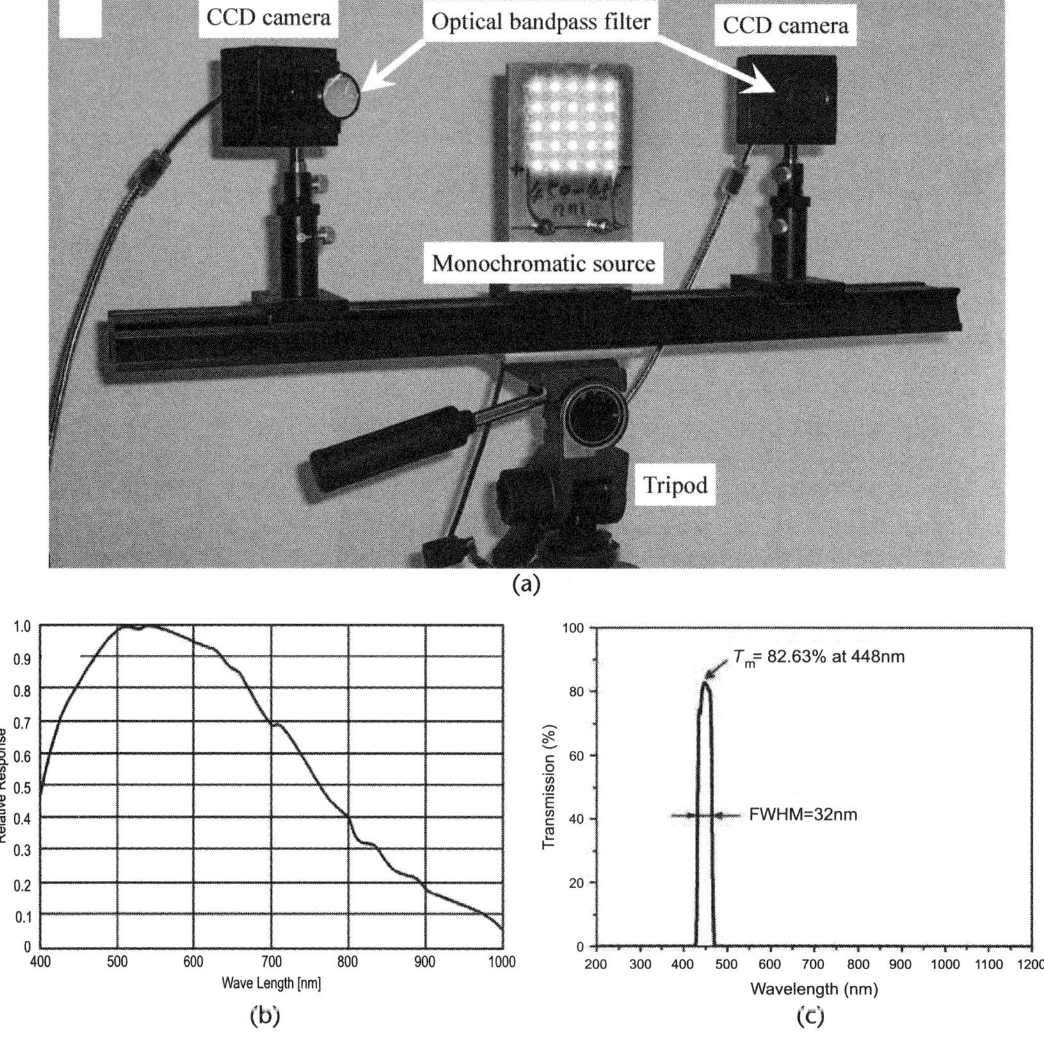

Figure 6.24 (a) Photograph of the self-established active imaging 3D-DIC system; (b) quantum efficiency curve of the CCD camera; (c) transmission efficiency of the bandpass filter attached before the imaging lens.

Figure 6.24 (b) gives the quantum efficiency of the CCD camera being used. The camera is sensitive to light within a wavelength range from 400 nm to 1000 nm, whereas the quantum efficiency of the CCD camera exceeds 50% for the light emitted in the range of 400–750 nm. Clearly, ambient light or thermal radiation emitted by a hot object within this bandpass range has an effect on the brightness of the recorded image. In the established active imaging 3D-DIC system, a self-made monochromatic blue LED light source, emitting at 450–455 nm, is employed to illuminate the test object surface to provide even and stable illumination, as shown in Figure 6.24(a). Two optical bandpass filters, with a center wavelength of 450±2 nm and a full-width at half-maximum (FWHM) value of approximately 32 nm, were attached just before the two imaging lenses. Figure 6.24(c) shows the transmission efficiency curve of the bandpass filter. The function of the optical bandpass filter is to allow all of the active illumination monochromatic lighting to enter the camera sensor and cut off all the light wavelengths shorter than 428 nm or longer than 470 nm. As shown in Figure 6.24(c), for the active illuminated monochromatic light emitting at 450–455 nm, the bandpass filter shows a very high transmission efficiency exceeding 80%. However, for the light with wavelengths outside of the bandpass range, the bandpass filter shows a very low transmission efficiency approaching zero. Because only a very limited portion of the ambient light or self-radiated light within the bandpass range of the filter can pass through the filter, the undesirable ambient light makes a negligible contribution to the intensity of the recorded image compared with the actively illuminated monochromatic light. As will be shown later, the images acquired by the proposed active imaging 3D-DIC method demonstrate constant image contrast, even though the energy of the ambient illumination is intentionally changed or the hot test object emitted dazzling light. Thus, high accuracy and high-fidelity shape and deformation measurements for the object in out-of-laboratory environments or extreme high-temperature environments using 3D-DIC is possible, and the application range of 3D-DIC is therefore greatly extended.

To measure 3D deformation in a high-temperature environment, a 2-mm-thick curved plate sample made of chromium-nickel austenite stainless steel (material type No 1Cr18Ni9Ti) is used. To ensure reliable and accurate measurements using 3D-DIC, white speckle granules capable of sustaining 1700°C were applied to the specimen surface to form artificial random speckle patterns suitable for DIC applications. The specimen is placed vertically on the specimen stage approximately 200 mm before the quartz lamps of the infrared heating device used to heat the specimen, as shown in Figure 6.25(a). In this experiment, the specimen is first heated to 20°C and the image pair of the specimen surface is recorded as reference images using the active imaging 3D-DIC system. Then the specimen is heated from 100°C to 1200°C with a temperature increment of 100°C. When the preset temperature is reached, an image pair is captured after a dwell time of 15 seconds. In total, 12 deformed image pairs were recorded. Figure 6.25(c)–(f) show the four images captured by the left camera at temperatures of 20°C, 500°C, 1000°C, and 1200°C, respectively. Figure 6.25(b) shows a photograph of the specimen surface at 1200°C captured by a common color digital camera. Figure 6.26 shows the resulting x-directional, y-directional, and z-directional displacement fields at 1200°C. The in-plane rotation has been removed.

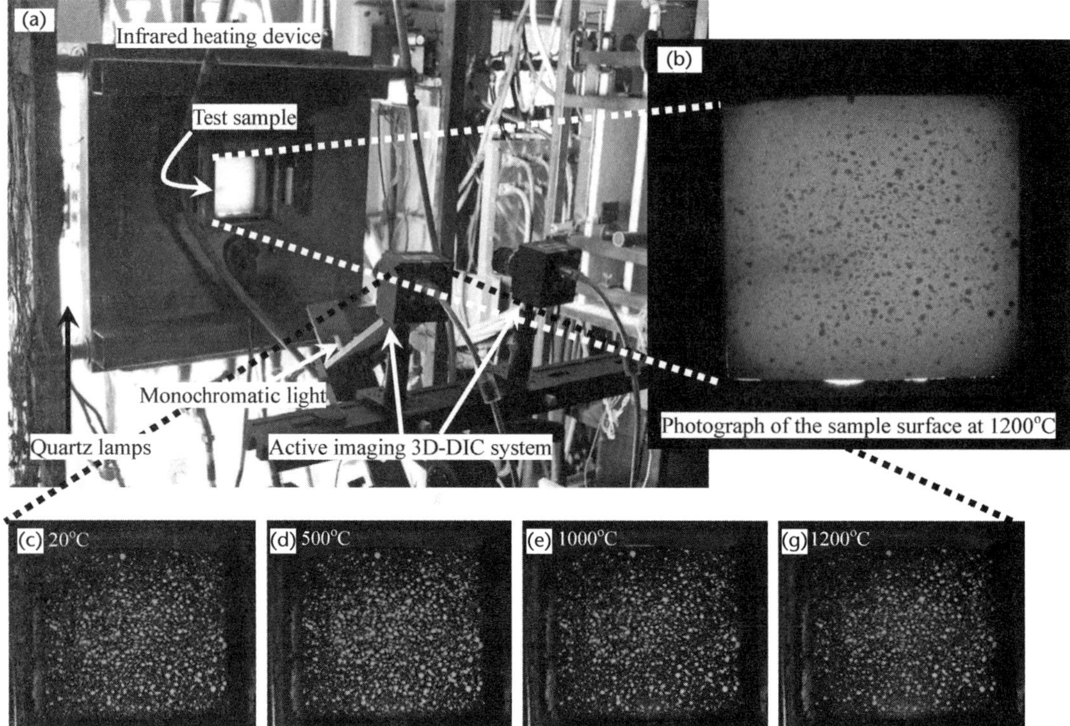

Figure 6.25 (a) In situ photograph of the experimental setup used for the high-temperature deformation measurement; (b) photograph of the specimen surface at 1200°C captured by a common digital camera; (c–f) surface images captured by the left camera of the active imaging 3D-DIC system at temperatures of 20°C, 500°C, 1000°C, and 1200°C.

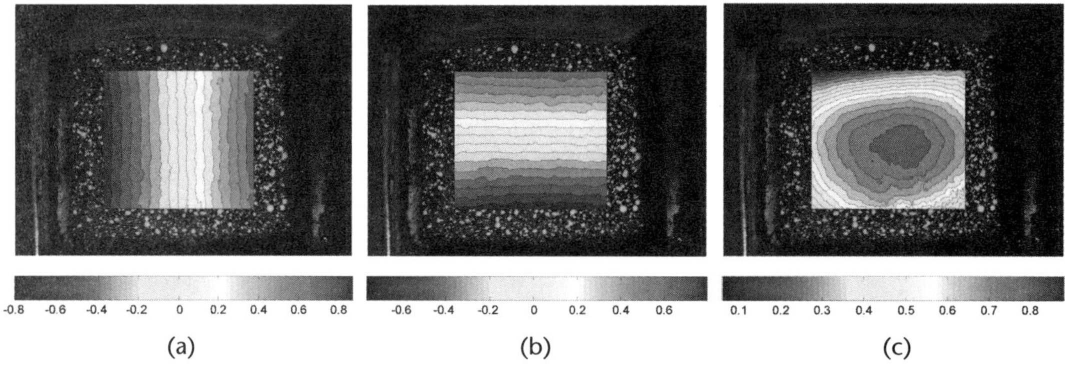

Figure 6.26 Thermal deformation of the specimen at 1200°C measured by the active imaging 3D-DIC system after removing in-plane rigid-body motion: (a) *u*-field displacement, (b) *v*-displacement, and (c) *w*-displacement (unit: mm).

Figure 6.27 plots the detected x-directional and y-directional thermal expansion, which increases almost linearly as the temperature increases. The measured thermal strains are compared with ideal thermal strains, which are calculated by the *Chinese Aeronautical Materials Handbook*. Figure 6.27 confirms the accuracy of the proposed active imaging 3D-DIC system for high-temperature strain measurements.

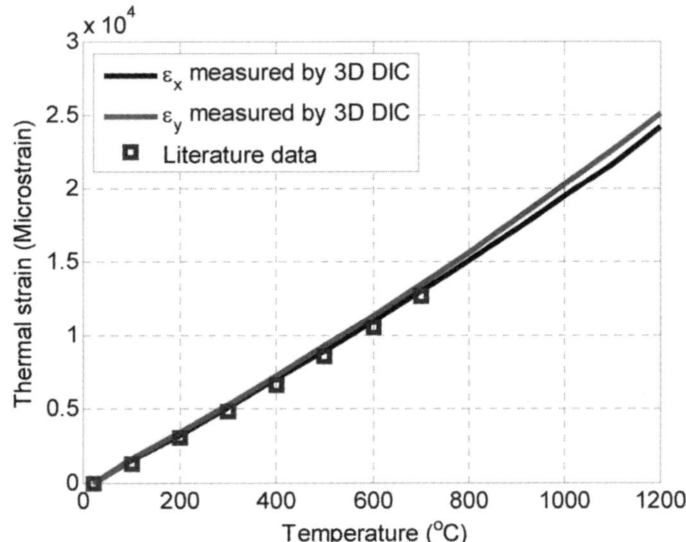

Figure 6.27 Comparison of the thermal strains measured by the active imaging 3D-DIC system and literature data measured by 2D-DIC.

6.4 Concluding Remarks and Future Work

In this chapter, the basic principles and implementation procedures for the simple but useful 2D and 3D digital image correlation techniques for surface deformation measurement are described. Representative applications of these techniques are also presented to demonstrate their effectiveness and practicality. With the development and popularity of the high-accuracy, easy-to-implement stereovision calibration technique, it is clear that 3D-DIC will gain more and more broad applications for deformation measurement of macroscopic objects and structures, considering its higher accuracy and wider applicability. However, 2D-DIC remains a simple, cost-effective yet practical tool for in-plane deformation measurement of nominally planar objects at various temporal and spatial scales. Particularly for observing and quantitatively measuring micro scale deformation at reduced size, the 2D-DIC combined with high-spatial-resolution imaging device (especially, SEM and AFM) will still be the most effective and irreplaceable tool.

Over the past thirty years, the DIC technique has been continuously improved by many researchers and gradually evolved to the most famous and popular tool in the field of experimental mechanics. Commercial DIC systems developed by various companies with friendly interface, powerful function, and high efficiency are available in the market. However, the ultimate objective of developing an intelligent DIC technique for fully automatic, real-time, high-accuracy measurements remains to be achieved. Some important challenges to this end involve, for example, automatic, adaptive, and optimal selection of various calculation parameters (in particular, a proper subset size and a displacement mapping function for displacement tracking and an optimal strain calculation window for strain field estimation) and fully automatic, high-accuracy calibration of a stereovision system.

References

[1]   Rastogi, P. K. *Photomechanics (Topics in Applied Physics)*, Springer Verlag, 2000.

[2]   Schnars U., Jueptner W. *Digital holography.* Springer, 2005.

[3]   Post, D., Ifju, P., Han, B. T. *High Sensitivity Moiré.* Springer, New York, 1994.

[4]   Sirkis, J. S., Lim, T. J. Displacement and strain-measurement with automated grid methods. *Exp. Mech.* 1991, 31(4): 382–388.

[5]   Sutton, M. A., Orteu, J. J., Schreier, H. W. *Image correlation for shape, motion and deformation measurements.* Springer, 2009.

[6]   Pan, B., Qian K. M., Xie H. M., Asundi A. Two-dimensional Digital Image Correlation for In-plane Displacement and Strain Measurement: A Review. *Measurement Science and Technology.* 2009, 20: 062001.

[7]   Pan, B., Recent Progress in Digital Image Correlation. *Exp. Mech.* 2011, 51, 1223–1235.

[8]   Peters, W. H., Ranson, W. F. Digital Imaging Techniques in Experimental Stress Analysis. *Opt. Eng.* 1981, 21: 427–431.

[9]   Peters, W. H., Ranson, W. F., Sutton, M. A., et al. Application of digital correlation methods to rigid body mechancis. *Opt. Eng.* 1983, 22(6): 738–742.

[10]   Chu, T. C., Ranson, W. F., Sutton, M. A., et al. Applications of digital-image-correlation techniques to experimental mechanics. *Exp. Mech.*1985. 25(3): 232–244.

[11]   Sutton, M. A., Yan, J. H., Tiwari, V., Schreier, W. H., Orteu, J. J. The effect of out-of-plane motion on 2D and 3D digital image correlation measurements. *Opt. Lasers Eng.* 2008, 46:746–757.

[12]   Pan, B., Yu, L. P., Wu, D. F. High-accuracy 2D digital image correlation measurements with bilateral telecentric lens: error analysis and experimental verification. *Exp Mech.* 2013, 53(9): 1719–1733.

[13]   Luo, P. F., Chao, Y. J., Sutton, M. A., et al. Accurate measurement of three-dimensional displacement in deformable bodies using computer vision. *Exp. Mech.* 1993, 33(2): 123–132.

[14]   Garcia, D., Orteu, J. J., Penazzi, L. A combined temporal tracking and stereo-correlation technique for accurate measurement of 3D displacements: application to sheet metal forming. *Journal of Materials Processing Technology*, 2002, 125: 736–742.

[15]   Bay, B. K., Smith, T. S., Fyhrie, D. P., et al. Digital volume correlation: Three-dimensional strain mapping using X-ray tomography. *Exp. Mech.* 1999, 39(3): 217–226.

[16]   Bay, B. K. Methods and applications of digital volume correlation. *J. Strain Anal. Eng. Des.*, 2008, 43(8): 745–760.

[17]   Helm, J. D., McNeil, S. R., Sutton, M. A. Improved three-dimensional image correlation for surface displacement measurement. *Opt. Eng.* 1996, 35(7): 1911–1920.

[18]   Pan, B., Xie, H. M., Gao, J. X., Asundi, A. Improved Speckle Projection Profilometry for Out-of-plane Shape Measurement. *Applied Optics*, 2008, 47(29): 5527–5533.

[19]   Sun, Z. L., Lyons, J. S., McNeill, S. R. Measuring microscopic deformations with digital image correlation. *Opt. Lasers Eng.* 1997; 27: 409–428.

[20]   Zhang, D. S., Luo, M., Arola, D. D. Displacement/strain measurements using an optical microscope and digital image correlation. *Opt. Eng.* 2006, 45(3): 033605.

[21]   Berfield, T. A., Patel, H. K., Shimmin, R. G., Braun, P. V., Lambros, J., Sottos, N. R. Fluorescent image correlation for nanoscale deformation measurements. *Small* 2006; 2: 631–635

[22]   Sabate, N., Vogel, D., Gollhardt, A., et al. Measurement of residual stresses in micromachined structures in a microregion. *Applied Physics Letters*, 2006, 88(7), Art. No. 071910.

[23]   Kang, J., Jain, M., Wilkinson, D. S., et al. Microscopic strain mapping using scanning electron microscopy topography image correlation at large strain. *Journal of Strain Analysis for Engineering Design*, 2005, 40(6): 559–570.

[24] Wang, H., Xie, H., Ju, Y., et al. Error analysis of digital speckle correlation method under scanning electron microscope. *Exp. Tech.* 2006, 30(2): 42–45.

[25] Sutton, M. A., Li, N., Garcia, D., et al. Metrology in a scanning electron microscope: theoretical developments and experimental validation. *Meas. Sci. Technol.* 2006, 17(10): 2613–2622.

[26] Chasiotis, I., Knauss, W. G. A New Microtensile Tester for the Study of MEMS Materials with the aid of Atomic Force Microscopy. *Exp. Mech.* 2002, 42(1): 51–57.

[27] Knauss, W. G., Chasiotis, I., Huang, Y. Mechanical Measurements at the Micron and Nanometer Scales. *Mechanics of Materials*, 2003, 35(3–6): 217–231.

[28] Chang, S., Wang, C. S., Xiong, C. Y., et al. Nanoscale in-plane displacement evaluation by AFM scanning and digital image correlation processing. *Nanotechnology* 2005, 16(4): 344–349.

[29] Li, X. D., Xu, W. J., Sutton, M. A., et al. Nanoscale deformation and cracking studies of advanced metal evaporated magnetic tapes using atomic force microscopy and digital image correlation techniques. *Meas. Sci. Technol.* 2006, 22(7): 835–844.

[30] Siebert, T., Becker, T., Spiltthof, K., et al. High-speed digital image correlation: error estimations and applications. *Opt. Eng.* 2007, 46(5): 051004.

[31] Tiwari, V., Sutton, M. A., McNeill, S. R. Assessment of high speed imaging systems for 2D and 3D deformation measurements: Methodology development and validation. *Exp. Mech.* 2007, 47(4): 561–579.

[32] Pankow, M., et al., Three-dimensional digital image correlation technique using single high-speed camera for measuring large out-of-plane displacements at high framing rates. *Appl. Opt.* 2010.

[33] Yoneyama, S., Kikuta, H., Kitagawa, A., Kitamura, K. Lens distortion correction for digital image correlation by measuring rigid body displacement. 2006, *Opt Eng.* 45 (2): 023602.

[34] Pan, B., Yu, L. P., Wu, D. F., Tang, L. Q. Systematic errors in two-dimensional digital image correlation due to lens distortion. *Opt Lasers Eng.* 2013, 51(2): 140–147.

[35] Ma, S. P., Pang, J. Z., Ma, Q. W. The systematic error in digital image correlation induced by self-heating of a digital camera. *Meas Sci Technol.* 2011, 23: 025403.

[36] Pan, B., Yu, L. P., Wu, D. F. High-accuracy 2D digital image correlation measurements with low-cost imaging lenses: implementation of a generalized compensation method. *Measurement Science & Technology*, 2014, 25(2): 025001.

[37] Schreier, H. W., Sutton, M. A. Systematic Errors in Digital Image Correlation Due to Undermatched Subset Shape Functions. *Exp. Mech.* 2002, 42(3): 303–310.

[38] Lu, H., Cary, P. D. Deformation Measurement by Digital Image Correlation: Implementation of a Second-order Displacement Gradient. *Exp. Mech.* 2000, 40(4): 393–400.

[39] Pan, B., Xie, H. M., Wang, Z. Y. Equivalence of Digital Image Correlation Criteria for Pattern Matching. *Appl Opt.* 2010, 49(28): 5501–5509.

[40] Tong, W. An evaluation of digital image correlation criteria for strain mapping applications. *Strain*, 2005, 41(4): 167–175.

[41] Pan, B., Asundi, A., Xie, H. M., Gao, J. X. Digital Image Correlation using Iterative Least Squares and Pointwise Least Squares for Displacement Field and Strain Field Measurements. *Optics and Lasers in Engineering.* 2009, 47(7–8): 865–874.

[42] Huang, J. Y., Zhu, T., Pan, X. Y., Qin, L., Peng, X. L., Xiong, C. Y., Fang, J. A high-efficiency digital image correlation method based on a fast recursive scheme. *Meas. Sc. Technol.* 2010, 21, 035101.

[43] Press, W. H., C++ *Numerical algorithms*. Beijing : Publishing House of Electronics Industry, 2003.

[44] Schreier, H. W., Braasch, J. R., and Sutton, M. A. Systematic errors in digital image correlation caused by intensity interpolation. *Opt Eng.* 2000; 39(11): 2915–2921.

[45] Pan, B., Li, K. A fast digital image correlation method for deformation measurement. *Optics and Lasers in Engineering*. 2011; 49(7): 841–847.

[46] Bruck, H. A, McNeil, S. R., Sutton, M. A., Peters, W. H. Digital Image Correlation Using Newton-Raphson Method of Partial Differential Correction. *Exp. Mech.* 1989, 29(3): 261–267.

[47] Vendroux, G., Knauss, W. G. Submicron Deformation Field Measurements: Part2. Improved Digital Image Correlation. *Exp. Mech.* 1998, 38(2): 86–92.

[48] Pan, B., Xie, H. M., Xu, B. Q., Dai, F. L. Performance of Sub-pixel Registration Algorithms in Digital Image Correlation. *Measurement Science and Technology*, 2006, 17 (6): 1615–1621.

[49] Sun, Y., Pang, J. H. L., Wong, C. K., et al. Finite element formulation for a digital image correlation method. *Applied optics*, 2005, 44(34): 7357–7363.

[50] Besnard, G., Hild, F., Roux, S. "Finite-element" displacement fields analysis from digital images: application to Portevin–Le Châtelier bands. *Experimental Mechanics*, 2006, 46(6): 789–803.

[51] Ma, S., Zhao, Z., Wang, X. Mesh-based digital image correlation method using higher order isoparametric elements. *The Journal of Strain Analysis for Engineering Design*, 2012, 47(3): 163–175.

[52] Pan, B. An evaluation of convergence criteria for digital image correlation using inverse compositional Gauss-Newton algorithm. *Strain*, 2014, 50(1): 48–56.

[53] Pan, B. Reliability-guided digital image correlation for image deformation measurement. *Appl Opt* 2009; 48(28): 1535–1542.

[54] Pan, B., Wang, Z. Y., Lu, Z. X. Genuine full-field deformation measurement of an object with complex shape using reliability-guided digital image correlation. *Opt Express* 2010; 18(2): 1011–1023.

[55] Zhang, Z. F., Kang, Y. L., Wang, H. W., et al. A novel coarse-fine search scheme for digital image correlation method. *Measurement*, 2006, 39(8): 710–718.

[56] Chen, D. J., Chiang, F. P., Tan, Y. S., Don, H. S. Digital speckle-displacement measurement using a complex spectrum method. *Appl Opt* 1993; 32(11): 1839–1849.

[57] Pan, B., Xie, H. M. Digital Image Correlation Method with Differential Evolution. *Journal of Optoelectronics & Laser.* 2007, 18(1):100–103. (in Chinese)

[58] Zhang, J. Q., Zeng, P., Lei, L.P., Ma, Y. Initial guess by improved population-based intelligent algorithms for large inter-frame deformation measurement using digital image correlation. *Optics and Lasers in Engineering*. 2012, 50(3):473–490.

[59] Zhou, Y. H., Pan, B., Chen, Y. Q. Large deformation measurement using digital image correlation: a full automatic approach." *Applied Optics*, 2012, 51(31): 7674–7683.

[60] Zhou, Y. H., Chen, Y. Q. Propagation function for accurate initialization and efficiency enhancement of digital image correlation. *Optics and Lasers in Engineering*. 2012, 50(12): 1789–1797.

[61] Pan, B., Xie, H. M., Guo, Z. Q., Hua, T. Full-field Strain Measurement using a Two-dimensional Savitzky-Golay Digital Differentiator in Digital Image Correlation. *Optical Engineering*, 2007, 46(3): 033601.2.

[62] Sutton, M. A., Turner, J. L., Bruck, H. A., Chao, T. A. Full-field Representation of Discretely Sampled Surface Deformation for Displacement and Strain Analysis. *Exp. Mech.* 1991; 31(2): 168–177.

[63] Meng, L. B., Jin, G. C., Yao, X. F. Application of iteration and finite element smoothing technique for displacement and strain measurement of digital speckle correlation. *Opt. Lasers Eng.* 2007; 45: 57–63.

[64] Yoneyama, S. Smoothing Measured Displacements and Computing Strains Utilising Finite Element Method. *Strain* 2011; 47: 258–266.

[65] Tong, W. Detection of plastic deformation patterns in a binary aluminum alloy. *Exp. Mech.* 1997; 37(4): 452–459.

[66] Wang, C. C. B., Deng, J. M., Ateshian, G. A., et al. An automated approach for direct measurement of two-dimensional strain distributions within articular cartilage under unconfined compression. *J. Biomech. Eng-T ASME* 2002; 124(5): 557–567.

[67] Xiang, G. F., Zhang, Q. C., Liu, H. W., et al. Time-resolved deformation measurements of the Portevin-Le Chatelier bands. *Scripta Materialia*, 2007, 56(8): 721–724.

[68] Han, Y. G., Kim, D. W., Kwon, H. J. Application of digital image cross-correlation and smoothing function to the diagnosis of breast cancer. *J. Mech. Behav. Biomed.* 2012; 14: 7–18.

[69] Wattrisse, B. C., Muracciole, A., Nemoz-Gaillard, J. M. Analysis of strain localization during tensile tests by digital image correlation. *Exp. Mech.* 2001, 41(1): 29–39.

[70] Schreier, H. W., Garcia, D., Sutton, M. A. Advances in light microscope stereo vision. *Exp. Mech.* 2004; 44: 278–288.

[71] Hu, Z. X., Luo, H. Y., Du, Y. J., Lu, H. B. Fluorescent stereo microscopy for 3D surface profilometry and deformation mapping. *Opt. Express*, 2013, 20,11808–11818.

[72] Pankow, M., Justusson, B., Waas, A. M. Three-dimensional Digital Image Correlation Technique Using Single High-speed Camera for Measuring Large Out-of-plane Displacements at High Framing Rates. *Applied Optics*, 2010, 49(17): 3418–3427.

[73] Genovese, K., Casaletto L., Rayas, J. A., Flores, V., Martinez, A. Stereo-Digital Image Correlation (DIC) measurements with a single camera using a biprism. *Optics and Lasers in Engineering.* 2013, 51 (3): 279–285.

[74] Xia, S., Gdoutou, A., Ravichandran, G. Diffraction Assisted Image Correlation: A Novel Method for Measuring Three-Dimensional Deformation using Two-Dimensional Digital Image Correlation. *Experimental Mechanics*, 2013: 1–11.

[75] Pan, B., Wang, Q. Single-camera microscopic stereo digital image correlation using a diffraction grating. *Optics Express*, 2013: 21, 25056–25068.

[76] Zhang, Z. Y. A flexible new technique for camera calibration. *IEEE Transactions on Pattern Analysis and Machine Intelligence*, 2000, 22(11):1330–1334.

[77] Xue, T., Wu, B., Zhu, J. G., et al. Complete calibration of a structure-uniform stereovision sensor with free-position planar pattern. *Sensors and Actuators A-Physical*, 2007, 135 (1): 185–191.

[78] Vo, M., Wang, Z. Y., Luu, L., Ma, J. Advanced geometric camera calibration for machine vision. *Optical Engineering*, 2011, 50(11): 110503.

[79] Zhang, H., Zhang, L. Y., Chen, J., Zhao, Z. P. Field Calibration of Binocular Stereo System Based on Planar Template and Free Snapping. *Acta Aeronautica & Astronautica Sinica*, 2007, 28(3): 695–701 (In Chinese).

[80] Pan, B., Xie, H. M., Yang, L. H., Wang, Z. Y. Accurate measurement of satellite antenna surface using 3D digital image correlation technique. *Strain*, 2009, 45, 194–200.

[81] Pan, B., Wu, D. F., Yu, L. P. Optimization of a three-dimensional digital image correlation system for deformation measurement in extreme environments. *Applied Optics*, 2012, 51, (19): 4409–4419.

CHAPTER 7

Digital Fringe Projection Profilometry

Lujie Chen, Chenggen Quan, and Cho Jui Tay

7.1 Introduction

Historically, fringe projection profilometry is essentially based on data-processing methods developed for interferometry. However, unlike interferometry, which relies on an optical path difference for information, it utilizes the principles of triangulation. In earlier works, a projection pattern is often generated using a grating and when the projected pattern is slightly defocused, the intensity of the pattern exhibits a sinusoidal distribution. In recent years, digital fringe projection, which belongs to a family of triangulation methods called structured light projection [1, 2] is employed. Its deployment is largely owing to the availability of digital devices, such as digital projectors and CCD cameras, which enable the projection and recording of a wide variety of different patterns. Digital fringe projection [3] as a branch of the triangulation methods has unique strength in many aspects.

This chapter describes the principles and development of digital fringe projection profilometry. The motivation for using fringe projection and the basic principles of triangulation are first described. This is followed by a discussion on the applications of the digital fringe projection method to shape measurement. A number of exercises with solutions are also included.

7.2 Principles of Method

Digital fringe projection profilometry is essentially based on the principles of triangulation from which quantitative phase evaluation of resulting fringe patterns are conducted. Subsequent calibration of the phase values against standard results would yield the actual profile of an object.

7.2.1 Triangulation

Do you remember seeing a straight, vertical line projected onto a flat surface (e.g., a wall in a classroom or a screen in a cinema)? Probably yes, and you remember that the line remains straight on the surface. Do you remember seeing such a line being projected onto a curved surface? Perhaps no, but it is a simple experiment to do: connect a digital projector to your computer, load an image with a straight, vertical line on the screen, direct the projector to a surface that is not flat, walk a few steps from the projector, and observe the projected line. What do you see? Most likely, you will no longer see a straight line. It may be a curved line along the curvature of the surface; or it may also be a broken line if the surface has a sharp discontinuity. Now, walk behind the projector and orientate your viewing direction so that your view is in the direction of the projected beam. What do you see? The line is now straight.

You have just experienced the principles of triangulation in action. A straight line appears to be curved when there is an angle between the projection and viewing directions; the larger the angle, the larger is the curvature. In optical terms, the depth sensitivity increases with the angle of triangulation. Hence, the principles of triangulation can be applied to the measurement of depth, a quantity that is not directly measurable on a 2D image.

How exactly can the depth information be quantitatively measured apart from the fact that it is indeed observed as a curved line? Consider the following experiment: a flat reference plane is imaged at two predetermined positions: Plane 1 (z_1) and Plane 2 (z_2), and a curved object is placed roughly midway in between the two planes. A straight line is projected onto the flat plane and the object. Figure 7.1(a) shows the projected image of the straight line on the three planes and Figure 7.1(b) shows the images recorded by a camera. The x coordinates of the line on planes 1 and 2 are x_1 and x_2 respectively. As the line on the object is curved due to the object's curved surface; its coordinates x_o vary with y.

We are interested in finding the depth of a point on an object, or more precisely, the z coordinate of the point on the projected line. It is not difficult to see that the z coordinate of such a point is related to x_o, x_1, x_2, z_1 and z_2 by

$$\frac{z_o - z_1}{z_2 - z_1} = \frac{x_o - x_1}{x_2 - x_1} \tag{7.1}$$

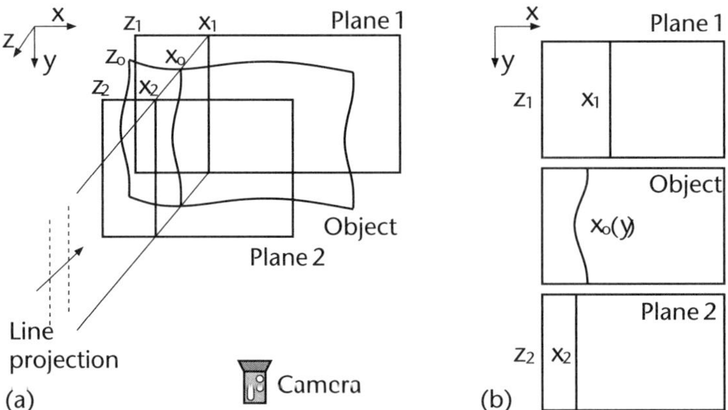

Figure 7.1 (a) A straight vertical line is projected onto a reference plane at two positions, z_1 and z_2, and onto an object at z_o. (b) The line remains straight on planes 1 and 2 but appears curved on the object.

Since $z_2 - z_1$ is known and x_1, x_2 and x_o can be determined from the images, the depth of the point can be calculated by

$$z_o - z_1 = \frac{x_o - x_1}{x_2 - x_1}\left(z_2 - z_1\right) \tag{7.2}$$

To calculate the depth of other points on the line, we would just have to utilize the coordinate $x_o(y)$ of each point. Subsequently, the depth information of the line can be obtained. This completes a quantitative measurement of the depth information based on the x coordinates of the projected line. As you may have suspected, in this experiment the accuracy of depth measurement depends on the accuracy of the line measurement method.

7.2.2   Fringe Projection

A question arising from the previous experiment would be: How do we obtain a whole-field depth map of an object? The simple answer is: Scan the entire object surface by projecting many lines at different locations on the object and calculate the depth of each line to reconstruct a whole-field depth map.

EXERCISE 7.1

Using a digital projector, suggest two methods for the projection of a pattern of lines on an object.

Solution

The first method is to generate a series of images, each consisting of a line. The line on each image is at a specific distance from the line on the next image. The images are then projected onto the test object. The second method is to project a line image on an object. The object is then shifted multiple times and each time the image is recorded. The recorded images are subsequently combined to form a pattern of lines.

The approach is known as the line-scan method, widely used in many practical applications; however, it has some limitations, such as the duration of a scan and sampling gap (the gap between two projected lines). In applications where these issues could cause serious problems, alternative solutions have to be found. Figure 7.2(a) illustrates a solution to address some of the issues. Instead of projecting one line at a time, two lines with a reasonable separation are projected at once. Due to the separation, the two lines as shown in Figure 7.2(b) can be easily identified. Subsequently, two line profiles can be obtained in one set of images. As a result, in a dual-line projection, to obtain a whole-field depth map, the scanning time is reduced by half. To further reduce the time, one could increase the number of lines in one projection but this could cause a new problem. As the number of lines increases, the separation between them decreases; it becomes increasingly harder to distinguish the lines in the image, especially in regions where the surface profile

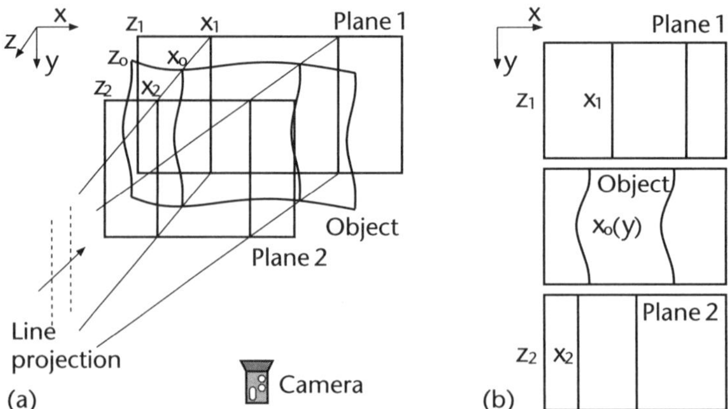

Figure 7.2 (a) Two vertical lines are simultaneously projected. (b) The projected lines are recorded in each image.

changes drastically. If the order of a line is identified wrongly, large errors would be introduced.

An insightful solution to these issues is to project a sinusoidal fringe pattern, as shown in Figure 7.3. The fringe pattern has a sinusoidally changing intensity distribution and hence the line detection approach to locate the x coordinates is no longer applicable; instead, a phase detection approach is employed. In the previous method, coordinates x_1, x_2, and x_o correspond to the x coordinates of the projected line in the three images; now, they correspond to the x coordinates of a point with a similar phase. The remaining calculations to obtain the depth information are similar.

The advantages of using a fringe pattern are (1) one single image would provide information on the entire surface and hence scanning is not needed, (2) the problem of a gap between two projected lines no longer exists as every point on the surface is sampled, (3) the order of a phase value at each point can be detected accurately by various methods (discussed in the next section), and (4) the accuracy is related

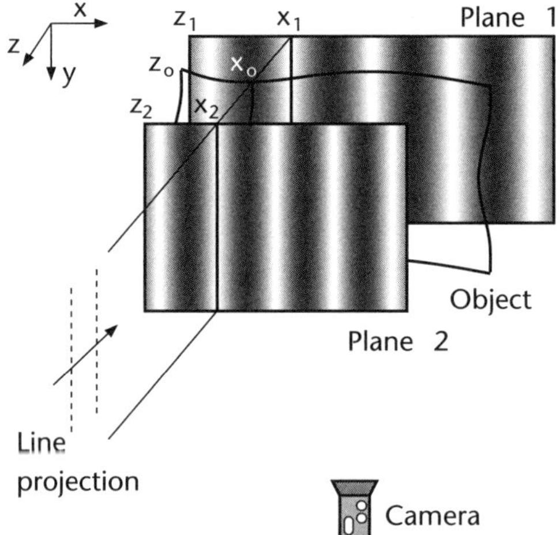

Figure 7.3 Sinusoidal fringe pattern projected on the reference and object planes.

to that of phase measurement, which is much higher than the accuracy of the line detection method.

EXERCISE 7.2

In measuring the profile of an object, explain when should a vertical and horizontal fringe pattern be employed.

Solution

If a projector and a camera optical centers both lie in a horizontal plane, a vertical fringe pattern should be used. In this case, a horizontal fringe pattern is not able to detect the depth variation. On the other hand, if they lie in a vertical plane, a horizontal fringe pattern should be used. In general, the direction of the fringe pattern should be perpendicular to the plane at which both centers lie to achieve the highest sensitivity.

7.2.3   Phase Evaluation

In the fringe projection method, it is important to obtain the correct phase for each pixel of a fringe pattern. This is carried out in two steps: wrapped phase extraction and phase unwrapping. The former focuses on retrieving phase values wrapped in 2π (i.e., in the range of $[-\pi, \pi]$) from one or more fringe patterns. The latter focuses on recovering a continuous phase distribution from a phase map wrapped in a value of 2π. There are many wrapped-phase-extraction and phase-unwrapping methods. In the following sections, we will introduce a few particularly useful methods.

7.2.3.1   Fourier Transform

As mentioned in Section 7.2.2, the spatial intensity distribution of a fringe pattern is sinusoidal in the direction of progressing phase (e.g., in the x direction of Figure 7.3). The intensity distribution can be expressed as

$$I(x,y) = a(x,y) + b(x,y)\cos\theta(x,y) \tag{7.3}$$

where I is the recorded intensity, a and b are background and modulation intensity respectively, and θ is the phase value. In (7.3), either a sine or cosine function can be used. A cosine function is used hereafter. The phase values θ is a near-periodic function with a dominant frequency of f_0, called carrier frequency, and information related to the profile. Equation (7.3) can be rewritten using Euler's formula:

$$I = a + \frac{be^{j\theta}}{2} + \frac{be^{-j\theta}}{2} \tag{7.4}$$

where j is $\sqrt{-1}$ (note that the function notation (x, y) is omitted for simplicity). Fourier transform of a fringe pattern projected on a specimen (Figure 7.4(a)) results in three spectrum concentrated areas [4, 5], as illustrated in Figure 7.4(b). The three concentrated areas correspond to the three terms in (7.4). The separation between the areas is f_0.

To retrieve the phase values θ, either the -1 or $+1$ frequency component can be used. First, the -1 or $+1$ component as indicated in Figure 7.4(b) is selected and the real and imaginary parts of the Fourier transform outside the region are set to zero. An inverse Fourier transform is then performed on the modified data and a single frequency component is obtained as follows:

$$c = \frac{be^{j\theta}}{2} = \frac{b}{2}(\cos\theta + j\sin\theta) \tag{7.5}$$

In [4, 5], a spectrum shift was applied before the inverse Fourier transform; this is equivalent to removing a carrier phase component. This process will be discussed in a Section 7.2.4. The final phase values θ are calculated using

$$\theta = \arctan\frac{\text{Im}(c)}{\text{Re}(c)} \tag{7.6}$$

where Re and Im denote the real and imaginary part of a complex number respectively and the values of θ are determined in a 2π range. In general, to obtain the values of θ using the Fourier transform method only one fringe pattern is needed.

7.2.3.2 Phase Shifting

Phase-shifting method [6] requires the input of at least three phase-shifted fringe patterns as in a sinusoidal intensity distribution there are three unknowns, a, b and θ (see (7.3)). Generating a phase-shifted fringe pattern is straightforward: this is done by changing the initial phase of a fringe pattern (see Figure 7.5) and projecting it sequentially onto the object.

It is important to note that phase shifting refers to the intensity change of a particular pixel due to the phase change and not the spatial shift of the fringe pattern, although the latter is a visual consequence. The phase-shifting algorithm is applied to a point's phase-shifted intensities, not to a spatial section of a fringe pattern, although the latter is also sinusoidal.

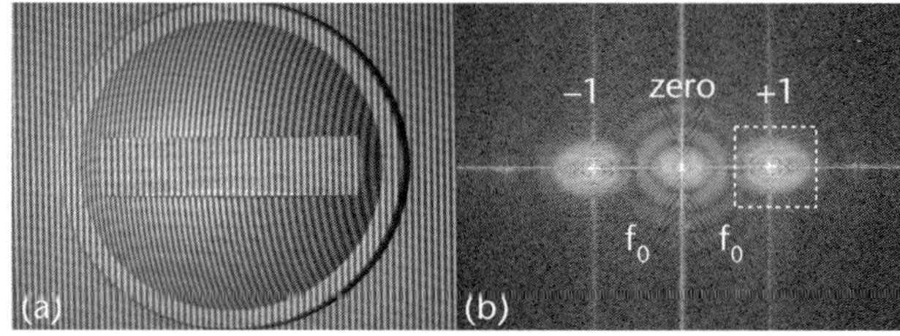

Figure 7.4 (a) A fringe pattern projected on a specimen. (b) Fourier transform spectrum of the fringe pattern in (a). The spectrum concentrated areas are zero order, −1, and +1 frequency components.

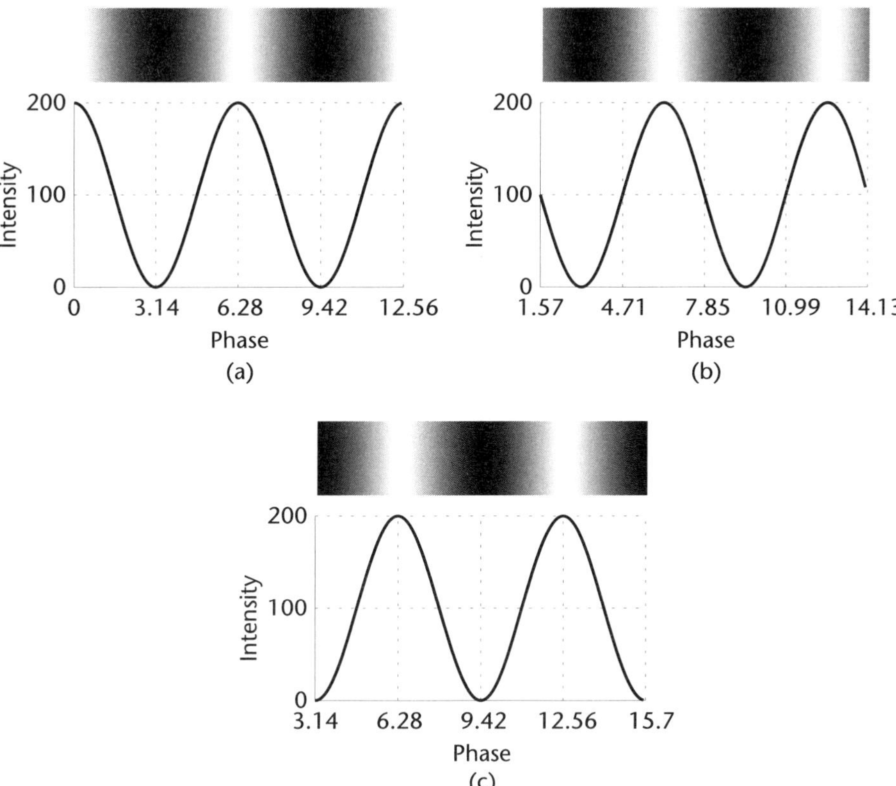

Figure 7.5   Three 90-degree phase-shifted fringe patterns.

The intensity of a pixel in a fringe pattern, which is phase-shifted n times, is expressed as

$$I_i = a + b\cos\left(\theta + \delta_i\right) \tag{7.7}$$

where $i \in [1, n]$ and δ_i is a known phase shift introduced at each step. If n is 3, the unknowns a, b and θ can be determined. If n is more than 3, the unknowns can be obtained using least-squares error. Rewriting I_i as

$$I_i = a + b\cos\theta\cos\delta_i - b\sin\theta\sin\delta_i \tag{7.8}$$

Let $B = b\cos\theta$ and $C = -b\sin\theta$. Define an error function

$$E(a,B,C) = \sum_{i=1}^{n}\left(I_i - a - B\cos\delta_i - C\sin\delta_i\right)^2 \tag{7.9}$$

Taking partial derivative of E with respect to a, B, and C, and equating the resulting expressions to zero to solve for B and C, the values of θ in the range of 2π can then be obtained from the following equation:

$$\theta = \arctan\frac{-C}{B} \tag{7.10}$$

EXERCISE 7.3

Express $\partial E/\partial a = 0$, $\partial E/\partial B = 0$, and $\partial E/\partial C = 0$ in a matrix form with a, B, and C as unknowns in a vector.

Solution

The matrix is given by:

$$\partial E/\partial a = -2\sum_{i=1}^{n}\left(I_i - a - B\cos\delta_i - C\sin\delta_i\right) = 0$$

$$\partial E/\partial B = -2\sum_{i=1}^{n}\left(I_i - a - B\cos\delta_i - C\sin\delta_i\right)\cos\delta_i = 0$$

$$\partial E/\partial C = -2\sum_{i=1}^{n}\left(I_i - a - B\cos\delta_i - C\sin\delta_i\right)\sin\delta_i = 0$$

$$\sum_{i=1}^{n}\begin{bmatrix} 1 & \cos\delta_i & \sin\delta_i \\ \cos\delta_i & \cos^2\delta_i & \sin\delta_i\cos\delta_i \\ \sin\delta_i & \sin\delta_i\cos\delta_i & \sin^2\delta_i \end{bmatrix}\begin{bmatrix} a \\ B \\ C \end{bmatrix} = \sum_{i=1}^{n}\begin{bmatrix} I_i \\ I_i\cos\delta_i \\ I_i\sin\delta_i \end{bmatrix}$$

7.2.3.3 Spatial Phase Unwrapping

After obtaining a wrapped phase map, the next step is to unwrap the phase values to recover a continuous phase distribution without a 2π jump. A simple 1D phase unwrapping as illustrated in Figure 7.6 involves scanning from left to right a wrapped phase map and adding or subtracting multiples of 2π to or from the phase values at each location to obtain a continuous curve that corresponds to the actual surface profile (Figure 7.6(b)).

The problem of phase unwrapping becomes significantly complicated in 2D due to the many possible unwrapping paths (as oppose to only one path in the 1D case). In 2D phase unwrapping, an unwrapping path may change from the x to y direction

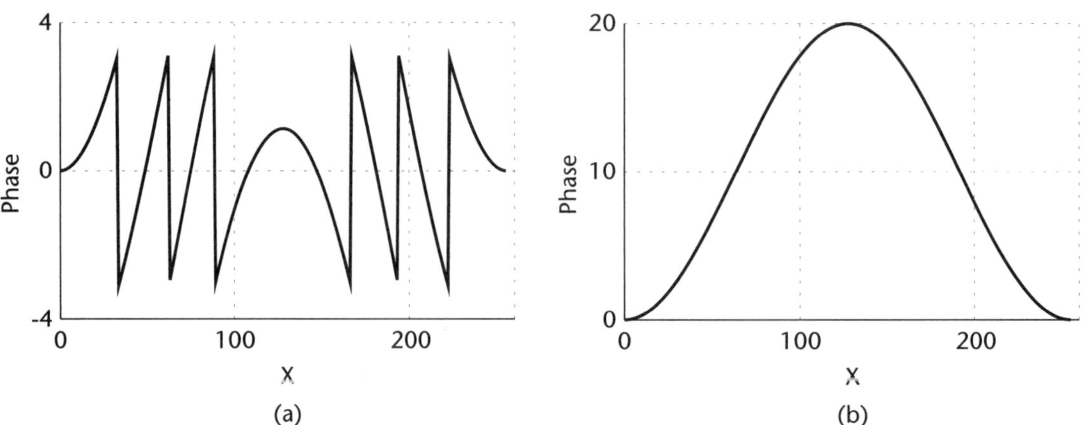

Figure 7.6 1D phase unwrapping: (a) A wrapped phase map and (b) unwrapped phase profile.

or vice versa at every step and depending on the chosen path the outcome will differ. For example, if a noisy region is encountered early in a chosen path, the error will be carried forward to subsequent pixels; and hence, the error would accumulate. However, if the noisy region is unvisited until near the end of a path, the error will be confined only to that region without affecting other pixels. Algorithms of spatial phase unwrapping and how to generate a good path that leads to acceptable results are discussed in detail in [7].

BOX 7.1   Quality-Guided Phase Unwrapping

A particularly effective approach for phase unwrapping is known as quality-guided phase unwrapping. It relies on a quality map to determine an unwrapping path. At each step, after the current pixel is processed, the next pixel to be processed is determined by a pixel with the highest quality on the propagation boundary. This method ensures the unwrapping path moves quickly to a higher-quality region before moving to a lower-quality region where errors are more likely to be encountered. The method is effective for phase fringes where the fringe contrast, which is defined as the ratio between the modulation and background intensity: b/a, is an accurate quality indicator. The modulation and background intensity can be obtained from the phase-shifting algorithm. Low fringe contrast is usually due to a dark region where a light beam is not able to reach, resulting in a low signal-to-noise ratio, consequently inaccurate results. The quality-guided phase unwrapping algorithm would process these low fringe contrast regions at the last stage to minimize error accumulation.

EXERCISE 7.4

Given that ψ is an unwrapped phase value, show how the corresponding wrapped phase value can be obtained.

Solution

Let ψ be an unwrapped phase value, its corresponding wrapped phase value is $\theta = \arctan(\sin \psi / \cos \psi)$. By checking the sign of $\sin \psi$ and $\cos \psi$, θ can be determined within a 2π range.

7.2.3.4   Temporal Phase Unwrapping

Though spatial phase unwrapping is speedy and requires only one wrapped phase map, there are certain limitations. For example, if an object profile is complicated, spatial phase unwrapping often produces suboptimal results. In cases where a field of view contains multiple objects whose profiles are disconnected or of different heights, spatial approach may not be able to produce the intended results.

Unlike spatial phase unwrapping that relies on a pixel's neighboring pixels to determine its phase value, temporal phase unwrapping deals with each pixel individually [8]. The method relies on different calculated phase values of a pixel to determine its actual phase value. Multifrequency fringe projection [9, 10] that employs temporal phase unwrapping is such a technique. Figure 7.7(a) shows the fringe patterns of four frequency levels, which vary from the lowest frequency L1 to the highest frequency L4. If the fringe patterns are phase-shifted and subsequently projected onto a flat plane, the wrapped phase map would be similar to that shown in Figure 7.7(b). Prior to phase unwrapping, the phase data of each level are scaled by a factor of 2^{n-i}, where n is the number of levels and i is the level order of a phase map. A section of such scaled phase data is shown in Figure 7.7(c).

For phase unwrapping, an appropriate value of 2π multiples is added to or subtracted from the phase values at level L4. The appropriate multiples to be used are determined using the phase data of the other levels. The process involves comparing the phase value $\theta_4(x, y)$ of a pixel in L4 with its corresponding pixel phase value $\theta_3(x, y)$ in L3 and the following procedure is carried out: (1) $\theta_4(x, y)$ is to remain unchanged, if $|\theta_4(x, y) - \theta_3(x, y)| \in [-\pi, \pi]$, (2) 2π is subtracted from $\theta_4(x, y)$, if $\theta_4(x, y) - \theta_3(x, y) > \pi$, and (3) 2π is added to $\theta_4(x, y)$, if $\theta_4(x, y) - \theta_3(x, y) < -\pi$.

The revised phase value $\theta'_4(x, y)$ is then compared with its corresponding pixel phase value $\theta_2(x, y)$ in L2. Again a similar procedure to that described above is carried out. However, the conditional threshold values are -2π and/or 2π instead of π. If conditions (2) or (3) are encountered, a 4π value is subtracted from or added to $\theta'_4(x, y)$ to obtain another revised phase value $\theta''_4(x, y)$, which is then compared with the corresponding pixel phase value $\theta_1(x, y)$ in L1. The conditional threshold

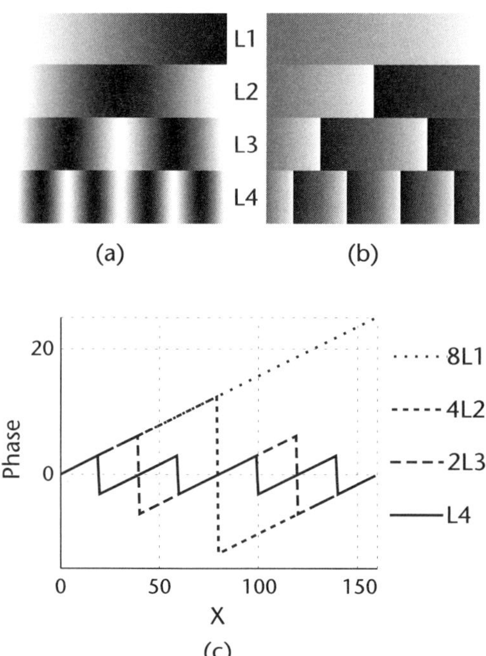

Figure 7.7   (a) Fringe patterns of four frequencies, L1 to L4. (b) Corresponding wrapped phase map of the fringe pattern in (a). (c) A section of the wrapped phase map in (b) scaled by a different coefficient.

is again doubled and the same procedure of subtracting or adding the resulting multiples of 2π to $\theta''_4(x, y)$ is applied. Unlike spatial phase unwrapping, temporal phase unwrapping does not require comparison of a pixel phase value with any of its neighbors.

Note that in the phase-unwrapping procedure described, the unwrapped phase map always contains the highest frequency of the fringes. The reason is that the higher the fringe frequency, the higher is the sensitivity of the unwrapping process to a change in the surface profile.

BOX 7.2   Absolute-Phase

Although multifrequency fringe projection method relies on many more wrapped phase maps than spatial phase unwrapping, it has a distinct advantage: the unwrapped phase map does not contain a random phase offset as long as the fringe pattern in L1 does not vary. Thus the phase value obtained can be referred to as the absolute phase as it is determined from the fringe pattern in L1 and not from the phase unwrapping process. The random offset exists in spatial phase unwrapping as the offset is related to the phase of the starting pixel and that the offset value may vary from one experiment to another. The offset value would affect the whole phase map and in situations where the absolute phase is of interest, the multifrequency fringe projection method would be most suitable.

EXERCISE 7.5

In the use of multifrequency fringe projection technique, if the frequency of a fringe pattern from one level to the next is not doubled but tripled or quadrupled, is the temporal phase unwrapping procedure still valid? What changes have to be made accordingly?

Solution

The temporal phase unwrapping procedure is still valid. The changes to be made are: instead of multiplying the phase values by a factor of 2^{n-i}, a factor of 3^{n-i} (tripled) or 4^{n-i} (quadrupled) should be used, where n is the number of levels and i is the level order. Figure 7.8 shows an example of a quadruple increase in the frequency of fringes projected on an electric plug specimen.

7.2.4   Phase Calibration

Calibration of a fringe projection system may vary according to its applications. If a phase-to-height relationship is required, one may measure an object with a known height and obtain a phase-to-height coefficient by removing the carrier phase component from the unwrapped phase map. If a 3D reconstruction of an object is the goal, a full 3D calibration is required. Three methods, which cover a good spectrum of various purposes, are described in the following sections.

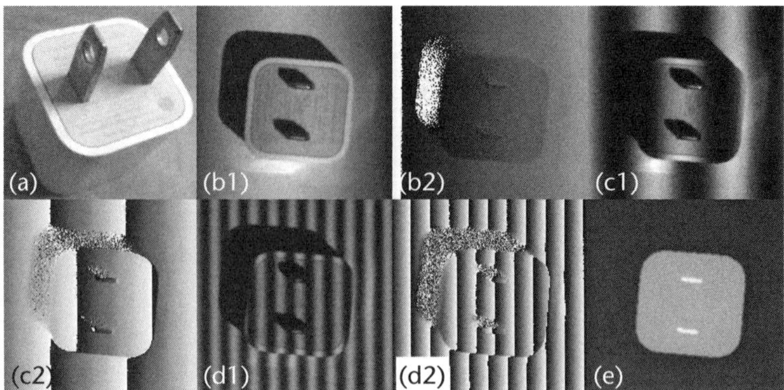

Figure 7.8 Measurement of an electrical plug specimen with drastic discontinuities in surface profile by multi-frequency fringe projection. The frequency increment is quadrupled from one level to the next. (a) An electrical plug specimen. (b1) Fringe pattern and (b2) wrapped phase map of level L1. (c1) Fringe pattern and (c2) wrapped phase map of level L2. (d1) Fringe pattern and (d2) wrapped phase map of level L3. (e) An unwrapped phase map with carrier phase component removed.

7.2.4.1 Carrier Phase Removal

Carrier fringes are a fringe pattern generated by fringe projection. The carrier fringe deformation that results in phase variations of the fringes would provide the surface profile information of an object. As it is the phase variation and not the original phase of the carrier fringes that is of interest, the latter should be removed from an unwrapped phase map [11]. A convenient method to remove the carrier phase component is to determine the component by a 2D polynomial function [12] given as follows:

$$\theta_p(x,y) = \begin{array}{lll} a_{0,0} & + \; a_{0,1}x & + \cdots + \; a_{0,n}x^n \\ a_{1,0}y & + \; a_{1,1}xy & + \cdots \\ + \cdots & & \\ a_{n-1,0}y^{n-1} & + \; a_{n-1,1}xy^{n-1} & \\ a_{n,0}y^n & & \end{array} \tag{7.11}$$

where n is the order of the polynomial, $a_{j,i}$ are unknown coefficients, and i and j are the power of x and y, respectively. Estimation of the coefficients of $\theta_p(x, y)$ is based on the least-squares technique. To obtain the unknown coefficients, an error function as shown below is employed:

$$E\left(a_{0,0}, \ldots, a_{0,n}, \ldots, a_{n,0}\right) = \sum_{(x,y)\in U} \left[\theta_p(x,y) - \theta_e(x,y)\right]^2 \tag{7.12}$$

where U is a region that contains the carrier phase and θ_e are the experimentally obtained unwrapped phase values. To minimize the error, partial derivatives of E with respective to each coefficient $a_{j,i}$ are set to zero. This produces $(n + 1)(n + 2)/2$ linear equations, from which the unknowns can be obtained.

The choice of n largely depends on the uniformity of the carrier fringes (i.e., the fringe spacing). If the projection system emits parallel beams, the fringe spacing on a reference plane will be uniform. In this case, a first-order polynomial function ($n = 1$) would be able to estimate the carrier phase component accurately. If a convergent or divergent beam is projected, a higher-order polynomial is required. It is rare to have to resort to a fourth- or higher-order polynomial. Figure 7.9 shows the effect of removing the carrier phase. As the projected fringes are not fully parallel, a first-order polynomial function would not be able to estimate the carrier phase component accurately and any remaining residual carrier phase would distort the originally flat plane, as shown in Figure 7.9(b). A second-order polynomial would improve the carrier phase removal process, as indicated by the flat plane in Figure 7.9(c).

After the carrier phase component is removed, a linear phase-to-height relationship is obtained and a step of a known height can be measured. The ratio between the actual step height and the phase difference between two consecutive steps is represented by the coefficient.

7.2.4.2   Multiple-Reference-Plane Calibration

The intensity of a fringe pattern projected on an object changes with the distance of the object. Although global intensity variation does not severely affect the resultant

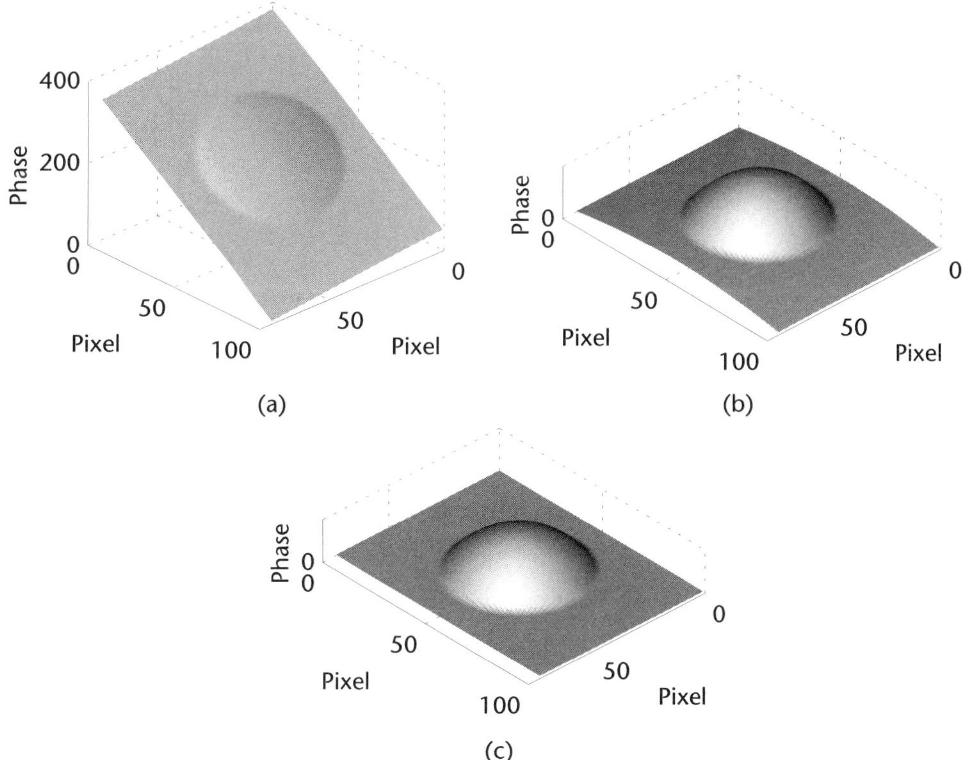

Figure 7.9   (a) Unwrapped phase map that contains a hump and a carrier phase component. (b) Removal of the carrier phase component estimated by a first-order polynomial function. (c) Removal of the carrier phase component estimated by a second-order polynomial function.

phase values, it does reduce the measurement accuracy to some extent. Other factors, such as lens distortion, and focus, also have different effects at different distances. To deal with them effectively, a multiple-reference-plane calibration can be applied. The simplest version is the two-plane setup approach shown in Figure 7.3.

A more effective version measures a reference plane at several axial distances (along the z-axis) with each incremental distance kept reasonably small. An unwrapped phase map (without phase removal) at each position is then recorded. In this case, the absolute phase is recorded and the phase values of a pixel at different distances directly indicate the phase change.

After the unwrapped phase values of an object are calculated, each pixel phase (θ_o) is compared with the corresponding pixel phase in the reference planes. The nearest two reference planes are obtained (phase θ_1 and θ_2) and each pixel position (along the z-axis) is interpolated from two reference positions based on the ratio of $(\theta_o - \theta_1)/(\theta_o - \theta_2)$. The process is repeated for all the points on the object to obtain the entire surface profile. The method ensures that only the most suitable (closest) reference phase values are used to calculate the object surface profile. The intensity variation, lens distortion, focus, etc. have all been taken into consideration; hence, the results are largely free from their influence.

7.2.4.3 3D Calibration

A full 3D calibration would establish the relationship between a 2D image and a 3D global coordinate. Various calibration methods, which rely mainly on techniques developed in computer vision, have been proposed [13, 14]. One of the key steps is to obtain the imaging parameters of a camera used. The camera imaging parameters are represented by a 3×4 matrix, which contains parameters such as the focal length and other parameters based on the global coordinates.

The actual calibration of the camera imaging parameters is beyond the scope of this chapter. However, it suffices to mention that the matrix is a bridge between a 2D image point and a 3D global point [15]. The calibrated equation is given by:

$$\left[\lambda x_i, \lambda y_i, \lambda\right]^T = M\left[x_w, y_w, z_w, 1\right]^T \tag{7.13}$$

where $\mathbf{M}$ is a 3×4 matrix, (x_i, y_i) is an image point, (x_w, y_w, z_w) is a global point, λ is a scaling factor, and T denotes a transpose. The equation is based on a homogeneous coordinate system to comply with the convention in computer vision: a 2D or 3D point is represented by a 3D or 4D vector. The last element of the vector is a scaling factor.

Calibration involves calibrating the camera imaging parameters and establishing the phase-to-height relationship using the methods described earlier. In the measuring process, a point on the object surface z_w is first retrieved and its corresponding coordinates x_w, y_w and the scaling factor λ are then calculated based on (7.13). Note that the coordinate (x_i, y_i) of the point in the 2D image and the matrix $\mathbf{M}$ are known quantities. Other points on the surface are then obtained in a similar manner and subsequently the entire 3D profile is determined.

7.3   Applications of Method

A number of objects are examined and are included here to demonstrate the afore-mentioned principles. Various data-processing methods used are also included.

7.3.1   Line Scan and Fringe Projection

An object in the form of a mask is evaluated by the line-scan method and a four-step, 90-degree phase-shifting fringe projection technique. Figure 7.10 shows the experimental setup. Figure 7.11(a) shows a line-scan image obtained by a projected line that scans over the entire mask surface. An edge detection algorithm retrieves the x coordinates of each line with subpixel accuracy and the depth of each point on the mask is calculated using (7.2). Figure 7.11(b) shows the retrieved surface profile of the mask. In the measurement only 150 lines were used and the mask appears distorted. This is a typical result in line-scan technique as a surface is usually undersampled. To retrieve a nondistorted surface profile, interpolation is required.

Figure 7.12 shows the mask evaluated by the fringe projection technique. Data-processing involves wrapped phase extraction by phase-shifting, phase unwrapping, locating points of equal phase on a near and a far reference planes, and calculating the height of each point based on (7.2). The accuracy of phase extraction is in general an order higher than that of edge detection; hence more accurate results

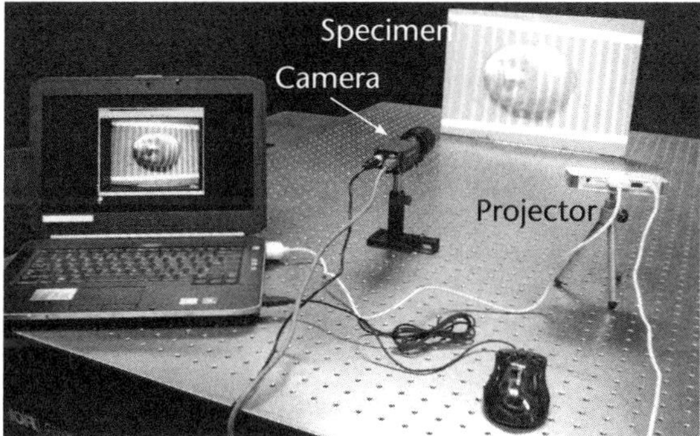

Figure 7.10   Experimental setup.

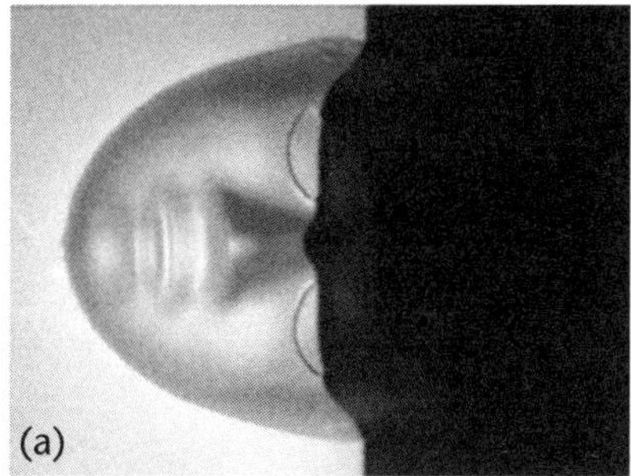

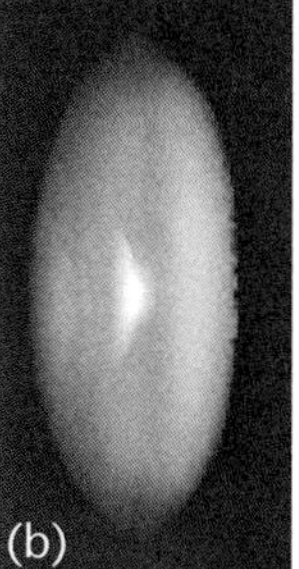

Figure 7.11   (a) A line-scan image of a mask. (b) Surface profile of the mask retrieved from 150 line-scan images.

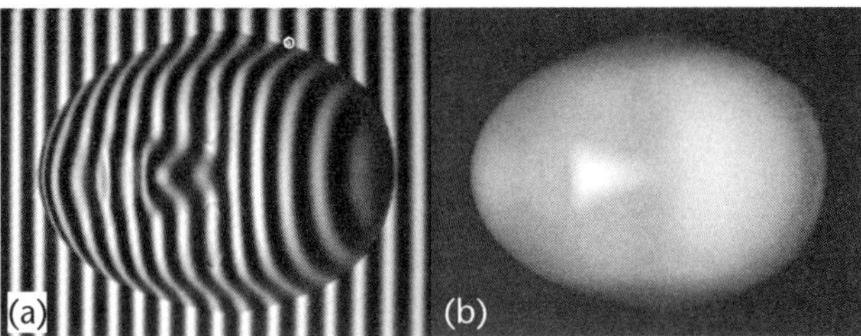

Figure 7.12 (a) A fringe pattern projected on a mask. (b) Surface profile of the mask retrieved by the fringe projection method.

can be obtained. The sampling size is dependent on the resolution of the camera and there is no necessity for interpolation. Hence to obtain a whole-field surface measurement, the fringe projection method is preferred over the line-scan method.

7.3.2 Quality-guided Phase Unwrapping

A fringe pattern projected on the model of a fish, as shown in Figure 7.13, is used to illustrate the principles of quality-guided phase unwrapping. During phase unwrapping, the fringe contrast is set as the quality indicator. As can be seen, regions at the tail end of the fish model are covered by shadows. They correspond to dark pixels in the contrast map and drastically varying phase values in the unwrapped phase map, which indicate erroneous results. This is due to poor signal-to-noise ratios. However, the errors are confined to the tail end, which means that they are unwrapped at the last stage in the unwrapping process.

It is seen from Figure 7.13(a) that the shadow regions are irregular. How does the unwrapping process avoid the shadow until the last stage of processing? To answer this question, we have to refer to the unwrapping sequence illustrated in Figure 7.14. The unwrapping process starts from Figure 7.14(a) and progresses to Figure 7.14(e); regions that show high quality (high intensity) are first processed. As indicated in the contrast map of Figure 7.13(b), the base plane is of high quality; hence it would be processed with priority while the boundary of the fish model is of relatively low quality and hence would be processed at a later stage. The model body region is processed in stages in Figure 7.14(f) to 7.14(i), with the processing path entering from the lower portion of the model boundary. Regions of the lowest quality are left until the last stage of processing.

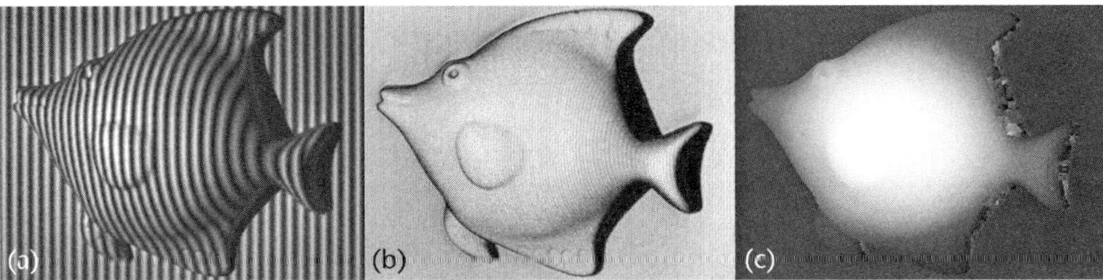

Figure 7.13 (a) A fringe pattern projected on a fish model. (b) A fringe contrast map, where brightness indicates good contrast. (c) Unwrapped phase map with carrier phase removed.

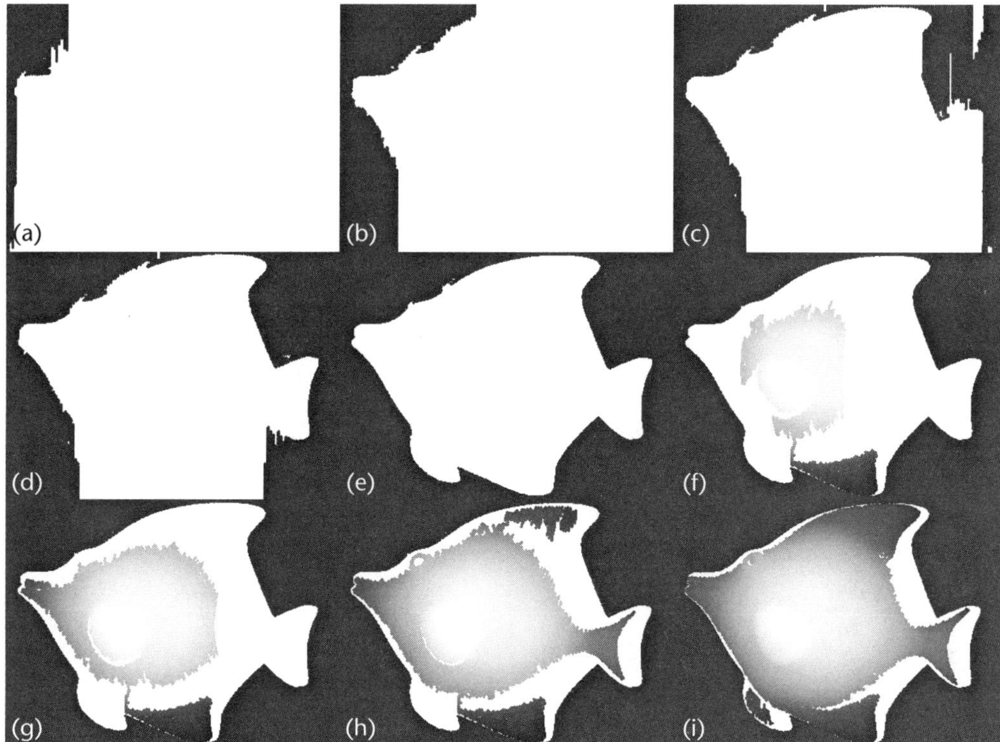

Figure 7.14   Unwrapping process of a fish model, guided by fringe contrast.

7.3.3   Multifrequency Fringe Projection

When it comes to the evaluation of an extremely irregular surface, such as the dragon wall specimen as shown in Figure 7.15, spatial phase unwrapping would be unsuitable. In this case the surface profile is evaluated by the multifrequency fringe projection technique. At each frequency, an eight-step, 45-degree phase-shift is applied to extract a wrapped phase map. An unwrapped phase map is obtained from the wrapped phase maps by the multifrequency unwrapping algorithm. As mentioned in Section 7.2.3.4, phase values in multiples of 2π are added to or subtracted from the wrapped phase values at the highest-frequency level and the multiples are determined by the corresponding wrapped phase values at the lower-frequency levels.

Figure 7.16 shows a comparison between quality-guided and multifrequency phase unwrapping. Results of the former approach guided by fringe contrast contain many erroneous regions, as indicated by the arrows shown in Figure 7.16(b). The errors are difficult to eliminate as they are widespread. The results using multifrequency unwrapping (Figure 7.16(c)) also contain errors, albeit mostly in the low-contrast regions (as indicated by the arrows). Since multifrequency unwrapping does not rely on spatial information, the issue of error accumulation is eliminated. Using the fringe contrast map, phase values of low contrast are replaced by the surrounding high-quality phase data and a filtered noise-free unwrapped phase map is obtained as shown in Figure 7.16(d).

Figure 7.15 Top image (first row): a dragon wall. (Source: S. Dritsas, SUTD. Reprinted with permission.) Bottom images (second and third rows): multifrequency fringe patterns projected on the dragon wall.

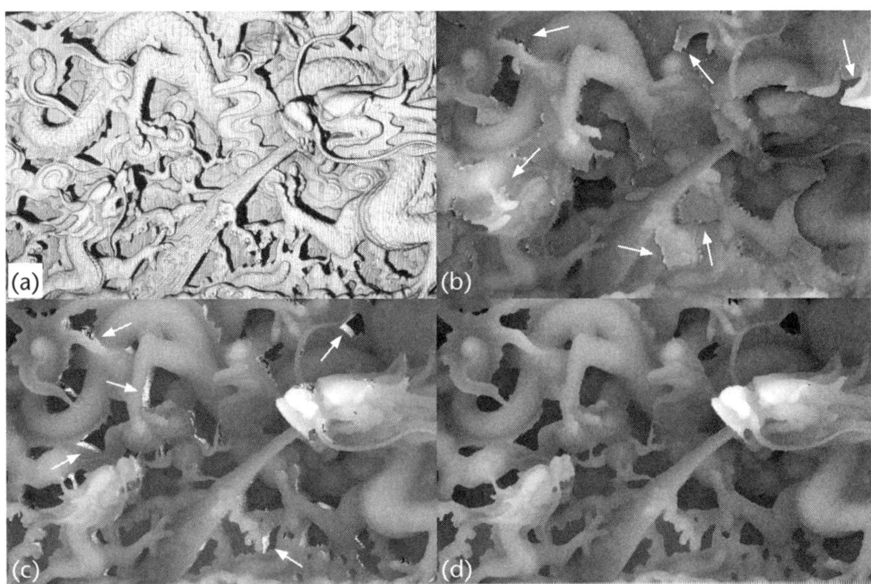

Figure 7.16 (a) Fringe contrast map of the highest frequency. Unwrapped phase map (carrier phase removed) obtained by (b) quality-guided and (c) multifrequency unwrapping. (d) A filtered unwrapped phase map of (c).

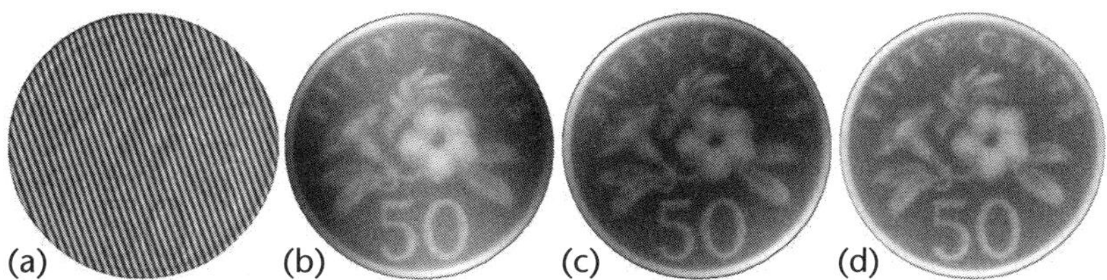

Figure 7.17   A 3D reconstruction of the dragon wall specimen.

Figure 7.17 shows a 3D reconstruction of the dragon wall specimen. As can be seen, the multifrequency fringe projection technique assisted by digital devices and data-processing algorithms is able to image most of the details with excellent results.

7.3.4   Carrier Phase Removal

Carrier phase removal method is normally employed in the evaluation of a surface profile with small variation in depth, such as the surface of a coin as shown in Figure 7.18. Owing to the small range in depth, distortion of image pertaining to large depth range can be ignored. Data processing in this case is reduced to retrieving the relationship between phase and height and the removal of the carrier phase component also offers a convenient shortcut in contrast to the more complicated calibration system mentioned earlier.

The fringe pattern on the coin specimen (Figure 7.18(a)) is not parallel to the x- or y-axis; hence an x and y variables-dependent polynomial function as shown

(a)   (b)   (c)   (d)

Figure 7.18   (a) Fringe pattern projected on a coin specimen. Surface profile of the coin specimen after removal of the carrier phase component using (b) first-order, (c) second-order, x and y variables-independent polynomial function, and (3) second-order, x and y variables dependent polynomial function.

Figure 7.19 (a) Magnified 3D surface of Figure 7.18(b). (b) Magnified 3D surface of Figure 7.18(c). (c) Magnified 3D surface of Figure 7.18(d).

in (7.11) is required for the phase removal. Although the fringe spacing appears to be uniform, residual carrier phase in Figure 7.18(b) and distortion of the surface (Figure 7.19(a)) are observable even after removal of the carrier phase using a first-order polynomial. If a second-order polynomial independent of x and y variables is employed as shown in Figure 7.18(c), some residual carrier phase and distortion of the surface (Figure 7.19(b)) can still be observed. However, if a second-order, x and y variables-dependent polynomial is used, the results show an undistorted surface profile (Figures 7.18(d) and 7.19(c)).

7.3.5 360-Degree Fringe Projection

Fringe projection can also be employed to evaluate an object surface in 360 degrees [16], as the shown in Figure 7.20. The measurement process consists of the following procedure: (1) calibration of the camera and fringe projector, (2) calibration of the test object axis of revolution, (3) imaging of the entire object surface using a rotating stage, (4) transformation of captured images to a common global coordinate, and (5) integration of the images to create a 360-degree surface profile. A linear stage is incorporated to introduce accurate shift in the z direction while a rotating stage is used to rotate the test object.

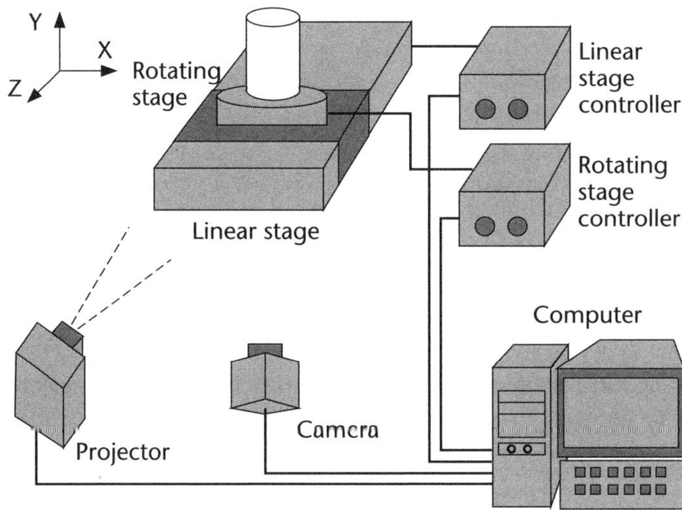

Figure 7.20 360-degree fringe projection system.

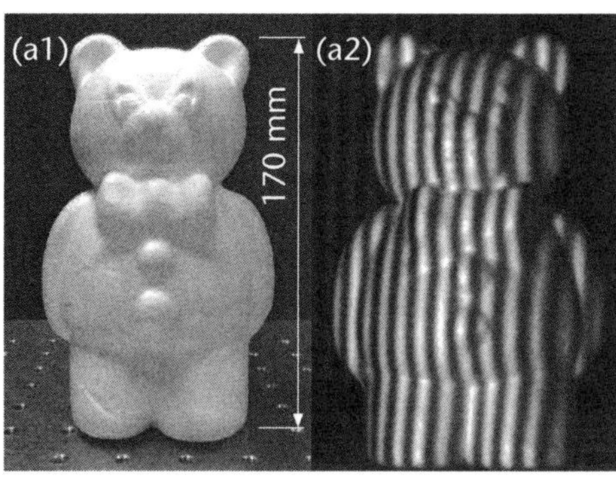
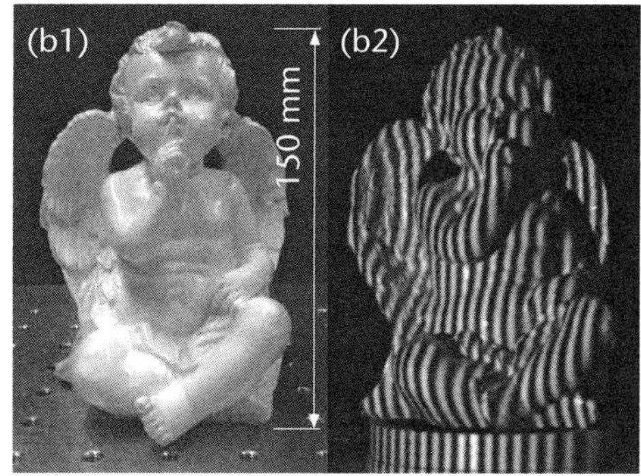

Figure 7.21   Test objects. (a1) Test model of a bear and (a2) a fringe pattern projected on the model. (b1) Test model of an angel and (b2) a fringe pattern projected on the model.

Figure 7.21 shows two test objects that are imaged at several viewing angles. At each viewing angle, the multifrequency fringe projection method is employed to capture the surface images. The recorded images are then combined to form a 360-degree surface profile.

Figure 7.22(a) shows a cloud image of the model of a bear captured at a 30-degree interval. For clarity not all patch images captured are included. Each image recorded at a 30-degree interval is transformed to a common coordinate system and a surface reconstruction algorithm is employed to generate a surface mesh on the model. As the top part of the model was not visible to the imaging system, no useful data was reconstructed as shown in Figure 7.22(b). In addition, phase errors are introduced due to the presence of shadow on the arm of the model as shown in Figure 7.22(c), while erroneous vertical stripes on the surface are introduced by non-sinusoidal fringe distribution. Such nonsinusoidal fringe induced phase errors are a topic that has received increasing interest in recent years [17].

Figure 7.23 shows a reconstructed image of the test model of an angel. The relatively complicated profile presented numerous problems in the reconstruction. As a result, sporadic erroneous spots on the model (dark spots) are observed. These are areas undetectable to the imaging system due to shadows or voids on the object. However, areas that are relatively smooth and visible to the imaging system are reconstructed with good accuracy.

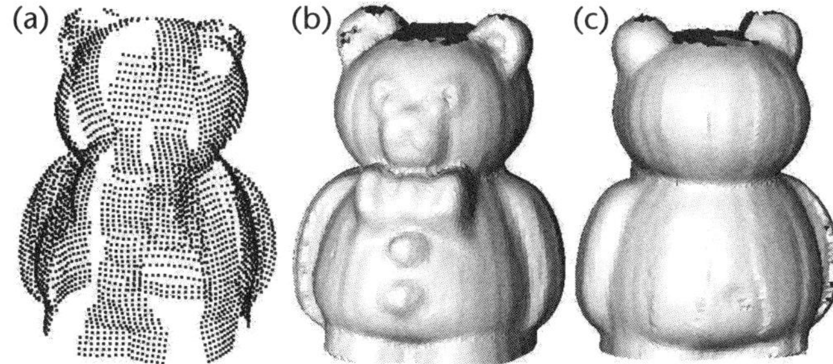

Figure 7.22   (a) A cloud image of a test bear model reconstructed at a 30-degree interval. (b) Reconstruction of the test model (front view). (c) Reconstruction of the test model (back view).

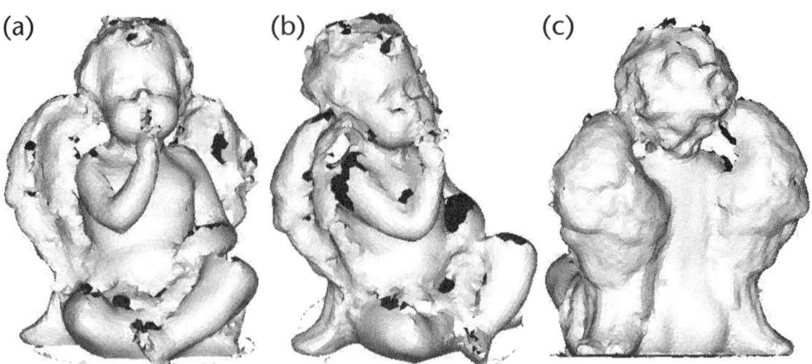

Figure 7.23 Reconstructed model of an angel in different views.

The case study shows that although fringe projection is primarily used in 3D measurement at a particular viewing direction, an ancillary technique can be incorporated to achieve a 360-degree 3D reconstruction.

7.4   Concluding Remarks

Fringe projection profilometry has seen many inspiring ideas, technological breakthroughs, and novel applications in over three decades of research and development. In recent years, it has become one of the most widely used digital measurement methods owing to the ease of access to digital devices. The advancement of fringe projection profilometry is also enhanced by data-processing development in the areas of optical interferometry and computer vision.

To date, numerous techniques based on fringe projection have achieved certain maturity; however, there are still challenges in this field that attract a great number of researchers to devote their time and effort. For instance, nonsinusoidal distribution of a fringe pattern introduces phase measurement errors. Eliminating such errors would lead to a significant increase in measurement accuracy. Various approaches have been proposed, albeit there is no consensus on the best method. Work in this area would warrant further investigations.

In this chapter, the main challenges encountered in digital fringe projection profilometry have been discussed. The various methods of phase evaluation involving Fourier transform, phase shifting, phase unwrapping, and phase calibration are presented. These methods have shown to be effective and accurate on the profile evaluation of various test objects.

References

[1]   Chen, F., Brown, G. M., and Song, M., "Overview of Three-Dimensional Shape Measurement Using Optical Methods," *Optical Engineering*, Vol. 39, No. 1, 2000, pp. 10–22.

[2]   Geng, J., "Structured-Light 3D Surface Imaging: A Tutorial," *Advances in Optics and Photonics*, Vol. 3, 2011, pp. 128–160.

[3]   Gorthi, S. S., and Rastogi, P., "Fringe Projection Techniques: Whither We Are?" *Optics and Lasers in Engineering*, Vol. 48, No. 2, 2010, pp. 133–140.

[4]   Takeda, M., Ina, H. and Kobayashi, S., "Fourier-Transform Method of Fringe-Pattern Analysis for Computer-Based Topography and Interferometry," *Journal of Optical Society of America A*, Vol. 72, No. 1, 1982, pp. 156–160.

[5]   Takeda, M., and Mutoh, K., "Fourier Transform Profilometry for the Automatic Measurement of 3-D Object Shapes," *Applied Optics*, Vol. 22, No. 24, 1983, pp. 3977–3982.

[6]   Morgan, C. J., "Least-Squares Estimation in Phase-Measurement Interferometry," *Optics Letters*, Vol. 7, No. 8, 1982, pp. 368–370.

[7]   Ghiglia, D. C., and M. D. Pritt, *Two-Dimensional Phase Unwrapping: Theory, Algorithms, and Software*, New York: Wiley, 1998.

[8]   Huntley, J. M., and Saldner, H., "Temporal Phase-Unwrapping Algorithm for Automated Interferogram Analysis," *Applied Optics*, Vol. 32, No. 17, 1993, pp. 3047–3052.

[9]   Zhao, H., Chen, W., and Tan, Y., "Phase-Unwrapping Algorithm for the Measurement of Three-Dimensional Object Shapes," *Applied Optics*, Vol. 33, No. 20, 1994, pp. 4497–4500.

[10]   Wang, Z., Du, H., Park, S., and Xie, H., "Three-Dimensional Shape Measurement with a Fast and Accurate Approach," *Applied Optics*, Vol. 48, No. 6, 2009, pp. 1052–1061.

[11]   Chen, L., and Quan, C., "Fringe Projection Profilometry with Nonparallel Illumination: A Least-Squares Approach," *Optics Letters*, Vol. 30, No. 16, 2005, pp. 2101–2103.

[12]   Chen, L., and Tay, C. J., "Carrier Phase Component Removal: A Generalized Least-Squares Approach," *Journal of Optical Society of America A*, Vol. 23, No. 2, 2006, pp. 435–443.

[13]   Zhang, S., and Huang, P. S., "Novel Method for Structured Light System Calibration," *Optical Engineering*, Vol. 45, No. 8, 2006, pp. 083601.

[14]   Zhang, S., "Recent Progresses on Real-Time 3D Shape Measurement Using Digital Fringe Projection Techniques," *Optics and Lasers in Engineering*, Vol. 48, 2010, pp. 149–158.

[15]   Hartley, R., and A. Zisserman, *Multiple View Geometry in Computer Vision*, Cambridge, UK: Cambridge University Press, 2000.

[16]   Dai, M., Chen, L., Yang, F., and He, X., "Calibration of Revolution Axis for 360 deg Surface Measurement," *Applied Optics*, Vol. 52, No. 22, 2013, pp. 5440–5448.

[17]   Wang, Z., Nguyen, D. A., and Barnes, J. C., "Some Practical Considerations in Fringe Projection Profilometry," *Optics and Lasers in Engineering*, Vol. 48, 2010, pp. 218–225.

CHAPTER 8

Digital Photoelasticity

Krishnamurthi Ramesh

8.1 Introduction

Photoelasticity comes under the category of common path interferometers, and the coherence requirements are not stringent. Hence, photoelasticity has gained wider acceptance and established itself as an excellent tool in visualizing/quantifying stress fields and as a teaching aid for stress analysis. Though the phenomenon of temporary birefringence, which is the physical principle on which photoelasticity is based, was discovered by Brewster in 1816, it is only in 1930s through the works of Coker and Filon in the UK that the technique gained popularity. For a wide variety of design problems, photoelasticity provides direct quantitative information and for phenomenological understanding, photoelasticity has always been in the forefront—the cases that are addressed include need for multiparameters for crack-tip stress field modeling, various issues related to stress wave propagation, crack-stress wave interactions, and so on.

Modern photoelasticity no longer looks at the basic data as fringe patterns but as intensity data. In the initial phase of developments, the use of digital cameras replaced the human eye and digital hardware is then used to mimic conventional data augmentation methods such as fringe multiplication and fringe thinning for improved accuracy. Though such developments simplified human effort and is quite useful in certain applications, whole field automated photoelasticity has been possible only with the development of phase shifting techniques. Several innovative techniques to use color information for quantitative analysis have also been developed. A detailed description of various methods could be found in [1–3]. In this chapter, for brevity only those techniques that are time-tested and in use widely are presented.

8.2 Basics of Photoelasticity

In photoelasticity, typically for a known incident polarized light the emergent light is analyzed for stress/strain information. The formation of fringes in all variants of photoelasticity [4] is due to the phenomenon of birefringence exhibited by the models/coatings.

8.2.1   Birefringence

The crystalline media are optically anisotropic and are characterized by the fact that a single incident ray will give rise to two refracted rays, ordinary "o" and extraordinary "e." The ordinary and extraordinary rays are plane polarized and their planes of polarization are perpendicular to each other. The extraordinary ray violates Snell's Law under suitable circumstances. When the incident light is perpendicular to the optic axis, the ordinary and extraordinary rays travel in the same direction but with different velocities. This aspect is exploited in photoelasticity for formation and the interpretation of fringe patterns.

8.2.2   Retardation Plates

Consider a crystalline plate of thickness h. Let a plane-polarized light of a single wavelength be incident normally as shown in Figure 8.1(a). Let the polarizing axes of the plate be oriented at angles θ and $\theta + \pi/2$ with the horizontal. Since the incident ray is perpendicular to the optic axes (polarizing axes); two rays, namely, the ordinary and the extraordinary, travel in the same direction but with different velocities of v_1 and v_2. These two rays will have a net phase difference of δ upon emergence from the crystal plate. Since the velocities of propagation within the crystal are different for these two rays, they will take, respectively, h/v_1 and h/v_2 seconds to traverse the plate. This time difference contributes to the phase difference. Let the frequency of the light be f, then

$$\delta = 2\pi f\left(\frac{h}{v_1} - \frac{h}{v_2}\right) = 2\pi h \frac{c}{\lambda}\left(\frac{1}{v_1} - \frac{1}{v_2}\right)$$
$$= \frac{2\pi h}{\lambda}\left(n_1 - n_2\right) \tag{8.1}$$

The emerging light is in general elliptically polarized. If the thickness is such as to produce a phase difference of $\pi/2$ radian, then it is a quarter-wave plate ($\lambda/4$), and if it is π radians, it is a half-wave plate. If the retardation is 2π, then one gets a full-wave plate and the incident light is unaltered. Wave-plates or retarders have two polarizing axes, and one of these axes is labeled as a fast axis (F) and the other as the slow axis (S). Using wave plates, a linearly polarized light can be changed to a circular or elliptically polarized light.

8.2.3   Stress-Optic Law

Certain noncrystalline transparent materials, notably some polymeric plastics, are optically isotropic under normal conditions but become doubly refractive or birefringent when stressed. This effect normally persists while the loads are maintained but vanishes almost instantaneously or after some interval of time depending on the material and conditions of loading when the loads are removed. This is the physical characteristic on which photoelasticity is based.

The photoelastic model behaves like a retarder with different retarder characteristics at different points in the model governed by the induced stress field.

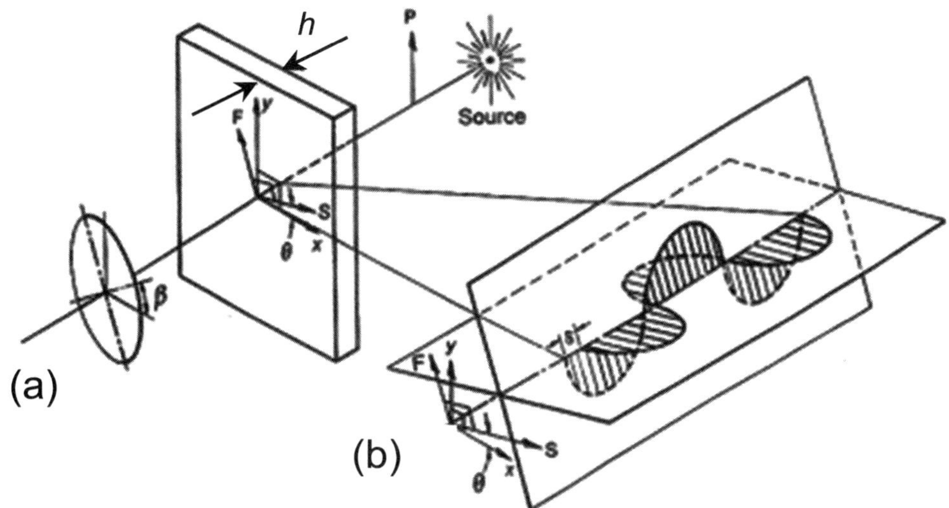

Figure 8.1 (a) Linearly polarized light incident on a retarder; (b) inside the retarder two plane polarized lights, whose planes of polarization coincide with the fast and slow axes of the retarder, travel with different velocities. (Courtesy [1].)

Consider a transparent model made of a high polymer subjected to a plane state of stress. Let the state of stress at a point be characterized by the principal stresses σ_1, σ_2, and their orientation θ with reference to a set of axes. Let n_1 and n_2 be the refractive indices for vibrations corresponding to these two directions. Following Maxwell's formulation, the relative retardation in terms of principal stress difference can be written as [1]

$$\delta = \frac{2\pi h}{\lambda}(n_1 - n_2) = \frac{2\pi h}{\lambda}C(\sigma_1 - \sigma_2) \tag{8.2}$$

where C is the relative stress-optic coefficient usually assumed to be a constant for a material. However, various studies have shown that this coefficient depends on wavelength and should be used with care. Equation (8.2) can be rewritten in terms of fringe order N as

$$N = \frac{\delta}{2\pi} = h\frac{C}{\lambda}(\sigma_1 - \sigma_2) \tag{8.3}$$

which can be recast as

$$\sigma_1 - \sigma_2 = \frac{NF_\sigma}{h} \tag{8.4}$$

where

$$F_\sigma = \frac{\lambda}{C} \tag{8.5}$$

which is known as the material stress fringe value with the units N/mm/fringe. Equation (8.4) is known as stress-optic law as it relates the stress information to optical measurement. The principal stresses are labeled such that σ_1 is always algebraically greater than σ_2. In view of it, the L.H.S of (8.4) is always positive. Thus, in photoelasticity fringe order N is always positive.

If the fringe order and the material stress fringe values are known, then one can obtain the principal stress difference. Equation (8.5) shows the dependence of F_σ on the wavelength. Equation (8.4) implicitly gives the indication that F_σ and $(\sigma_1 - \sigma_2)$ are linearly related. However, at higher stress levels, the relationship is nonlinear and (8.4) should be used with care.

8.2.4   Optical Arrangements in Conventional Photoelasticity

One of the simplest optical arrangements possible is a plane polariscope (Figure 8.2(a)). Let the light source be a monochromatic one. A circular disk under diametral compression is kept in the field of view. The incident light on the model is

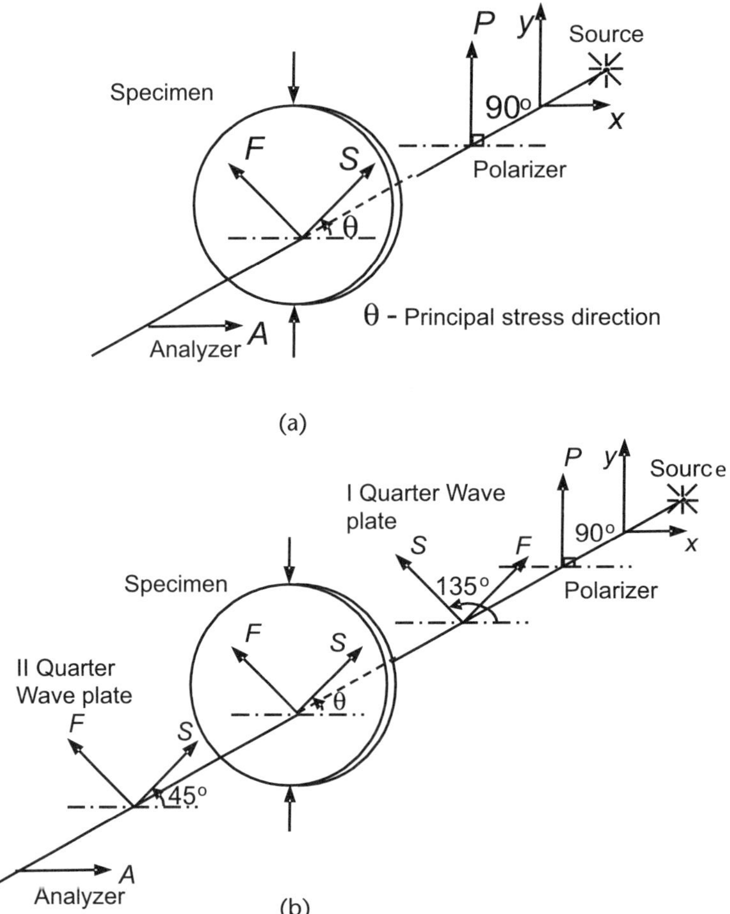

Figure 8.2 Optical arrangements in conventional photoelasticity: (a) plane polariscope; (b) circular polariscope.

plane polarized. As it passes through the model, the state of polarization changes from point to point depending on the magnitude of principal stress difference and the principal stress direction. The information about the stress field can be obtained if the state of polarization of the emergent light is studied. This is easily achieved by introducing a polarizer at 0°, which is termed as an analyzer as it analyzes the state of polarization of the exit light. In the absence of a model, no light will be transmitted after the analyzer, and the optical arrangement is known as *dark field* plane polariscope. Another optical arrangement commonly employed is a circular polariscope in which a circularly polarized light is incident on the model. Placement of a quarter-wave plate after the polarizer at 45°/135° helps to get a circularly polarized light. The elements after the model are complementary to the one kept before and helps to analyze the state of polarization of the exit light. The basic optical arrangement shown is for a dark field, and if the analyzer is kept at 90°, one would observe that entire light will be transmitted by the analyzer in the absence of a model and such an arrangement is termed as *bright field* circular polariscope. Although, bright and dark fields could be achieved through different relative orientations of the optical elements [1, 2], usually only those arrangements wherein the quarter-wave plates kept crossed are preferred as the error is reduced due to mismatch of quarter-wave plates.

8.3   Jones Calculus

In general, an optical element in a polariscope introduces a rotation and retardation. In Jones calculus, these basic operations are represented as matrices. Rotation matrix is useful to find the components of a vector if the reference axes are rotated by an arbitrary angle θ. It is given by

$$\begin{bmatrix} \cos\theta & \sin\theta \\ -\sin\theta & \cos\theta \end{bmatrix} \tag{8.6}$$

The usual convention of measuring angles is valid for constructing a rotation matrix and the angle is positive if it is measured counterclockwise.

A doubly refracting medium introduces a relative retardation δ between the vibrating components along its axes. Using complex number notations, the emerging light can be obtained as a matrix operation of the incident light as [1]

$$\begin{Bmatrix} u' \\ v' \end{Bmatrix} = \Re \begin{bmatrix} e^{-i\delta/2} & 0 \\ 0 & e^{i\delta/2} \end{bmatrix} \begin{Bmatrix} a_1 e^{i\alpha_1} \\ a_2 e^{i\alpha_2} \end{Bmatrix} e^{i\omega t} \tag{8.7}$$

The first matrix on the right side of (8.7) is called the retardation matrix. On the explicit understanding that we deal with real parts only, the notation $\Re$ is usually left out. It is to be noted that in this representation, the u' axis is considered as the slow axis.

As many elements of a generic polariscope can be considered as retarders, it is desirable to obtain a matrix representation of the retarder. Let the slow axis of the retarder be oriented at an angle θ with the x-axis and let the total relative retardation introduced by the retarder be δ. In order to trace the transformation of light through the retarder, the incident light has to be rotated by an angle θ when it enters the retarder, then the relative retardation of δ is to be introduced, and finally the emerging light is to be represented with respect to the original reference axes. These operations are represented in matrix form as

$$\begin{bmatrix} \cos\theta & -\sin\theta \\ \sin\theta & \cos\theta \end{bmatrix} \begin{bmatrix} e^{-i\delta/2} & 0 \\ 0 & e^{i\delta/2} \end{bmatrix} \begin{bmatrix} \cos\theta & \sin\theta \\ -\sin\theta & \cos\theta \end{bmatrix} \tag{8.8}$$

The product of the matrices in (8.8) gives the matrix as follows

$$\begin{bmatrix} \cos\dfrac{\delta}{2} - i\sin\dfrac{\delta}{2}\cos 2\theta & -i\sin\dfrac{\delta}{2}\sin 2\theta \\ -i\sin\dfrac{\delta}{2}\sin 2\theta & \cos\dfrac{\delta}{2} + i\sin\dfrac{\delta}{2}\cos 2\theta \end{bmatrix} \tag{8.9}$$

Equation (8.9) is a very useful one to quickly write the Jones matrix for retardation plates. If the values of the retardation and the orientation (of the slow axis with the x-axis) are known, then one gets a very simple matrix for computations.

EXERCISE 8.1

Figure 8.2(b) shows the quarter-wave plates used in a conventional circular polariscope. Determine the Jones matrices for them.

Solution

From Figure 8.2(b), it can be seen that the angle θ (orientation of the slow axis) for the first and second quarter-wave plates are 135° and 45°, respectively. The retardation introduced by a quarter-wave plate is $\pi/2$. Substituting these in (8.9) for quarter-wave plate I one gets

$$\begin{bmatrix} \cos\dfrac{\pi}{4} - i\sin\dfrac{\pi}{4}\cos\dfrac{3\pi}{2} & -i\sin\dfrac{\pi}{4}\sin\dfrac{3\pi}{2} \\ -i\sin\dfrac{\pi}{4}\sin\dfrac{3\pi}{2} & \cos\dfrac{\pi}{4} + i\sin\dfrac{\pi}{4}\cos\dfrac{3\pi}{2} \end{bmatrix} = \begin{bmatrix} \dfrac{1}{\sqrt{2}} & i\dfrac{1}{\sqrt{2}} \\ i\dfrac{1}{\sqrt{2}} & \dfrac{1}{\sqrt{2}} \end{bmatrix} = \dfrac{1}{\sqrt{2}}\begin{bmatrix} 1 & i \\ i & 1 \end{bmatrix}$$

Similarly, the Jones matrix for quarter-wave plate II can be written as

$$\begin{bmatrix} \cos\dfrac{\pi}{4} - i\sin\dfrac{\pi}{4}\cos\dfrac{\pi}{2} & -i\sin\dfrac{\pi}{4}\sin\dfrac{\pi}{2} \\ -i\sin\dfrac{\pi}{4}\sin\dfrac{\pi}{2} & \cos\dfrac{\pi}{4} + i\sin\dfrac{\pi}{4}\cos\dfrac{\pi}{2} \end{bmatrix} = \begin{bmatrix} \dfrac{1}{\sqrt{2}} & -i\dfrac{1}{\sqrt{2}} \\ -i\dfrac{1}{\sqrt{2}} & \dfrac{1}{\sqrt{2}} \end{bmatrix} = \dfrac{1}{\sqrt{2}}\begin{bmatrix} 1 & -i \\ -i & 1 \end{bmatrix}$$

8.3.1   Analysis of Plane Polariscope by Jones Calculus

Using Jones calculus, the components of light vector along the analyzer axis and perpendicular to the analyzer axis for the plane polariscope arrangement are obtained as

$$\left\{ \begin{array}{c} E_x \\ E_y \end{array} \right\} = \left[\begin{array}{cc} \cos\dfrac{\delta}{2} - i\sin\dfrac{\delta}{2}\cos 2\theta & -i\sin\dfrac{\delta}{2}\sin 2\theta \\ -i\sin\dfrac{\delta}{2}\sin 2\theta & \cos\dfrac{\delta}{2} + i\sin\dfrac{\delta}{2}\cos 2\theta \end{array} \right] \left\{ \begin{array}{c} 0 \\ 1 \end{array} \right\} ke^{i\omega t} \qquad (8.10)$$

where $ke^{i\omega t}$ is the incident light vector, and E_x and E_y are the components of light vector along the analyzer axis and perpendicular to the analyzer axis, respectively. In (8.10), polarizer is represented as a vector, and the model is represented as a retarder. The intensity of light transmitted is the product $E_x E_x^*$ where E_x^* denotes the complex conjugate of E_x and the intensity of light transmitted is

$$I_p = I_a \sin^2\frac{\delta}{2}\sin^2 2\theta \qquad (8.11)$$

8.4   Fringe Contours in Conventional Photoelasticity

The intensity equation (8.11) is a function of both the magnitude of the principal stress difference (δ) and its orientation (θ). The intensity of light emerging from the model is zero when the retardation δ, which is related to the principal stress difference ($\sigma_1 - \sigma_2$) is such as to cause a relative phase difference of $2m\pi(m = 0, 1, 2, \ldots)$, where m is an integer. Since stress is continuous, one observes a collection of points forming contours that satisfy this condition and the respective fringe field is known as *isochromatics*. In conventional photoelasticity, only the dark field is valid and the isochromatic fringes are counted as 0, 1, 2, ... , and so on.

The term *isochromatics* is more appropriate to use when white light is used as a source. It has been pointed out earlier that the wave plates are wavelength dependent. Hence, when white light is incident on the model, at any point only a single wavelength is cut off. In view of this one observes white light minus the extinct color over the field. *Iso* means constant, and *chroma* means color. Thus, isochromatics represent contours of constant color.

Another possibility, wherein the intensity of light is zero is, when the polarizer axis coincides with one of the principal stress directions at the point of interest. In this case, light extinction is not wavelength dependent and one observes a dark fringe even in white light. These are known as *isoclinics* meaning contours of constant inclination. Isoclinics are usually numbered with the angles they denote such as 0°, 10°, 15°, and so on. The principal stress direction on all points lying on an isoclinic is a constant. Thus, in a plane polariscope, one has two sets of contours, namely, *isochromatics* and *isoclinics* superposed over each other (Figure 8.3(a)).

The intensity of light transmitted in dark (I_d) and bright (I_ℓ) fields for a circular polariscope are obtained as [1]

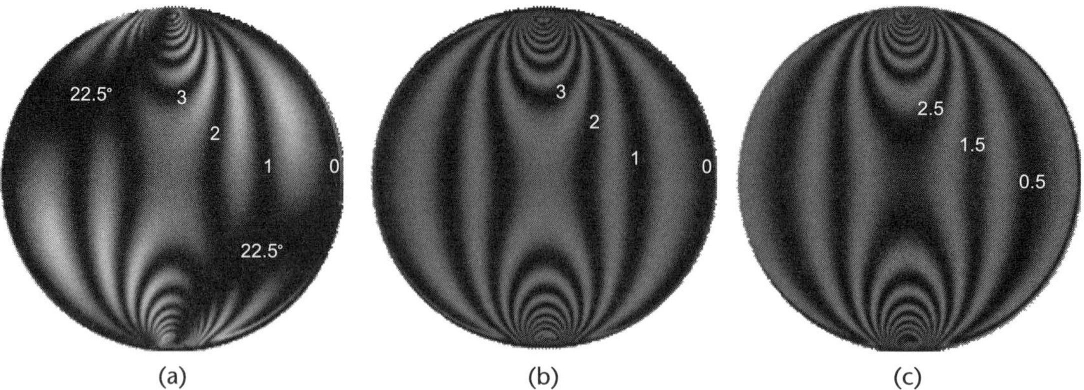

Figure 8.3 Fringe patterns in conventional photoelasticity for the problem of a disk under diametral compression: (a) plane polariscope; (b) dark field circular polariscope; (c) bright field circular polariscope.

$$I_d = I_a \sin^2\frac{\delta}{2}; \quad I_\ell = I_a \cos^2\frac{\delta}{2} \tag{8.12}$$

It is to be noted that for both dark and bright fields, the intensity equations are independent of θ and hence the extinction condition is only a function of δ; thus, only isochromatics will be seen. This is a significant achievement in separating isoclinics and isochromatics.

In dark field arrangement, intensity is zero when $\delta = 2m\pi(m = 0, 1, 2, \ldots)$ and fringes correspond to 0, 1, 2, ... and so on (Figure 8.3(b)). In bright-field arrangement, intensity is zero when $\delta = (2m + 1)\pi$, that is, when the retardation is an odd multiple of half-wave lengths. The fringes correspond to 0.5, 1.5, 2.5, and so on (Figure 8.3(c)).

8.5   Calibration of Model Materials

Determination of the material stress fringe value is known as calibration of a photoelastic material. The stress-fringe values of model materials vary with time and also from batch to batch. Hence, it is necessary to calibrate each sheet or casting at the time of the test. Calibration is performed on simple specimens for which closed form stress field solution is known. Although the stress fields for simple tension or beam under pure bending are known, the use of a circular disk under diametral compression is preferred for calibration. This is because the specimen is compact, easy to machine, and easy to load. The principle stress difference $(\sigma_1 - \sigma_2)$ at any point in the disk can be expressed as [1]

$$\left(\sigma_1 - \sigma_2\right) = \frac{4PR}{\pi b} \frac{R^2 - \left(x^2 + y^2\right)}{\left(x^2 + y^2 + R^2\right)^2 - 4y^2 R^2} \tag{8.13}$$

In the conventional method fringe order at the center of the disk alone is measured. For this data, using (8.13) and (8.4), the material stress fringe value is obtained as

$$F_\sigma = \frac{8P}{\pi DN} \tag{8.14}$$

Material stress fringe value has to be evaluated with at least two to three decimal places accuracy, as it is the only parameter that links the optical information to stresses. The load is first increased and then decreased incrementally. A graph is drawn between the load P and fringe order N. A best-fit straight line is then drawn through the points (graphical approach to least squares) and the slope of the line is substituted for P/N in (8.14) to evaluate the material stress fringe value.

BOX 8.1   Stress Field in a Circular Disk Under Diametral Compression

The stress field at any point in a circular disk under diametral compression is given as [1],

$$\left\{ \begin{array}{c} \sigma_x \\ \sigma_y \\ \tau_{xy} \end{array} \right\} = -\frac{2P}{\pi h} \left\{ \begin{array}{c} \dfrac{(R-y)x^2}{r_1^4} + \dfrac{(R+y)x^2}{r_2^4} - \dfrac{1}{D} \\[2ex] \dfrac{(R-y)^3}{r_1^4} + \dfrac{(R+y)^3}{r_2^4} - \dfrac{1}{D} \\[2ex] \dfrac{(R+y)^2 x}{r_2^4} - \dfrac{(R-y)^2 x}{r_1^4} \end{array} \right\}$$

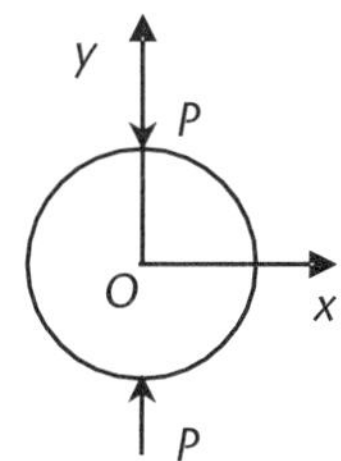

where

$$r_1^2 = x^2 + (R-y)^2$$
$$r_2^2 = x^2 + (R+y)^2$$

These equations are referred with respect to a Cartesian coordinate system with the disk center as the origin. R denotes the radius of the disk, D represents its diameter, h is the thickness of the disk, and P is the compressive load applied along its vertical diametral line.

EXERCISE 8.2

The values of fringe order at the cener of a disk of diameter 60 mm under diametral compression during loading and unloading phases is summarized in the table in Exercise 8.2. For a load of 400 N, the table gives only the fringe order that has moved to the center while performing Tardy's method of compensation. During the loading phase, a higher fringe order has moved to the center of the disk and the corresponding analyzer angle β_1 is 108° and during the unloading phase for the same load the sense of analyzer rotation is such that a lower fringe order has moved to the point of interest and the corresponding analyzer angle β_2 is 81°. With this data, determine the material stress fringe value of the photoelastic model in a least squares sense.

Load (N)	Fringe Order, N	
	Loading	Unloading
100	0.35	0.36
200	0.72	0.72
300	1.06	1.1
400	2	1
500	1.79	1.82
600	2.14	2.2

Solution

If the angle of rotation of the analyzer is β, then Tardy's method of compensation (Box 8.2), gives the fractional fringe order as, $\delta_N = \pm |\beta|/180°$. During the loading phase since a higher fringe order has moved to the point, one has to select the sign as negative for δ_N and for the unloading phase as the lower fringe order has moved, one has to select it as positive. This gives the total fringe orders as

$$N_{\text{loading}} = 2 - \frac{108}{180} = 1.4;$$

$$N_{\text{unloading}} = 1 + \frac{81}{180} = 1.45$$

It is instructive to note that $\beta_1 + \beta_2 \approx 180°$. The fringe order at each load is taken as the average of the measurements in loading and unloading phases. Instead of using (8.4) for each load and then taking the average, one can perform a graphical least squares analysis by plotting a graph between the fringe order and the load. The slope of the least squares line gives the best value of P/N, which is 276.1. The material stress fringe value is determined as

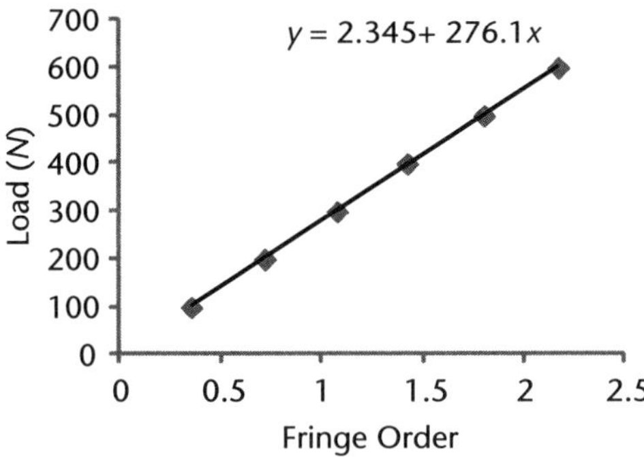

$$F_\sigma = \frac{8 \times 276.1}{\pi \times 60} = 11.72 \text{ N/mm/fringe}$$

In digital photoelasticity, as the fringe data can be easily obtained over the model (details of which can be found later in the chapter), use of a large number of data points from the field is the preferred approach for calibration. It is desirable that the final results are nearly independent of the choice of data points from the field. In order to achieve this, the least squares technique is usually combined with a random sampling process. It is worthwhile to account for residual birefringence also in the analysis. Let the residual birefringence expressed in fringe orders be assumed as a linear function in x and y as [1]

$$N_r(x,y) = Ax + By + C \tag{8.15}$$

The material stress fringe value is evaluated in a least squares sense by solving the following equation:

$$[b]^T[b]\{u\} = [b]^T\{N\} \tag{8.16}$$

where

$$[b] = \begin{bmatrix} S_1 & x_1 & y_1 & 1 \\ S_2 & x_2 & y_2 & 1 \\ \vdots & \vdots & \vdots & \vdots \\ S_M & x_M & y_M & 1 \end{bmatrix}, \{u\} = \begin{Bmatrix} 1/F_\sigma \\ A \\ B \\ C \end{Bmatrix} \{N\} = \begin{Bmatrix} N_1 \\ N_2 \\ \vdots \\ N_M \end{Bmatrix} \tag{8.17}$$

$$S(x,y) = \frac{4PR}{\pi} \frac{R^2 - \left(x^2 + y^2\right)}{\left(x^2 + y^2 + R^2\right)^2 - 4y^2R^2} \tag{8.18}$$

and

The unknown coefficient vector $\{u\}$ can be easily evaluated using the standard Gaussian elimination procedure. The construction of (8.16) is very simple from experimental data, and hence the method has found wide acceptance.

Least squares technique by itself does not guarantee that the result is physically admissible. Evaluation of the parameter values is complete only if the theoretically reconstructed fringe patterns agree well with the experimentally obtained ones [1].

8.6   Digital Fringe Multiplication

Techniques for fringe multiplication use the digital image processing (DIP) hardware as a paperless camera. A simple digital subtraction of bright and dark field images could result in fringe multiplication by an order of two. In such a case one gets

$$g(x,y) \approx I_a \cos\delta \tag{8.19}$$

In (8.19), the extinction of light will occur when $\delta = (2n + 1)\pi/2$. The resultant image is termed as a *mixed image* and the fringe orders are N = 0.25, 0.75, 1.25, ... Figure 8.4 shows the fringe multiplication obtained for the problem of a disk under three radial loads.

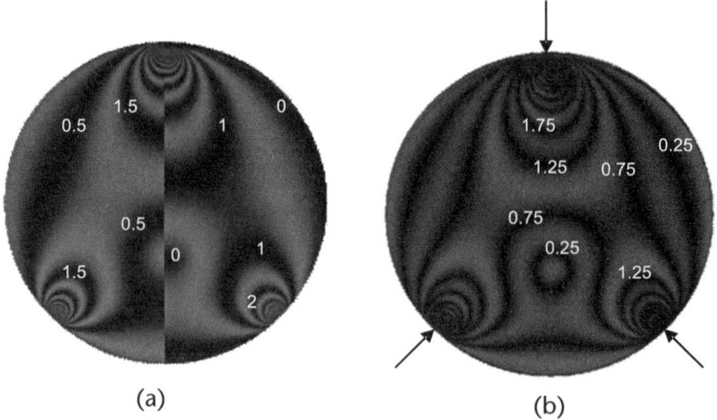

(a)   (b)

Figure 8.4 Fringe multiplication by image subtraction of bright and dark field images: (a) bright field (left) and dark field (right) image and (b) mixed image.

8.7   Digital Fringe Thinning

One of the simplest methods to fringe thinning is to treat the fringe patterns as a binary image, and the fringe center lines are determined by a process of erosion [5]. However, in general the center lines may not be the actual fringe skeletons. This is because only those points where the intensity of light transmitted is zero actually depicts a fringe. Due to experimental difficulties, it is difficult to extract the fringe skeletons by collecting only those points for which the intensity is zero. Identifying the fringe areas in a given image greatly simplifies the development of fringe thinning methodologies. As photoelastic images have high contrast, a simple global thresholding operation can identify the fringe areas. The choice of the threshold is problem dependent.

Figure 8.5 shows the fringe areas for different threshold values for a disk under three radial loads. If the thresholding operation is such that the intensity variation below the threshold is retained, then it is called *semithresholding*. Figure 8.5(d) gives a semithresholded image.

In the binary-based fringe thinning algorithm, the binary image obtained after thresholding is scanned left to right, right to left, top to bottom, and bottom to top sequentially to eliminate border pixels forming the fringe band. During each

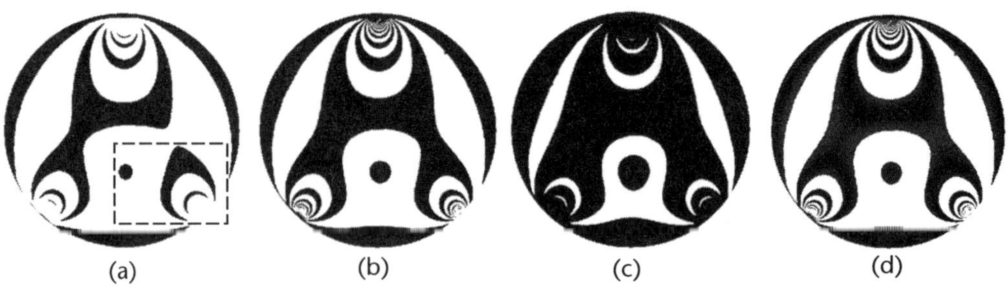

(a)   (b)   (c)   (d)

Figure 8.5 Fringe areas identified for various threshold values: (a) 60, (b) 80, (c) 110, (d) same as (b) but semithresholded.

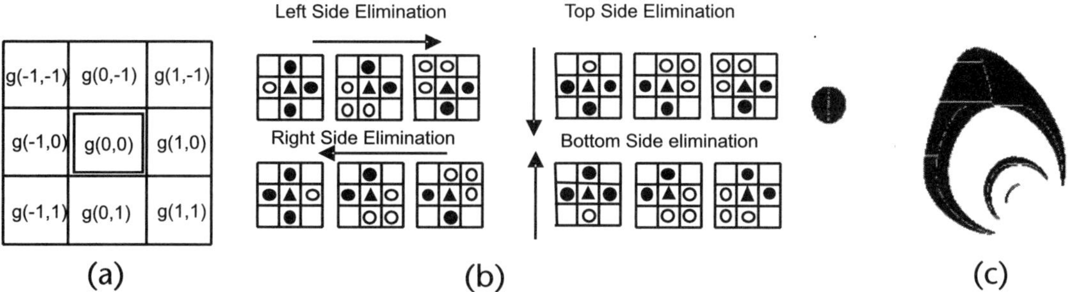

Figure 8.6   (a) 3 × 3 pixel mask. (b) Pictorial representation of the eliminative conditions needed for scanning the image left to right, right to left, top to bottom, and bottom to top. In the figure, filled circles are fringe pixels, the circles are nonfringe pixels, and the filled triangles are the pixels that are considered for elimination. (c) Fringe skeletons superimposed on the binary image for the tile shown in Figure 8.5(a).

such scan, for every pixel with a gray-level value below the threshold (a point on a fringe), a 3 × 3 pixel mask is considered (Figure 8.6(a)) to eliminate the border pixel.

In view of progressive thinning, the approach is basically iterative in nature. It is to be noted that in broad fringes the center lines may not be the actual fringe location. Figure 8.6(c) gives the fringe skeleton superimposed on original image, obtained after 10 iterations for a portion of the disk under three radial loads identified in Figure 8.5(a). Spurious skeletons are identified in zones where the fringe is thicker.

Among the various intensity-based fringe thinning algorithms, the one proposed by Ramesh and Pramod [6] is the fastest and assures fringe skeletons of one pixel width [7]. This has two steps: the first one is edge detection followed by minimum intensity identification. Since photoelastic fringes have high contrast, a simple global thresholding can help in finding the fringe areas. Once the fringe areas are identified, the image is scanned to locate the minimum intensity points forming the fringe skeleton. In order to account for various fringe curvatures, a unique process of logical operators is proposed in which the edge detected image is scanned for

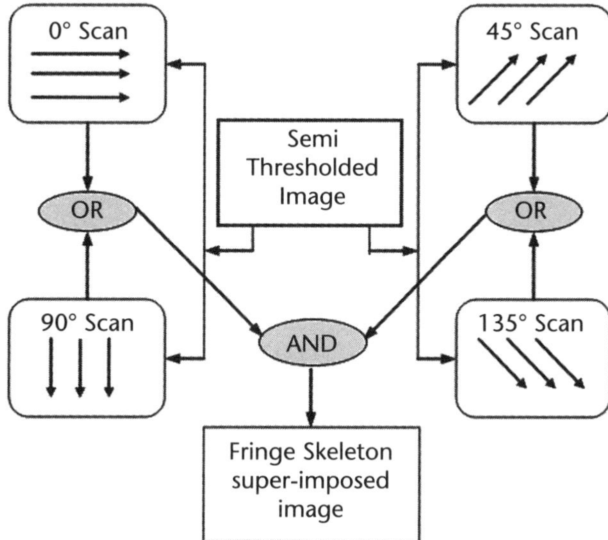

Figure 8.7   Global fringe thinning algorithm of Ramesh and Pramod.

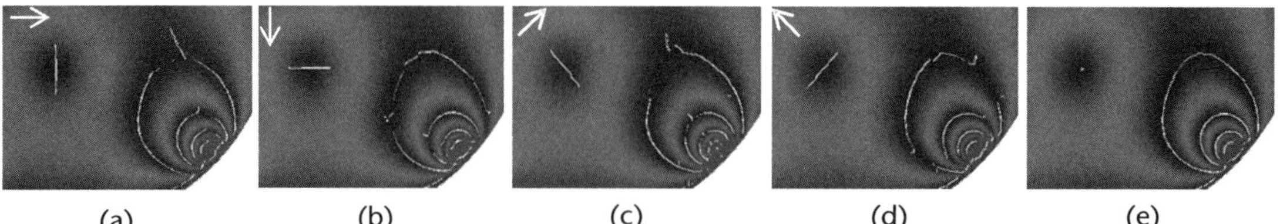

(a)   (b)   (c)   (d)   (e)

Figure 8.8 Fringe skeleton of the tile in Figure 8.5(a) obtained for various scan directions: (a) 0-degree scan; (b) 90-degree scan; (c) 45-degree scan; (d) 135-degree scan; (e) final fringe skeleton superimposed on the original image after the application of logical operators.

four scanning directions to determine points of minimum intensity. The scheme is illustrated in Figure 8.7. Figure 8.8 summarizes the skeletons obtained for various scans. One can see in each scan when the scan direction is tangential to a fringe, information is lost and in addition noise is generated. These are scan direction dependent and the use of orthogonal scans and the logical operators has helped to get a smooth fringe skeleton. It can be noted that the OR operation of orthogonal scans removes the gaps in the fringe skeleton and the final AND operation removes the noise. This algorithm has correctly identified the fringe skeleton of the circular patch on the left as a point (Figure 8.8(e)), whereas the binary based algorithm has not identified it (Figure 8.6(c)).

Binary-based algorithms come in handy if the fringe pattern available is of a poor quality. However, if the fringe patterns are carefully recorded by a digital camera, ihe intensity-based method will give the best results. For the evaluation of fracture parameters/contact stress parameters, the use of fringe multiplication in conjunction with fringe thinning techniques have been found to be quite useful [2].

8.8 Phase Shifting in Photoelasticity

With the advent of affordable high-quality digital image acquisition systems, a paradigm shift in processing of optical fringe patterns has taken place. Use of intensity information for extracting the phase information has come into place, which require recording of a few images with known phase shifts. Phase-shifting technique (PST) has been applied to many classical interferometric techniques [8]. In most classical interferometers, the path length of the two interfering beams is distinct and separate. In general, phase differences can be introduced by altering the optical path length of any one of the light beams. Usually the phase of the reference light beam is altered in known steps. Photoelasticity falls into a special category, in that the two light beams cannot be treated separately but always go together. Use of an external mirror as used in other interferometric techniques to introduce a phase shift cannot be applied here. In practice, a change in phase between the beams involved is achieved by appropriately rotating the optical elements of the polariscope.

In conventional photoelasticity, one is concerned with very simple optical arrangements that provide a dark field in a plane polariscope and both bright and dark fields in a circular polariscope. However, if the optical elements are kept at arbitrary positions a great degree of flexibility in utilizing the intensity information for proposing newer methodologies for data reduction is possible. A study on the intensity of light transmitted for arbitrary orientations of the optical elements is necessary to appreciate the methodology of phase shifting in photoelasticity.

8.8.1   Intensity of Light Transmitted for Generic Arrangements of Plane and Circular Polariscopes

In the subsequent discussions, δ is used to represent the retardation (in radians) introduced by the model, θ is the orientation of the slow axis of the model (one of the principal stress directions at the point of interest) with respect to the horizontal, $ke^{i\omega t}$ is the incident light vector, and E_β and $E_{\beta+\pi/2}$ are the components of light vector along the analyzer axis and perpendicular to the analyzer axis, respectively. The angular orientations of the various optical elements are referred with respect to the x-axis.

Figure 8.9(a) shows a photoelastic specimen kept in a plane polariscope with the polarizer and the analyzer kept at arbitrary angles of α and β, respectively.

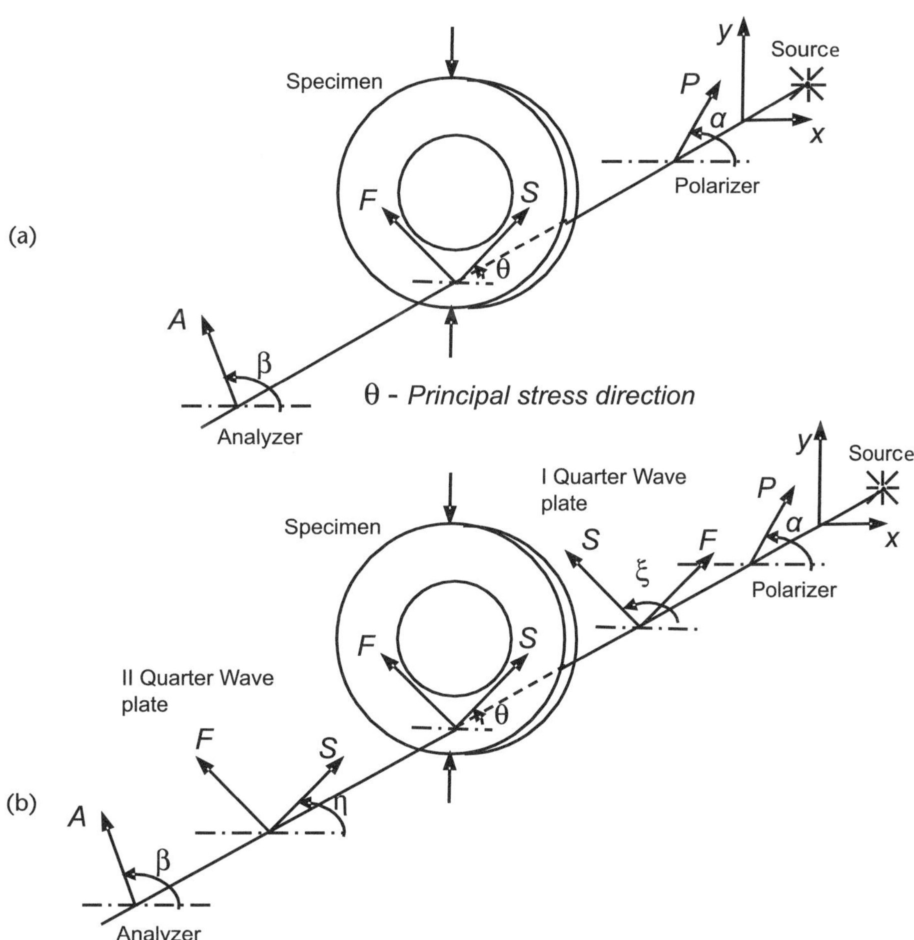

Figure 8.9   Generic arrangements of (a) plane polariscope, and (b) circular polariscope.

Using Jones calculus, the components of light vector along the analyzer axis and perpendicular to the analyzer axis are obtained as [1, 9]

$$\begin{Bmatrix} E_\beta \\ E_{\beta+\pi/2} \end{Bmatrix} = \begin{bmatrix} \cos\beta & \sin\beta \\ -\sin\beta & \cos\beta \end{bmatrix} \begin{bmatrix} \cos\dfrac{\delta}{2} - i\sin\dfrac{\delta}{2}\cos 2\theta & -i\sin\dfrac{\delta}{2}\sin 2\theta \\ -i\sin\dfrac{\delta}{2}\sin 2\theta & \cos\dfrac{\delta}{2} + i\sin\dfrac{\delta}{2}\cos 2\theta \end{bmatrix}$$

$$\times \begin{Bmatrix} \cos\alpha \\ \sin\alpha \end{Bmatrix} ke^{i\omega t}$$

$$(8.20)$$

The intensity of light transmitted I_p (the subscript "p" denotes that the incident light is plane polarized) is obtained as

$$I_p = I_a\left[\cos^2\frac{\delta}{2}\cos^2(\beta - \alpha) + \sin^2\frac{\delta}{2}\cos^2(\beta + \alpha - 2\theta) \right] \qquad (8.21)$$

where I_a accounts for the amplitude of incident light vector and the proportionality constant. It is interesting to note that (8.21) is not altered when the model slow axis orientation is changed from θ to $\theta + \pi/2$. In other words, θ can represent either the major or minor principal stress direction at the point of interest.

If $\alpha = \beta + \pi/2$, (8.21) reduces to

$$I_p = I_a\left[\sin^2\frac{\delta}{2}\sin^2 2(\theta - \beta) \right] \qquad (8.22)$$

Equation (8.22) represents the intensity of light transmitted for crossed positions of polarizer-analyzer combination.

In most data acquisition algorithms that use a circular polariscope, only the second quarter-wave plate and the analyzer are kept at arbitrary positions of η and β, respectively (Figure 8.9(b)). The polarizer is kept at 90 degrees and the first quarter-wave plate is kept at either $\xi = 135$ degrees or 45 degrees. Using Jones calculus, the components of light vector along the analyzer axis and perpendicular to the analyzer axis (for $\xi = 135$ degrees) are obtained as [1, 4],

$$\begin{Bmatrix} E_\beta \\ E_{\beta+\pi/2} \end{Bmatrix} = \frac{1}{2}\begin{bmatrix} \cos\beta & \sin\beta \\ -\sin\beta & \cos\beta \end{bmatrix}\begin{bmatrix} 1 - i\cos 2\eta & -i\sin 2\eta \\ -i\sin 2\eta & 1 + i\cos 2\eta \end{bmatrix}$$

$$\times \begin{bmatrix} \cos\dfrac{\delta}{2} - i\sin\dfrac{\delta}{2}\cos 2\theta & -i\sin\dfrac{\delta}{2}\sin 2\theta \\ -i\sin\dfrac{\delta}{2}\sin 2\theta & \cos\dfrac{\delta}{2} + i\sin\dfrac{\delta}{2}\cos 2\theta \end{bmatrix}\begin{bmatrix} 1 & i \\ i & 1 \end{bmatrix}\begin{Bmatrix} 0 \\ 1 \end{Bmatrix}ke^{i\omega t}$$

$$(8.23)$$

The combination of polarizer at 90 degrees and the quarter-wave plate at $\xi = 135$ degrees produces a left circularly polarized light. Denoting the intensity of light transmitted as I_ℓ, the intensity expression is obtained as

$$I_\ell = \frac{I_a}{2} + \frac{I_a}{2}\left[\sin 2(\beta - \eta)\cos\delta - \sin 2(\theta - \eta)\cos 2(\beta - \eta)\sin\delta\right] \qquad (8.24)$$

where I_a accounts for the amplitude of light vector and the proportionality constant. If in Figure 8.9(b), the model fast axis is kept at θ (i.e., slow axis at $\theta + \pi/2$), then the intensity of light transmitted is

$$I_\ell = \frac{I_a}{2} + \frac{I_a}{2}\left[\sin 2(\beta - \eta)\cos\delta + \sin 2(\theta - \eta)\cos 2(\beta - \eta)\sin\delta\right] \qquad (8.25)$$

Unlike for the generic arrangement of a plane polariscope, inspection of (8.24) and (8.25) show that the intensity of light transmitted depends on the orientation of the model fast or slow axis for the cases when the optical elements after the model are kept at arbitrary orientations. As the model fast or slow axes are not known a priori, this leads to an ambiguity in the sign of fractional retardation δ. Hence, this aspect needs special attention in automated techniques based on intensity measurement.

Many investigators have explicitly added a term I_b to the intensity expression to account for the stray light/background illumination. In view of the optical inhomogenity and nonuniform characteristics of the light source and the diffuser, the quantities I_a and I_b are functions of the spatial coordinates viz., $I_a(x, y)$ and $I_b(x, y)$. With this explicit understanding, these will be represented as I_a and I_b further in the text.

8.8.2   Brief Historical Development of Phase Shifting Techniques

The concept of phase-shifting in photoelasticity was first introduced by Hecker and Morche [10] in 1986. They recorded five phase-shifted images and demonstrated the possibilities for the determination of isochromatic parameter over the whole domain using a circular polariscope arrangement. Patterson and Wang [11] extended the work of Hecker and Morche [10] and proposed a six-step PST for the determination of both isoclinic and isochromatic parameters, which gave a new impetus to digital photoelasticity. They were the first to obtain both isoclinics and isochromatic parameters over the model domain. Ramesh and Ganapathy in 1996 [9] performed a Jones calculus analysis of the six-step phase-shifting method of Patterson and Wang [11] and simplified the evaluation of the expression for intensity of light transmitted. This approach opened up the possibility of exploring new optical arrangements conveniently. In 1998, Ajovalasit et al. [12] proposed a slight but important modification of the basic six-step phase-shifting algorithm that uses both left and right circularly polarized light incident on the model judiciously. They showed that such a modification has reduced the influence of mismatch of quarter-wave plates on the accuracy of the results, provided both the quarter-wave plates have uniform and identical mismatch error. In reality, the mismatch error is not uniform over the field and also this variation could be different for the second quarter-wave plate. All these influence the final result from experimentally recorded images. Though several different phase-shifting algorithms have been proposed [1], the final expressions of intensity of light transmitted in many of these are similar. Ramesh in 2000 [1] reported that there could be multiple optical arrangements and comprehensively presented these arrangements for the popular six-step PST. He further improved the six-step PST proposed by Ajovalasit et al. [12] by incorporating the use of crossed quarter-wave plates wherever possible to minimize the error due to mismatch of quarter-wave plates. Around the year 2002, researchers realized that though isoclinic evaluation is simple in conventional photoelasticity, it is quite involved in digital photoelasticity. To address this, composite six-step PST have been reported by Barone et al. and Ramji and Ramesh [3].

Ramji and Ramesh in 2008 proposed a ten-step PST [13, 14]. The methodology intelligently combines a four-step approach of Brown and Sullivan [15] for isoclinic data evaluation and a six-step circular polariscope-based method for isochromatic data evaluation (Table 8.1). The optical arrangements are carefully selected to minimize the influence of quarter-wave plate mismatch error. This has become a standard approach if one is interested in evaluating both isoclinics and isochromatics accurately [3, 16]. For a detailed discussion on the historical developments, refer to [1, 2].

8.8.3 Evaluation of Photoelastic Parameters Using Intensity Information

The number of unknowns at each point in a model are four, which are I_a, I_b, δ, and θ. In principle, just four optical arrangements should be sufficient to evaluate these. However, research in the past three decades has established that ten images as in Table 8.1 are needed for accurate evaluation of the parameters. This will become

Table 8.1 Optical Arrangement for Ten-Step PST

α	ξ	η	β	Intensity Equation
$\pi/2$	—	—	0	$I_1 = I_b + I_a \sin^2 \dfrac{\delta}{2} \sin^2 2\theta$
$5\pi/8$	—	—	$\pi/8$	$I_2 = I_b + \dfrac{I_a}{2} \sin^2 \dfrac{\delta}{2}(1 - \sin 4\theta)$
$3\pi/4$	—	—	$\pi/4$	$I_3 = I_b + I_a \sin^2 \dfrac{\delta}{2} \cos^2 2\theta$
$7\pi/8$	—	—	$3\pi/8$	$I_4 = I_b + \dfrac{I_a}{2} \sin^2 \dfrac{\delta}{2}(1 + \sin 4\theta)$
$\pi/2$	$3\pi/4$	$\pi/4$	$\pi/2$	$I_5 = I_b + \dfrac{I_a}{2}(1 + \cos\delta)$
$\pi/2$	$3\pi/4$	$\pi/4$	0	$I_6 = I_b + \dfrac{I_a}{2}(1 - \cos\delta)$
$\pi/2$	$3\pi/4$	0	0	$I_7 = I_b + \dfrac{I_a}{2}(1 - \sin 2\theta \sin\delta)$
$\pi/2$	$3\pi/4$	$\pi/4$	$\pi/4$	$I_8 = I_b + \dfrac{I_a}{2}(1 + \cos 2\theta \sin\delta)$
$\pi/2$	$\pi/4$	0	0	$I_9 = I_b + \dfrac{I_a}{2}(1 + \sin 2\theta \sin\delta)$
$\pi/2$	$\pi/4$	$3\pi/4$	$\pi/2$	$I_{10} = I_b + \dfrac{I_a}{2}(1 - \cos 2\theta \sin\delta)$

clear in the following sections. Figure 8.10 shows the ten phase-shifted images for the problem of a ring under diametral compression (outer dia = 80 mm, inner dia = 40 mm). The diametral load is 504 N, thickness of the specimen is 6 mm, and the material stress fringe value is 11.54 N/mm/fringe.

Using the first four equations in Table 8.1, the isoclinic parameter is obtained as

$$\theta_c = \frac{1}{4}\tan^{-1}\left(\frac{I_4 - I_2}{I_3 - I_1}\right) = \frac{1}{4}\tan^{-1}\left(\frac{I_a \sin^2\frac{\delta}{2}\sin 4\theta}{I_a \sin^2\frac{\delta}{2}\cos 4\theta}\right) \qquad \text{for } \sin^2\frac{\delta}{2} \neq 0 \qquad (8.26)$$

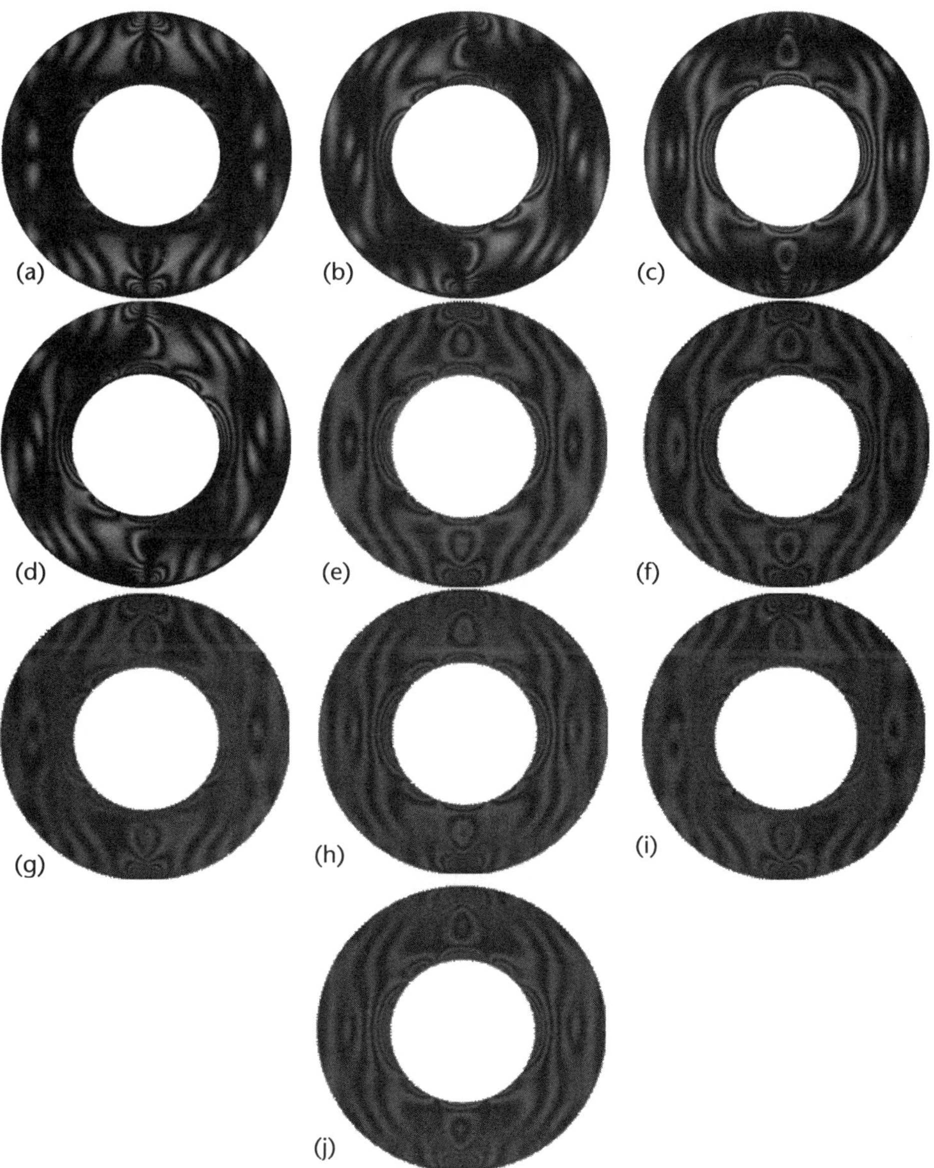

Figure 8.10   Ten-step phase-shifted images of ring under diametral compression. (Courtesy [2].)

where, the subscript "c" refers to the principal value of the inverse trigonometric function in this equation as well as in subsequent equations. Equation (8.26) also gives the mathematical expressions of the intensity difference, which helps to visualize certain issues on parameter estimation. Representing equations in this manner is adopted for some of the key equations discussed in this chapter. Equation (8.26) is valid only when $\sin^2(\delta/2) \neq 0$. This implies that the isoclinic parameter is undefined when $\delta = 0, 2\pi, 4\pi, \ldots$. In other words, the isoclinic contour will not be continuous over the domain. When δ is not exactly equal to 0 or 2π or 4π, and so on, but very close to these values, θ_c does not become indeterminate but the value of θ_c determined will be unreliable and appears as noise in isoclinic fringe when plotted.

It is important to note that the first four equations of Table 8.1 are recorded in a crossed plane polariscope and the next six equations are recorded in a generic circular polariscope. The isoclinics can also be determined by using the equations recorded in a generic circular polariscope as [1, 2]

$$\theta_c = \frac{1}{2}\tan^{-1}\left(\frac{I_9 - I_7}{I_8 - I_{10}}\right) = \frac{1}{2}\tan^{-1}\left(\frac{I_a \sin\delta \sin 2\theta}{I_a \sin\delta \cos 2\theta}\right) \quad \text{for } \sin\delta \neq 0 \quad (8.27)$$

Equation (8.27) shows that the isoclinic value is indeterminate at those points where $\delta = 0, \pi, 2\pi, 3\pi, \ldots$. Here again, the isoclinic contour will not be continuous over the domain. From the last six equations of Table 8.1, the isochromatics can be obtained in two different ways as [1, 11]

$$\delta_c = \tan^{-1}\left(\frac{I_8 - I_{10}}{\left(I_5 - I_6\right)\cos 2\theta_c}\right) \quad \text{for } \cos 2\theta_c \neq 0 \quad (8.28)$$

$$\delta_c = \tan^{-1}\left(\frac{I_9 - I_7}{\left(I_5 - I_6\right)\sin 2\theta_c}\right) \quad \text{for } \sin 2\theta_c \neq 0 \quad (8.29)$$

Inspection of (8.28) and (8.29) shows that the evaluation of retardation (isochromatic fringe order) is interlinked to the evaluation of θ (isoclinic value). Further, in (8.28) and (8.29), δ_c has low-modulation regions when $\cos 2\theta_c$ or $\sin 2\theta_c$ are small. In these regions, the intensity differences $(I_8 - I_{10})$ or $(I_9 - I_7)$ will also be small. The evaluation of δ_c will involve the ratio of two small discrete numbers and will be in considerable error in such regions. This issue was ably addressed by Quiroga and Gonzàlez-Cano [17], and they suggested the following form of intensity processing:

$$\delta_c = \tan^{-1}\left(\frac{\left(I_9 - I_7\right)\sin 2\theta_c + \left(I_8 - I_{10}\right)\cos 2\theta_c}{\left(I_5 - I_6\right)}\right) = \tan^{-1}\left(\frac{I_a \sin\delta}{I_a \cos\delta}\right) \quad (8.30)$$

Albeit simple, it is a significant development noted by Ramesh [1] in 2000, and it took some time for the researchers to see its advantage.

It is to be noted that (8.26), (8.27), and (8.30) involve inverse trigonometric operations and thus one has multivalued solutions, and this poses a problem in correctly evaluating δ and θ over the domain. To evaluate the inverse trigonometric function, one can use atan function if one knows only the value of the ratio and it returns the value in the range $-\pi/2$ to $\pi/2$. If, however, the numerator and denominator can be independently evaluated including the sign, then the atan2 function is suitable for evaluation, and it returns the value in the range $-\pi$ to π. Equations (8.26) and (8.30) are amenable for using the atan2 function, and for (8.27) one can use only the atan function. Nevertheless, both (8.26) (using atan2) and (8.27) (using atan) give θ_c in the range of $-\pi/4$ to $\pi/4$. Physically θ is in the range of $-\pi/2$ to $\pi/2$. Use of atan2 function for δ_c gives it in the range of $-\pi$ to π. For further processing, it is desirable that δ_c is expressed in the range of 0 to 2π.

8.8.4   Intricacies in Digital Photoelasticity

A distinction has to be made between the nature of phase information recorded in photoelasticity and other interferometric techniques. In most techniques, the intensity information is related to a single physical information. For example, in holography one gets out-of-plane displacements, and in Moiré one records either u or v displacement from a single experiment. On the other hand, in photoelasticity the intensity information is affected by both the difference in principal stresses (δ) and the orientation of the principal stress (θ). This influences the evaluation of isochromatics and isoclinics differently and needs to be addressed properly for accurate estimation of these parameters.

The optical arrangements used in conventional photoelasticity are insensitive to the orientation of the model fast or slow axis, which are linked to the principal stress directions at the point of interest, which are not known a priori. This makes the interpretation of data simple in conventional photoelasticity. The only case where the model axis influences the interpretation is while applying Tardy's method of compensation at a point of interest. The sign of the fractional retardation is ambiguous with just the record of value of the rotation by the analyzer such that the intensity at the point of interest is zero. The observer usually takes care of the ambiguity by noting which fringe order moves to the point of interest when the analyzer is rotated clockwise/counterclockwise. If a higher fringe order moves then the sign is negative, and if a lower fringe order moves then it is positive. This is very important information, which is recorded by the user in conventional photoelasticity for every point the user analyzes. One of the challenges in digital photoelasticity is to find the fractional fringe order with the correct sign at every point in the domain with minimal human intervention.

BOX 8.2   Tardy's Method of Compensation

Compensation techniques are basically point-by-point techniques. The basic principle is that, by external means, the retardation provided by the model is compensated such that a fringe passes through the point of interest. The additional retardation added or subtracted is known as fractional retardation. The use of the analyzer as a compensator is known as Tardy's method of compensation.

Tardy's method of compensation is quite a useful technique, and the fringe order can be measured with a high degree of accuracy. Following are its steps:

1. Initially, the principal stress direction at the point of interest is determined using a plane polariscope.
2. A circular polariscope is then formed such that the polarizer is kept at the isoclinic angle and all the other optical arrangements are appropriately arranged. At this stage, if the optical elements are correctly aligned, there should be no difference in the isochromatic field compared to the conventional arrangement.
3. The analyzer alone is then rotated until a fringe passes through the point of interest. Let the angle of rotation be β
4. If β is measured in degrees, the fractional fringe order δ_N is obtained as

$$\delta_N = \pm \frac{|\beta|}{180°}$$

Unlike the full fringe order N, the fractional fringe order has a sign. The sign convention is not decided by the sign of β but by the physics of the problem. The fractional fringe order δ_N is taken as positive if a lower fringe order moves to the point of interest, and vice versa.

In conventional photoelasticity, the optical arrangement is usually set based on the background field. The relative orientations of the optical elements are more important, and their absolute orientations are less significant in most studies. However, in digital photoelasticity, the absolute orientations of the optical elements do play a significant role. Thus, every effort must be made to correctly align the polariscope such that one-to-one correspondence exists between the arrangement used for theoretical development of intensity equations and the arrangement used for experimental recording of intensities [1, 2]. This is important to develop suitable phase unwrapping algorithms in digital photoelasticity, which helps in getting the total fringe order.

As the basic equations used for parameter estimation are inverse trigonometric functions, they are multivalued as noted in Section 8.8.3. Special care has to be taken in finding the correct solution. If not handled properly, it can lead to certain *inconsistencies* in isoclinic evaluation and also cause the formation of *ambiguous* zones in isochromatic phasemaps.

8.8.5   Evaluation of Isoclinics

It is seen in Section 8.8.3 that isoclinics can be evaluated by processing the first four images recorded in plane polariscope arrangement or independently by using the last six equations of Table 8.1, which are recorded in a generic arrangement of circular polariscope. Before proceeding further, one should be clear which of these two equations is to be used for accurate evaluation of the isoclinics. In order to facilitate this comparison, it is desirable to plot the isoclinics over the domain as contours of 10-degree steps. This can be done by the following code [2]:

$$\text{If } \big((-2.5 + \theta) < \theta < (2.5 + \theta)\big)\, g_b(x,y) = 0 \ \text{ else } g_b(x,y) = 255 \qquad (8.31)$$

where $\theta = -80°, -70°, ..., 0°, ..., 10°, 20° ...$

The contour plot is binary in nature and would mimic the assembly of isoclinic contours as seen in conventional books on photoelasticity. Figure 8.11 shows a comparison of isoclinics plotted in steps of 10 degrees by (8.26) and (8.27) on both theoretically simulated phase shifted images and experimentally recorded phase shifted images for a ring under diametral compression. For experimentally recorded images, the plot obtained by (8.26) is closer to the theoretical plot of isoclinics (Figure 8.11(e)). Though the algorithm of Ajovalasit using six step PST based on a circular polariscope has worked well for theoretically phase shifted images, it has failed badly when applied to experimentally recorded phase shifted images (Figure 8.11(d)). Ajovalasit has pointed out that the quarter-wave plate error is different for left and right circularly polarized lights. If one considers ε as error for left circularly polarized light and as ε' for right circularly polarized light then [12]

$$\theta_c = \frac{1}{2}\tan^{-1}\left(\frac{\sin 2\theta(\cos\varepsilon' + \cos\varepsilon)\sin\delta + \big(\cos^2 2\theta + \sin^2 2\theta\cos\delta\big)(\sin\varepsilon' - \sin\varepsilon)}{\cos 2\theta(\cos\varepsilon' + \cos\varepsilon)\sin\delta - \big(1 - \cos\delta\big)\sin 2\theta\cos 2\theta(\sin\varepsilon' - \sin\varepsilon)}\right)$$

$$(8.32)$$

Equation (8.32) shows that there is significant error due to quarter-wave plate mismatch. Further, the equation is obtained with the premise that the mismatch error is uniform over the field, which is not the case in practice. The error changes from point to point, and this further complicates experimental evaluation. Hence use of circular polariscope is not quite suitable for evaluating isoclinics [18].

As the isoclinics are not defined at the isochromatic fringe order, one observes gaps in Figure 8.11(a). and in Figure 8.11(c) these are observed as kinks. Further improvements to Figure 8.10 is desirable for comparison with Figure 8.11(e), which is discussed in Section 8.8.10.

8.8.6   Phasemaps in Photoelasticity

Phasemap is a grayscale plot of the phase information. In photoelasticity, phasemaps can be plotted for isoclinics (θ) and isochromatics (δ). Isoclinic value physically lies in the range $-\pi/2$ to $+\pi/2$. A grayscale theta plot of isoclinic can be obtained by defining [2]:

$$g(x,y) = INT\left[\frac{255}{\pi}\left(\theta + \frac{\pi}{2}\right)\right] = INT[R] \qquad (8.33)$$

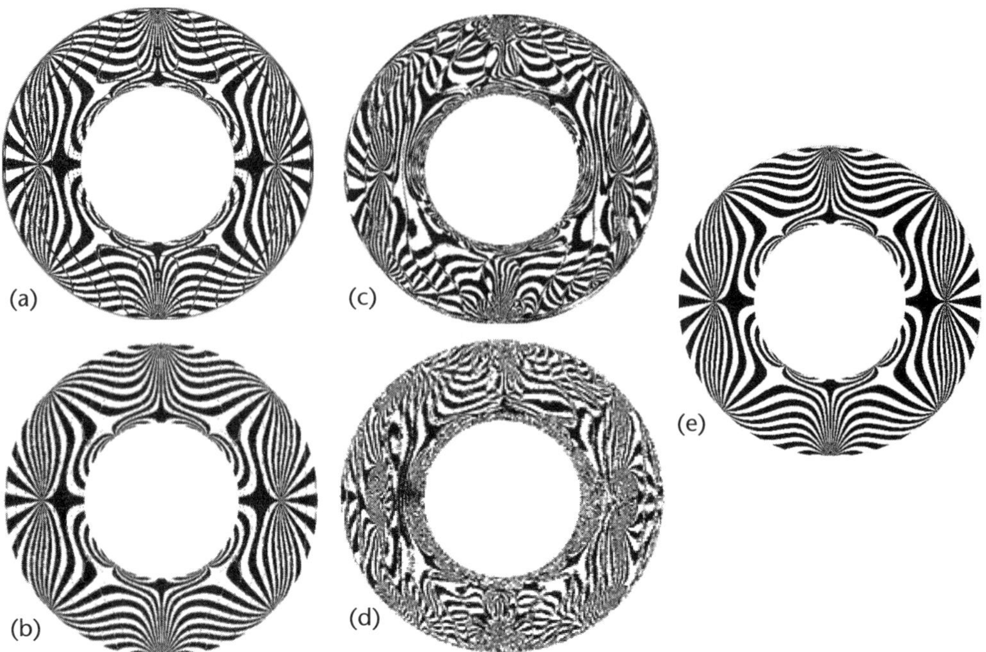

Figure 8.11 Isoclinics in steps of 10 degrees for the problem of a ring under diametral compression using theoretically simulated phase shifted images: by (a) (8.26); (b) (8.27) experimentally recorded images; (c) (8.26); (d) (8.27); (e) theory. (Adapted from [18].)

where $g(x, y)$ is the gray level value at the point (x, y), INT $[R]$ is the nearest integer of R. Such a representation will plot black for $\theta = -\pi/2$ and white for $\theta = \pi/2$, and is known as unwrapped phasemap. If the same plotting scheme is adopted for the principal value of the isoclinic (θ_c) calculated in (8.26), which lies in the range $-\pi/4$ to $+\pi/4$, one would see specific zones in the model domain rather than smooth variation of the grayscale values (Figure 8.12(c)). The reason and their nature will be discussed in the next section.

Fractional retardation is obtained in the range $-\pi \leq \delta_c \leq \pi$. To obtain wrapped phasemap, fractional retardation has to be first expressed in the range of 0 to 2π. This is then converted into gray levels between 0 and 255 for graphical plotting. The zero fractional retardation corresponds to pitch black, and 2π fractional retardation corresponds to pure white (i.e., 255). The phase map is plotted using the following relations [1]:

$$\delta_p = \begin{cases} \delta_c & \text{for } \delta_c > 0 \\ 2\pi + \delta_c & \text{for } \delta_c \leq 0 \end{cases}$$

$$g(x,y) = \frac{255}{2\pi}\delta_p \tag{8.34}$$

For getting the total fringe order, the wrapped phase has to be unwrapped suitably.

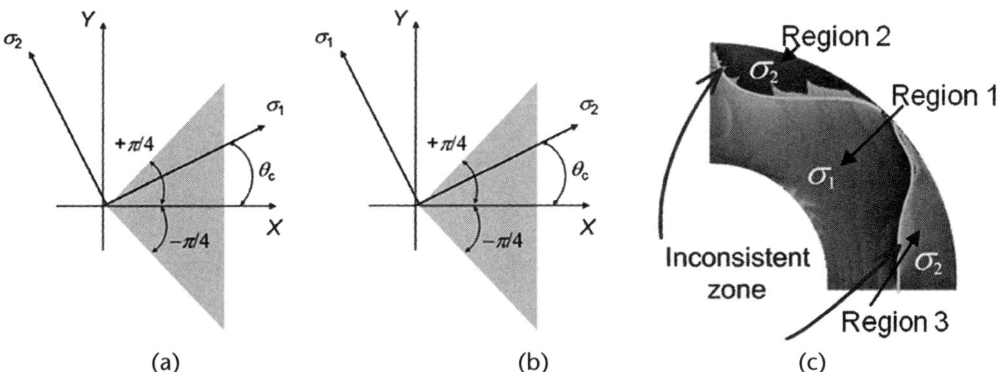

Figure 8.12   Principal value of theta can represent either (a) σ_1 direction or (b) σ_2 direction; (c) plot of wrapped isoclinic phasemap for a quarter of a ring under diametral compression. (Courtesy [2].)

8.8.7   Inconsistent and Ambiguous Zones in Wrapped Phasemaps

Calculation of fractional retardation requires first the evaluation of isoclinic angle. It is to be noted that the principal value of θ_c lies in the range of $-\pi/4 \leq \theta_c \leq \pi/4$. Since one does not know the principal stress direction a priori, there exists an inconsistency on whether θ_c corresponds to σ_1 or σ_2 direction over the domain. This is schematically illustrated in Figures 8.12(a) and (b).

Figure 8.12(c) shows the grayscale plot of wrapped isoclinics for quarter of a ring under diametral compression. Inspection of the figure shows three regions as well as observation of isochromatic skeletons. The isochromatic skeletons are actually noise, as (8.26) is not defined at isochromatic fringes. In the figure, the values of theta in region 1 corresponds to maximum principal stress, and in regions 2 and 3 it correspond to minimum principal stress. One is able to identify that these regions correspond to a specific principal stress, as the problem on hand has a theoretical solution. If one wants to find only the orientation of the major principal stress over the domain, then regions 2 and 3 could be classified as *inconsistent zones*. One can also represent for the entire domain the orientation of the minimum principal stress and if done then region 1 becomes an inconsistent zone. An inconsistent zone is labeled in a relative perspective, and what is important is to obtain over the model domain the isoclinic value in the range $-\pi/2 \leq \theta \leq \pi/2$ for one of the principal stresses consistently. This process is called unwrapping of isoclinic phasemap, which is discussed in Section 8.8.9.

If the principal value of θ_c is used in (8.30) to get δ_c and if the wrapped phase is plotted using (8.35), then one obtains the isochromatic phasemap as shown in Figure 8.13(a). There is a striking similarity of boundaries of various regions between Figures 8.12(c) and 8.13(a). The regions that are labeled as *inconsistent zones* in isoclinic phasemap are labeled as *ambiguous zones* in isochromatic phasemap. This interdependence is first noted by Ramesh [1].

Unlike the total fringe order, which is always positive, the fractional fringe order has a sign, too. While performing Tardy's method of compensation, additional heuristic information is noted by the user at each point to assign a sign to the fractional

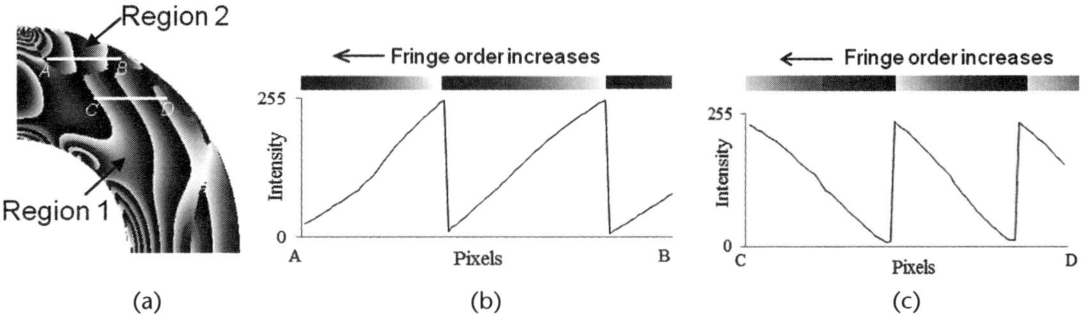

Figure 8.13 (a) Wrapped isochromatic phasemap with ambiguous zones; (b) variation of fractional fringe order along line *AB*; (c) variation of fractional fringe order along line *CD*.

retardation. Figures 8.13(b) and (c) show the variation of fractional fringe order in regions 1 and 2 for two lines in which the total fringe order increases from right to left in both. The sign of the fractional retardation in ambiguous zones is opposite to that in the other zones. Referring to (8.34) when the fractional retardation is negative, for the purpose of plotting the phasemap, it is added with 2π. This essentially provides the complementary value of fractional retardation. In view of this the gradient is reversed in normal and ambiguous zones. For isochromatic unwrapping, the isochromatic phasemap has to be free of ambiguous zones [19]. Ambiguous zone is again a relative term like an inconsistent zone.

8.8.8 Unwrapping Methodologies

In the case of isoclinic phasemaps, unwrapping refers to the process of obtaining the direction of either σ_1 or σ_2 consistently over the domain, while in the case of isochromatic phasemaps, unwrapping refers to suitable addition of the integral value to the fractional retardation values to get continuous fringe order data. The two focal issues are how to avoid propagation of errors and how to handle complex geometries with cutouts.

The choice of the basic algorithm for phase unwrapping influences its sensitivity to propagation of errors. Simplest phase unwrapping methodology is to start from a seed point and scan the image horizontally, scan it vertically, or use these scans judiciously. This algorithm can work well if it does not encounter a noise point, geometric discontinuity, or material discontinuity.

Another approach is to guide the path of unwrapping through a quality measure. In quality guided phase unwrapping, the pixels in the phase map are graded depending on its quality where "1" refers to highest quality and "0" denotes the lowest quality. Consequently in the quality map, the black areas are of low quality, the white areas are of high quality, and the gray areas are of intermediate quality. Unwrapping starts from the highest quality pixel and then proceeds toward lowest quality pixels, thereby assuring the best possible results. It is computationally intensive and several quality measures are reported in the literature. Among these, phase derivative variance is recommended for photoelasticity [20, 21].

Phase derivative variance gives the badness of the pixel as

$$\frac{\sqrt{\sum\left(\Delta^{x}_{i,j} - \overline{\Delta}^{x}_{m,n}\right)^{2}} + \sqrt{\sum\left(\Delta^{y}_{i,j} - \overline{\Delta}^{y}_{m,n}\right)^{2}}}{k^{2}} \tag{8.35}$$

where for each sum, the indices (i, j) range over the $k \times k$ neighborhood of each center pixel (m, n) and $\Delta^{x}_{i,j}$ and $\Delta^{y}_{i,j}$ are the partial derivatives of the phase (i.e., the wrapped phase differences). The terms $\overline{\Delta}^{x}_{m,n}$ and $\overline{\Delta}^{y}_{m,n}$ and are the averages of these partial derivatives in the $k \times k$ windows. Equation (8.35) is a root-mean-square measure of the variances of the partial derivatives in the x and y directions. The variance is negated consequently to represent goodness. As the integration path is dictated by quality, in principle, accumulation of error is minimized in this approach.

Phase unwrapping of complex model boundaries with cut outs is a challenge by itself. Simply connected models are in general easy to handle. For multiply connected bodies, one can choose either domain delimiting or boundary masking approaches. In domain delimiting, the multiply connected models are unwrapped by dividing them into an assembly of simply connected domains. Madhu and Ramesh [22] proposed an approach to unwrap complex geometries having cut outs. The boundary coordinates information required for delimiting is obtained using boundary extraction techniques. In boundary masking approach, the image area is identified by a binary image, and by storing it as a separate image, it is possible to handle any geometric shape. Quality guided algorithms for phase unwrapping in conjunction with boundary masking would be a user-friendly phase unwrapping strategy.

8.8.9 Unwrapping of Isoclinics

In order to remove inconsistencies in wrapped phasemap (Figure 8.12(c)), the algorithm should first identify the boundaries of these zones. From the discussion in Section 8.8.7, the boundaries of the inconsistent zones are characterized by a $\pi/2$-jump. This is not the only issue that needs to be addressed. Unwrapping of the isoclinic phasemap is complicated by the presence of isotropic points and π-jumps. At an isotropic point all isoclinics merge, while a π-jump is a sudden jump of 180° in the isoclinic values. which starts from an isotropic point [2]. These are marked on theoretically simulated unwrapped isoclinic phasemap for clear understanding (Figure 8.14). In complex photoelastic models, many isotropic points/π-jumps may be present in the isoclinic phasemap.

Ramji and Ramesh [14] proposed an adaptive quality guided phase unwrapping (AQGPU) algorithm for unwrapping isoclinics that accommodates the presence of isotropic points as well as π-jumps suitably. The model domain is identified using a boundary mask image (Figure 8.14(a)). Wrapped isoclinic phasemap with inconsistent zones is shown in Figure 8.14(b). Figure 8.14(c) shows the quality map obtained using phase derivative variance (8.35). A seed point is selected to initiate the phase unwrapping, and the integration path is dictated by the quality map.

Phase unwrapping is done by applying suitable checking conditions for the immediate neighborhood to the current pixel such as the left pixel, right pixel,

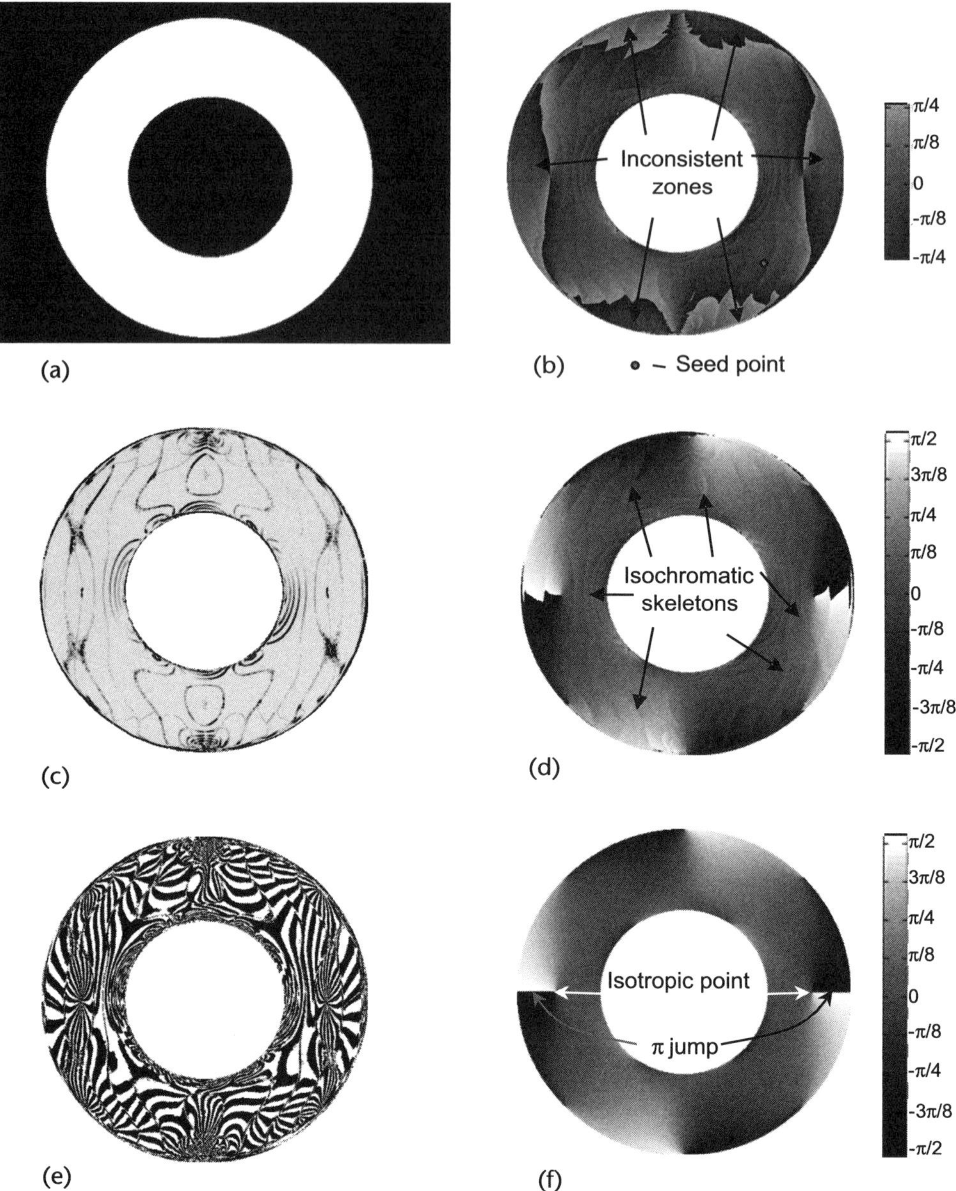

Figure 8.14 Steps involved in unwrapping isoclinic phasemap obtained for the problem of a ring under diametral compression: (a) boundary mask; (b) wrapped isoclinic (with inconsistent zones and seed points marked); (c) phase derivative variance quality map; (d) unwrapped isoclinic phasemap obtained by phase derivative variance algorithm; (e) binary plot of isoclinics ((8.31)); (f) Unwrapped isoclinic phasemap obtained using analytical solution. (Courtesy [2].)

bottom pixel, and top pixel positions. The checking conditions for two pixel positions are shown in Table 8.2. The checking conditions autonomously take care of the isotropic point(s)/π-jumps occurring in the isoclinic phasemap while unwrapping. For identifying the inconsistent zones and a π-jump, a variable theta tolerance (*theta_tol*) is needed. A value of $\pi/3$ for *theta_tol* is found to be sufficient for a variety of problems. To explain the role of these conditions, in Table 8.2 for the

Table 8.2 Checking Conditions for Isoclinic Unwrapping Based upon Pixel Position for AQGPU Algorithm

S No.	Pixel position	Checking condition
1	Left Pixel (LP)	$abs(\theta_{i,j} - \theta_{i-1,j}) > theta\_tol$, then $$\theta_{i-1,j} = \begin{cases} \theta_{i-1,j} + \dfrac{\pi}{2} & \theta_{i,j} <= \dfrac{-\pi}{2} \;\&\; \theta_{i-1,j} <= 0 \;\&\; (\theta_{i-1,j} - \theta_{i,j}) > theta\_tol \quad (1a) \\[2ex] \theta_{i-1,j} - \dfrac{\pi}{2} & \theta_{i,j} > \dfrac{-\pi}{2} \;\&\; (\theta_{i-1,j} - \theta_{i,j}) > theta\_tol \quad (2a) \end{cases}$$ $$\theta_{i-1,j} = \begin{cases} \theta_{i-1,j} - \dfrac{\pi}{2} & \theta_{i,j} >= \dfrac{\pi}{2} \;\&\; \theta_{i-1,j} >= 0 \;\&\; (\theta_{i-1,j} - \theta_{i,j}) <- theta\_tol \quad (3a) \\[2ex] \theta_{i-1,j} + \dfrac{\pi}{2} & \theta_{i,j} < \dfrac{\pi}{2} \;\&\; (\theta_{i-1,j} - \theta_{i,j}) <- theta\_tol \quad (4a) \end{cases}$$
2	Right Pixel (RP)	$abs(\theta_{i,j} - \theta_{i+1,j}) > theta\_tol$, then $$\theta_{i+1,j} = \begin{cases} \theta_{i+1,j} + \dfrac{\pi}{2} & \theta_{i,j} <= \dfrac{-\pi}{2} \;\&\; \theta_{i+1,j} <= 0 \;\&\; (\theta_{i+1,j} - \theta_{i,j}) > theta\_tol \\[2ex] \theta_{i+1,j} - \dfrac{\pi}{2} & \theta_{i,j} > \dfrac{-\pi}{2} \;\&\; (\theta_{i+1,j} - \theta_{i,j}) > theta\_tol \end{cases}$$ $$\theta_{i+1,j} = \begin{cases} \theta_{i+1,j} - \dfrac{\pi}{2} & \theta_{i,j} >= \dfrac{\pi}{2} \;\&\; \theta_{i+1,j} >= 0 \;\&\; (\theta_{i+1,j} - \theta_{i,j}) <- theta\_tol \\[2ex] \theta_{i+1,j} + \dfrac{\pi}{2} & \theta_{i,j} < \dfrac{\pi}{2} \;\&\; (\theta_{i+1,j} - \theta_{i,j}) <- theta\_tol \end{cases}$$

left pixel position (LP), the checking conditions are marked as 1a, 2a, 3a, and 4a, respectively. The conditions (1a and 3a) respectively take care of both the π-jumps and the presence of isotropic points while unwrapping. These conditions ensure that the isoclinic values lie in the range $-\pi/2 < \theta < +\pi/2$ so that at the interface of π-jump it exactly reverses (from $-\pi/2$ to $+\pi/2$, or vice versa). The conditions (2a and 4a) respectively take care of the $\pi/2$-jump denoting an inconsistent zone depending upon the sign of the gradient of the adjacent pixels. Proceeding in this manner, unwrapping is done in a single step for the entire model domain without any further human intervention.

The semblance of isochromatic skeleton on isoclinic phasemap (Figure 8.14(d)) is actually noise and smoothing of isoclinic data is mandatory for their use in stress separation studies. This noise is clearly evident when one looks at the binary representation of isoclinics (Figure 8.14(e)).

EXERCISE 8.3

The following table summarizes the intensities observed for two points in a circular disk under diametral compression (dia = 60 mm, thickness = 6 mm, load = 574 N, Fsigma = 10.9 N/mm/fringe), corresponding to the optical arrangements of Table 8.1. Determine the principal stress direction at these points of interest. Also discuss the role of unwrapping of theta and atan, atan2 functions in the calculations. Compare the experimental results with theoretical results and comment.

Coordinate (x, y) mm	I_1	I_2	I_3	I_4	I_5	I_6	I_7	I_8	I_9	I_{10}
0,0	27	61	101	60	88	67	84	117	82	44
−10.53, 25.7	116	42	60	106	66	93	70	93	71	49

Solution

For point 1, using a plane polariscope, theta can be found by (8.26) as

$$\theta_c = \frac{1}{4}\tan^{-1}\left(\frac{I_4 - I_2}{I_3 - I_1}\right) = \frac{1}{4}\tan^{-1}\left(\frac{60 - 61}{101 - 27}\right)$$

Theta values can be obtained using atan function (θ_{c1}) or atan2 function (θ_{c2}), which are given as

$$\theta_{c1} = \frac{1}{4}\tan^{-1}(-0.0135) = -0.19°; \quad \theta_{c2} = \frac{1}{4}\tan^{-1}\left(\frac{-7}{74}\right) = -0.19°$$

Since the value of inverse tangent function lies in the fourth quadrant, there is no difference between the use of atan or atan2 for this point.

Based on circular polariscope, theta can be found by (8.27) as

$$\theta_c = \frac{1}{2}\tan^{-1}\left(\frac{I_9 - I_7}{I_8 - I_{10}}\right) = \frac{1}{2}\tan^{-1}\left(\frac{82 - 84}{117 - 44}\right)$$

The values obtained using atan and atan2 functions are

$$\theta_{c1} = \frac{1}{2}\tan^{-1}(-0.0274) = -0.785°; \quad \theta_{c1} = \frac{1}{2}\tan^{-1}\left(\frac{-2}{73}\right) = -0.785°$$

At the center of the disk, theoretically the isoclinic value is zero. However, the evaluation of theta by both plane or circular polariscope give a small negative value, which may be due to a small misalignment of the model with respect to the optical system.

For both the points, theoretical values of principal stress and their associated directions are determined using the equation in Box 8.1, which are summarized in the following table

Coordinates (x, y) mm	σ_1 (MPa)	σ_2 (MPa)	θ_1 (deg)	θ_2 (deg)
0, 0	1.02	−3.05	0	90
−10.53, 25.7	0.041	−1.092	−57.07	32.93

For point 2, from plane polariscope arrangements, one gets

$$\theta_c = \frac{1}{4}\tan^{-1}\left(\frac{106 - 42}{60 - 116}\right)$$

The theta values calculated using atan and atan2 functions are

$$\theta_{c1} = \frac{1}{4}\tan^{-1}(-1.143) = -12.2°; \quad \theta_{c2} = \frac{1}{4}\tan^{-1}\left(\frac{64}{-56}\right) = 32.8°$$

The values obtained are different and the value of inverse tangent function is in the second quadrant. This is correctly evaluated only by the atan2 function. As it is divided by a factor of 4, θ lies in first quadrant.

From circular polariscope, the values obtained using atan and atan2 functions are

$$\theta_{c1} = \frac{1}{2}\tan^{-1}(0.023) = -0.651°; \quad \theta_{c1} = \frac{1}{2}\tan^{-1}\left(\frac{1}{44}\right) = -0.651°$$

These are far different from the theoretical values. It has already been pointed out that the results from the circular polariscope are not reliable and this is established for this point.

The value of θ_{c2} from plane polariscope arrangement coincides with the θ_2 (i.e., the orientation of principal stress σ_2). For determining the retardation, it is essential that one unwraps θ such that it always represents one of the principal stress directions consistently over the domain. Thus, θ_{c2} has to be added or subtracted by $\pi/2$. If $\pi/2$ is added, one gets $\theta > \pi/2$ and as θ has to be in the range $-\pi/2 < \theta < \pi/2$, it is desirable to subtract $\pi/2$, which gives $\theta_{unwrapped} = -57.2°$. which is very close to the theoretical value.

Although in this example, discussion on unwrapping for point 2 is based on theoretical calculation, in a generic case, the condition given in Table 8.2 takes care of it systematically. Alternately, if θ phasemap is plotted using (8.33), one can easily say whether or not a point needs unwrapping by observing visually whether or not the point lies in the inconsistent zone.

The solution to this problem supports the use of atan2 function as well as (8.26) for theta calculation.

8.8.10   Smoothing of Isoclinics

Mathematically, isoclinic values are undefined at isochromatic skeletons. The isoclinic contour has a discontinuity and the number of such discontinuities increase with increased load; this is a major source of error in evaluating the isoclinic parameter.

8.8.10.1   Smoothing by Robust Outlier

In this, the data points lying outside the trend are omitted and a local curve fitting is done by least squares analysis [14]. The smoothing procedure is termed local because each smoothed value is determined by the neighboring data points defined within a span. The span defines a window of neighboring points to be included in the smoothing calculation for each data point. The larger the span, the better the smoothed curve will follow the trend.

The omission of data points that are present outside the trend is done by a weighting process. The weights are calculated based upon the statistical parameter of median absolute deviation (MAD) and scaled between 0 and 1. Data points with weights nearer to one are those that are lying outside the data trend locally, and those with weights nearer to zero are those that exactly follow the data trend. The algorithm appropriately chooses the data points closer to zero for local least squares curve fitting. An important factor here is the selection of the span width and the order of the polynomial for least squares curve fitting. It varies from problem to problem but usually a longer span width is better.

8.8.10.2   Adaptive Smoothing Algorithm for Isoclinics

The presence of π-jumps and isotropic points do affect smoothing, and in a complex problem there could be many and an autonomous approach is desirable. Checking conditions as in Table 8.3 are introduced to segment the line in case of a π-jump and in case of an isotropic point to exclude it [23]. Robust outlier is then used to reduce the discontinuities in the isoclinic contour. A variable theta_tol is used to identify π-jump and for smoothing it varies between 80° and 175°.

Figure 8.15 shows the smoothed isoclinic phasemap obtained using the robust outlier algorithm. One can clearly notice that a semblance of isochromatic fringes is not seen here. A span length of 40 pixels and a linear polynomial for the least squares curve fitting is used along the horizontal direction. For smoothing along the vertical direction, a span length of 30 pixels was found to be sufficient.

8.8.11   Unwrapping of Isochromatics

In isochromatic phasemap unwrapping, the integral value should be added to the fractional fringe order appropriately. It is essential that the wrapped isochromatic phasemap is free of ambiguous zones. It has been demonstrated by Ramji and Ramesh [13, 14] that if the unwrapped isoclinic value (smoothed one is still better) is used in (8.30), one gets a phasemap free of ambiguous zones (Figure 8.16(a)). It is surprising that although the consistency aspect of isoclinics has influenced the evaluation of isochromatic data significantly, the noise present in the isoclinics

Table 8.3   Checking Conditions Used by Adaptive Smoothing Algorithm to Take Care of Isotropic Points and π-Jumps

S No.	Category	Smoothing Direction	Checking Condition
1	π-jump	Top to Bottom (Vertical)	$abs(\theta_{i,j} - \theta_{i,j+1}) >$ theta_tol
		Left to Right (Horizontal)	$abs(\theta_{i,j} - \theta_{i+1,j}) >$ theta_tol
			$\mathrm{int}(\theta_{i,j}) = 0$
2	Isotropic point	Top to Bottom (Vertical)	$\theta_{i,j+1},\ \theta_{i,j+2} < 0\ \&\ \theta_{i,j-1},\ \theta_{i,j-2} > 0$ (or)
			$\theta_{i,j+1},\ \theta_{i,j+2} > 0\ \&\ \theta_{i,j-1},\ \theta_{i,j-2} < 0$
			$\mathrm{int}(\theta_{i,j}) = 0$
		Left to Right (Horizontal)	$\theta_{i+1,j},\ \theta_{i+2,j} < 0\ \&\ \theta_{i-1,j},\ \theta_{i-2,j} > 0$ (or)
			$\theta_{i+1,j},\ \theta_{i+2,j} > 0\ \&\ \theta_{i-1,j},\ \theta_{i-2,j} < 0$

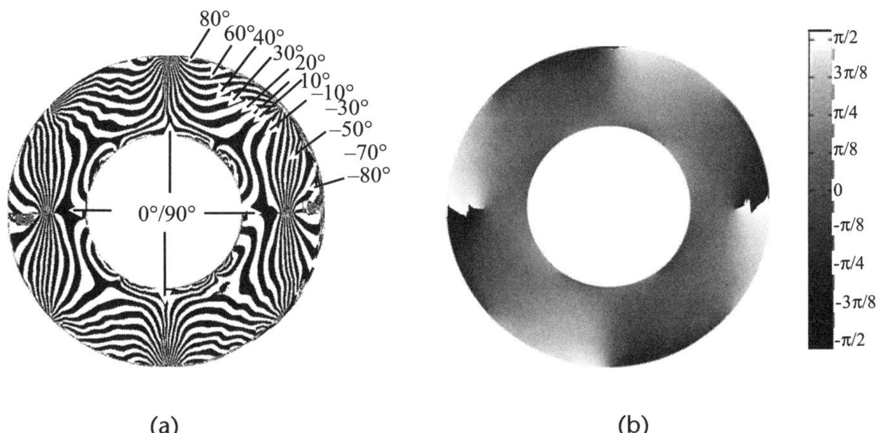

(a)   (b)

Figure 8.15 Smoothed isoclinic plots for the problem of a ring under diametral compression: (a) binary plot in steps of 10 obtained by adaptive smoothing; (b) grayscale plot. (Courtesy [2].)

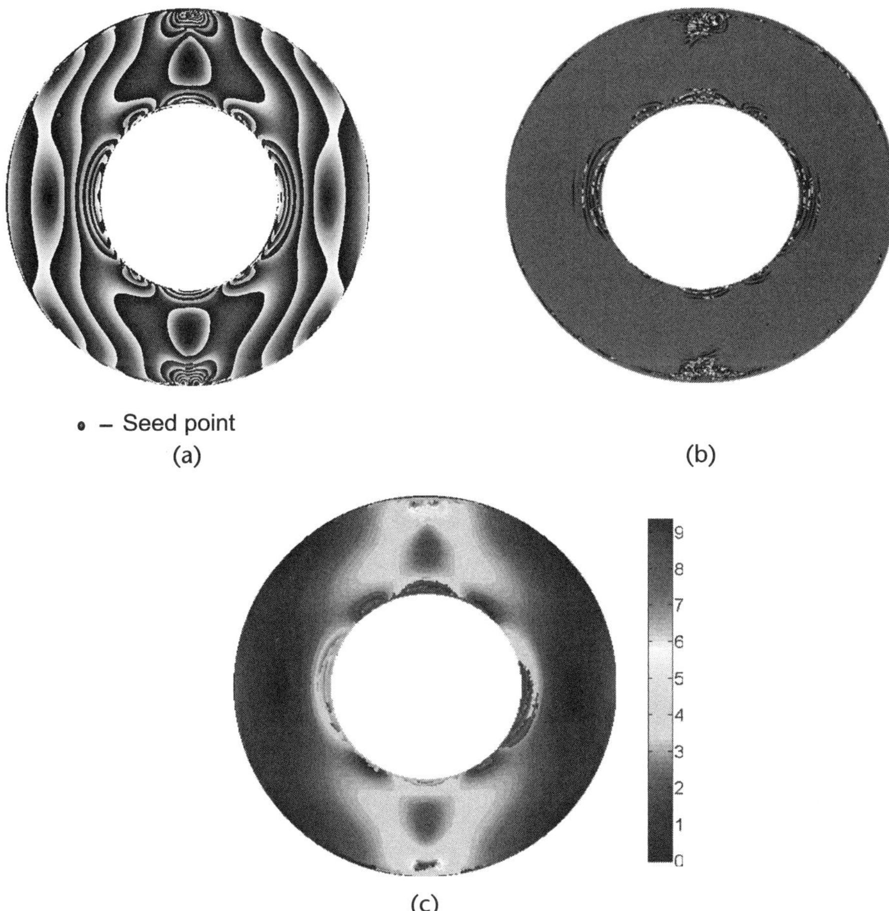

(a)   (b)

(c)

Figure 8.16 (a) Wrapped phase of fractional retardation obtained by using unwrapped isoclinic value in (8.30); (b) quality map obtained using phase derivative variance; (c) Matlab plot of unwrapped phase. (Courtesy [2].)

did not affect it [1]. In the phasemap, if one notices the variation of the value of fractional fringe order, it increases in the direction of increasing total fringe order. Such an observation greatly simplifies the algorithm for phase unwrapping. This is so if the optical elements are aligned exactly (i.e., the slow axis of the quarter-wave plates) as in Figure 8.9(b). It is interesting to note that the phasemap is similar to the dark field image (Figure 8.10(f)). This is one indirect check that the polariscope is correctly aligned.

A suitable seed point has to be selected for which the total fringe order needs to be given. In practice, as the phasemap has high contrast, usually the integer fringe points are easily selectable and the user has to identify one of these and provide the corresponding integer fringe order determined through auxiliary means. The unwrapping process starts from the seed point and proceeds further based upon the quality map. The sign of the gradient of the adjacent pixels decides the addition or subtraction of integral value. If it is positive, then integer value is subtracted (k); otherwise, it is added. Identification of the transition point to add or subtract integers is a crucial step, and it is selected through a threshold value, and a value of 123 is normally used. The checking condition is given as

$$W\left(\delta_{i,j}\right) = \begin{cases} \delta_{i,j} - k & \Delta\delta_{i,j} > 123 \\ \delta_{i,j} + k & \Delta\delta_{i,j} < -123 \end{cases} \tag{8.36}$$

where $\delta_{i,j}$ is the fractional fringe order at corresponding pixel position (i, j), W is unwrapping operator, and $\Delta\delta_{i,j}$ is the gradient of the fractional fringe order. The AQGPU algorithm discussed in Section 8.8.9 is used with slight modification for isochromatic phasemap unwrapping for accommodating the condition given in (8.37). Figure 8.16(c) gives the quality map, and Figure 8.16(d) gives the unwrapped phase plotted in Matlab.

8.8.12 Summary of Ten-Step Method

One of the key aspects of the ten-step method as proposed by Ramji and Ramesh is to use unwrapped theta values in the calculation of the fractional retardation. For getting high-quality theta values, use of a plane polariscope method is suggested. Yao-Ting et al. [24] have supported this idea analytically in 2012. The (8.30) is recast by substituting the value of the intensity differences in terms of expressions involving δ and unwrapped θ.

$$\delta_c = \tan^{-1}\left(\frac{I_b \sin 2\theta \sin \delta \sin 2\theta_c + I_b \cos 2\theta \sin \delta \cos 2\theta_c}{I_b \cos \delta}\right) \tag{8.37}$$

$$\delta_c = \tan^{-1}\left(\frac{\sin \delta}{\cos \delta}\left(\sin 2\theta \sin 2\theta_c + \cos 2\theta \cos 2\theta_c\right)\right) \tag{8.38}$$

Equation (8.38) indicates that the principal value of the retardation is different if θ_c or $\theta_c + \pi/2$ is used as both are possible solutions. However, if unwrapped isoclinic value is used, then one gets a unique value for fractional retardation.

$$\delta_c = \tan^{-1}\left(\frac{\sin\delta}{\cos\delta}(\sin 2\theta \sin 2\theta + \cos 2\theta \cos 2\theta)\right)$$
$$= \tan^{-1}\left(\frac{\sin\delta}{\cos\delta}\right) \tag{8.39}$$

Among the various PSTs, the ten-step phase shifting has demonstrated its capability to obtain photoelastic parameters of high accuracy for a variety of problems [2, 3]. Several other methods reported are a subset of this. The last six arrangements correspond to the six-step phase shifting technique [1, 11, 12]. For dynamic problems, one would like to minimize the number of images, and four step methods are proposed. Among these the one proposed by Ajovalasit et al. is the best [3], and it is the images 5–8 of the ten-step method.

8.9   Color Image Processing Techniques

Among the various experimental techniques, photoelasticity has a unique distinction of providing fringe data in color. Use of a color code to identify fringe gradient direction and to assign the total fringe order approximately has been in use in conventional photoelasticity. In digital photoelasticity, a color image processing system is basically used as an image acquisition device. Use of RGB model is quite common, and in this the image is seen as a superposition of image planes of red, green, and blue.

8.9.1   Intensity of Light Transmitted in White Light for Various Polariscope Arrangements

Most systems that use color image processing hardware for data acquisition use white light as the illuminating source. The color camera does not have a flat response for all wavelengths, and its spectral response $F(\lambda)$ is a function of the wavelength. Further, the amplitudes of light corresponding to each of the wavelengths of the light source may not be the same and will be different for different sources. In addition to these effects, one may have to include a function $T(\lambda)$ that is the sum total of the transmission behavior of all the optical elements as a function of wavelength forming the particular polariscope arrangement. Finally, in view of a spectrum of colors, the perceived intensity will be an integration of the individual light intensity corresponding to each of the wavelengths in the spectrum. For a plane polariscope, in a dark field arrangement, with the analyzer at an angle β, (8.22) could be modified to include these aspects as [1, 25]

$$I_p = I_b + \left[\frac{1}{(\lambda_2 - \lambda_1)}\int_{\lambda_1}^{\lambda_2} T(\lambda)F(\lambda)I_a(\lambda)\sin^2\frac{\delta(\lambda)}{2}\,d\lambda\right]\sin^2 2(\theta - \beta) \tag{8.40}$$

where I_b is the total background light intensity for the entire spectrum of colors and the factor $(\lambda_2 - \lambda_1)$ in the denominator is the normalizing factor [1, 25] such that the intensity varies in the range 0–1 from dark to bright field arrangement.

Normally, one would like to express the retardation $\delta(\lambda)$ with respect to a reference wavelength λ_{ref}.

$$\delta(\lambda) = \frac{\delta_{ref}F_{\sigma_{ref}}}{F_{\sigma}(\lambda)} = \frac{2\pi N_{ref}F_{\sigma_{ref}}}{F_{\sigma}(\lambda)} \tag{8.41}$$

In most studies, $T(\lambda)$ is assumed to be unity. Substituting (8.41) in (8.40), one gets

$$I_p = I_b + \left[\frac{1}{(\lambda_2 - \lambda_1)}\int_{\lambda_1}^{\lambda_2} F(\lambda)I_a(\lambda)\sin^2\frac{\pi N_{ref}F_{\sigma_{ref}}}{F_{\sigma}(\lambda)}\,d\lambda\right]\sin^2 2(\theta - \beta) \tag{8.42}$$

where N_{ref} is the total fringe order corresponding to reference wavelength.

A circular polariscope is normally constructed with quarter-wave plates meant for a specific wavelength. When white light is used, the quarter-wave plate no longer provides a phase shift of $\pi/2$ for all wavelengths and in general behaves like a retarder. If the reference wavelength for which the quarter-wave plate is matched is, say, λ_{ref}, the error introduced for all other wavelengths (λ) is

$$\varepsilon = \frac{\pi}{2}\left(\frac{\lambda}{\lambda_{ref}} - 1\right) \tag{8.43}$$

Considering the quarter-wave plate as a retarder with a retardation of $(\pi/2 + \varepsilon)$, the intensity of light transmitted for dark field circular polariscope, with crossed quarterwave plates, for a particular wavelength is

$$I_d(\lambda) = I_a(\lambda)\sin^2\left(\frac{\pi N_{ref}F_{\sigma_{ref}}}{F_{\sigma}(\lambda)}\right)\left[1 - \cos^2 2\theta\sin^2\varepsilon\right] \tag{8.44}$$

For the complete spectrum of white light, including the background light intensity, (8.44) changes to

$$I_d = I_b + \frac{1}{(\lambda_2 - \lambda_1)}\int_{\lambda_1}^{\lambda_2} F(\lambda)I_a(\lambda)\sin^2\left(\frac{\pi N_{ref}F_{\sigma_{ref}}}{F_{\sigma}(\lambda)}\right)\left[1 - \cos^2 2\theta\sin^2\varepsilon\right]d\lambda \tag{8.45}$$

Though the intensity equations presented in this section appear quite complex, judicious approximations are invoked while using them for practical applications. One of the simplest approaches is to approximate each image plane as a monochrome image for image processing. Ramesh and Deshmukh [26] proposed the use of the green plane of the color image recorded by a color CCD camera for phase shifting. Although approximate, the methodology can give both isoclinics and isochromatics for a given problem.

8.9.2   TFP, RTFP, and Window Search Method

Equations (8.42) and (8.45) show that the mathematical modeling of the RGB intensity variation of the output image requires accurate consideration of various factors such as quarter wave plate error, dispersion of the stress-optic coefficient, spectral response of the camera, spectral composition of the light source, and the transmission response of the polariscope components [25]. It is very difficult to accurately estimate all these parameters. Nevertheless, for each data point, one can digitally extract the relevant grayscale value of the three image planes R, G, and B. From a model of known fringe order variation, such as a beam under four point bending, it is possible to associate a total fringe order for each combination of R, G, and B values. Hence in practice an approach based on a calibration table is generally followed for finding the total fringe order. Isoclinic values cannot be obtained through this method. Nevertheless, a large class of problems exist wherein isochromatics themselves yield rich data for further analysis.

The RGB values recorded digitally are sensitive to fringe gradient or fringe density. The fringe order gradient variation is accounted for by using a merged calibration table [27], which is prepared by combining three different calibration tables corresponding to 0–1 (low gradient), 0–2 (medium gradient), and 0–3 (high gradient) that are obtained by using the same calibration specimen under different loading conditions.

The technique is variedly known as three-fringe photoelasticity (TFP) [27], as beyond three fringe orders colors tend to merge, or RGB photoelasticity (RGBP) [25] as information of R, G, and B planes are used for fringe order evaluation. In TFP/RGBP for any test data point, the fringe order is obtained by comparing the R, G, and B values of the data point with those in the calibration table. The most direct approach is to calculate the least squares error term (e_i) for each row i in the calibration table and find the fringe order for which the e_i is the minimum using the following equation [27]:

$$e_i = \sqrt{\left(R - R_i\right)^2 + \left(B - B_i\right)^2 + \left(G - G_i\right)^2} \qquad (8.46)$$

The fringe order data thus obtained can be converted into a set of gray level values using the following equation:

$$g(x,y) = \mathrm{INT}\left[\frac{255}{N_{\max}} \times f(x,y)\right] = \mathrm{INT}[R] \qquad (8.47)$$

where $f(x, y)$ is the fringe order at point (x, y), $N_{\max}$ is the maximum fringe order of the calibration table (three in most cases), $g(x, y)$ is the gray level value at the point (x, y), and INT $[R]$ is the nearest integer of R.

Although, the least squares error method using the calibration table is very simple in implementation, it is prone to error at several locations in the model domain due to repetition of colors [28]. It leads to false estimation of fringe order (N) at some locations, and spatial continuity of N is lost. Figure 8.17(a) shows the color image of the isochromatics recorded for a disk under diametral compression, and Figure 8.17(b) shows the grayscale representation of the total fringe order evaluated by (8.46) and plotted using (8.47). In view of the repetition of colous, one observes several contours in Figure 8.17(b) rather than a smooth variation of grayscale values.

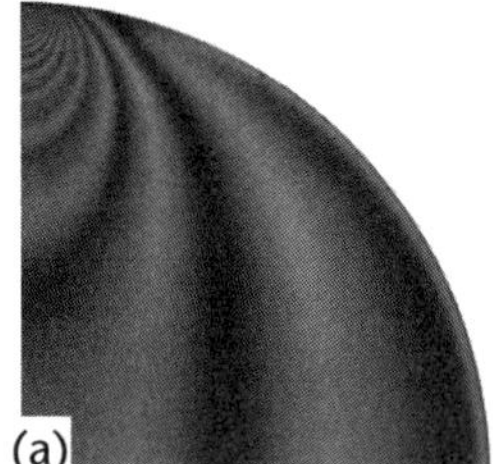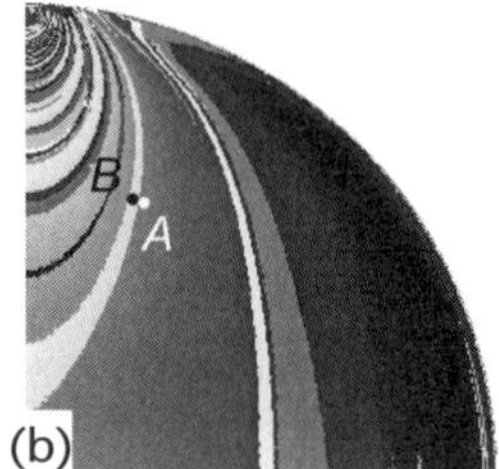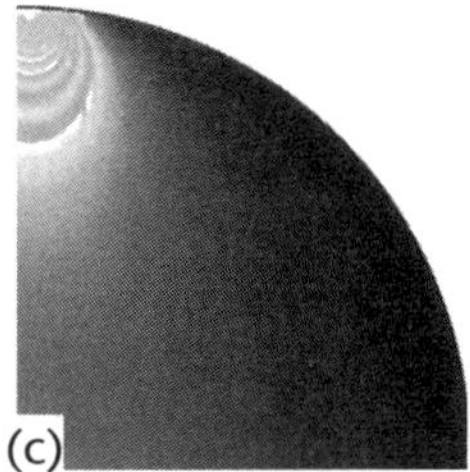

Figure 8.17 (a) Dark field isochromatics for one quarter of a disk under diametral compression; (b) total fringe order evaluated using TFP represented as an image; (c) smooth variation of fringe order by RTFP.

This problem is successfully addressed by Madhu and Ramesh [28] through refined TFP (RTFP) method in which the error e_i is calculated as

$$e_i = \sqrt{\left(R - R_i\right)^2 + \left(B - B_i\right)^2 + \left(G - G_i\right)^2 + \left(N_p - N\right)^2 K^2} \qquad (8.48)$$

The additional term N_p in (8.48) is the fringe order of the neighboring resolved pixel, and N is the fringe order at the current checking point of the calibration table. The parameter K is determined interactively for refining the results and lies between 20 and 200 [2]. Figure 8.17(c) shows a smooth variation of the grayscale values obtained using RTFP. The averaged value of the adjacent neighboring resolved pixels as N_p in (8.48) would enhance the accuracy further.

Fringe order continuity can also be addressed by the window search method [29]. In this, only a small window centered on the previously resolved pixel is searched in the calibration table, thus assuring the spatial continuity of the fringe order. Fringe order based window is recommended for use with merged calibration tables to account for fringe gradient issues and is given as follows:

$$e_i = \sqrt{\left(R - R_i\right)^2 + \left(B - B_i\right)^2 + \left(G - G_i\right)^2}\,;\; N(i) \in \left[N_p - \Delta N, N_p + \Delta N\right] \qquad (8.49)$$

where ΔN is the search window defined in terms of fringe order and its value is usually taken as 0.4 [30].

EXERCISE 8.4

A circular disk of 60 mm diameter and 6 mm thickness is subject to diametral compression of 492 N (Figure 8.17(a)). Total fringe order at point A with coordinates (7.19, 16.65) has to be determined by TFP. The RGB values at A are 157, 71, and 35, respectively. Also find the fringe order of point B (7.07, 16.65) next to A whose RGB values are 156, 63, and 38, respectively. Verify the results obtained with theory. Take the material stress fringe value to be 12.16 N/mm/fringe. If necessary, refine the results obtained using RTFP. Use the calibration table given for some selected data to base your computations.

S. No.	R	G	B	N
1	17	19	11	0.00
2	68	94	54	0.30
3	105	95	28	0.71
4	33	37	46	1.01
5	57	121	50	1.30
6	135	74	23	1.70
7	132	64	27	1.73
8	55	31	48	2.00
9	22	111	19	2.29
10	134	52	41	2.75
11	50	76	21	3.00

Solution

At first the error "e" is found out using (8.46) for both the points. The fringe order corresponding to minimum error term is assigned to the point. The following figure shows the variation of error terms for the two points.

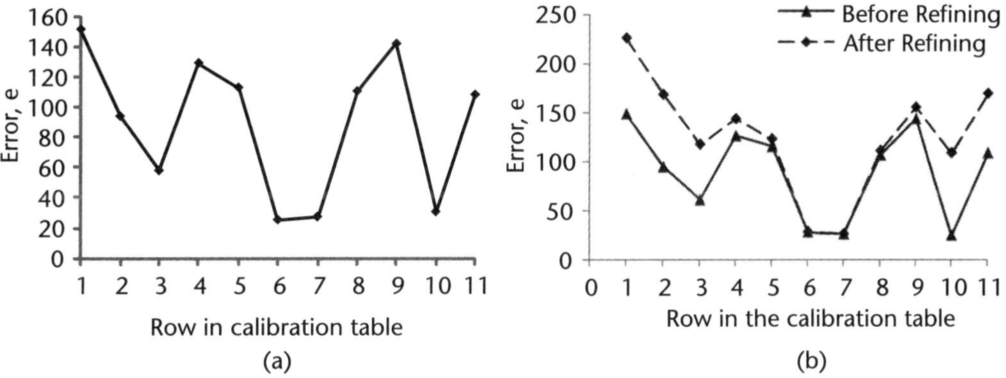

the plot of error "e" for all the rows of the calibration table for (a) point A (b) point B before and after refining.

It can be seen from the figure that the error term for point A is minimum at sixth row and the corresponding fringe order is 1.70. For point B, the error term is minimum at row 10 and the fringe order assigned by TFP is 2.75, which can't be.

Referring to Figure 8.17(b), point A is in the noise free zone and point B is in the noisy zone; hence RTFP needs to be applied for point B. In (8.48) the fringe order at point A is taken as N_p and for a value of K of 100, the error is replotted, which identifies the minimum fringe order at B as 1.73.

Theoretically, the fringe orders using (8.13) and (8.4) for points A and B are obtained as 1.72 and 1.74. The comparison is good.

8.9.3 Color Adaptation

The variation of color tint between the calibration specimen and application specimen and/or difference between lighting conditions during calibration procedure and testing can result in erroneous fringe order evaluation [31]. Using color adaptation, the same calibration table after suitable modifications can be used for various specimens and for different lighting conditions. A simple approach to do this is to assess the tint of the calibration and application specimens first. In order to do that, the RGB variation of the bright field image corresponding to zero-load for both the calibration and application specimens are obtained initially. Let the values for the calibration specimen be R_{zc}, G_{zc}, and B_{zc} and for the application specimen be R_{ze}, G_{ze}, and B_{ze}. The calibration table is revised by modifying the RGB intensities for each color as [5]

$$R_{mi} = \frac{R_{ze}}{R_{zc}} R_{ci}; \quad G_{mi} = \frac{G_{ze}}{G_{zc}} G_{ci}; \quad B_{mi} = \frac{B_{ze}}{B_{zc}} B_{ci} \tag{8.50}$$

where R_{mi}, G_{mi}, and B_{mi} are the RGB values in the modified calibration table at ith row and R_{ci}, G_{ci}, and B_{ci} are the RGB values at ith row of the original calibration table. The fringe order N for each row is not modified in the equivalent table. While using stress-frozen slices, this approach may not be useful as an unloaded slice may not be available. This can be overcome if two-point color adaptation is used. In this, the maximum and minimum RGB intensities from the application dark field image (R_{ImageMax}, R_{ImageMin}, G_{ImageMax}, G_{ImageMin}, B_{ImageMax}, B_{ImageMin}) and the calibration table (R_{TableMax}, R_{TableMin}, G_{TableMax}, G_{TableMin}, B_{TableMax}, B_{TableMin}) are used to find modified calibration table (R_{mi}, G_{mi}, B_{mi}) from the original calibration table (R_{ci}, G_{ci}, B_{ci}) as follows [32]:

$$R_{mi} = \frac{R_{\text{ImageMax}} - R_{\text{ImageMin}}}{R_{\text{TableMax}} - R_{\text{TableMin}}} \left(R_{ci} - R_{\text{TableMin}} \right) + R_{\text{ImageMin}}$$

$$G_{mi} = \frac{G_{\text{ImageMax}} - G_{\text{ImageMin}}}{G_{\text{TableMax}} - G_{\text{TableMin}}} \left(G_{ci} - G_{\text{TableMin}} \right) + G_{\text{ImageMin}} \tag{8.51}$$

$$B_{mi} = \frac{B_{\text{ImageMax}} - B_{\text{ImageMin}}}{B_{\text{TableMax}} - B_{\text{TableMin}}} \left(B_{ci} - B_{\text{TableMin}} \right) + B_{\text{ImageMin}}$$

EXERCISE 8.5

The calibration table for TFP is available for epoxy material. In order to relieve the machining stresses, the model made of epoxy material is subjected to a suitable annealing cycle, which has slightly changed the tint of the model. The no load RGB values of the calibration and model specimens are, respectively: $R_{zc} = 146$, $G_{zc} = 135$, $B_{zc} = 94$ and $R_{ze} = 183$, $G_{ze} = 165$, $B_{ze} = 119$. The RGB values for a row of the calibration table are 129, 59, and 29 respectively. Determine the modified values after performing single point color adaptation.

Solution

The modified gray scale values as per (8.50) are

$$R_{mi} = \frac{R_{ze}}{R_{zc}} R_{ci} = \frac{183}{146} \times 129 = 161.69; \quad G_{mi} = \frac{G_{ze}}{G_{zc}} G_{ci} = \frac{165}{135} \times 59 = 72.11;$$

$$B_{mi} = \frac{B_{ze}}{B_{zc}} B_{ci} = \frac{119}{94} \times 29 = 36.71$$

The fractional grayscale values are used for calculating the least squares error.

Figure 8.18 shows the robustness of the use of two-point color adaptation scheme in the presence of ambient illumination, too. Direct use of two point color adaptation scheme gives a better variation of grayscale representation of the total fringe order (Figure 8.18(e)), which is further improved by RTFP (Figure 8.18(f)) that is comparable with the total fringe order evaluated by the ten-step phase shifting technique (Figure 8.18(b)).

8.9.4   Advancing Front Scanning Method

One of the biggest challenges in solving industrial problems is how to handle complex geometric shapes (Figure 8.19(a)). This requires the development of a flexible

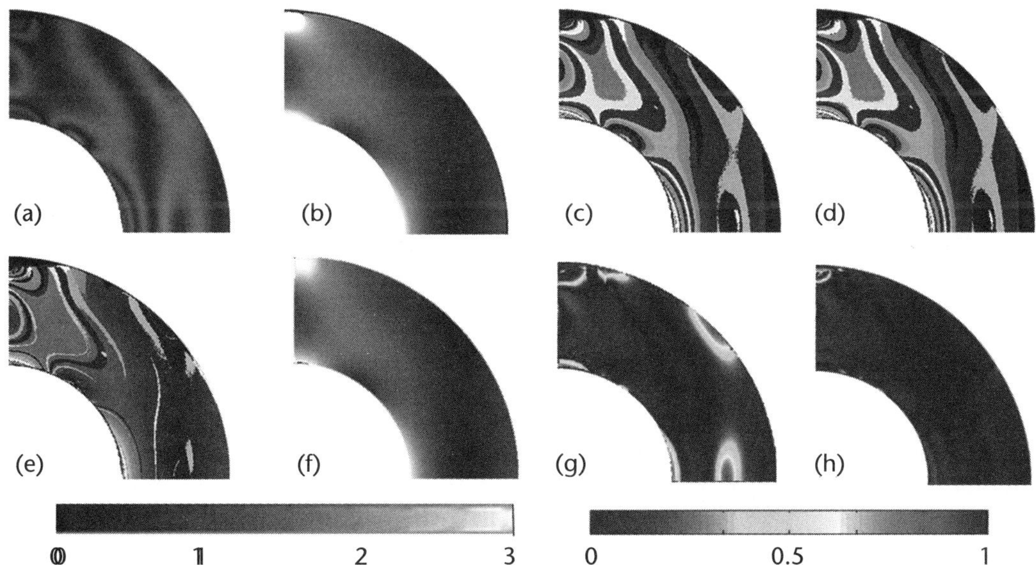

Figure 8.18 One quarter of a ring under diametral compression with ambient illumination: (a) dark field isochromatics; (b) total fringe order evaluated by ten-step PST; (c) TFP without color adaptation; (d) TFP with one-point color adaptation; (e) TFP with two-point color adaptation; (f) RTFP after two-point color adaptation. Whole-field error compared with ten-step PST results with ambient illumination when fringe order is evaluated by (g) RTFP without color adaptation, and (h) RTFP with two-point color adaptation. (Adapted from [32].)

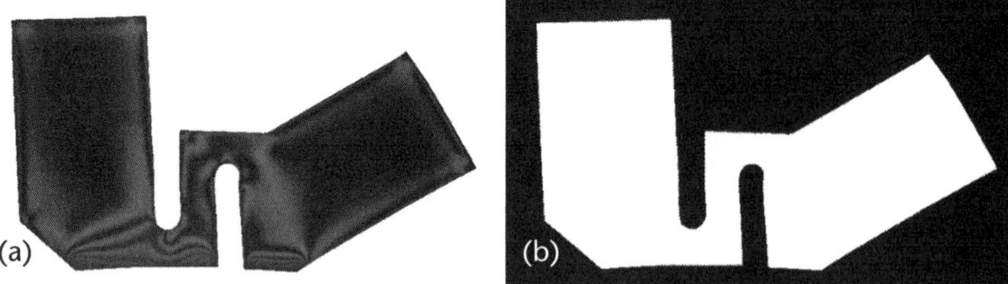

Figure 8.19   (a) Dark field image of a complex simply connected stress frozen slice; (b) image domain mask. (Courtesy [30].)

scanning procedure. Another issue is how the model boundary has to be incorporated in the algorithm. This is easily achieved by using image domain mask.

The image domain mask is generated by carefully selecting the boundaries of the specimen and assigning the value 255 to all pixels inside the boundary, while the value 0 is assigned for all pixels falling outside the boundary (Figure 8.19(b)). Accurate image domain mask generation is important, as inclusion of some background pixels within the mask may lead to erroneous regions in the final result.

The advancing front scanning method [30] progresses by resolving a pixel that is surrounded by maximum number of resolved pixels within a 3×3 window centered on it at each step. First, using the least squares method (8.46), the fringe orders are calculated within the model domain. Though the result obtained using the least squares method could be erroneous in some regions, one could identify several zones with correctly resolved fringe orders. Then a seed point(s) is selected interactively within the correctly resolved zone. A 3×3 window surrounding the selected seed point is considered for advancing the scanning. As all eight pixels within the 3×3 window adjacent to the resolved seed point have only one resolved neighboring pixel, any one of the eight pixels can be resolved. The upper left corner pixel within the 3×3 window is usually chosen for resolving. For each pixel resolved in the subsequent steps, the list labeled unresolved_pixel_list is expanded to include all the unresolved pixels that are adjacent to the entire set of resolved pixels. For managing the scanning procedure, another list labeled surrounding_resolved_pixel_count is created, which stores the number of resolved pixels within the 3×3 window centered on a given pixel within the image domain mask. At each step, the scanning procedure searches for a pixel in the unresolved_pixel_list having the maximum number of adjacent neighboring resolved pixels, using the surrounding_resolved_pixel_count list. The pixel is then resolved using either (8.48) or (8.49). Scanning of the image domain is continued until the unresolved_pixel_list is empty.

The working of advancing front is very similar to eight-way queue based flood fill method with adjacency condition. An analogy can also be made with the process of filling a mold shaped as the specimen, being filled with molten metal entering through the seed point. The major difference between the advancing front scanning and flood fill methods is that the flood fill method will just take the first pixel in the queue for resolving, while the advancing front scanning method searches for pixels having maximum resolved neighbors.

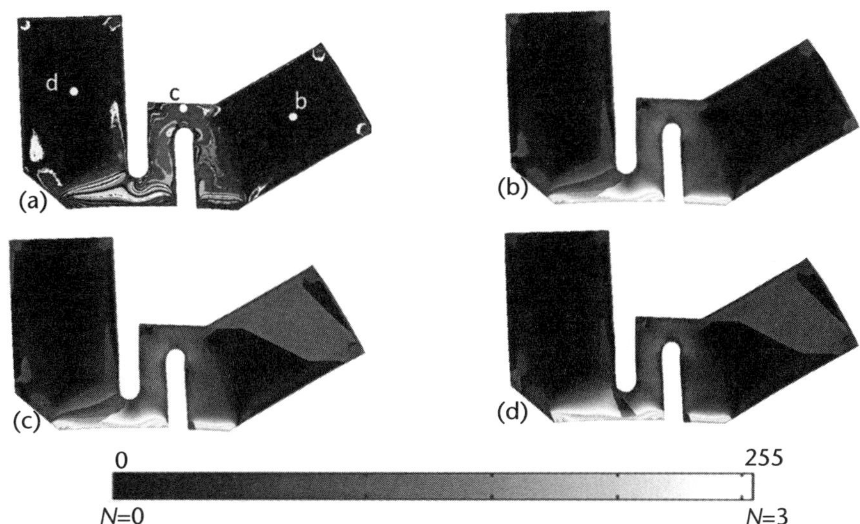

Figure 8.20 Fringe order evaluation of the complex specimen using single seed point advancing front scanning with window search method. Different seed point locations b, c, and d are shown in (a), with correspondingly labeled solutions in (b), (c), and (d). (Courtesy [30].)

The fringe orders obtained for specimen shown in Figure 8.19(a) using advancing front type scanning with single seed points specified at three different locations are shown in Figure 8.20. Although the advancing front scanning algorithm is able to evaluate fringe orders in the whole image domain mask, one can clearly observe patches where fringe order is spatially discontinuous for all three single seed points (Figure 8.20 (b)–(d)). Resolving complex shaped specimen in Figure 8.19(a) can be comparatively easy with multiple seed points than just one.

The simple structure of an advancing front scanning algorithm allows an easy incorporation of the multiple seed points feature. The mold filling analogy for advancing front scanning can be extended for multiple seed points, by having multiple gates for pouring the molten metal into the mold. The working of the algorithm is the same, except that each one of the multiple seed points provided is independently processed until all the unresolved pixels within the image domain mask are resolved.

It is to be noted that seed points are nothing but the points at which the fringe order is correctly known, and they act like reference points for resolving the fringe field in the model domain. The region around each seed point is resolved based on the neighborhood results. Thus, each seed point acts similar to a point of crystal growth. In order to capture the local variations, if there are resolved regions near the stress concentration or high fringe gradient zones, select them as possible seed points.

With carefully selected multiple seed points (Figure 8.21(a)), the entire complex domain is resolved without any zones with fringe order discontinuities (Figures 8.21(e) and (f)). One can see in the intermediate figures, Figures 8(b)–8(d), the seed points act like centers of crystal growth and its immediate neighborhood is resolved.

With discrete spectrum fluorescent light source, fringe orders beyond three could also be resolved [29].

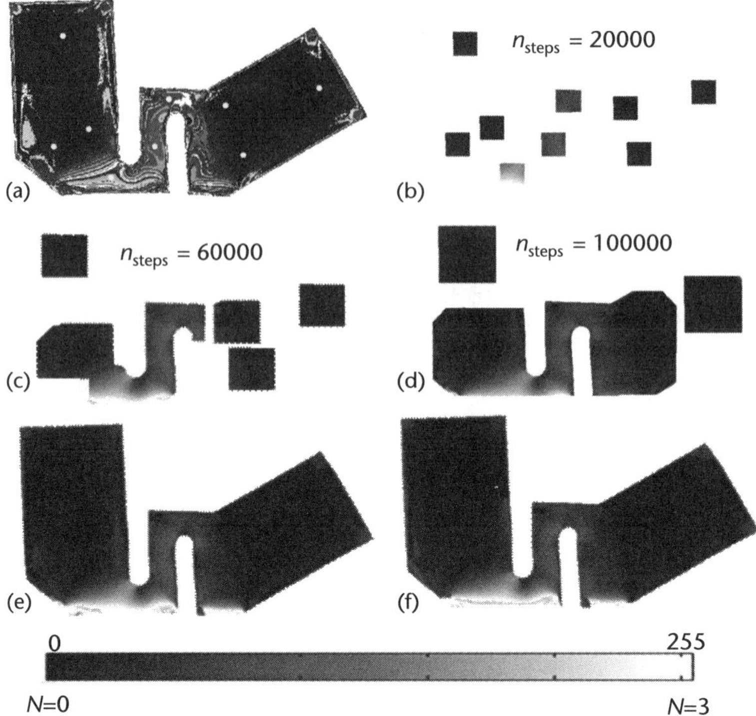

Figure 8.21 Fringe order evaluation of the complex specimen using multiple seed points (a) seed point locations on least squares error method solution; (b)–(d) intermediate steps at each 40,000 pixels resolved. Final solution obtained using (e) window search method, and (f) RTFP. (Adapted from [30].)

8.10 Digital Reflection Photoelasticity

Reflection photoelasticity is an extension of transmission photoelastic analysis for the analysis of opaque prototypes. A thin temporarily birefringent coating is pasted on the prototype with a reflective backing at the interface. The prototype is loaded appropriately, and a reflection polariscope is used for collecting the optical information. The birefringence in the coating is introduced through the surface deformations at the interface and it is useful to represent the photoelastic phenomena to the strains developed. The strain-optic law is

$$\varepsilon_1^c - \varepsilon_2^c = \frac{NF_\varepsilon}{2h_c} \tag{8.52}$$

where F_ε is the strain-optic coefficient, and $(\varepsilon_1^c - \varepsilon_2^c)$ gives the principal strain difference in the coating.

The optical response achievable is an order of magnitude less than what is possible in transmission photoelasticity. This has necessitated the use of white light for reflection photoelastic analysis and with developments in color image processing techniques, digital reflection photoelasticity has become a reality. One of the initial goals is to obtain the isoclinic and isochromatic information over the entire field. This can be achieved by a five-step method discussed next.

8.10.1 Five-Step Method [33]

It is seen in Section 8.9 how isochromatics data can be obtained by processing a single color image recorded in conventional dark field circular polariscope. The same methodology can be extended in reflection photoelasticity too for isochromatic fringe order evaluation. Further, for the evaluation of isoclinics, it is seen in Section 8.8.5 that plane polariscope-based methods are ideal and a record of four isoclinics in steps of 22.5° can help in finding the data over the field. Instead of recording the isoclinics with a monochrome light source, it is recorded using white light as the source. The polarization-stepped images are preferably recorded using a 3CCD color camera for high fidelity. The isoclinic angle is determined by [34]

$$\theta = \frac{1}{4}\tan^{-1}\left(\frac{\left(I_{4,R} + I_{4,G} + I_{4,B}\right) - \left(I_{2,R} + I_{2,G} + I_{2,B}\right)}{\left(I_{3,R} + I_{3,G} + I_{3,B}\right) - \left(I_{1,R} + I_{1,G} + I_{1,B}\right)}\right) \tag{8.53}$$

where $I_{i,j}$ ($i = 1, 2, 3, 4;\ j = R, G, B$) correspond to pixel gray levels of R-, G-, and B-planes for the first four analyzer positions of Table 8.1.

One of the challenging problems in industrial environments is due to assembly stresses. The use of the five-step method for evaluating the stresses developed due to assembly of a flange coupling is discussed next.

Two circular plates made of aluminum material (inner diameter = 35 mm, outer diameter = 100 mm, four holes of 8.4-mm diameter at PCD 75 mm, and thickness = 7 mm) are assembled to simulate a flange coupling using four M8 Allen bolts and nuts.

A photoelastic coating of thickness 3.07 mm made of commercially available polycarbonate sheet (PS-1A; Measurements Group, USA) is machined and pasted on one side of the flange using the epoxy-hardener mixture as bonding material. The assembled flange coupling with the reflective coating bonded on one side of the aluminum plate is shown in Figure 8.22. A mechanical torque spanner is used to apply a uniform torque of 14.7 Nm. Only the right half of the coupling is taken up

Figure 8.22 Flange coupling pasted with reflective coating.

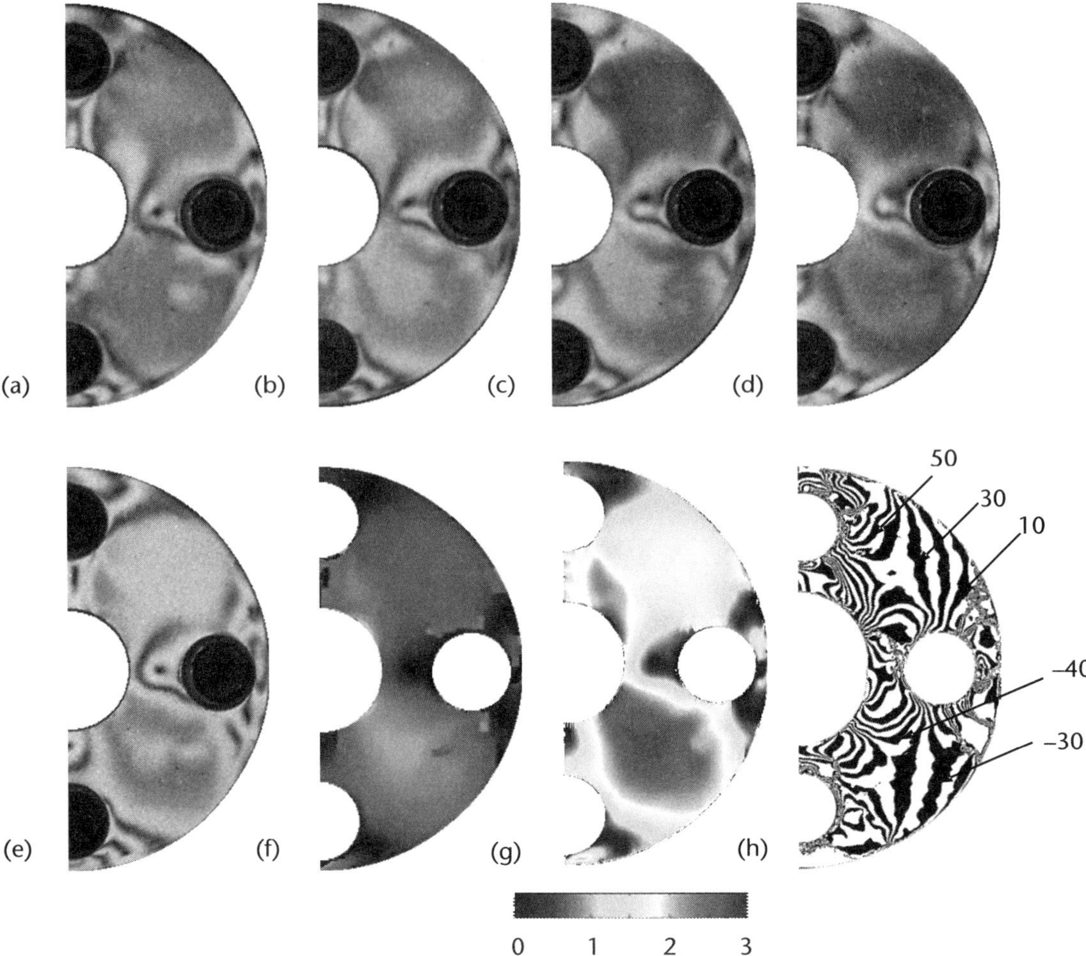

Figure 8.23   Flange coupling, experimentally obtained dark field: (a) 0° isoclinics; (b) 22.5° isoclinics; (c) 45° isoclinics; (d) 67.5° isoclinics; (e) isochromatics; (f) RTFP after two-point color adaptation; (g) smoothed isochromatics in color plot; (h) smoothed isoclinics in binary plot.

for analysis. The strain field developed in the reflective coating is recorded and the five-step images are shown in Figures 8.23(a)–(e). The grayscale representation of the total fringe order using two-point color adaptation in conjunction with RTFP, smoothed with outlier smoothing, is shown in Figure 8.23(f). The smoothed isochromatics are shown as a color plot in Figure 8.23(g). The isoclinc value is calculated by (8.53) and unwrapped and smoothed by using the methodologies mentioned in Sections 8.8.9 and 8.8.10 is shown as a binary plot in Figure 8.23.

8.11   Some Applications

8.11.1   Role of RTFP in Evaluating Transient Thermal Stress Intensity Factors

Most engineering components in use today, such as electronic packages, bimetal thermostats, and gas turbine blades, consist of materials with different thermal and mechanical properties bonded together. A major concern in such bimaterials is the

Table 8.4 Material Properties

S.No	Property	Units	PSM-1[13]	Aluminium
1.	Density, ρ	g/cc	1.17	2.823
2.	Expansion coefficient, α	/°C	146×10^{-6}	22.7×10^{-6}
3.	Thermal conductivity, k	W/m/K	0.365	142
4.	Specific heat capacity, C_P	J/g/°C	1.1052	0.963
5.	Young's modulus, E	GPa	2.39	71
6.	Poisson's ratio, v	—	0.38	0.33
7.	Material stress fringe value, F_σ	N/mm/fringe	7	—

mismatch in thermal coefficients of expansion of the constituent materials, which gives rise to thermal stress in addition to the stress due to mechanical loading. Knowledge of the physical phenomenon behind edge heating of bimaterial system is of key importance because, in several engineering applications, components come into contact with one another at different temperatures and are heated or cooled by heat conduction from the surfaces. In this section, transient thermal analysis of an edge-heated PSM-1 (polycarbonate, Measurements Group Inc., USA)—aluminium bimaterial specimen with symmetric interface edge cracks is studied at 50°C [35]. It is to be noted that PSM-1 has been chosen as the photothermoelastic material as its material properties remain constant between −10°C and 55°C. The relevant mechanical properties are given in Table 8.4. The expansion coefficient of PSM-1 is about six times greater than that of aluminum, while its thermal conductivity and Young's modulus are several orders of magnitude less than those of aluminum. For the determination of stress intensity factors (SIF), it is enough one records the information of isochromatic fringe order. As the problem under study is slowly varying with time, use of ten-step PST is not feasible, and, as the focus is only on recording isochromatic data, TFP is the ideal choice.

A temperature-controlled heating plate is maintained at a temperature of 50°C and the bottom face of the aluminum part of the bimaterial specimen (room temperature is maintained at 28°C) is suddenly placed on it (Figure 8.24(a)). The transient phenomenon until thermal equilibrium is achieved is of interest. Color dark field isochromatics are automatically recorded at time intervals of 15 seconds successively using a color 3CCD camera (Sony XC-003P) having a resolution of 752 × 576 pixels, with a white light source. The transient phenomenon is recorded for about 12 minutes (Figure 8.24).

As soon as the specimen is kept on the hot plate, the aluminum gets heated from the bottom face by surface conduction, and very quickly the temperature becomes fairly uniform because of its high thermal conductivity (Table 8.4). Although the thermal coefficient of aluminum is lower than that of PSM-1, aluminum heats to a higher temperature and expands more in relation to PSM-1, causing the crack to open. Because the thermal conductivity of PSM-1 is very low, it expands slowly, only after the heat flows through the aluminum. Because of the high expansion coefficient, even the small temperature rise occurring in the polycarbonate portion causes it to expand considerably, and the magnitudes of stresses present along the bond begin to reduce and reach a steady state. After about 12 minutes from the

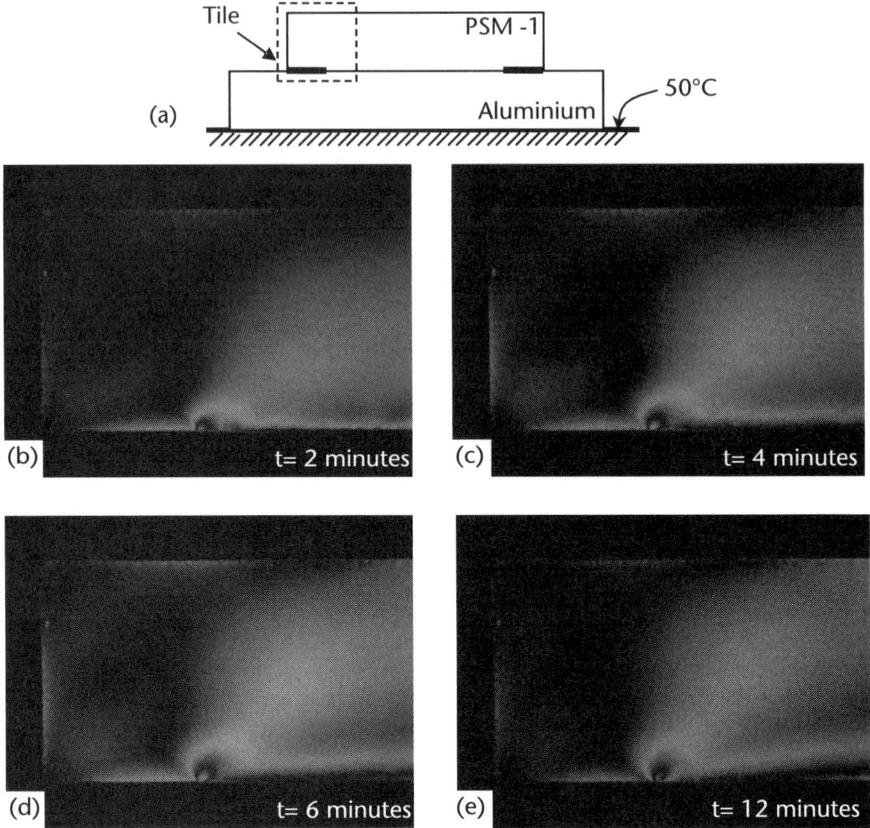

Figure 8.24 Polycarbonate-aluminum bimaterial specimen kept on hot plate at 50°C: (a) schematic; (b)–(e) transient dark field isochromatics recorded for the rectangular tile shown in (a). (Adapted from [35].)

commencement of the experiment, the temperature field of the specimen and the thermal stresses therein approach a steady state.

Data extraction and analysis is discussed for the time step at 6 minutes (Figure 8.24(d)). Using color adaptation followed by RTFP, the fringe data over the field is extracted. For representation, instead of using (8.47), a *composite field* image consisting of bright, dark, and mixed field fringes (fringes in steps of 0.25) is generated from the whole-field fringe order data, in order to have sufficient number of fringes for data extraction for SIF studies (Figure 8.25(a)).

The method of least squares [36] is used to evaluate the multiparameters governing the stress field iteratively. For easy convergence, it has been proposed [36] that the fringe order and the corresponding positional coordinates need to be collected such that, when plotted, they capture the basic geometric features of the fringe field. The zone for SIF evaluation (usually expressed as a ratio of the radial distance r to the crack length a) is typically in the range $0.05 < r/a < 0.65$, but it varies slightly from one time step to another. The *composite field* image is scanned at intervals of 1°, and only those points that are closer ($N \pm 0.02$) to fringe orders of 0.25, 0.5, 0.75, and so on that lie in the range $0.06 < r/a < 0.59$ (for this time step) are selected, which totals to 336 data points, for SIF evaluation. The automated data

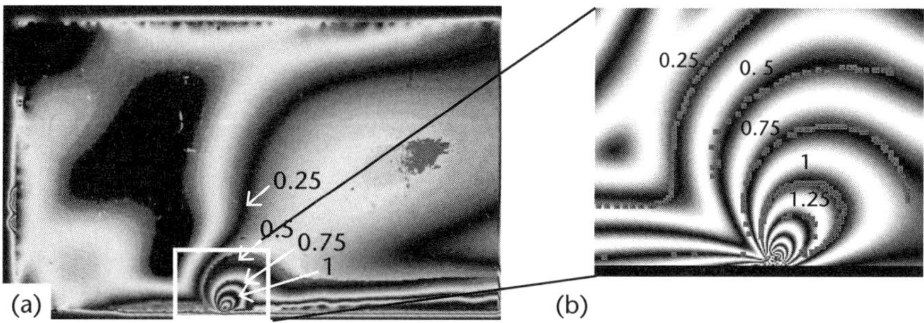

Figure 8.25 Total fringe order evaluation for bimaterial specimen kept on hot plate at 50°C after $t = 6$ minutes: (a) composite field image generated after data smoothing; (b) theoretically reconstructed fringe patterns with data points echoed for six mode I and six mode II parameter solution of the zoomed portion of the tile in (a). (Adapted from [35].)

collection software has an interactive module to remove outliers. A six-parameter solution matches the fringe field well (Figure 8.25(b)). From the six-parameter solution, the values of K_I and K_{II} are 0.064 MPa√m and 0.036 MPa√m, respectively. The variations in the SIFs with time are plotted as a graph in Figure 8.26, which clearly indicates that the opening mode is dominant. The peak in K_I and K_{II} occurs after about 7 minutes and 2 minutes 45 seconds, respectively.

This example has demonstrated the utility of digital photoelasticity, particularly the use of RTFP in handling even a transient problem so effectively.

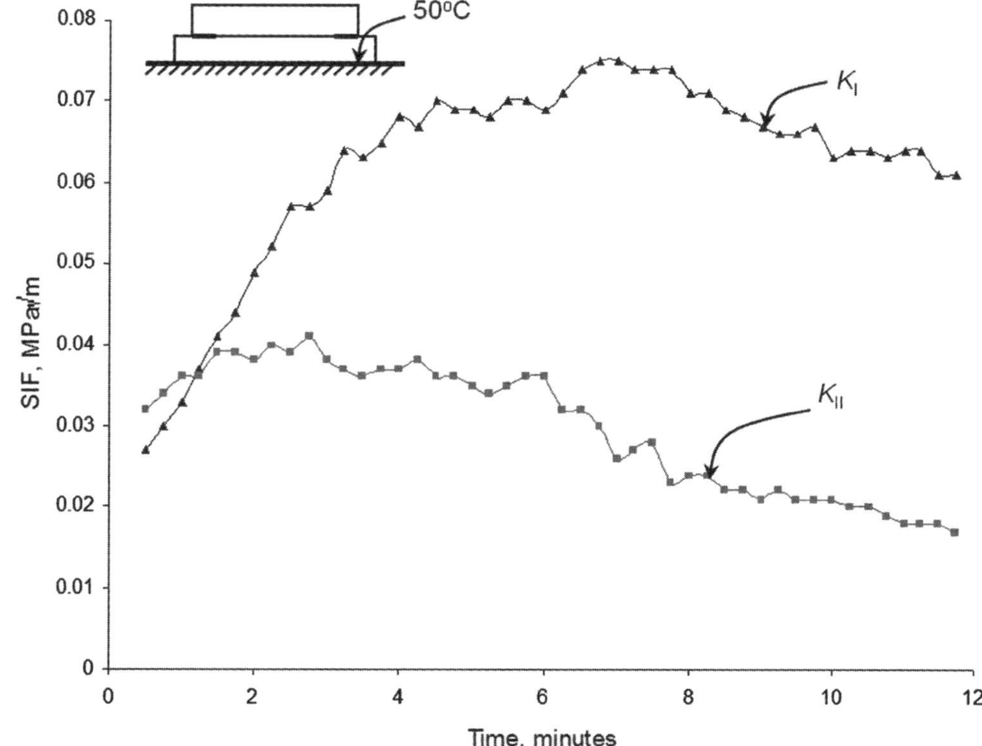

Figure 8.26 Graph showing the variation of SIFs for bimaterial specimen with time when bottom face of aluminum is maintained at 50°C. (Courtesy [35].)

8.11.2   Role of Digital Photoelasticity in Residual Stress Measurements of Glass

Photoelasticity has been widely used to measure the residual stresses in glass articles since glass is birefringent. The residual retardation is usually low as glass is weakly birefringent. One important class of problems is the measurement of residual stresses in glass plates, and the other most fundamental one is the calibration of glass itself for quantification of the results. Unlike in conventional photoelasticity, where material stress fringe value (F_σ, N/mm/fringe) is widely used for quantifying the photoelastic behavior, in glass literature one finds the use of photoelastic constant C (TPa^{-1}), which is related to F_σ as in (8.5). For most glasses, the value of C lies in the range of 2 to 4 TPa^{-1}. The challenge is how to measure such a small value accurately. It is interesting to note that the use of carrier fringe in conjunction with fringe thinning can provide high-quality results for these classes of problems.

Ramesh et al. [37] have illustrated the use of carrier fringe method coupled with fringe thinning to calibrate glass. A rectangular beam of glass is subject to four point bending, and the isochromatics are shown in Figure 8.27(a). A carrier with linear variation of fringes is placed such that the principal stress directions in the carrier and glass are perpendicular to each other. The composite fringes obtained by the superposition of the carrier and the glass beam are shown in Figure 8. 27(b).

The retardation in the glass can be obtained by measuring the horizontal deviation of the composite fringes using the equation [37]

$$\delta_g(y) = \left| \frac{x - x_i}{p} \right| \tag{8.54}$$

where $(x - x_i)$ is the deviation of the fringes at an ordinate y (Figure 8.27(c)), and p is the pitch of the carrier. The load of the glass beam is increased in steps and the images are recorded from which the fringe skeletons are extracted. Figure 8.27(c) shows the fringe skeleton extracted using the global fringe thinning algorithm (Section 8.7) using an image recorded at a load of 47 N. The retardation is calculated using (8.54) for a large number of points along the height of the glass beam. Since the glass beam is subjected to pure bending, the stresses vary linearly, and hence the fringe order varies linearly with height. Once the fringe order is measured for several data points along the height of the beam, a plot is made between fringe order versus distance from the neutral axis of the beam. Let S_1 be the slope of the least squares line plotted for this data for a particular load. The experiment is repeated for a few loads. For each of these loads, slope S_1 is calculated. Next, another plot

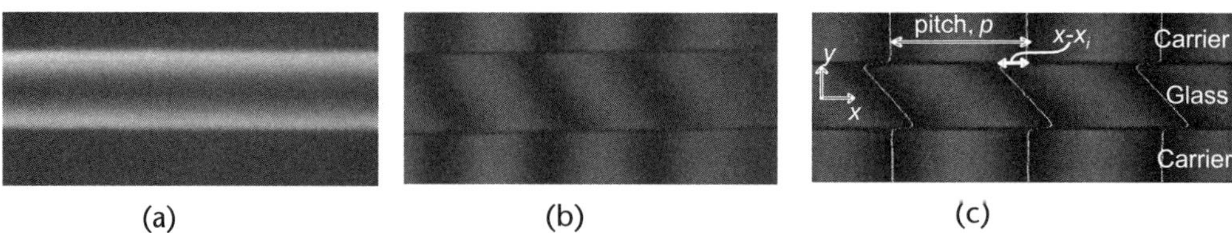

        (a)   (b)   (c)

Figure 8.27   (a) Glass beam subjected to a load of 47 N; (b) composite fringes in color obtained by superimposing with a carrier; (c) composite fringe thinned by global fringe thinning algorithm with various features marked. (Adapted from [37].)

is made between these slopes and the corresponding loads. As the beam is loaded within its elastic limit, the plot between the slope S_1 and the corresponding load is expected to be a straight line. A least squares line is drawn for this plot and its slope (S_2) is used for obtaining the value of C as

$$C = \frac{2S_2 \lambda I_{zz}}{at} \tag{8.55}$$

where, λ is the wavelength of light, I_{zz} is the moment of inertia of the beam cross section, a is the moment arm, and t is the thickness of the glass beam. Ramesh et al. [37] have demonstrated this technique by calibrating LB2000 glass material and have reported its C value as 2.79 TPa^{-1} with an uncertainity of 0.026.

In glass plates, the residual stresses are introduced due to thermal gradients between its various regions during manufacturing. The residual stress in the glass plates can be generally divided into two kinds—thickness stresses and membrane stresses. Thickness stresses are the most prominent of all the stresses in a glass plate and are found to depend on the cooling process and the dimensions of the glass plate. The thickness stresses could be easily seen in a transmission polariscope by placing the glass plate such that the light passes through its thickness (Figure 8.28(a)). The variation of the thickness stresses along the thickness is parabolic in nature with compression in the surfaces and tension in the central region. Interestingly when a linear carrier is placed, one observes the composite fringes to be parabolic (Figure 8.28(b)). The tensile stress in the mid plane is approximately half the magnitude of the surface compression. The compressive stress near the surface increases the bending strength of glass, whereas the tension in the mid-plane affects the fragmentation properties. The stress components σ_z and σ_y constitute thickness stresses, and usually they are assumed to be equal for a large plate.

Membrane stresses are introduced in a glass plate due to some cooling conditions when there is a thermal gradient along a plane of the glass plate. They are generally present in tempered glasses and absent in annealed glasses since the cooling process is very slow. The membrane stress at the edge of the glass plate is called edge stress

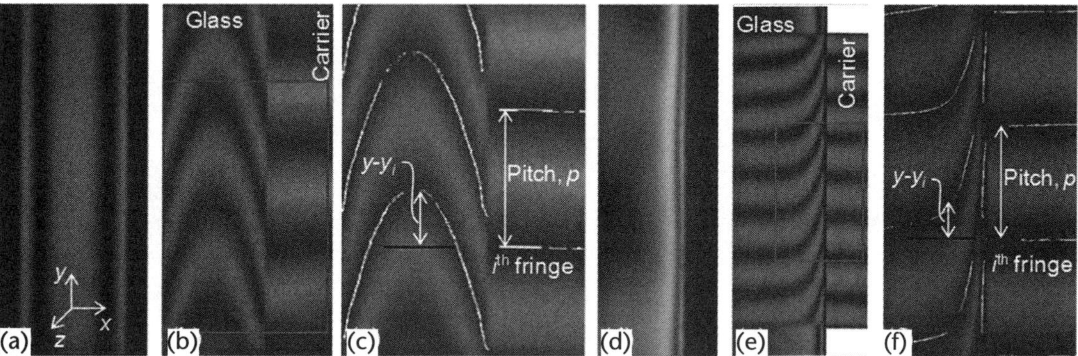

Figure 8.28   (a) Dark field isochromatics showing the thickness stress in a glass plate; (b) thickness stress with the superposition of carrier fringes; (c) magnified view of the rectangular portion with the fringe skeletons extracted using global fringe thinning algorithm; (d) dark field isochromatics showing the edge stress in a tempered glass plate; (e) edge stress with carrier fringes; (f) same as 'c' for (e).

and is caused because the edges cool quickly compared to the central part of the glass plate. They are generally compressive in nature and its measurement is very important from practical considerations. The edge stresses in glass can be seen in a conventional circular polariscope (Figure 8.28(c)). When a linear carrier is placed, deviations of the fringes are seen (Figure 8.28(d)). As the edge is a free surface, σ_y stress component is the edge stress.

For quantitative measurements, the recipe is simple: just identify the fringe skeletons and measure the vertical deviations in both cases [38, 39] (Figure 8.28(c) and (f)).

$$\delta^g(x) = \frac{|y - y_i|}{p} \tag{8.56}$$

where p is the pitch of the carrier, and $y - y_i$ gives the deviation in the carrier. The variation of thickness stress in the glass plate is shown in Figure 8.29(a). Figure 8.29(b) shows the variation of stresses near the edge of the tempered glass, and since the edge of the glass plate is ground, the edge stress is obtained by extrapolating the curve to the edge of the glass plate. The edge stress in the tempered glass plate is measured to be −88 MPa.

8.12 Closure

In this chapter, starting from the basic optical principle exploited in photoelasticity, the current developments that include early use of digital hardware for automation of conventional photoelasticity to the fully automated methods involving phase shifting and color image processing techniques have been discussed. The role of a ten-step method for accurate estimation of photoelastic parameters by phase shifting techniques has been dealt with systematically. It is shown that all the other phase shifting methods in digital photoelasticity are subset to this. Elegance of using color information for isochromatic fringe evaluation using a single photograph is brought with the discussion of TFP and RGB photoelasticity.

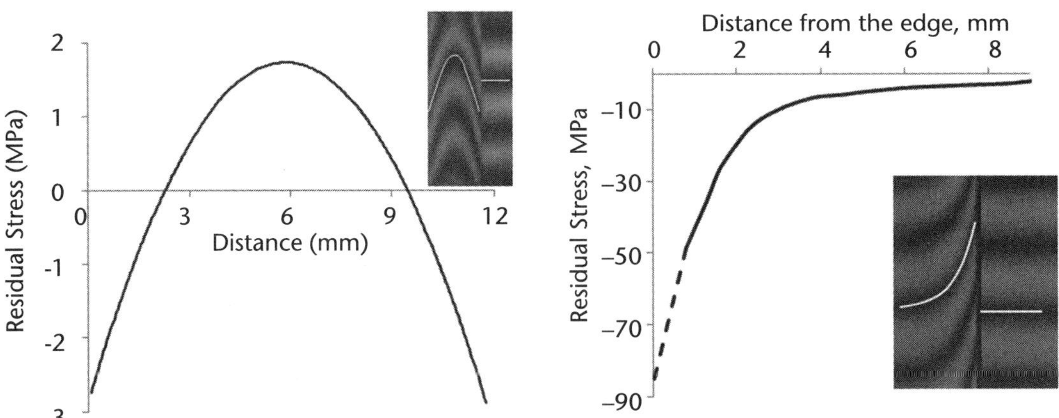

Figure 8.29 (a) The plot of thickness stress in the glass plate shown in 8.28(b); (b) the variation of stress near the edge of the tempered glass plates shown in Figure 8.28(d).

Problems that are most challenging for stress field evaluation are those due to assembly stresses or residual stresses. Numerical modeling of these problems is quite difficult, and experimental analysis can help to develop a suitable numerical model for further parametric study. The use of a five-step method to solve industrial problems using reflection photoelasticity is brought out by the example of the fringe field in a flange coupling due to assembly stresses. The role of digital photoelasticity in evaluating the residual stresses in glass plates is discussed, and it is shown that the use of carrier fringe in conjunction with fringe thinning still has a role to play. Transient stress field developed in the neighborhood of a crack is solved by employing the right combination of various digital photoelastic techniques such as RTFP in conjunction with color adaptation for data acquisition, fringe plotting for data extraction, and over-deterministic least squares method for the evaluation of stress intensity factors.

Application of photoelasticity to solve industrial problems had a temporary setback between mid-1960s and the 1980s due to the euphoria generated by the development of the numerical method of finite elements. It is clearly demonstrated in this chapter how the developments in computer hardware and low-cost image acquisition have significantly influenced the methodology of photoelasticity, too. With advances in digital photoelasticity, newer and sleek equipment for photoelasticity has been developed. The advances on rapid prototyping have made a significant impact on photoelastic model making. With these developments, photoelasticity is now quite competitive to numerical methods and is being used more widely in industries.

For color versions of some of the figures in this chapter, please visit www.artechhouse.com/static/Downloads/RastogiColorFigs.zip.

Problems

1. What is meant by thresholding an image? What are the different criteria that are usually employed to determine the threshold value? What is meant by semi-thresholding, and where is it used?
2. The isochromatics in the neighborhood of a crack is shown in dark field, bright field, and mixed fields. Based on the dark and bright field images, label the fringes in the mixed-field image.

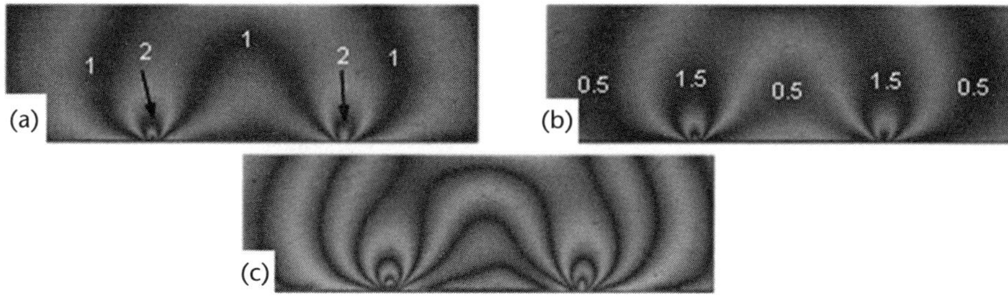

3. How is fringe thinning followed by fringe ordering is still relevant even after the development of fully automated procedures like phase shifting interferometry?

4. Mention the various steps involved in the global fringe thinning algorithm. What is the source of noise and how can logical operators help to remove them?

5. Write a software program to evaluate the material stress fringe value from just one photograph of the circular disk under diametral compression. The circular disk shown in Figure 8.3 corresponds to a load of 665 N. The diameter of the disk is 60 mm and the thickness is 6 mm. Determine its material stress fringe value using your software.

6. What has influenced the paradigm shift in data acquisition and processing of data in optical methods? Explain phase shifting in photoelasticity by comparing it to Tardy's method of compensation.

7. Write briefly the genesis of the ten-step method.

8. A circular polariscope is constructed using various optical arrangements of the elements. Using Jones Calculus, establish which of the arrangements correspond to bright and dark fields.

α	ξ	η	β
$\pi/2$	$3\pi/4$	0	$\pi/4$
$\pi/2$	$3\pi/4$	0	$3\pi/4$
$\pi/2$	$3\pi/4$	$\pi/4$	$\pi/2$
$\pi/2$	$3\pi/4$	$\pi/4$	0

9. How are the fractional retardation and the isoclinic angle calculated in the ten-step method? Discuss the range of these values obtained directly from the trigonometric equations.

10. What do you understand by *ambiguous zones* in isochromatic phasemaps and *inconsistent zones* in isoclinic phasemaps? Are they interrelated? What is the criterion to determine the ambiguous zones? Identify the ambiguous zones and inconsistent zones for the angle bracket

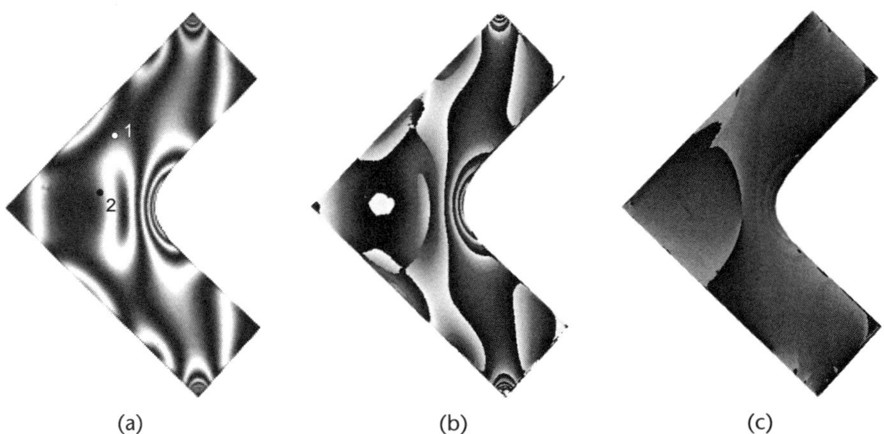

(a)   (b)   (c)

Angle bracket under diametral compression: (a) dark field isochromatics; (b) wrapped isochromatic phasemap; (c) wrapped isoclinic phasemap.

11. What is the difference in the phase unwrapping of phasemaps of isochromatic and isoclinic data?

12. Using the intensity data given in Exercise 8.3, determine the total fringe order at the two points and compare them with theoretical calculations.

13. The following table summarizes the intensities observed for two points (marked in "a" given in Problem 10) in an angle-bracket under compression, corresponding to the optical arrangements of Table 8.1. Determine the principal stress direction and the total fringe order at these points of interest by the ten-step method.

Point	I_1	I_2	I_3	I_4	I_5	I_6	I_7	I_8	I_9	I_{10}
1	114	49	69	151	107	85	162	109	52	112
2	46	112	114	48	115	95	131	165	95	60

14. Explain briefly the various quality measures and of these which quality measure is ideally suitable for handling photoelastic images?

15. One can observe in the grayscale representation of the unwrapped isoclinic phasemap a faint appearance of isochromatics. What are they? Is it desirable to have them?

16. Is it necessary to calibrate a digital polariscope? If so how do you do that and in what aspect of data processing is this essential?

17. Solve the Exercise Problem 8.4 using window search method. Generate the calibration table in steps of 0.005 fringe orders from 0 to 3 using a 15th order polynomial for RGB values.

$$I = p_{15}.N^{15} + p_{14}.N^{14} + p_{13}.N^{13} + p_{12}.N^{12} + p_{11}.N^{11} + p_{10}.N^{10} + p_9.N^9 + p_8.N^8 + p_7.N^7 + p_6.N^6 + p_5.N^5 + p_4.N^4 + p_3.N^3 + p_2.N^2 + p_1.N^1 + p_0$$

The coefficients of the polynomial are summarized in the following tables.

Coeff.	p_{15}	p_{14}	p_{13}	p_{12}	p_{11}	p_{10}	p_9	p_8
R	4.51	−93.28	887.96	−5171.42	20527.28	−58092.12	118522.22	−172751.78
G	-0.89	−28.91	774.90	−7353.55	38452.61	−125523.21	268321.20	−380866.23
B	13.59	−209.17	1034.11	919.15	−34314.61	184656.03	−545892.74	1026549.17

Coeff.	p_7	p_6	p_5	p_4	p_3	p_2	p_1	p_0
R	176693.72	−125638.93	63627.58	−23347.60	4549.28	312.81	−12.85	17.03
G	357921.74	−222666.28	95243.31	−29025.50	3826.54	975.56	−33.98	19.20
B	−1274991.36	1044410.98	−550973.37	181165.69	−36906.91	4780.88	−208.94	12.79

18. The maximum and minimum R, G, B values from the calibration table of epoxy material are 169, 200, 105 and 22, 25, 18, respectively. A model made of epoxy material is subjected to stress freezing and in the process, the tint

of the material has changed. The maximum and minimum R, G, B values of the stress frozen specimen are 168, 166, 81 and 17, 20, 12, respectively. It is proposed to use the same calibration table available and one of the R, G, B entries in the calibration table are known to be 25, 48, and 53. Determine the modified R, G, B values using two-point color adaptation.

References

[1] Ramesh, K. *Digital Photoelasticity: Advanced Techniques and Applications*, 2000 (Springer-Verlag, Berlin, Heidelberg).

[2] Ramesh, K. *e-book on Experimental Stress Analysis*, 2009, (IIT Madras) http://apm.iitm.ac.in/smlab/kramesh/book_5.htm.

[3] Ramesh, K., Kasimayan, T., and Neeti Simon, B. Digital photoelasticity—A comprehensive review, *J. Strain analysis for Engng. Design*, 2011, 46(1), 245–266.

[4] Ramesh, K. *Photoelasticity*, in Sharpe, W.N. (ed.), *Springer Handbook of Experimental Solid Mechanics*, pp. 701–742, 2008, (Springer, NY).

[5] Chen, T. Y., and Taylor, C. E. Computerised fringe analysis in photomechanics, *Experimental Mechanics*, 1989, 29(3), 323–329.

[6] Ramesh, K., and Pramod, B. R. Digital image processing of fringe patterns in photomechanics, *Opt. Engineering*, 1992, 31(7), 1487–1498.

[7] Ramesh, K., and Singh R. K. Comparative performance evaluation of various fringe thinning algorithms in photomechanics. *Electronic Imaging*, 1995, 4(1), 71–83.

[8] Sharpe, W. N. (ed.) *Springer Handbook of Experimental Solid Mechanics*, 2008 (Springer, NY).

[9] Ramesh, K., and Ganapathy, V. Phase-shifting methodologies in photoelastic analysis—the application of Jones calculus. *J. Strain analysis for Engng. Design*, 1996, 31(6), 423–432.

[10] Hecker, F. W., and Morche, B. Computer-aided measurement of relative retardations in plane photoelasticity. In: Wieringa, H. (ed.) Experimental stress analysis, 1986, pp. 535–542 (Martinus Nijhoff, Dordrecht, The Netherlands).

[11] Patterson, E. A., and Wang, Z. F. Towards full field automated photoelastic analysis of complex components. *Strain*, 1991, 27(2), 49–56.

[12] Ajovalasit, A., Barone, S., and Petrucci G. A method for reducing the influence of the quarterwave plate error in phase-shifting photoelasticity. *J. Strain Analysis for Engng. Design*, 1998, 33(3), 207–216.

[13] Ramji, M., and Ramesh, K. Whole-field evaluation of stress components in digital photoelasticity—issues, implementation and application. *Optics and Lasers in Engng.*, 2008, 46(3), 257–271.

[14] Ramji, M., and Ramesh, K. Adaptive quality guided phase unwrapping algorithm for whole-field digital photoelastic parameter estimation. *Strain*, 2010, 46(2), 184–194.

[15] Brown, G. M., and Sullivan, J. L. The computer-aided holophotoelastic method. *Experimental Mechanics*, 1990, 30(2), 135–144.

[16] Ramji, M., Prasath, R. G. R., and Ramesh, K. A generic error simulation in digital photoelasticity by Jones calculus. *J. of Aerospace Sciences & technologies*, 2009, 61(4), 475–481.

[17] Quiroga, J. A. and González-Cano, A. Phase measuring algorithm for extraction of isochromatics of photoelastic fringe patterns. *Applied Optics*, 1997, 36(32), 8397–8402.

[18] Ramji, M., Gadre, V. Y., and Ramesh K. Comparative study of evaluation of primary isoclinic data by various spatial domain methods in digital photoelasticity. *J. Strain Analysis for Engng. Design*, 2006, 45(5), 333–348.

[19] Prasad, V. S., Madhu, K. R., and Ramesh, K. Towards effective phase unwrapping in digital photoelasticity. *Optics and Lasers in Engng.*, 2004, 42(4), 421–436.

[20] Siegmann, P., Backman, D., and Patterson, E. A. A robust approach to demodulating and unwrapping phase-stepped photoelastic data. *Experimental Mechanics*, 2005, 45(3), 278–289.

[21] Ramji, M., Nithila, E., Devvrath, K., and Ramesh, K. Assessment of autonomous phase unwrapping methodologies in digital photoelasticity. *Sādhanā*, 2008, 33(1), 27–44.

[22] Madhu, K. R., and Ramesh, K. New boundary information encoding for effective phase unwrapping of specimens with cut outs. *Strain*, 2007, 43(1), 54–57.

[23] Kasimayan, T., and Ramesh, K. Adaptive smoothing for isoclinic parameter evaluation in digital photoelasticity, *Strain*, 2011, e371–e375.

[24] Yao-Ting, Zhang, Huang, Min-Jui, Liang, Hua-Rong, and Lao, Fu-You. Branch cutting algorithm for unwrapping photoelastic phase map with isotropic point. *Optics and Lasers in Engineering*, 2012, 50, 619–631.

[25] Ajovalasit, A., Barone, S., and Petrucci, G. Towards RGB photoelasticity: full-field automated photoelasticity in white light. *Experimental Mechanics*, 1995, 35(3), 193–200.

[26] Ramesh, K., and Deshmukh, S. S. Automation of white light photoelasticity by phase-shifting technique using colour image processing hardware. *Optics and Lasers in Engng.*, 1997, 28(1), 47–60.

[27] Ramesh, K., and Deshmukh, S. S. Three fringe photoelasticity—use of colour image processing hardware to automate ordering of isochromatics, *Strain*, 1996, 32(3), 79–86.

[28] Madhu, K. R., and Ramesh, K. Noise removal in three fringe photoelasticity by adaptive colour difference estimation. *Optics and Lasers in Engng.*, 2007, 45(1), 175–182.

[29] Ajovalasit, A., Petrucci G., and Scafidi, M. RGB photoelasticity: Review and improvements, *Strain*, 2010, 46(2), 137–147.

[30] Kale, S., and Ramesh, K. Advancing front scanning approach for three-fringe photoelasticity, *Optics and Lasers in Engng.*, 2013, 51, 592–599.

[31] Madhu, K. R., Prasath, R. G. R., and Ramesh, K. Colour adaptation in three fringe photoelasticity. *Experimental Mechanics*, 2007, 47(2), 271–276.

[32] Neethi Simon, B., Kasimayan, T., and Ramesh, K. The influence of ambient illumination on colour adaptation in three fringe photoelasticity, *Optics and Lasers in Engng.* 2011, 49, 258–264.

[33] Kasimayan, T., and Ramesh, K. Digital reflection photoelasticity using conventional reflection polariscope. *Experimental Techniques*, 2010, 34(5) 45–51.

[34] Petrucci, G. Full-field automatic evaluation of an isoclinic parameter in white light. *Experimental Mechanics*, 1997, 37(4), 420–426.

[35] Neethi Simon, B., Prasath, R. G. R., and Ramesh, K. Transient thermal SIFs of bi-material interface cracks using RTFP. *J. of Strain Analysis for Engng. Design*, 2009, 44(6), 427–438.

[36] Ravichandran, M., and Ramesh, K. Evaluation of stress field parameters for an interface crack in a bimaterial by digital photoelasticity. *J. of Strain Analysis for Engng. Design*, 2005, 40(4), 327–343.

[37] Ramesh, K., Vivek, R., Tarkes Dora, P., and Dipayan Sanyal, A simple approach to photoelastic calibration of glass using digital photoelasticity, *Journal of non-crystalline solids*, 2013, 378, 7–14.

[38] Ajovalasit A, Petrucci G, and Scafidi M. Photoelastic Analysis of Edge Residual Stresses in Glass by Automated Test Fringes Methods, *Experimental Mechanics*, 2011, 10(7).

[39] Vivek, R., and Ramesh, K. Residual Stress Analysis of Commercial Float Glass using Digital Photoelasticity. *International Journal of Applied Glass Science*, 2014. doi: 10.11/ijag.12106.

CHAPTER 9

Digital Particle Image Velocimetry

Julio Soria

9.1 Introduction to the Principles of Digital Particle Image Velocimetry

Laser-based fluid velocity measurement techniques can be be broadly divided into point measurement techniques and field measurement techniques. An example of the former is laser Doppler anemometry (LDA) [1] and of the latter laser speckle velocimetry (LSV) [2]. The former uses photodetectors to measure the timing of a known displacement at a point whereas in the latter imaging is used to measure a displacement field that has occurred during a known time period. In this chapter we will only be concerned with a measurement technique that falls into the latter category and is known as digital particle image velocimetry (DPIV).

In general, particle image velocimetry (PIV), sometimes also referred to as particle image displacement velocimetry (PIDV), which was historically derived from LSV, can be used to estimate instantaneous three-component three-dimensional (3C-3D) velocity fields by recording the 3D location of markers in the fluid at multiple times. There is a subtle distinction between PIV and the traditional particle tracking methods that suffer from the complication of having to identify and follow individual particles in the flow field; it is in the processing of the signal in PIV, where it is not required to track individual particles because the 3C-3D velocity components in the flow are determined using digital processing of a finite 3D interrogation volume (IV) or measurement volume. It is worth elaborating on the concept of a signal used in this chapter and how it relates to PIV. A signal in its most general form is understood to be an instantaneous record of a 3D intensity field that represents the 3D locations of the markers in the fluid.

As with most optical field and point velocity measurement techniques, PIV uses as its cornerstone the fundamental definition of velocity to arrive at an estimate of the local velocity u; that is,

$$\mathbf{u}(\mathbf{x},t) = \lim_{\delta t \to 0} \frac{\delta \mathbf{x}(\mathbf{x},t)}{\delta t} \tag{9.1}$$

where $\delta \mathbf{x}$ represents the average displacement over a short time period δt of the marker particles that are located within the finite IV centered at $\mathbf{x}$ at time t. The principle behind the implementation of (9.1) in PIV is simple: the flow is seeded with small tracer particles, also referred to as marker particles, which faithfully follow the flow, the flow is illuminated by a powerful light source, typically high energy pulsed lasers, and the signal in the form of the 3D particle positions are recorded via digital imaging of the scattered light from the tracer particles in the light sheet, or volume in the case of 3D, onto a digital recording medium (i.e., one or multiple electronic photo-array detectors, typically CCD or CMOS arrays). The 3D displacement and hence, 3C velocity is then estimated from sequential 3D recordings of particle positions (i.e., the signal) within the *IV*. As a consequence of the recording stage of PIV using one or more electronic photoarray detectors and the processing of the data being carried out using digital computing techniques the method is referred to today as digital particle image velocimetry (DPIV). The roots of DPIV lie in the works of [3–5].

DPIV can be implemented to measure: two velocity components in a two-dimensional plane (2C-2D), three velocity components in a two-dimensional plane (3C-2D), and three components in a three-dimensional volume (3C-3D) with the complexity of the experimental setup and the DPIV postprocessing of the experimental data increasing, as well as the dimensional measuring capacity. The main focus of this chapter is not the experimental acquisition of the raw particle location data, (i. e., the signal), which includes 2D imaging used to acquire that data for 2C-2D DPIV [4, 6, 7], stereo imaging used to acquire the data for 3C-2D DPIV [8–10], and holographic imaging [11–14] or tomographic imaging [15–18] used to acquire the data for 3C-3D DPIV or even the reconstruction of the 3D signal in the holographic imagining or the tomographic imaging applications. Rather, this chapter develops the theoretical framework for the general 3C-3D DPIV analysis where the starting point is the signal in the form of the instantaneous 3D intensity distribution of the raw particle location data. Note that 3C-2D DPIV and 2C-2D DPIV are reduced cases of the general 3C-3D DPIV analysis.

BOX 9.1   Advanced DPIV Analysis Techniques

A number of matters such as advanced DPIV techniques that can overcome some of the limitations due to fluid mechanic phenomena such as vorticity and strain rate are not covered as they are beyond the introductory nature of this chapter. Neither are experimental aspects such as illumination, image acquisition, tracer particle selection, and so forth of DPIV, which are also not within the scope of this chapter; and the reader is referred to [19–21] for in-depth presentations for all of these aspects.

9.2   Preliminaries: Fluid Flow Kinematics

Our study of DPIV starts with some important preliminary considerations of the fluid dynamics and kinematics. It is generally understood that under the continuum assumption, the motion of a Newtonian, uniform density fluid is governed by the

Navier-Stokes and continuity equations, which in Cartesian tensor notation are given as

$$\frac{\partial u_i}{\partial t} + u_j \frac{\partial u_i}{\partial x_j} = -\frac{1}{\rho}\frac{\partial p}{\partial x_i} + v\frac{\partial^2 u_i}{\partial x_j \, \partial x_j} \tag{9.2}$$

$$\frac{\partial u_i}{\partial x_i} = 0 \tag{9.3}$$

Here u_i is the velocity in the x_i coordinate direction, ρ is the fluid density, p is the pressure, and v is the kinematic viscosity. If it is assumed that the solution of the uniform-density incompressible Navier-Stokes equations are regular at all points, then following [22] at any arbitrary point **O** in the fluid flow field a Taylor series can be used to expand the velocity u_i in terms of the space coordinates x_j, with the origin for x_j located at **O**. This gives the velocity at the point **x** in tensor notation as

$$u_i(\mathbf{x}) = u_i(\mathbf{O}) + \frac{\partial u_i}{\partial x_j}x_j + \frac{1}{2}\frac{\partial^2 u_i}{\partial x_j \, \partial x_k}x_j x_k + \ldots \tag{9.4}$$

Note that although not explicitly shown in (9.4), the velocity is time-dependent and is assumed to be so unless otherwise stated. The first term in the expansion represents the velocity at **O** and the spatial derivatives of u_i in the subsequent terms of the expansion are evaluated at **O**. The first term and the spatial derivatives of u_i are functions of time in unsteady flow. One can think of the order of the truncation of the series as representing the degree of modeling of the local velocity field in the neighborhood of the arbitrary point **O**. How well the model represents the local velocity field depends on the local flow structure. The size of the domain where the model applies also depends on the local flow structure. These two statements have immediate application in DPIV, because the size of the domain is related to the 3D size of the measurement volume, that is the *IV*. Therefore, the motion of the tracer particles as a consequence of the fluid particle kinematics inherent within an *IV* as described by (9.4) are dependent on the size of the *IV*. As clearly shown by (9.4), during a short time interval, δt, the displacement, $\delta x_i(\mathbf{x})$, of tracer particles within the *IV* centred at **O** is given by

$$\delta x_i(\mathbf{x}) = u_i(\mathbf{x})\delta t = u_i(\mathbf{O})\delta t + \delta t\frac{\delta u_i}{\delta x_j}x_j + \delta t\frac{1}{2}\frac{\delta^2 u_i}{\delta x_j \, \delta x_k}x_j x_k + \ldots \tag{9.5}$$

If the velocity is uniform within the *IV* (i.e., the velocity gradient tensor (VGT) is zero), then all points within the *IV* have the same uniform displacement and velocity at the centre of the *IV* (i.e., at **O**):

$$\delta x_i(\mathbf{x}) = u_i(\mathbf{x})\delta t = u_i(\mathbf{O})\delta t \tag{9.6}$$

No physical flows of any scientific or practical interest are described by (9.6), yet this is the working assumption of DPIV and PIV in general, which by reference to (9.5) is therefore only an approximation, and furthermore only a good approximation if

$$\frac{\left|\dfrac{\partial u_i}{\partial x_j} x_j\right|}{\left|u_i(O)\right|} \ll 1, \quad \frac{\left|\dfrac{1}{2}\dfrac{\partial^2 u_i}{\partial x_j\,\partial x_k} x_j x_k\right|}{\left|u_i(O)\right|} \ll 1, \ldots \tag{9.7}$$

In practice the leading order term dominates and thus, the first assumption given in (9.7) is the only one that must be satisfied in order for (9.6) to be a good approximation and a good working principle for DPIV. Equation (9.7) indicates that this can be achieved if the VGT components have small values and/or if the size of the IV is small as it dictates the size of x_j in (9.4), (9.5), and (9.7).

If we define the linear size of the IV by l_{IV}, then $|x_j| \leq l_{IV}/2$ and the assumption that needs to be satisfied for (9.6) to be a good working principle for DPIV is thus

$$\frac{l_{IV}\left|\dfrac{\partial u_i}{\partial x_j}\right|}{2\left|u_i(O)\right|} \ll 1 \tag{9.8}$$

In the following mathematical analysis of the 3D cross-correlation of single-exposed IV pairs, which lies at the heart of DPIV, no assumption regarding the magnitude of the VGT component will be made unless otherwise indicated, but a description of the local velocity or more appropriately the displacement field in the form of (9.5) will be assumed to be possible.

EXERCISE 9.1

For an interrogation volume with a linear dimension of 1 mm and a mean velocity dynamic range of 1–10 m/s, what is the maximum permissible velocity gradient size allowable so that the fundamental assumption in PIV is satisfied?

Solution

Rearranging (9.8) yields

$$\left|\frac{\partial u_i}{\partial x_j}\right| \ll \frac{2\left|u_i(O)\right|}{l_{IV}}$$

which for the minimum velocity of 1 m/s requires

$$\left|\frac{\partial u_i}{\partial x_j}\right| \ll \frac{2 \times 1}{10^{-3}} = 2 \times 10^3 \ 1/s$$

and for the maximum velocity of 10 m/s requires

$$\left|\frac{\partial u_i}{\partial x_j}\right| << \frac{2 \times 10}{10^{-3}} = 2 \times 10^4 \ 1/s$$

The reader should note that although our preliminary consideration of the fluid flow kinematics and the link to DPIV in this section is based on the restrictive case of a Newtonian fluid, this is not a restriction as far as the analysis presented in this chapter of DPIV and the application of DPIV to measurements in non-Newtonian fluids is concerned. Non-Newtonian fluids are governed by the Cauchy momentum equations complemented with the appropriate constitutive equations. The special case of the Newtonian fluid constitutive equations lead to the Navier-Stokes equations (9.2). Thus, the discussion and analysis described above also applies to non-Newtonian fluids.

9.3   Mathematical Analysis of 3D Cross-Correlation DPIV of Single-Exposed Interrogation Volume Pairs

We will start the mathematical description and analysis of DPIV by stating the aim of DPIV and a detailed specification of the data that is available for the DPIV analysis. The aim of DPIV is simply stated as the desire to determine instantaneously $\mathbf{u}(O)$ in an IV centered at an arbitrarily located position O in a general 3D fluid flow field and to do this at the very same instant for any number of IVs in the fluid flow field of interest where the neighboring IV centers are separated by a desired spatial separation $\Delta \mathbf{x}_s$. The data for DPIV analysis that is available are two signals, and each signal consists of a 3D intensity field that corresponds to the information of the location and size of the tracer particles located within the IV of interest extracted from within the fluid flow field of interest.

For all subsequent analysis unless otherwise stated we will assume that they follow the flow field exactly (i.e., the slip velocity between the fluid and the tracer particles is zero). Furthermore, the two signals are separated in time by a known time interval δt. The two signals can also be separated in space by a known finite distance $\Delta \mathbf{x}_O$ so that the respective centers of the IV separated in time by δt do not coincide spatially. Separation in space and time of the IV is the basis of advanced DPIV methods such as multigrid DPIV [6]. However, unless otherwise stated our analysis will assume that $\Delta \mathbf{x}_O = 0$.

9.3.1   The Signal in DPIV

As stated above the signal in DPIV analysis consists of two components that are themselves referred to as signals and are separated in time by a time interval δt. Each signal represents the location and size of tracer particles within the IV at the respective time. Assume that each tracer particle within the 3D flow field of interest can be represented by an intensity distribution with a functional form

$$I_i(\mathbf{x},t;d_i) = f\left(\mathbf{x} - \mathbf{x}_i(t);d_i\right) \tag{9.9}$$

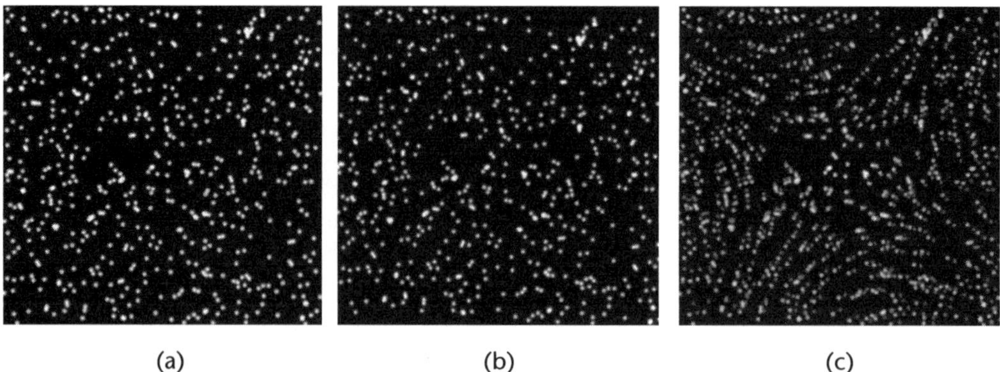

(a)   (b)   (c)

Figure 9.1   Example of a 2D tracer particle field recorded at (a) time t and (b) $t + \delta t$, and (c) the summed fields showing the fluid motion.

The subscript i in (9.9) refers to an ith arbitrary particle in the fluid flow volume of interest, whose centroid is located at time t at $\mathbf{x}_i(t)$. The particle is assumed to be spherical in nature and is characterized by its diameter d_i. The 3D functional form on the right-hand side of (9.9) applies to all particles within the fluid flow volume. This now allows us to represent the 3D intensity distribution of all N particles within the fluid flow volume of interest by

$$I(\mathbf{x},t) = \sum_{i=1}^{N} I_i(\mathbf{x},t;d_i) = \sum_{i=1}^{N} f\left(\mathbf{x} - \mathbf{x}_i(t);d_i\right) \tag{9.10}$$

Figure 9.1 illustrates these relationships for a 2D case of planar flow around a saddle point. Figure 9.1(a) is the 2D tracer particle images that represent $I(\mathbf{x}, t)$ and Figure 9.1(b) the corresponding tracer particle images $I(\mathbf{x}, t + \delta t)$ at a time δt later after they have displaced due to the fluid motion. Figure 9.1(c) shows the superposition of both $I(\mathbf{x}, t)$ and $I(\mathbf{x}, t + \delta t)$, which now also shows the fluid motion via the flow pattern depicted in this image.

The IV is assumed to be a cubic domain of linear length l_{IV}.[1] In order to arrive at the intensity distribution within the IV we need to define a sampling function that will operate on $I(\mathbf{x}, t)$ to extract the sampled signal. We start by defining the the sampling function as

$$\Pi(\mathbf{x}) = \begin{cases} 1, & |x_k| \le \dfrac{l_{IV}}{2} \\[2mm] 0, & |x_k| > \dfrac{l_{IV}}{2} \end{cases} \tag{9.11}$$

The jth signal at time t corresponding to the jth IV within the fluid flow domain of interest is thus given by extracting this IV with its center located at $\mathbf{x}_{s_m}$ using the following relationship

---

[1] Note that other domains such as a rectangle, a sphere, and so forth could also be used.

$$S_1(\mathbf{x},t) = I(\mathbf{x},t) \times \Pi\left(\mathbf{x} - \mathbf{x}_{s_m}\right) \tag{9.12}$$

If we assume that the 3D functional form describing the intensity distribution of a particle i given by (9.9) is a compact function with spatial extend much less than l_{IV}, then the effect of (9.12) is that of extracting the particle intensities that reside within the domain defined by (9.11) centered at $\mathbf{x}_{s_m}$. We will assume that those particles have been clearly identified and number N_{s_m} with $N_{s_m} \leq N$, then the signal $S_1(\mathbf{x}, t)$ can be written as

$$S_1(\mathbf{x},t) = \sum_{i=1}^{N_{s_m}} I_i\left(\mathbf{x},t;d_i\right) = \sum_{i=1}^{N_{s_m}} f\left(\mathbf{x} - \mathbf{x}_i(t);d_i\right) \tag{9.13}$$

Note that in general (9.13) is not exact and there is an error term on the right-hand side to make it an equality; only the compact nature of the functional describing the intensity distribution of the particles may result in an essential zero error term and hence the applicability of (9.13). In general this needs to be analyzed for specific cases to ascertain whether (9.13) is valid. We will assume that all necessary conditions are satisfied for this relationship to be valid.

In order to proceed with the analysis we assume that the N_{s_j} particles located within an arbitrary jth IV move due to a velocity field that is accurately described by (9.4) and furthermore, that during the time interval of interest δt: (1) the spatial velocity distribution is time-invariant so that (9.5) applies with negligible error and (2) the particles remain within the spatial domain defined by the jth IV during the time period δt. This allows us to write the mathematical description for the signal at time $t + \delta t$ as

$$S_2(\mathbf{x},t + \delta t) = \sum_{i=1}^{N_{s_m}} I_i\left(\mathbf{x},t + \delta t;d_i\right) = \sum_{i=1}^{N_{s_m}} f\left[\mathbf{x} - \left(\mathbf{x}_i(t) + \delta\mathbf{x}\left(\mathbf{x}_i(t)\right)\right);d_i\right] \tag{9.14}$$

This process is illustrated in Figure 9.2 by the example of a typical 2C-2D DPIV image, which illustrates the flow from left to right over a blunt leading edge flat plate. Figure 9.2(a) illustrates the two single-exposed images containing image records of the tracer particles superimposed to render the fluid motion. Figure 9.2(b) shows the sampling function located at an arbitrary sampling point $\mathbf{x}_{s_m}$ indicating the domain of the IV in white, which has a value of 1, and black, which has a value of 0. The multiplication of the images in Figure 9.2(a) with the sampling function shown in Figure 9.2(b) results in the two DPIV signal samples $S_1(\mathbf{x}, t)$ and $S_2(\mathbf{x}, t + \delta t)$ shown superimposed in Figure 9.2(c) and magnified in Figure 9.2(d). Note that the 3C-3D is analogous to this 2C-2D example.

Without loss of generality the origin of the coordinate system can be translated to the origin of the arbitrary jth IV denoted by $\mathbf{O}_j$ to be consisted with the fluid kinematics developed in Section 9.2. Thus, from now on unless otherwise stated, $\mathbf{x}$, $\mathbf{x}_i(t)$, $\delta\mathbf{x}(\mathbf{x}_i(t)) \equiv \delta\mathbf{x}_i(t)$ are assumed to be measured from the centroid of the jth IV, i.e., $\mathbf{O}$.

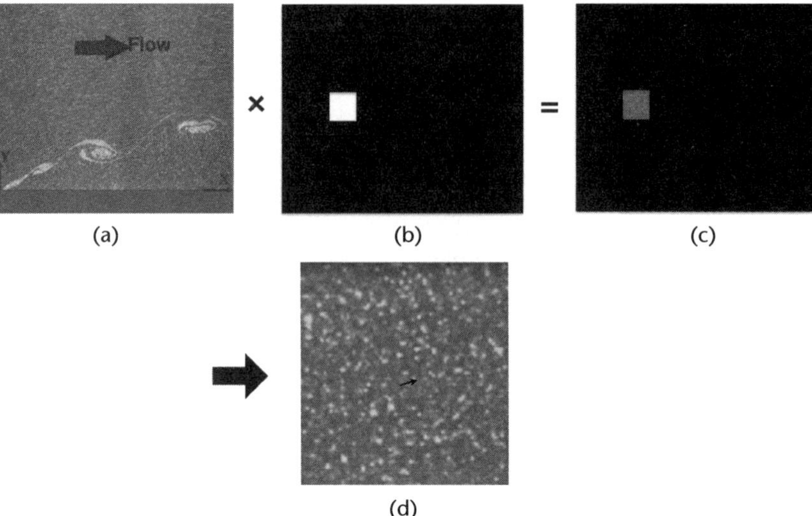

Figure 9.2 Illustration of signal sampling process in 2C-2D DPIV. (a) The superposition of particles recorded at two times of a typical flow (i.e., it represents $I(\mathbf{x}, t) + I(\mathbf{x}, t + \delta t)$), (b) represents the 2D sampling function given by (9.11) with its center located at a certain $\mathbf{x}_{s_m}$, (c) represents the process represented by (9.12) applied to both $I(\mathbf{x}, t)$ and $I(\mathbf{x}, t + \delta t)$ and superimposed here, and (d) shows an enlarged view of the resultant superimposed IV that result from this process and the expected 2C-2D velocity within this IV. Note that in practice the two signal fields are not superimposed and here this has only been done to help with the illustration.

9.3.1.1 The Digital Nature of the Signal in DPIV

We now introduce some of the practical aspects of the signals used in DPIV analysis that have been described in Section 9.3.1. We will do this via illustration in relation to 2C-2D DPIV. However, all aspects apply equally to the general 3C-3D DPIV case. In 2C-2D DPIV the signal is an intensity map that represents the light scattered by the tracer particles, and thus, it represents the projected planar image of the particles onto a recording medium. Today, the recording media in DPIV are either 2D CCD arrays or CMOS arrays. Both these electronic digital recording devices are made up of an $N \times M$ array of small finite size detectors that integrate the intensity of the light projected on it over a given time and quantises the collected electronic signal into an integer number that typically have a range of 8 bits (0, ..., 255), 10 bits (0, ..., 1023) up to 16 bits (0, ..., 65535), with most commonly available array having 8, 12, or 14 bits dynamic range.

The optical imaging scenario can be viewed as a mapping from physical space to the recording array. In the case of 2C-2D PIV a planar rectangular area of dimensions $H \times L$ maps onto the $N \times M$ array of small detectors, which are usually referred to as a *pixel* and have an effective physical light integration area of $h \times l$ (e.g., typical arrays can be 1200 × 1200 pixels with a pixel size of 7.4 μm × 7.4 μm with a 14 bit signal dynamic range).

There are a number of factors that affect the quality of the signal from each pixel, most prominent among them are the noise due to Brownian electron motion and the quantum efficiency of the detector to the wavelength of the imaging light. These differ from recording device to device and the practitioner needs to be aware of these factors and how they may affect their signal in an adverse manner.

In this section we will concentrate on the two more general crucial factors, namely spatial resolution and quantization. Spatial resolution is important because the higher it is, the more details and features of the image can be recorded. The only way to increase the spatial resolution when imaging a fixed physical area is to increase the digital $N \times M$ array size. From an optical point of view there is a limit to the smallest feature that can be optically resolved. This limit it is known as the diffraction limit of the optical system [23], which is usually represented by the diffraction-limited spot size. If the diffraction limited spot size is significantly larger than the spacing between the pixels on the recoding array, the optical imaging components (i.e., the lens), is said to have empty magnification, which means that increasing the optical magnification, by for example using a larger focal length lens at the expense of a decrease in the numerical aperture, does not improve the spatial resolution no matter how much we increase the recording array size. Most PIV imaging is diffraction-limited.

In order to illustrate the effect of spatial resolution and quantization and their connection we are going to consider the cross-sectional ideal intensity distribution of a recorded Gaussian intensity distribution, which is a good approximation for the scattered light of a spherical particle. Let the effective diameter of the the particle be d, where $d = 6\sigma$ and σ usually referred to as the standard deviation of the Gaussian function that controls its width, then the normalized intensity of a particle represented in a Cartesian coordinate system is given by

$$I(x,y) = e^{-\frac{18(x^2+y^2)}{d^2}} \tag{9.15}$$

The integrated intensity within $r = \sqrt{x^2 + y^2}/d$ is 99.7% of the total intensity. Table 9.1 shows up to the effective radius relative to the particle diameter d that can be resolved if the maximum intensity for both quantization is the same for two digitization levels and two different definitions of the effective particle diameter d. The results in this table show that if particle light scattering results in Gaussian intensity distributions, then 8-bit intensity quantization with properly matched maximum intensity to the quantization range is sufficient to record the correct size of the particle image, which contains 99.7% of the scattered light intensity and hence the particle area.

This effect is also shown in Figure 9.3, which shows the 1D normalized cross-sectional intensity distribution of particles acquired at different spatial resolution for 8-bit and 16-bit intensity quantization. The Gaussian intensity distribution is shown

Table 9.1 Particle-Size Resolution Limits for Typical Intensity Dynamic Range Digitization and Different Particle Models

	Intensity Dynamic Range Digitization	
Particle Model	*8 bit*	*12 bit*
$d = 6\sigma$, $I(x,y) = e^{-\frac{18(x^2+y^2)}{d^2}}$	$r/d \leq 0.555$	$r/d \leq 0.785$
$d = 2\sigma$, $I(x,y) = e^{-\frac{2(x^2+y^2)}{d^2}}$	$r/d \leq 1.665$	$r/d \leq 2.355$

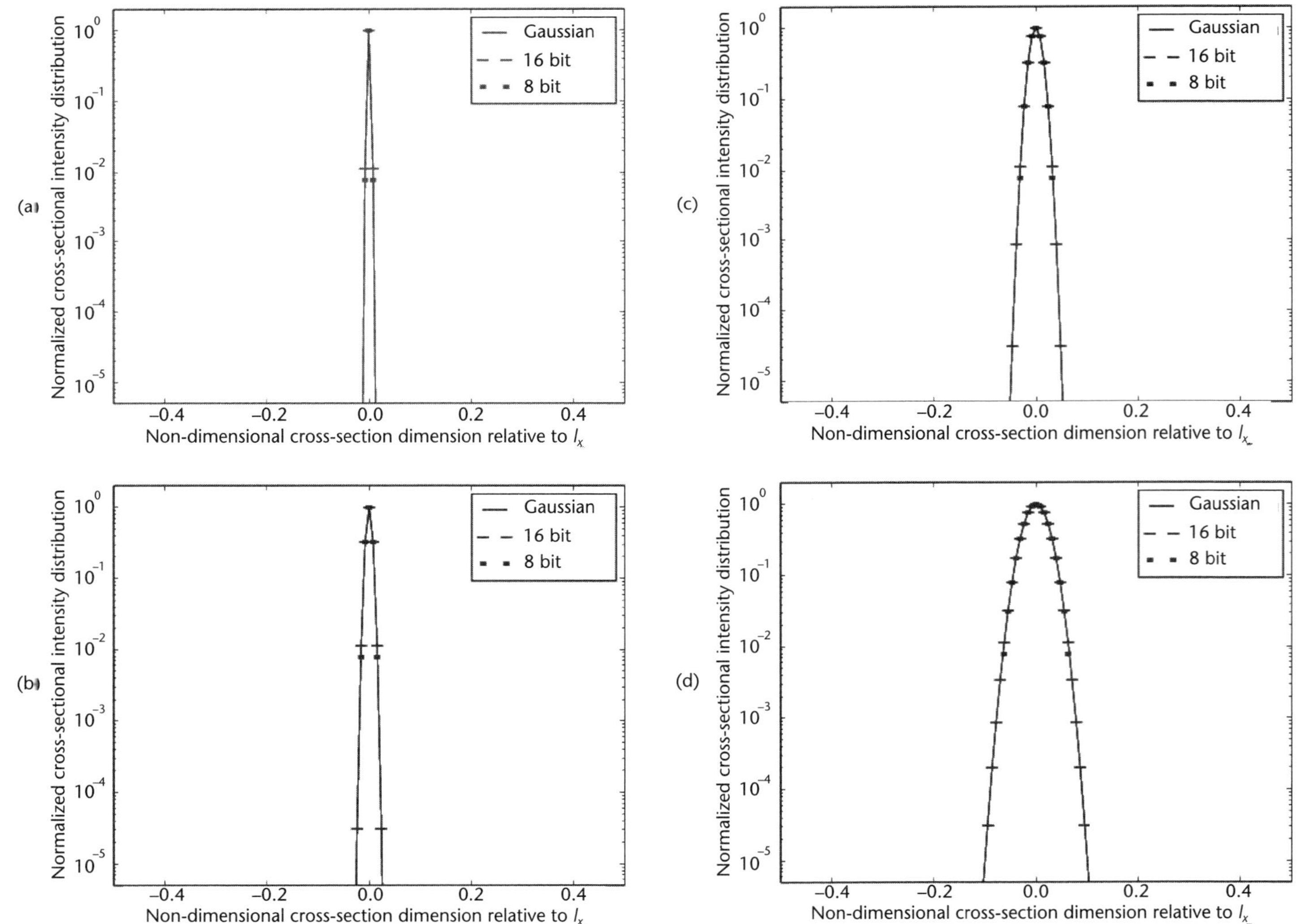

Figure 9.3 Normalized intensity cross-sectional intensity distribution for particles of different sizes and thus different spatial resolutions w th 8-bit and 16-bit intensity quantization. Note that $l_x = 64$ px.

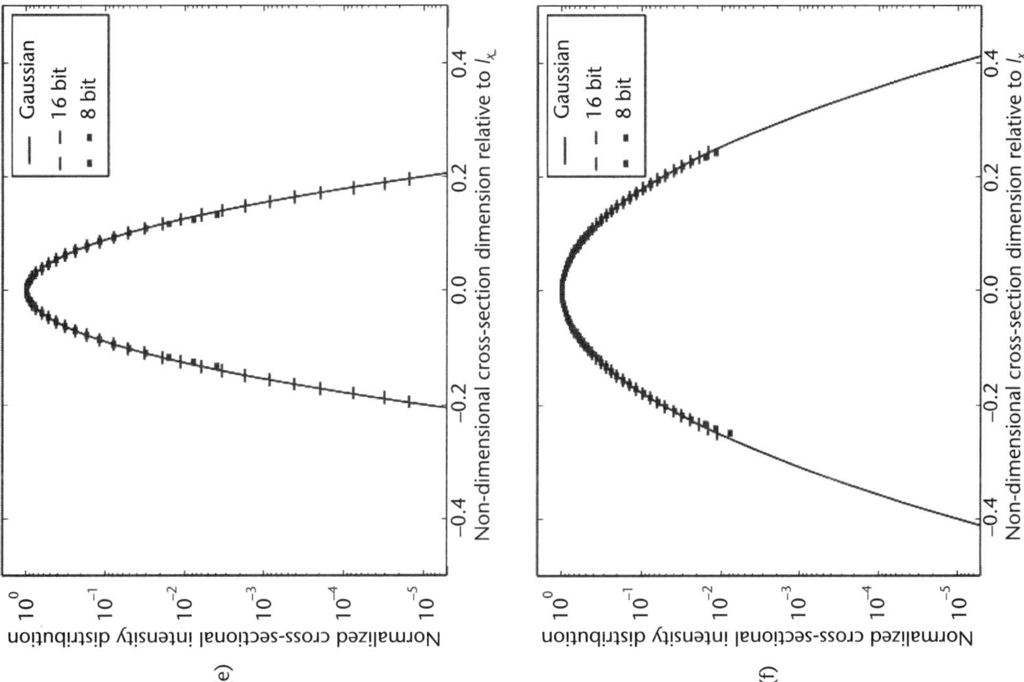

Figure 9.3   cont'd.

for comparison. All cases show that the effective diameter of the particle image is resolved with 8-bit digitization. Even though the lowest spatial resolution gives the correct digitized intensity for this ideal case, a good Gaussian function representation is only observed from Figure 9.3(c) onward. We also note that as the spatial resolution increases, 8-bit quantization compared to 16-bit quantization begins to show its limitation and inability to measure the intensity distribution to small levels.

Only for the cases in Figure 9.3(e) and (f) is the effect of lower 8-bit quantization noticeable, resulting in a measurement of a smaller particle size with the correct particle size only represented by 16-bit quantization. But these two cases represent extremely high spatial resolution that in practice are usually never achieved or used. Furthermore, the diffraction limit discussed above will most likely come into play at these high spatial resolutions. It must also be stressed that measurements of these very low levels of intensity at the edge of the particle play an insignificant part in the recording of the particle intensity in most practical DPIV measurements performed to date for the reason mentioned above.

EXERCISE 9.2

In the case of 2C-2D PIV, assume that a particle intensity distribution is represented by a 2D Gaussian distribution with an effective diameter equal to 6 standard deviations. If the sensor used for imaging has a 10-bit intensity dynamic range digitization, how much of the particle can be resolved? Assume that the entire digitization range is available to represent the particle intensity distribution in the digital image.

Solution

If $d = 6\sigma$, then the Gaussian particle intensity distribution distribution in terms of polar coordinates is given as

$$I(r) = e^{-\frac{18(r^2)}{d^2}}$$

For 10-bit digitization, the number of discrete intensity levels that can be resolved by the sensor are $2^{10} = 1024$. Without loss of generality we are assuming a maximum intensity of 1 and therefore the smallest level that can be resolved is $I_{min} = 1/1024 = 9.766 \times 10^{-4}$. The radius at which this intensity is found is given by

$$\frac{r}{d} = \sqrt{\frac{\ln\left(I_{min}\right)}{-18}} = \sqrt{\frac{\ln\left(1/1024\right)}{-18}} = 0.621$$

Given that it is possible to resolve a radial distance from the peak intensity that is $0.621d$, effectively all of the particle intensity distribution can be resolved.

9.3.2   3D Cross Correlation Analysis of a Single Exposed IV Pair

The two components that make up the single-exposed IV pair are $S_1(\mathbf{x}, t)$ and $S_2(\mathbf{x}, t + \delta t)$ given by (9.13) and (9.14), respectively. The cross-correlation function between $S_1(\mathbf{x}, t)$ and $S_2(\mathbf{x}, t + \delta t)$ is

$$R_{12}(\eta) = \int_{IV} S_1(\mathbf{x},t) S_2(\mathbf{x}+\eta, t+\delta t)\, d\mathbf{x} \tag{9.16}$$

Using the correlation theorem [24], the cross-spectrum $G_{12}(\mathbf{f}_x)$ is given as the 3D Fourier transform of $R_{12}(\eta)$

$$\begin{aligned} G_{12}(\mathbf{f}_x) &= \mathfrak{F}\big[R_{12}(\eta)\big] \\ &= \mathfrak{F}\big[S_1(\mathbf{x},t)\big]^* \mathfrak{F}\big[S_2(\mathbf{x},t+\delta t)\big] \end{aligned} \tag{9.17}$$

where $\mathfrak{F}[g]$ denotes the Fourier transform of g and $*$ denotes the complex conjugate. Conversely, the cross spectrum is related to the cross-correlation function through the inverse Fourier transform

$$\big[R_{12}(\eta)\big] = \mathfrak{F}^{-1}\big[G_{12}(\mathbf{f}_x)\big] \tag{9.18}$$

The Fourier transform pair definition used in this chapter follows the definition used by [23, 24]. We now define the Fourier transform of the functional describing the 3D intensity of particle i as

$$\hat{I}_i(\mathbf{f}_x) = \mathfrak{F}\big[I_i(\mathbf{x},t;d_i)\big] = \mathfrak{F}\big[f(\mathbf{x},t;d_i)\big] \tag{9.19}$$

which permits us to use the shift theorem [23] to write the Fourier transform of $S_1(\mathbf{x}, t)$ as

$$\begin{aligned} \mathfrak{F}\big[S_1(\mathbf{x},t)\big] &= \sum_{i=1}^{N_{s_m}} \mathfrak{F}\big[f(\mathbf{x} - \mathbf{x}_i(t); d_i)\big] \\ &= \sum_{i=1}^{N_{s_m}} \hat{I}_i(\mathbf{f}_x) \exp\big[-i2\pi\, \mathbf{f}_x \cdot \mathbf{x}_i(t)\big] \end{aligned} \tag{9.20}$$

and the corresponding Fourier transform of $S_2(\mathbf{x}, t + \delta t)$ as

$$\begin{aligned} \mathfrak{F}\big[S_2(\mathbf{x},t+\delta t)\big] &= \sum_{i=1}^{N_{s_m}} \mathfrak{F}\big[f(\mathbf{x} - (\mathbf{x}_i(t) + \delta x_i(\mathbf{x}_i(t))); d_i)\big] \\ &= \sum_{i=1}^{N_{s_m}} \hat{I}_i(\mathbf{f}_x) \exp\big[-i2\pi\, \mathbf{f}_x \cdot \{\mathbf{x}_i(t) + \delta\mathbf{x}(\mathbf{x}_i(t))\}\big] \end{aligned} \tag{9.21}$$

The cross-spectrum $G_{12}(\mathbf{f}_x)$, given by (9.17), can now be rewritten using (9.20) and (9.21) in the form

$$\begin{aligned} G_{12}(\mathbf{f}_x) = &\left(\sum_{i=1}^{N_{s_m}} \hat{I}_i^*(\mathbf{f}_x) \exp\big[i2\pi\, \mathbf{f}_x \cdot \mathbf{x}_i(t)\big] \right) \\ &\times \left(\sum_{j=1}^{N_{s_m}} \hat{I}_j(\mathbf{f}_x) \exp\big[-i2\pi\, \mathbf{f}_x \cdot \{\mathbf{x}_j(t) + \delta\mathbf{x}(\mathbf{x}_j(t))\}\big] \right). \end{aligned} \tag{9.22}$$

Expanding the product of the sums yields the following relationship for the 3D cross-spectrum $G_{12}(\mathbf{f_x})$

$$G_{12}(\mathbf{f_x}) = \sum_{i=1}^{N_{s_m}} \hat{I}_i^*(\mathbf{f_x})\hat{I}_i(\mathbf{f_x})\exp\left[-i2\pi\,\mathbf{f_x}\cdot\delta\mathbf{x}(\mathbf{x}_i(t))\right]$$

$$+ \sum_{\substack{i=1,j=1\\i\neq j}}^{N_{s_m}} \hat{I}_i^*(\mathbf{f_x})\hat{I}_j(\mathbf{f_x})\exp\left[-i2\pi\,\mathbf{f_x}\cdot\left\{\mathbf{x}_j(t)-\mathbf{x}_i(t)+\delta\mathbf{x}(\mathbf{x}_j(t))\right\}\right].$$

(9.23)

Taking the inverse Fourier transform of (9.23) yields the cross correlation function

$$R_{12}(\boldsymbol{\eta}) = \sum_{i=1}^{N_{s_m}} \mathfrak{F}^{-1}\left[\hat{I}_i^*(\mathbf{f_x})\hat{I}_i(\mathbf{f_x})\exp\left[-i2\pi\,\mathbf{f_x}\cdot\delta\mathbf{x}(\mathbf{x}_i(t))\right]\right]$$

$$+ \sum_{\substack{i=1,j=1\\i\neq j}}^{N_{s_m}} \mathfrak{F}^{-1}\left[\hat{I}_i^*(\mathbf{f_x})\hat{I}_j(\mathbf{f_x})\exp\left[-i2\pi\,\mathbf{f_x}\cdot\left\{\mathbf{x}_j(t)-\mathbf{x}_i(t)+\delta\mathbf{x}(\mathbf{x}_j(t))\right\}\right]\right]$$

(9.24)

Defining the following 3D particle intensity correlation functions

$$R_{ii}(\boldsymbol{\eta}) \equiv \mathfrak{F}^{-1}\left[\hat{I}_i^*(\mathbf{f_x})\hat{I}_i(\mathbf{f_x})\right]$$

$$R_{ij}(\boldsymbol{\eta}) \equiv \mathfrak{F}^{-1}\left[\hat{I}_i^*(\mathbf{f_x})\hat{I}_j(\mathbf{f_x})\right]$$

(9.25)

where $R_{ii}(\boldsymbol{\eta})$ is interpreted as the correlation of the ith particle intensity with itself when both are located at the same location in the IV and $R_{ij}(\boldsymbol{\eta})$ represents the correlation of the ith particle intensity with jth particle intensity when both are located at the same location in the IV, allows us to write (9.24) in a simplified form after the application of the shift theorem as

$$R_{12}(\boldsymbol{\eta}) = \sum_{i=1}^{N_{s_m}} R_{ii}\left(\boldsymbol{\eta}-\delta\mathbf{x}(\mathbf{x}_i(t))\right)$$

$$+ \sum_{\substack{i=1,j=1\\i\neq j}}^{N_{s_m}} R_{ij}\left(\boldsymbol{\eta}-\left\{\mathbf{x}_j(t)-\mathbf{x}_i(t)+\delta\mathbf{x}(\mathbf{x}_j(t))\right\}\right)$$

(9.26)

The elegant result given by (9.26) represents the general mathematical description of the digital cross-correlation function between two single-exposed IV containing N_{s_m} particles which have been displaced by virtue of a fluid flow within the IV. Note that no assumption of the spatial distribution of the displacement $\delta\mathbf{x}(\mathbf{x}_i(t))$ due to the fluid flow velocity during the time interval δt of each individual particle has been made in the derivation of (9.26), which points to the quite general nature of this result.

EXERCISE 9.3

In 2C-2D PIV the particle intensity distribution can be represented by a 2D Gaussian distribution with an effective diameter equal to 6 standard deviations. Compute $R_{ii}(\boldsymbol{\eta})$. Comparing the width of the function describing the particle intensity distribution and the function describing $R_{ii}(\boldsymbol{\eta})$, what can you conclude?

Solution

If $d = 6\sigma$ then the Gaussian particle intensity distribution in Cartesian coordinates is given as

$$I(x,y) = e^{-\frac{18(x^2+y^2)}{d^2}}$$

and its Fourier transform is given by

$$\hat{I}(\mathbf{f_x}) = \mathfrak{F}[I(x,y)] = \frac{d^2}{36}e^{\frac{-d^2}{72}\left((2\pi f_x)^2+(2\pi f_y)^2\right)}$$

Since $R_{ii}(\boldsymbol{\eta})$ is given by

$$R_{ii}(\boldsymbol{\eta}) = \mathfrak{F}^{-1}\left[\hat{I}_i^*(\mathbf{f_x})\hat{I}_i(\mathbf{f_x})\right]$$

this yields

$$R_{ii}(\boldsymbol{\eta}) = \frac{d^2}{72}e^{-\frac{9(x^2+y^2)}{d^2}}$$

This last expression can be rewritten as

$$R_{ii}(\boldsymbol{\eta}) = \frac{d^2}{72}e^{-\frac{18(x^2+y^2)}{(\sqrt{d^2})^2}}$$

which implies that the correlation function of the intensity distribution of a particle, $R_{ii}(\boldsymbol{\eta})$, has a larger diameter than the diameter of the particle intensity distribution, in this particular case it is larger by a factor of $\sqrt{2}$.

The cross correlation given by (9.26) is made up of two parts indicated by the two sums on the right-hand side of this relationship: (1) the first sum constitutes the superposition of the autocorrelation functions of each individual particle intensity, where the autocorrelation function origin (i.e., its peak location) is shifted to the position in $\boldsymbol{\eta}$ space, which corresponds to the displacement of that particular particle, $\delta\mathbf{x}(\mathbf{x}_i(t))$, and (2) the second sum is the result of the superposition of all possible cross-correlation functions of each individual particle in the first IV with each of the other particle in the second IV except itself, but in this case the peaks of each of the cross-correlation functions are located at a position in η space that is dependent on the position of the two particles involved in the cross-correlation and the displacement of the particle identified in the second IV. Therefore, the first sum, made up of the superposition of the autocorrelation functions, contributes to the average displacement of all the particles within the IV, while the second sum, made up of cross-correlation functions, contributes to noise and reduces the signal-to-noise ratio in the DPIV measurement. It will prove useful to define the first sum in (9.26) as the noiseless cross-correlation function [21, 25]

$$R_{12}^{s}(\eta) = \sum_{i=1}^{N_{s_m}} R_{ii}\left(\eta - \delta\mathbf{x}\left(\mathbf{x}_i(t)\right)\right), \tag{9.27}$$

as it contains all the information pertaining to the measurement that we are seeking. If the fluid within IV moves with constant velocity or (9.7) is satisfied, then $\delta\mathbf{x}(\mathbf{x}_i(t)) \approx \delta\mathbf{x}$, which is the same for all tracer particles. Dividing $\delta\mathbf{x}$ by the time between the two IV acquisition, δt yields the fluid velocity measurement $\mathbf{u}(\mathbf{O}, t)$ for the fluid position $\mathbf{O}$ at time t:

$$\mathbf{u}(\mathbf{O},t) \approx \frac{\delta\mathbf{x}}{\delta t}. \tag{9.28}$$

A typical example for 2C-2D digital PIV is shown in Figure 9.4. Figure 9.4(a) and (b) represent the 2D image signals $S_1(\mathbf{x}, t)$ and $S_2(\mathbf{x}, t + \delta t)$ at time t and $t + \delta t$, respectively, and Figure 9.4(c) is the corresponding cross-correlation function $R_{12}(\eta)$. In this example the very bright white region located around $(0.25, 0.07)$ in Figure 9.4(c) corresponds to the noiseless cross-correlation function $R_{12}^{s}(\eta)$ contribution to $R_{12}(\eta)$, while the fainter grey regions distributed within the cross-correlation function corresponds to the noise of the second term summation in (9.26).

With reference to this example we note that the number of tracer particles are less at time $t + \delta t$ shown in Figure 9.4(b) relative to the number of tracer particles that we started of with at time t in Figure 9.4(a). This is a consequence of certain

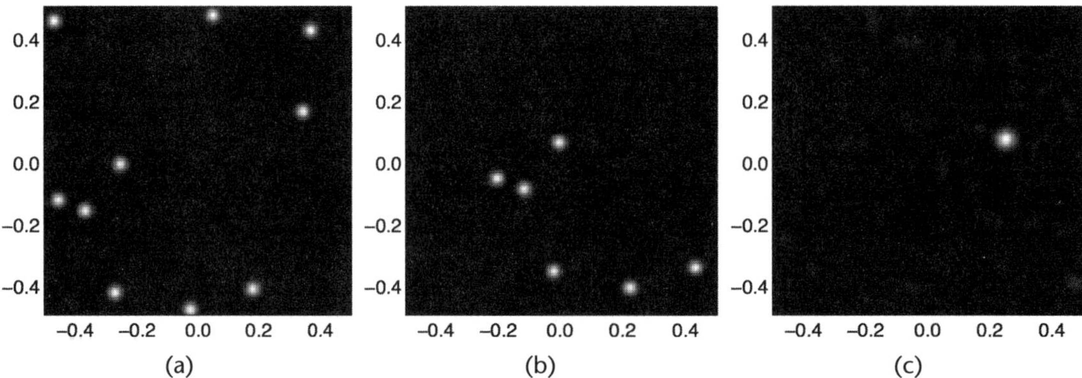

Figure 9.4 Typical 2D images of tracer particles acquired at t, (a) and $t + \delta t$, (b) and the resulting 2D cross-correlation function between the two images, (c) for a uniform displacement field $\delta\mathbf{x}_i(\mathbf{x}_i(t)) = (0.25, 0.07)$ within the IV.

tracer particles leaving the fixed IV domain for this given flow velocity and δt. This effect results in a loss of contribution to the noiseless cross-correlation function $R_{12}^s(\boldsymbol{\eta})$ and an increase in the contribution to the noise term in (9.26). Of course the converse effect of particles entering the IV between t and $t + \delta t$ is just as likely, but more likely is in fact the case that both possibilities occur in the experiment, which results in a reduction of the level of the noiseless cross-correlation function $R_{12}^s(\boldsymbol{\eta})$ to the noise term in (9.26) and hence, a reduction in the signal-to-noise level of the DPIV measurement. This can result in the incorrect detection of the noiseless cross-correlation function $R_{12}^s(\boldsymbol{\eta})$ contribution to $R_{12}(\boldsymbol{\eta})$ and hence, increases the potential for an erroneous DPIV measurement. One approach to minimize this possibility is to design the experiment so that the tracer particles do not move more than a small fraction of l_{IV}. This will be further discussed in Section 9.4.

Figure 9.5 shows the comparison between the theoretically computed cross-correlation function for signals similar to those shown in Figure 9.4 and the cross-correlation function computed directly from the signals given in a similar 2C-2D DPIV case by the two IV images shown in Figure 9.4(a) and (b). Note that the difference between the actual IV images used to produce the results in Figure 9.5 and those in Figure 9.4 is that the vertical movement is upward in Figure 9.4 and downward for Figure 9.5. The results in Figure 9.5 show that since all the details regarding the individual signals from the tracer particles are known, the theoretically computed cross-correlation function shown in Figure 9.5(a) has a higher signal-to-noise ratio than the cross-correlation function shown in Figure 9.5(b), which is computed directly from the resulting signals that are represented by image intensities similar to those as shown in Figure 9.4(a) and (b). This latter is the DPIV analysis used in practice. Mathematically this process is represented as

$$R_{12}(\boldsymbol{\eta}) = \mathfrak{F}^{-1}\left[\mathfrak{F}\left[S_1(\mathbf{x},t)\right]^* \mathfrak{F}\left[S_2(\mathbf{x},t + \delta t)\right]\right] \tag{9.29}$$

In practice the entire processing is described by (9.29), where $\mathfrak{F}[\,]$ and its inverse are carried out using the FFT to compute the discrete Fourier transform. Note that

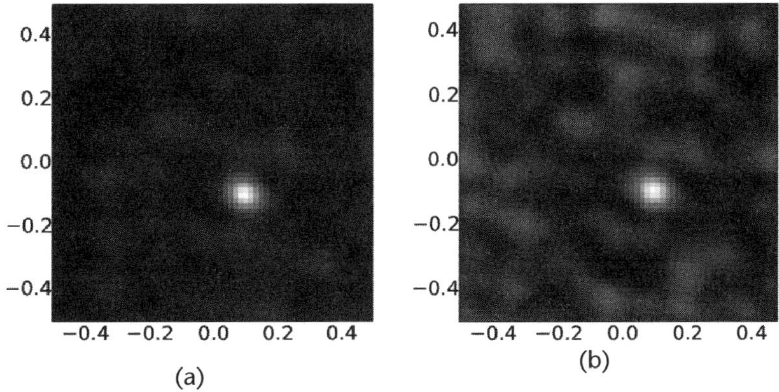

Figure 9.5   Cross-correlation functions for the example data shown in Figure 9.4 with (a) computed using the known particle information and application of (9.26), while (b) is calculated by computing the cross-correlation of the two signals given in Figure 9.4(a) and (b).

in (9.29) no specific knowledge of the individual tracer particle signals that produced $S_1(\mathbf{x}, t)$ and $S_2(\mathbf{x}, t + \delta t)$ are assumed. The reason why the signal-to-noise ratio is visibly lower in Figure 9.5(b) is as a result that some tracer particles have left the *IV* domain during the period δt of the signal acquisition and hence, there is a lack of signal pairing.

9.3.3　Determining the Expected Velocity in the IV from the Digital Cross-Correlation Function

Let us start by considering the situation that the velocity and hence, the displacement during δt, of the N_{sj} particles in the *j*th *IV* is known. Then a number of expected velocities for the *IV* can be defined, with the most straightforward being the arithmetic mean given by

$$\bar{\mathbf{u}}(\mathbf{O},t) = \frac{1}{N_s}\sum_{i=1}^{N_{s_j}}\mathbf{u}\left(\mathbf{x}_i,t\right) = \frac{1}{\delta t N_{sj}}\sum_{i=1}^{N_{s_j}}\delta\mathbf{x}\left(\mathbf{x}_i(t)\right). \tag{9.30}$$

An alternative expected velocity that can be computed is a weighted mean at the centroid of the *IV* given by

$$\tilde{\mathbf{u}}(\mathbf{O},t) = \frac{\sum_{i=1}^{N_{s_j}}W_i\mathbf{u}\left(\mathbf{x}_i,t\right)}{\sum_{i=1}^{N_{s_j}}W_i} = \frac{\sum_{i=1}^{N_{s_j}}W_i\,\delta\mathbf{x}\left(\mathbf{x}_i(t)\right)}{\delta t\sum_{i=1}^{N_{s_j}}W_i} \tag{9.31}$$

where the weights depend on the distance of the velocity measurement location, (i.e., the tracer particle location $\mathbf{x}_i$ from the origin of the *IV*, $\mathbf{O}$). A number of different approaches exist in the literature; here we adopt the well-established adaptive Gaussian window (AGW) weighting [26] where, recalling that $\mathbf{x}_i$ is measured from $\mathbf{O}$, the weights are given by

$$W_i = \exp\left(\frac{-|\mathbf{x}_i|^2}{H^2}\right). \tag{9.32}$$

The optimal H value in this scenario has been found to be of the order of $H = 1.24\Delta_P$, where Δ_P is the mean distance between particles. It should be noted that: (1) the sum of the weighting coefficients in fact add up to 1 and (2) (9.31) is also used as an interpolator and hence, the value of $\tilde{\mathbf{u}}(\mathbf{O}, t)$ can be considered to be the interpolation of the particle velocities within the *IV* at the centroid of the the *IV*.

The next step in DPIV is determining the velocity from the cross-correlation function. This being the result of the signal analysis of the two signals acquired a time separation δt apart as presented in Section 9.3.2. The starting point is the cross-correlation function between the two signals given by (9.26) or (9.29). We start by noting a number of points:

1. The space spanned by the independent variable $\boldsymbol{\eta}$ in $R_{12}(\boldsymbol{\eta})$ represents the range of the fluid particle displacements over the time period δt that are possible (i.e., $\mathbf{u}(\mathbf{O}, t)\,\delta t \in \boldsymbol{\eta}$);

2. $R_{ii}(\boldsymbol{\eta} - \delta\mathbf{x}(\mathbf{x}_i(t)))$ in (9.26) provides the location of the sample fluid particle displacement (or velocity once divided by δt) in the possible fluid displacement space given in Step 1 within the IV (i.e., the measurement volume);

3. $R_{ij}(\boldsymbol{\eta} - \{\mathbf{x}_j(t) - \mathbf{x}_i(t) + \delta\mathbf{x}(\mathbf{x}_j(t))\})$ is an unintended consequence of the formalism used to compute the cross-correlation function between the two signals $S_1(\mathbf{x}, t)$ and $S_2(\mathbf{x}, t + \delta t)$, which is considered noise, but is in fact correlated to the signals [27].

We proceed to the determination of the fluid particle displacement and hence, its velocity from the 3D cross-correlation function by using the result analytically deduced by [27] that the 3D noiseless crosscorrelation function, $R^s_{12}(\boldsymbol{\eta})$ given by (9.27), is related to the joint probability density function (JPDF) of the 3C fluid displacements (or velocity once the displacement is divided by δt). Denoting the JPDF of the 3C fluid displacements by $p_{\delta x}(\boldsymbol{\eta})$, we deduce using the results from [27] that

$$
\begin{aligned}
p_{\delta x}(\boldsymbol{\eta}) &= \lim_{N_s \to 100} \frac{R^s_{12}(\boldsymbol{\eta})}{N_s} \\
&= \lim_{N_s \to 100} \frac{\sum_{i=1}^{N_{sm}} R_{ii}\left(\boldsymbol{\eta} - \delta\mathbf{x}\left(\mathbf{x}_i(t)\right)\right)}{N_s}
\end{aligned}
\tag{9.33}
$$

Therefore formally, the mean fluid particle velocity within the IV is computed by taking the first moment with respect to each of the orthogonal velocity components of the JPDF of the fluid particle displacements and dividing the result by δt; that is

$$
\bar{u}_i(\mathbf{O},t) = \frac{1}{\delta t} \int_{-\infty}^{\infty} \eta_i p_{\delta x}(\boldsymbol{\eta})\, d\eta_i
\tag{9.34}
$$

Here the subscript is used to denote the vector component of the fluid velocity; that is, $\bar{u}_i(\mathbf{O},t)$ denotes the fluid velocity in the x_i-direction at the center of the IV denoted by the position vector $\mathbf{O}$, and $\eta_i = \delta x_i$ is the fluid particle displacement within the IV. It is worth noting that higher-order moments of the fluid velocity can be derived in a similar fashion as described and discussed in [27].

Following on from the relationship between the JPDF of the 3C fluid displacements and the 3D cross-correlation function, alternatively, instead of the mean velocity within the IV we may wish to measure the most likely fluid velocity within the IV. This is given by the location of the maximum of $R^s_{12}(\boldsymbol{\eta})$; that is,

$$
\hat{\mathbf{u}}(\mathbf{O},t) = \frac{\arg\max_{\eta} R^s_{12}(\boldsymbol{\eta})}{\delta t} = \frac{\arg\max_{\eta}\left(\sum_{i=1}^{N_{sm}} R_{ii}\left(\boldsymbol{\eta} - \delta\mathbf{x}\left(\mathbf{x}_i(t)\right)\right)\right)}{\delta t}
\tag{9.35}
$$

Typically the signal of the tracer particle can be modeled quite well as being represented by a Gaussian function, which results in $R_{ii}(\boldsymbol{\eta})$ functions that are a Gaussian function and in this case it can be shown that

$$
\bar{u}_i(\mathbf{O},t) = \hat{\mathbf{u}}(\mathbf{O},t)
\tag{9.36}
$$

which means that either method given by (9.34) or (9.36) can be used to deduce the fluid velocity within the IV provided the noiseless cross-correlation function is available. Other models for the particle signal such as the Dirac delta function yields the same conclusion. However, we must now step back and accept reality and the fact that in an experiment we have the two signals $S_1(\mathbf{x}, t)$ and $S_2(\mathbf{x}, t + \delta t)$ and no a priori detailed knowledge about the tracer particles and their motion that result in these two signals. The cross-correlation analysis of the two signals $S_1(\mathbf{x}, t)$ and $S_2(\mathbf{x}, t + \delta t)$ yields the cross-correlation function $R_{12}(\boldsymbol{\eta})$, which as (9.26) shows not only contains the noiseless $R_{12}^s(\boldsymbol{\eta})$ but also a correlated noise term, which we will denote by $R_{12}^n(\boldsymbol{\eta})$ and is given by

$$R_{12}^n(\boldsymbol{\eta}) = \sum_{\substack{i=1,j=1 \\ i \neq j}}^{N_{s_m}} R_{ij}\left(\boldsymbol{\eta} - \left\{\mathbf{x}_j(t) - \mathbf{x}_i(t) + \delta\mathbf{x}\left(\mathbf{x}_j(t)\right)\right\}\right) \tag{9.37}$$

such that

$$R_{12}(\boldsymbol{\eta}) = R_{12}^s(\boldsymbol{\eta}) + R_{12}^n(\boldsymbol{\eta}) \tag{9.38}$$

$R_{12}^n(\boldsymbol{\eta})$ introduces errors in either the determination of the velocity by taking moments of $R_{12}(\boldsymbol{\eta})$ or by finding the maximum likelihood velocity estimate from the location of the maximum of $R_{12}(\boldsymbol{\eta})$ in the fluid displacement plane $\boldsymbol{\eta}$ using the procedures given by (9.34) and (9.35).

EXERCISE 9.4

One measure to determine the signal-to-noise ratio of the crosscorrelation function, $R_{12}(\boldsymbol{\eta})$, is to determine the ratio of the maximum peak representing the most likely velocity of the particles with the IV to the second largest peak that is due to the displacement of uncorrelated particles. This ratio is also known as the peak-to-peak ratio (PPR). Show that this ratio is less than or equal to the number of particles found within the IV.

Solution

As mentioned previously particle intensity distributions can be well represented by a Gaussian function, which has a compact nature and a spatial extend given by the effective diameter d. The cross-correlation function, $R_{ii}(\boldsymbol{\eta})$, of this functional description of the particle intensity distribution also has a compact nature, as it is also a Gaussian function, which is $\sqrt{2}$ larger than the particle intensity distribution function. If the particle motion satisfies the uniform motion assumption within the IV, then the maximum within the IV, assuming that there are M monodispersed particles within the IV, is given by

$$R_{12_{max}} = \underset{\boldsymbol{\eta}}{\operatorname{argmax}}\left(\sum_{i=1}^{M} R_{ii}\left(\boldsymbol{\eta} - \delta\mathbf{x}\left(\mathbf{x}_i(t)\right)\right)\right) = M\,\underset{\boldsymbol{\eta}}{\operatorname{argmax}}\left(R_{ii}(\boldsymbol{\eta})\right).$$

The second largest maximum in $R_{12}(\eta)$ is related to the second summation term in (9.26). Each of the terms in this summation contributes potentially to the second largest maximum in $R_{12}(\eta)$. However, the location of the particles within the IV have a uniform probability and because of this fact and the compact nature of $R_{ij}(\eta)$, which for monodispersed particles is equal to the function $R_{ii}(\eta)$, this implies that the second largest maximum is given by

$$R_{12_{max_2}} = \underset{\eta}{\mathrm{argmax}}\left(R_{ii}(\eta)\right).$$

Thus, the ratio of the ratio of the maximum in $R_{12}(\eta)$ to the second largest maximum is given as

$$\mathrm{PPR} = \frac{R_{12_{max}}}{R_{12_{max_2}}} = \frac{M\,\mathrm{argmax}_\eta\left(R_{ii}(\eta)\right)}{\mathrm{argmax}_\eta\left(R_{ii}(\eta)\right)} = M$$

Due to noise and overlapping contributions of uncorrelated particles the denominator in the above equation may be larger and therefore in general

$$\mathrm{PPR} \le \mathrm{M};$$

that is, PPR is less than or equal to the number of particles within the IV.

In practice a two-step analysis process is used, whereby once $R_{12}(\eta)$ has been computed:

1. An estimate of the fluid particle displacement within the IV is found using either (9.34) or (9.35);
2. The estimate of Step 1 is then used to apply an assumed model for the functional description of the local region of the η space in the neighorhood around the location of the maximum of $R_{12}(\eta)$ to improve the precision of finding this location.

The precision to which the location of the maximum of $R_{12}(\eta)$ can be computed using the assumed model, remembering that this location represents the most likely velocity within the IV, is highly dependent on the quality of the model that is used to represent the shape of the $R_{12}(\eta)$ maximum. Typically a Gaussian function model is assumed for reasons already discussed. This model is used in a least-square sense and allows the fluid displacement and hence, the fluid velocity to be determined to better than the discrete spatial resolution and therefore the term subpixel resolution is used for this improvement in precision [6].

The computation of the location of the maximum of the crosscorrelation function is illustrated for a 2C-2D DPIV case in Figure 9.6, representing a uniform flow model case with $\delta \mathbf{x}(\mathbf{O}, t) = (3.2, -8.3)$. The signal images separated by the time interval δt are shown in Figure 9.6(a) and (b). Figure 9.6(c) is the noiseless cross-correlation function computed directly from the known particle velocity and its characteristics using (9.27), which using the method of finding the location of the maximum of the cross-correlation function just described yields an estimate for $\overline{\delta \mathbf{x}}(O,t) = (3.193, -8.320)$. Figure 9.6(d) shows the theoretically computed cross-correlation function computed directly from the known particle velocity and its characteristics using (9.26) and its maximum location is computed to yield an estimate for $\overline{\delta \mathbf{x}}(O,t) = (3.193, -8.318)$. Figure 9.6(d) shows the result of the method that is used in practice to compute $R_{12}(\boldsymbol{\eta})$ using (9.29) from the signals shown in Figure 9.6(a) and (b). Its maximum location is computed to yield an estimate for $\overline{\delta \mathbf{x}}(O,t) = (3.193, -8.319)$.

The relative error of these results based on this ideal model problem is approximately 0.22%, which is well within the experimental uncertainty of cross-correlation DPIV [6]. No statistically significant difference is found here between the theoretical measurements of the displacements and that computed from the signal information. This very low uncertainty using ideal signal information that was computer-generated (see Figure 9.6(a) and (b)), suggests that if careful DPIV experiments are undertaken, quite precise DPIV measurements can be achieved with low experimental uncertainty.

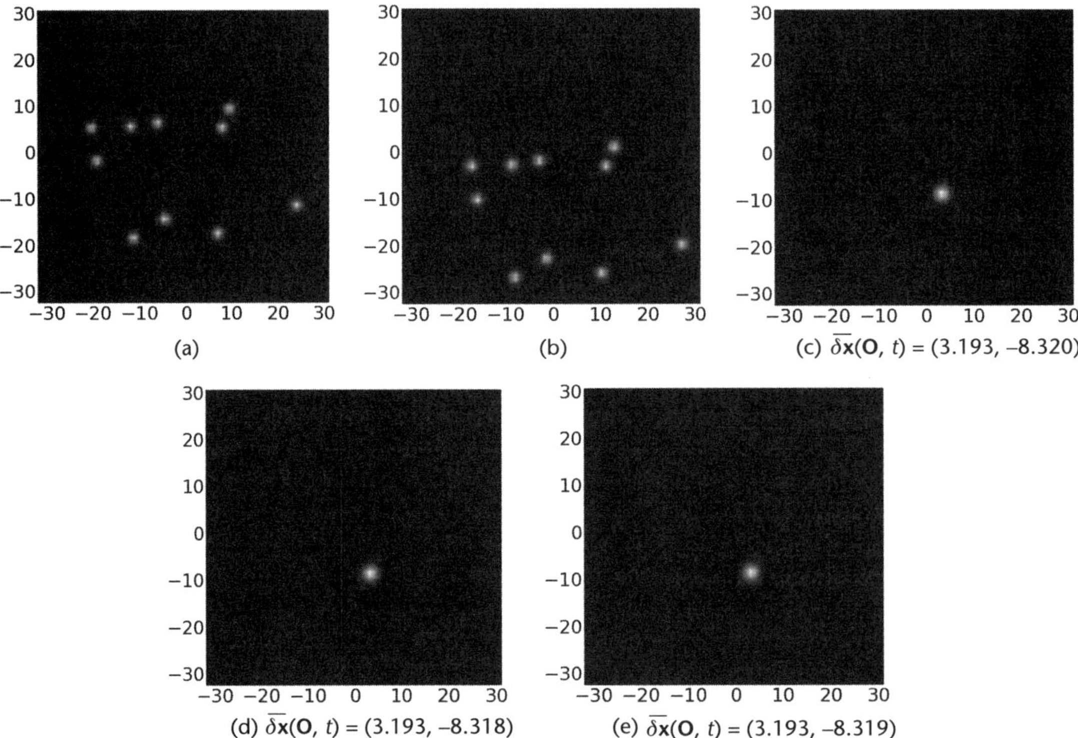

Figure 9.6 Example of 2C-2D DPIV images shown in (a) at time t, (b) at time $t + \delta t$ after a displacement of $\delta \mathbf{x}(O, t) = (3.2, -8.3)$ with particle whose diameter $d = 5$ px. Part (c) shows the noiseless crosscorrelation function, $R^s_{12}(\boldsymbol{\eta})$, (d) shows the theoretically computed cross-correlation function from the knowledge of the particle positions and their characteristics, $R_{12}(\boldsymbol{\eta})$, and (e) shows the cross-correlation function computed from the signal information shown in (a) and (b).

BOX 9.2   Experimental Imaging Variations of DPIV

There are a range of different experimental imaging variations that are used depending on the measurement sought. Two-dimensional imaging is used to acquire the data for 2C-2D DPIV [4, 6, 7], stereo imaging is used to acquire the data for 3C-2D DPIV [8–10], and holographic imaging [11–14] or tomographic imaging [15–18] is used to acquire the data for 3C-3D DPIV.

9.4   Increasing Velocity Dynamic Range via Colocated Different Size IV Cross-Correlation DPIV

An important characteristic of any measurement instrument is its measurement range, because we are concerned with measuring lengths, and in this case we want to specify the ratio of the maximum length and the minimum length that we can determine from the cross-correlation function using the methods described in Section 9.3.3. So let us start by defining the velocity dynamic range (VDR), which is equal to the displacement dynamic range (DDR), since in DPIV we measure the displacement within the IV as discussed in Section 9.2, as

$$\mathrm{VDR} \equiv \frac{|\mathbf{u}|_{max}}{|\mathbf{u}|_{min}} = \frac{|\delta\mathbf{x}|_{max}}{|\delta\mathbf{x}|_{min}} = \mathrm{DDR} \tag{9.39}$$

In the case where the interrogation volumes that are collocated, the theoretical $|\delta\mathbf{x}|_{max} = l_x/2$ where l_x is the characteristic length of the IV. However, in practice $|\delta\mathbf{x}|_{max} = l_x/4$ to minimize loss of particle pairs due to particles moving out of the IV as a consequence of a δt, which allows excessive particle displacements [28]. This can be generalized to a condition which states that

$$|\delta\mathbf{x}|_{max} = Kl_x, \text{ where typically } K \in (0.2, 0.35) \tag{9.40}$$

$|\delta\mathbf{x}|_{min}$, the smallest displacement that can be found depends on the quality of the data and the algorithm used to determine the velocity/ displacement from the cross-correlation function. For computer-generated particle images using Monte Carlo simulation, [6] found that the minimum resolvable DPIV displacement using the method described in Section 9.3.3 is 0.1 pixels (px)[2] with an uncertainty of 0.06 px at the 95% confidence level. Other studies using a similar methodology are available in the literature but for the the purpose of our discussion this minimum resolvable DPIV displacement serves its purpose and will be assumed to be the minimum resolvable DPIV displacement in px. All subsequent discussion will be in terms of this nondimensional length scale referred to as a px (e.g., l_x is given in terms of px). Therefore, we can state that

---

[2] A pixel is a nondimensional number associated to be equivalent to one discrete detector cell of the imaging sensor. It can converted into a physical size if the size of the cell is given and by taking any optical magnification into account.

$$\mathrm{VDR} = \mathrm{DDR} = 10\,Kl_x. \tag{9.41}$$

To illustrate this for 2C-2D DPIV using $K = 0.25$ with typical IV sizes of $l_x = 16$ px, 32 px, 64 px results in VDR = 40, 80, 160, respectively. The first two are very low dynamic ranges for the measurements of the fluid velocity field, particularly if it is a turbulent velocity field that we are interested in measuring. The obvious method to increase the VDR is to increase l_x (i.e., increase the IV). However, this has the potential to bring in undesirable effects due to the velocity gradients [21] and increases the averaging or filtering effect due to the enlarged measurement volume.

To this end in order to increase the VDR yet keep the good spatial resolution by keeping the IV small, advanced DPIV analysis methods have been developed that use adaptive interrogation volume and multigrid approaches [6]. The analysis of these methods is beyond the scope of this chapter and will not be presented here, but the interested reader can refer to [6] and subsequent works on this topic.

However, what we will consider here is the analysis of unequal sized IV^3 between the two sampling times and how the VDR is thereby increased while keeping the effective size of the sampling volume small. The modification to the analysis that will be presented also avoids the loss of useful signal in the noiseless cross-correlation function due to loss of image pair information as was illustrated in Figure 9.4.

We start our analysis with two IV such that

$$IV(t) < IV(t + \delta t), \tag{9.42}$$

where $IV(t + \delta t)$ sympolises the signal from all the tracer particles that contributed to the signal in the $IV(t)$. Denoting the number of individual tracer particle contained within $IV(t)$ as N_{s_m} as in Section 9.3.1, and the number of all the tracer particles that contribute to the signal from $IV(t + \delta t)$ as $N_{s_2} \equiv N_{s_m} + N_{s_n}$, where N_{s_n} are the number of tracer particles in $IV(t + \delta t)$ that are unpaired with the tracer particles in $IV(t)$. Symbolically the two signals $S_1(\mathbf{x},\ t)$ and $S_2(\mathbf{x},\ t + \delta t)$ are related to the two sampling volumes $IV(t)$ and $IV(t + \delta t)$, respectively, and can be mathematically expressed as

$$
\begin{aligned}
S_1(\mathbf{x},t) \quad &= \sum_{i=1}^{N_{s_m}} I_i\left(\mathbf{x},t;d_i\right) \\
&= \sum_{i=1}^{N_{s_m}} f\left(\mathbf{x} - \mathbf{x}_i(t);d_i\right) \\
S_2(\mathbf{x},t + \delta t) &= \sum_{i=1}^{N_{s_2}=N_{s_m}+N_{s_n}} I_i\left(\mathbf{x},t + \delta t;d_i\right) \\
&= \sum_{i=1}^{N_{s_2}=N_{s_m}+N_{s_n}} f\left[\mathbf{x} - \left(\mathbf{x}_i(t) + \delta\mathbf{x}\left(\mathbf{x}_i(t)\right)\right);d_i\right]
\end{aligned}
\tag{9.43}
$$

---

[3] Note that IV is also used here to denote in symbolic form the size of the interrogation.

Following the analysis presented in Section 9.3.2 but replacing the signals in (9.16) with the signals given by (9.43) yields the cross-correlation function

$$R_{12}(\eta) = \sum_{i=1}^{N_{s_m}} R_{ii}\big(\eta - \delta\mathbf{x}(\mathbf{x}_i(t))\big)$$

$$+ \sum_{i=1}^{N_{s_m}} \sum_{\substack{j=1, \\ j \neq 1, j \in N_{s_m}}}^{N_{s_m}+N_{s_n}} R_{ij}\Big(\eta - \big\{\mathbf{x}_j(t) - \mathbf{x}_i(t) + \delta\mathbf{x}(\mathbf{x}_j(t))\big\}\Big). \tag{9.44}$$

Letting l_{x_1} be the characteristic length of the $IV(t)$ and l_{x_2} be the characteristic length of the $IV(t + \delta t)$, with $l_{x_1} < l_{x_2}$, this cross-correlation function now allows the measurement of a maximum displacement that theoretically is $|\delta\mathbf{x}|_{\max} = l_{x_2}/2 > l_{x_1}/2$ and therefore results in a VDR that is increased to

$$\text{VDR} = \text{DDR} = 10\,Kl_{x_2} > 10\,Kl_{x_1}. \tag{9.45}$$

Yet the sampling volume size corresponding to this increased VDR is the smaller of $IV(t)$ (i.e., l_{x_1}).

The noiseless cross-correlation function that is still given by (9.27) can be recognized as the first summation term in (9.44), which contains the information pertaining to the velocity of the fluid within $IV(t)$. The second summation term in (9.44) is a noise term analogous to the second summation term given previously in (9.26), but with possibly increased noise contribution. The calculation of the velocity from the cross-correlation function proceeds exactly as before and follows the analysis presented in Section 9.3.3.

This method of increasing the VDR via colocated different-sized IV cross-correlation DPIV is illustrated by two 2C-2D examples with particle signals that correspond to $d = 2$ px and $d = 5$ px in Figure 9.7 and Figure 9.8 respectively. In both examples uniform flow is assumed with $\delta\mathbf{x}(\mathbf{O}, t) = (19.84, -10.56)$. This displacement required a 2D IV of $l_x = 64$ px as a minimum if equal-sized $IV(t)$ are used because the theoretical maximum displacement that can be measured with a 2D IV of $l_x = 32$ px is $|\delta x_i(\mathbf{O}, t)|_{\max} = 0.5\,l_x = 16$ px. However, as already pointed out in practice due to loss of particle pairs in the IV, this maximum is typically $|\delta x_i(\mathbf{O}, t)|_{\max} = 0.2 l_x = 6.4$ px. This maximum displacement that can be measured with a 2D IV of $l_x = 32$ px is certainly smaller than the uniform displacement of $\delta\mathbf{x}(\mathbf{O}, t) = (19.84, -10.56)$ that is to be measured and illustrated in the examples of Figures 9.7 and 9.8.

The unequal-sized IV cross-correlation DPIV of the signal data shown in Figure 9.7(a) and (c) and Figure 9.8(a) and (c) correspond to IV, which are centred at the origin of the domain shown and has a size of 64 px. The $IV(t)$ has a size $l_{x_1} = 32$ px, indicated by the red square in Figure 9.7(a) and (b) and Figure 9.8(a) and (b). Figure 9.7(b) and Figure 9.8(b) show only the particle signals that contribute to the signal $S_1(\mathbf{x}, t)$ in the cross-correlation analysis, while Figure 9.7(c) and Figure 9.8(c) show the particle signals which contribute to the signal $S_2(\mathbf{x}, t + \delta t)$ in the cross-correlation analysis and correspond to $IV(t + \delta t)$ with $l_{x_2} = 64$ px. Figure

9.7(d) and Figure 9.8(d) show the corresponding noiseless cross-correlation function, $R^s_{12}(\boldsymbol{\eta})$, which indicate as expected that the cross-correlation function is narrower for the $d = 2$ px case in the former than the $d = 5$ px case in the latter. The error in this single example, which must be noted is not statistically significant, is slightly larger for the $d = 5$ px case shown in Figure 9.8(d), but is still less than 0.6% of the measurement. Using this error as the smallest resolvable displacement, the VDR for this example based on $l_{x_2} = 64$ px but with an effective IV of $l_{x_1} = 32$ px and the noiseless cross-correlation function is of the order of VDR = 170 with a theoretical limit of VDR = 246. This shows that the theoretical VDR has been increased by a factor of 3 using unequal IV DPIV for this example.

The theoretically computed cross-correlation function based on the knowledge of the particle positions and their characteristics is given by $R_{12}(\boldsymbol{\eta})$ and shown in Figure 9.7(e) and Figure 9.8(e). The cross-correlation function computed from the signal samples (b) and (c) of Figure 9.7 and Figure 9.8 are shown in Figures 9.7(f) and 9.8(f), respectively. Both clearly show the noise introduced due to the unpaired correlations. Nevertheless, for these particular idealized examples the error in either case is of the same order as the error in the measurement using the noiseless

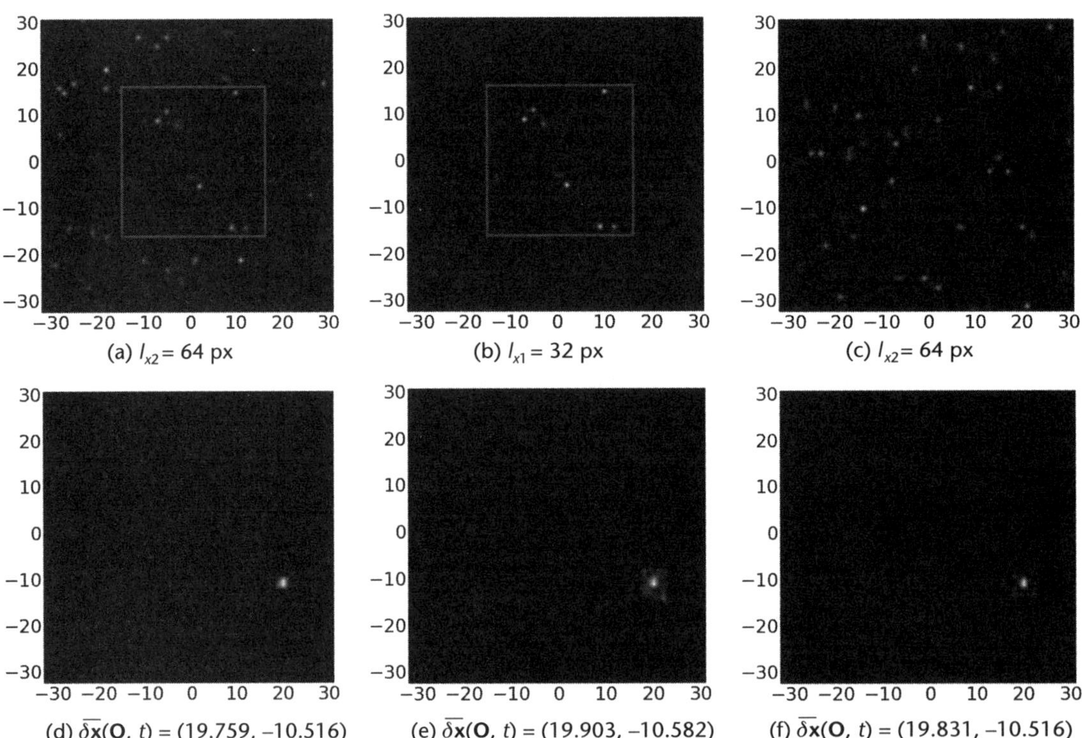

Figure 9.7 Example of 2C-2D DPIV images shown in (a) and (b) at time t, (c) at time $t + \delta t$ after a displacement of $\delta\mathbf{x}(\mathbf{O}, t)$ = (19.84, -10.56) with particle whose diameter $d = 2$ px. The $IV(t)$, which has an l_{x_1} = 32 px, is shown by the red square in (a) and extracted in (b). Part (d) shows the noiseless cross-correlation function, $R^s_{12}(\boldsymbol{\eta})$, (e) shows the theoretically computed cross-correlation function from the knowledge of the particle positions and their characteristics, $R_{12}(\boldsymbol{\eta})$, and (f) shows the cross-correlation function computed from the signal information shown in (b) and (c).

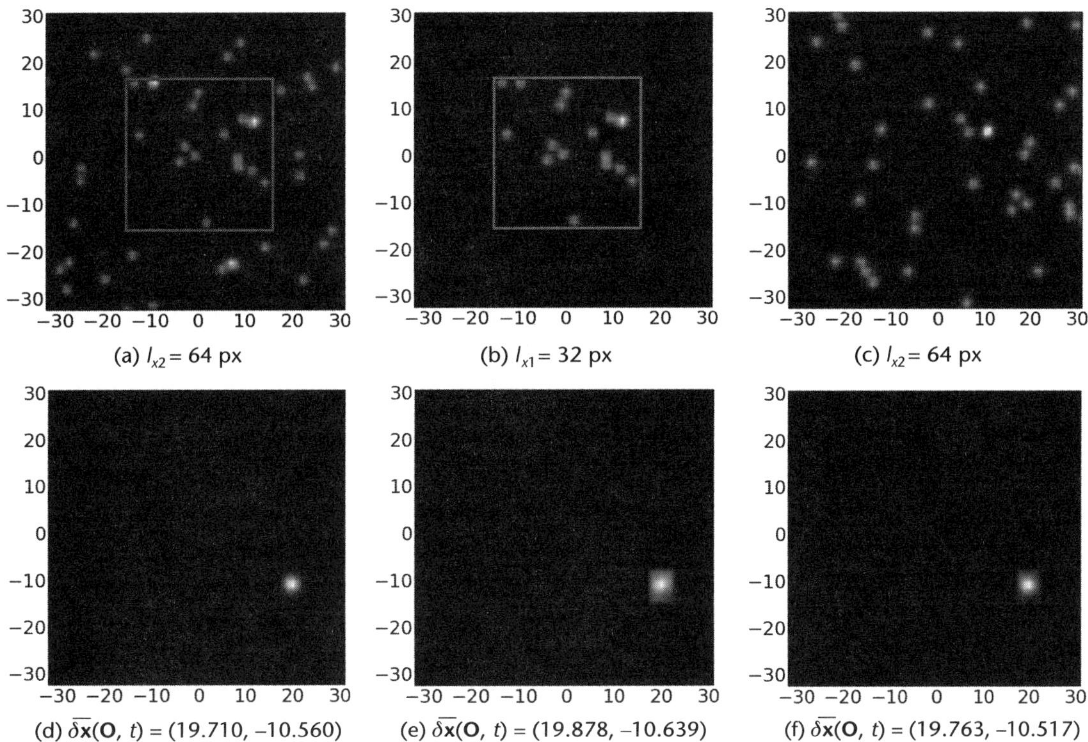

(a) $I_{x2} = 64$ px

(b) $I_{x1} = 32$ px

(c) $I_{x2} = 64$ px

(d) $\overline{\delta\mathbf{x}}(\mathbf{O}, t) = (19.710, -10.560)$

(e) $\overline{\delta\mathbf{x}}(\mathbf{O}, t) = (19.878, -10.639)$

(f) $\overline{\delta\mathbf{x}}(\mathbf{O}, t) = (19.763, -10.517)$

Figure 9.8 Example of 2C-2D DPIV images shown in (a) and (b) at time t, (c) at time $t + \delta t$ after a displacement of $\delta\mathbf{x}(\mathbf{O}, t) = (19.84, -10.56)$ with particle whose diameter $d = 5$ px. The $IV(t)$, which has an $I_{x_1} = 32$ px, is shown by the red square in (a) and extracted in (b). Part (d) shows the noiseless cross-correlation function, $R^s_{12}(\boldsymbol{\eta})$, (e) shows the theoretically computed cross-correlation function from the knowledge of the particle positions and their characteristics, $R_{12}(\boldsymbol{\eta})$, and (f) shows the cross-correlation function computed from the signal information shown in (b) and (c).

cross-correlation function and within the uncertainty of cross-correlation DPIV previously established [6]. However, the important benefit of the present approach is the increased VDR for a smaller IV size.

BOX 9.3   Increasing the Velocity Dynamic Range and Spatial Resolution of DPIV

There are a range of advanced techniques that have been developed over the last 20 years to increase the velocity dynamic range and the spatial resolution of DPIV. The most significant ones are the multigrid DPIV technique [6, 29, 30] that increase the velocity dynamic range and spatial resolution while reducing the experimental uncertainty. The particle image distortion approach to overcome the adverse effect of velocity gradients within the IV [31, 32] and super-resolution PIV that combines DPIV as a preprocessor with particle tracking velocimetry (PTV) [33, 34] to measure the velocity of tracer particles in high tracer particle seeded flows. Many of these techniques are found today in both open-source and commercial PIV software analysis computer programs.

References

[1] Durst, F., Melling, A., and Whitelaw, J. H., *Principles and Practice of Laser-Doppler Anemometry*, London: Academic Press, 1981.

[2] Dudderar, T. D., Meynart, R., and Simpkins, P. G., "Full-Field Laser Metrology for Fluid Velocity Measurement," *Optics and Lasers in Engineering*, Vol. 9, 1988, pp. 163–199.

[3] Cho, Y.- C., "Digital Image Velocimetry," *Applied Optics*, Vol. 28, No. 4, 1989, pp. 740–748.

[4] Willert, C. E., and Gharib, M., "Digital Particle Image Velocimetry," *Experiments in Fluids*, Vol. 10, 1991, pp. 181–193.

[5] Keane, R. D., and Adrian, R. J., "Theory of Cross-Correlation Analysis of PIV Images," *Applied Scientific Research*, Vol. 49, 1992, pp. 191–215.

[6] Soria, J., "An Investigation of the Near Wake of a Circular Cylinder Using a Video-Based Digital Cross-Correlation Particle Image Velocimetry Technique," *Experimental Thermal and Fluid Science*, Vol. 12, 1996, pp. 221–233.

[7] Silva, C. M., Gnanamanickam, E. P., Atkinson, C., Buchmann, N. A., Hutchins, N., Soria, J., and Marusic, I., "High Spatial Range Velocity Measurements in a High Reynolds Number Turbulent Boundary Layer," *Physics of Fluids*, Vol. 26, No. 2, 2014, p. 025117.

[8] Willert, C., "Stereoscopic Digital Particle Image Velocimetry for Application in Wind Tunnel Flows," *Measurement Science & Technology*, Vol. 8, No. 12, 1997, pp. 1465–1479.

[9] Parker, K., von Ellenrieder, K. D., and Soria, J., "Using stereo multigrid DPIV (SMDPIV) measurements to investigate the vertical skeleton behind a finite-span flapping wing," *Experiments in Fluids*, Vol. 39, No. 2, 2005, pp. 281–298.

[10] Herpin, S., Wong, C., Stanislas, M., and Soria, J., "Stereoscopic PIV Measurements Of A Turbulent Boundary Layer with a Large Spatial Dynamic Range," *Experiments in Fluids*, Vol. 45, 2008, pp. 745–763.

[11] Barnhart, D. H., Adrian, R. J., and Papen, G. C., "Phase-Conjugate Holographic System for High Resolution Particle Image Velocimetry," Applied Optics, Vol. 33, No. 30, 1994, pp. 7159–7170.

[12] Lozano, A., Kostas, J. and Soria, J., "Use of Holography in Particle Image Velocimetry Measurements of a Swirling Flow," *Experiments in Fluids*, Vol. 27, 1999, pp. 251–261.

[13] Palero, V., Arroyo, M. P., and Soria, J., "Digital Holography for Micro-Droplet Diagnostics," *Experiments in Fluids*, Vol. 43, No. 2, 2007, pp. 185–195.

[14] Sheng, J., Malkiel, J., and Katz, J., "Using Digital Holographic Microscopy for Simultaneous Measurements of 3D Near Wall Velocity and Wall Shear Stress in a Turbulent Boundary Layer," *Experiments in Fluids*, Vol. 45, 2008, pp. 1023–1035.

[15] Elsinga, G. E., Scarano, F., Wieneke, B., and van Oudheusden, B. W., "Tomographic Particle Image Velocimetry," *Experiments in Fluids*, Vol. 41, No. 6, 2006, pp. 933–947.

[16] Atkinson, C., and Soria, J., "An Efficient Simultaneous Reconstruction Technique for Tomographic Particle Image Velocimetry," *Experiments in Fluids*, Vol. 47, 2009, pp. 553–568.

[17] Atkinson, C. Coudert, S., Foucaut, J.-M., Stanislas, M., and Soria, J., "The Accuracy of Tomographic Particle Image Velocimetry for Measurements of a Turbulent Boundary Layer," *Experiments in Fluids*, Vol. 50, No. 4, 2011, pp. 1031–1056.

[18] Buchmann, N. A., Atkinson, C., Jeremy, M. C., and Soria, J., "Tomographic Particle Image Velocimetry Investigation of the Flow in a Modeled Human Carotid Artery Bifurcation," *Experiments in Fluids*, Vol. 50, No. 4, 2011, pp. 1131–1151.

[19] Adrian, R., "Particle-Imaging techniques for experimental fluid mechanics," *Annual Review of Fluid Mechanics*, Vol. 23, 1991, pp. 261–304.

[20] Raffel, M., Willert, C., Wereley, S., and Kompenhans, J., *Particle Image Velocimetry*, Berlin: Springer Verlag, 2007.

[21] Soria, J., "Particle Image Velocimetry—Application to Turbulence Studies," in *World Scientific Lecture Notes in Complex Systems:* Volume 4, Lecture Notes on Turbulence and Coherent Structures in Fluids, Plasmas and Nonlinear Media, World Scientific Publishers, 2006.

[22] Perry, A. E., and Chong, M. S., "A Description of Eddying Motions and Flow Patterns Using Critical-Point Concepts," *Annual Review of Fluid Mechanics*, Vol. 19, 1987, pp. 125–155.

[23] Goodman, J. W., *Introduction to Fourier Optics*, Second Edition, McGraw-Hill, 2006.

[24] Bracewell, R. N., *The Fourier Transform and its Applications*, Third Edition, McGraw-Hill, 2005.

[25] Mitchell, D., Honnery, D., and Soria, J., "Particle Relaxation and Its Influence on the Particle Image Velocimetry Cross-Correlation Function," *Experiments in Fluids*, Vol. 51, No. 4, 2011, pp. 933–947.

[26] Agüí, J. C. and Jiménez, J., "On the Performance of Particle Tracking," *Journal of Fluid Mechanics*, Vol. 185(-1), 1987, pp. 447–468.

[27] Soria, J., and Willert, C., "On Measuring the Joint Probability Density Function of Three-Dimensional Velocity Components in Turbulent Flows," *Measurement Science and Technology*, Vol. 23, No. 6, 2012, pp. 5301.

[28] Keane, R. D., and Adrian, R. J., "Optimization of Particle Image Velocimeters. I. Double Pulsed Systems," *Measurement Science and Technology*, Vol. 1, No. 11, 1990, pp. 1202–1215.

[29] Soria, J., "Multigrid Approach to Cross-Correlation Digital PIV and HPIV Analysis," In Proceedings of the 13th Australasian Fluid Mechanics Conference, 1998, pp. 381–384.

[30] Soria, J., Cater, J., and Kostas, J., "High Resolution Multigrid Cross-Correlation Digital PIV Measurements of a Turbulent Starting Jet Using Half Frame Image Shift Film Recording," *Optics & Laser Technology*, Vol. 31, No. 1, February 1999, pp. 3–12.

[31] Huang, H. T., Fiedler, H. E., and Wang, J. J., "Limitation and Improvement of PIV," *Experiments in Fluids*, Vol. 15-15, No. 4–5, September 1993, pp. 263–273.

[32] Scarano, F., and Riethmuller, M. L., "Advances in Iterative Multigrid PIV Image Processing," *Experiments in Fluids*, Vol. 29, No. 7, December 2000,pp. S051–S060.

[33] Keane, R. D., and Adrian, R. J., "Prospects for Super-Resolution with Particle Image Velocimetry," SPIE's 1993 International Symposium on Optics, Imaging, and Instrumentation, Vol. 2005, December 1993, pp. 283–293.

[34] Keane, R. D., Adrian, R. J., and Zhang, Y., "Super-Resolution Particle Imaging Velocimetry," *Measurement Science and Technology*, Vol. 6, 1995, pp.754–768.

CHAPTER 10

Optical Fiber Sensors

Brian Culshaw

10.1 Introduction

This chapter discusses a somewhat different topic from those that precede it in exploring the application of digital techniques in the context of optical fiber sensors. The overall conclusion is, however, very similar in that while fiber optic sensors are based on well-rehearsed physical principles, their effective implementation requires considerable exploitation of digital techniques. There is an important caveat on any discussion of these digital techniques in that they are often, perhaps invariably, the key to successfully operating the sensor system and in many, perhaps most, cases those who realize this inventive step are understandably somewhat reticent about making too many of the details too public. So while the objectives of the processing procedures are well rehearsed, the details of their implementation are often the subject for speculation.

Fiber sensors have been with us for approaching half a century, and indeed most of the physical principles upon which these sensors are based were elucidated well before the digital age. The basic idea of the fiber sensor is extremely simple and shown in Figure 10.1. The intent is that the output light be modulated in intensity,

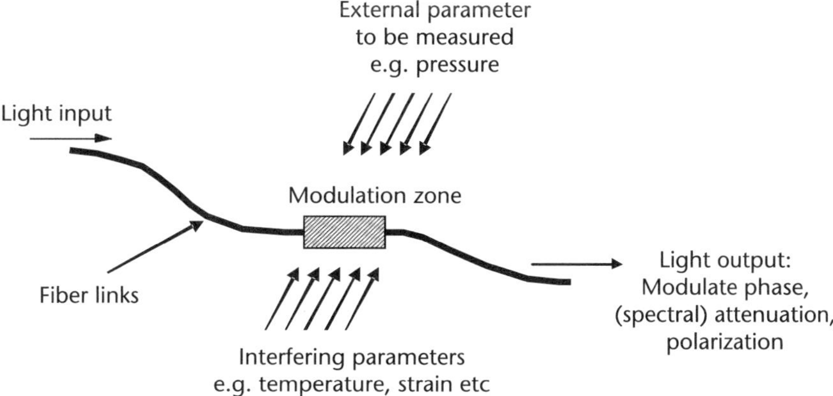

Figure 10.1 The basic idea of the fiber sensor—a parameter to be measured modulates a selected property of light propagating through a sensitive region either in the fiber or through some separate structure interfaced to the fiber. There are almost always interfering measurands, which need compensation!

spectral content, polarization, or phase in a manner that is uniquely related to the measurand of interest. Of course, this unique repeatable relationship is very rarely found in practice (sometimes for temperature), and this is one area in which signal processing, be it optical, electronic, analog, or electronic digital plays an important part [1].

Fiber sensors though have a number of particularly interesting features, the most important of which at a conceptual level is that they are unique in facilitating complex sensing networks capable of monitoring not only the measurand at a particular point but also its distribution over a large number of points extending into the hundreds or even thousands. This feature alone is interesting from an application perspective, and this interest is significantly enhanced by the fact that this network of sensing points can be realized without any electrical connections in the area to be monitored. Furthermore, thanks to the very low transmission losses of optical fibers, these networks can extend to the many tens of even the order of 100 km.

These networks (Figure 10.2) can themselves be realized in two basic configurations.

The first involves connecting together in a suitable architecture a network of point sensors, each connected individually; the second simply uses the fiber itself in a distributed format and exploits the fact that the external environment affects the detailed optical properties of the fiber in a potentially useful manner. These optical properties typically involve nonlinear processes such as Raman and Brillouin scatter (Figure 10.3). The former can be accurately and reliably configured to measure only temperature; the latter responds to both temperature and local strain but with careful interrogation techniques can in fact be configured to interrogate both phenomena independently. Distributed sensing operates essentially by injecting a short pulse of light, typically nanoseconds in duration, into the optical fiber and observing the time dependence of the backscattered light that has been modulated by the

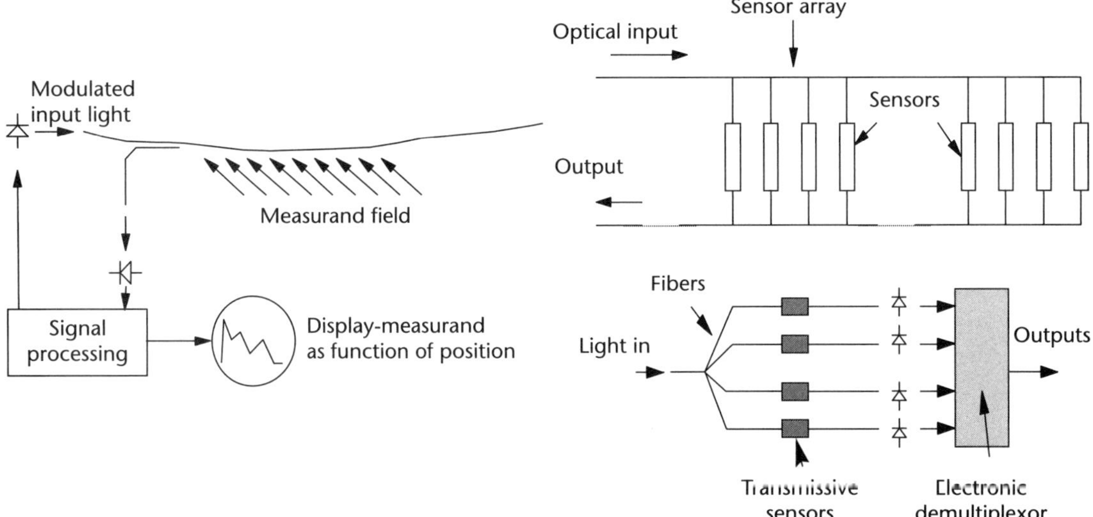

Figure 10.2 Fiber sensing networks: left, the very powerful distributed architecture using the fiber as the interface to the measurand; right, some entirely optical localized sensor networks facilitating large entirely passive systems.

phenomena in question. Some form of spectral filtering on the return light enables Brillouin and Raman scattering systems, and the design of these filters is critical to their success. There are, however, some systems that operate on Rayleigh scatter and that rely upon minute changes in the local phase of propagating light induced through modifications in mechanical strain (or pressure) and/or temperature. These changes are monitored either through some form of polarization analysis or effectively an interferometric realization at the receiver. While inherently such systems are quite complex compared to the Raman and Brillouin scatter versions, they do have the considerable advantage of operating on the Rayleigh backscatter signal, which is typically orders of magnitude higher than that from Raman or Brillouin. Recent progress from these systems has, for example, demonstrated real time operation of a distributed optical microphone capable of not only detecting a particular acoustic (or vibrational) signal but also determining its location to within a meter or better.

In the context of point sensing networks, those based on the fiber Bragg grating (FBG) [2] (Figure 10.4) are by far the most widely exploited. The Bragg grating is written into the core of the fiber using one of a variety of optical writing techniques based upon periodically varying ultraviolet induced damage modulating the core effective index. Depending on the period of this structure the reflected waveguide length changes with applied strain and/or varying temperature. Large networks of Bragg gratings (Figure 10.5) can be realized through appropriate combinations of wavelength multiplexing (assigning a particular spectral band to each measurement point and varying this spectral band for gratings written at different points along the fiber length) and coupling this wavelength multiplexing with appropriate

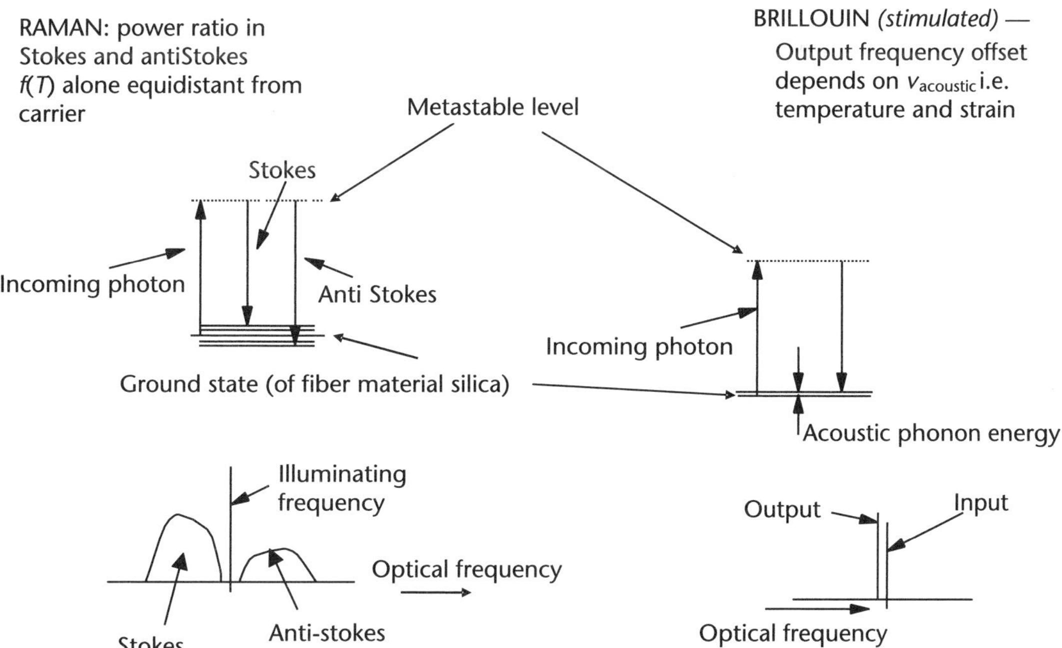

Figure 10.3   The basic nonlinear processes, expressed in terms of energy level diagrams, used in nonlinear distributed fiber optic sensor systems. All utilize these phenomena in standard silica based optical fibers.

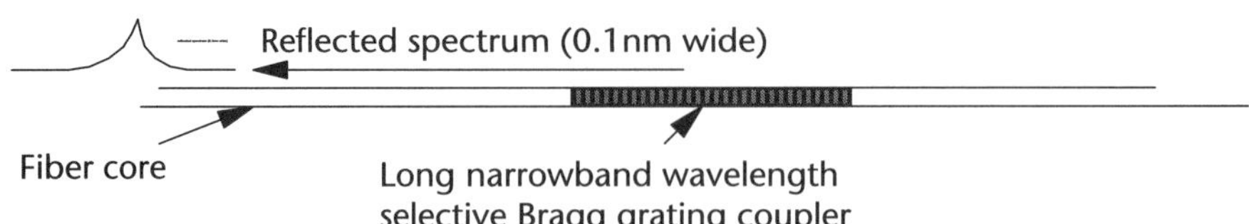

Figure 10.4 The optical fiber Bragg grating uses an photorefractively induced periodic structure in the core of the fiber as a wavelength sensitive reflector, the reflected wavelength depending on the optical path (and therefore temperature and strain) in the grating. The reflected spectrum from a broadband light source enables readings from a multitude of such gratings.

combinations of temporal and spatial multiplexing. The fundamental key to FBG operation clearly lies in accurately and repeatably measuring the peak wavelength of the reflected spectrum. The other important factor, not only to Bragg gratings but also to distributed systems, is the relationship (Figure 10.6) between the physical variations in the optical propagation properties of the optical fiber core and the physical variations to be measured, which clearly have to proceed through a complex set of interfaces.

This very brief introduction highlights two generic areas in which digital processing plays a critical role in sensor implementation. The first lies in extracting measurand data pertinent to a particular measurement point and ensuring that this processing algorithm is optimally matched to the optical physics problem that needs to be solved to extract the measurand data. The second lies in matching this data reliably and repeatably into useful information for the final system user. In effect, the user wants reliable answers to simple questions often reducing to "are things okay today?" If they are not okay, do we need to observe a particular section more closely, or do we shut down to avoid catastrophic consequences, or do we need to simply vary some or other parameter that is easily controlled?

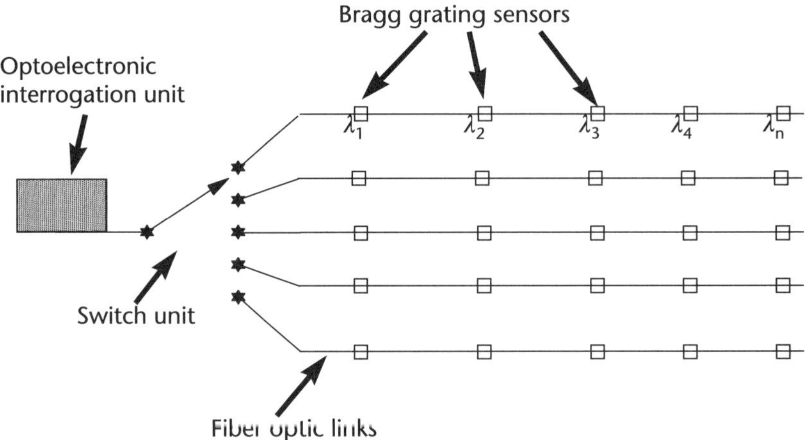

Figure 10.5 A generalized FBG system combining time and space multiplexing. Depending on system needs, up to several hundred sensors can be addressed from a single interrogator unit. Each fiber comprises numerous FBG sensors operating at different central wavelengths and arranged to be nonoverlapping when making measurements.

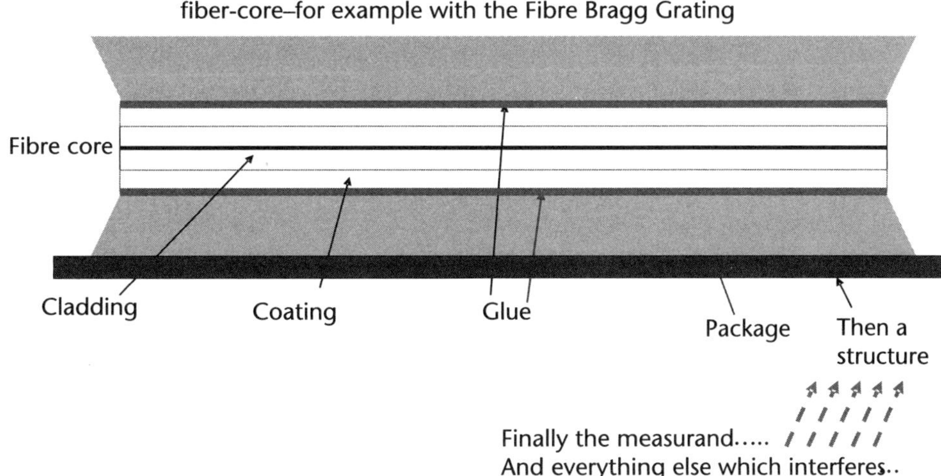

Figure 10.6   Demonstrating the numerous interfaces, all of which need to reliably transmit the parameter to be measured, en route from the measured sample to the core of the optical fiber applicable to both FBG-based and distributed networks. Temperature changes at rates slow compared to the thermal time constants involved are the only ones for which consistency is apparent.

In the sections that follow, we shall briefly discuss some examples of where signal processing has had a critical impact upon the operation of a fiber optic sensor or sensor system. We shall see that there do remain some fundamental issues to address which while they can be assisted by data processing also require a combination of agreed protocols and well-rehearsed physical installations (here the Bragg grating is the best example). There are also many where the objective is really to present useful information rather than a plethora of data to a final user, and here hard-won solutions are often jealously guarded.

10.2 Extracting Information from Data: Some Brief Examples

10.2.1 Single Point Sensors: Spectroscopy

Spectroscopic signal analysis is an extremely powerful tool. With care, a spectroscopic signature can be interpreted in terms of the chemical content of the material or material mixture from which the signature has been scattered or through which light has been transmitted (Figure 10.7). However, in order for this process to function reliably and effectively, the original light source needs to have stable spectral and spatial properties and the spectroscopic analysis tool typically, but not necessarily always, a prism or grating based spectrometer needs to be stable and reliably calibrated. These are simple statements but imply considerable care in the practical realization of these basic requirements.

The spectral signature is typically very complex, reflecting the mix of materials responsible for it and indeed the mix of processes—at the very least a combination of scatter and optical absorption—involved in producing the signature. The traditional "scientific" approach to interpreting this signature is to provide a menu

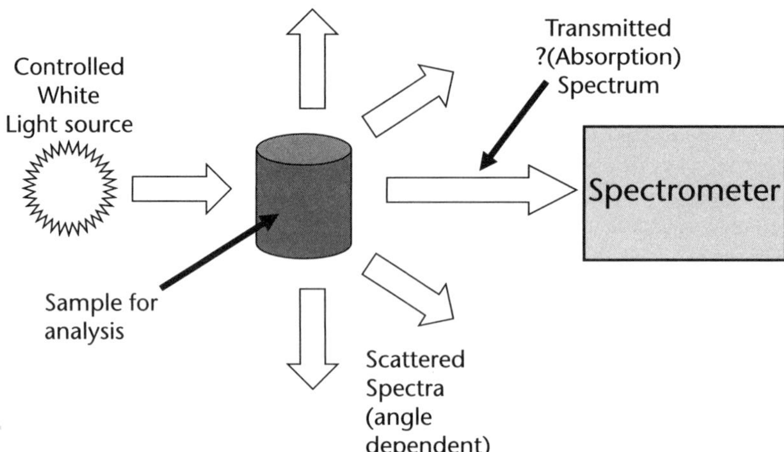

Figure 10.7 Illustrating the immense diversity of spectral analysis tools available, all of which present different information concerning the sample. The spectrometer and light source may be arranged to measure any one of these combinations.

of compounds thought to be candidates for contributors to the composition of the mixture under analysis and thereafter perform some complex curve fitting to extract a reasonable estimate of how much of which potential candidate compound is present in a particular sample. Frequently, this implies using calibration mixtures as a reference and check point, and the output is percentages of particular components (Figure 10.8). This may well be appropriate for some applications (e.g., looking at the content of the exhaust gases from an engine or a smoke stack [3]).

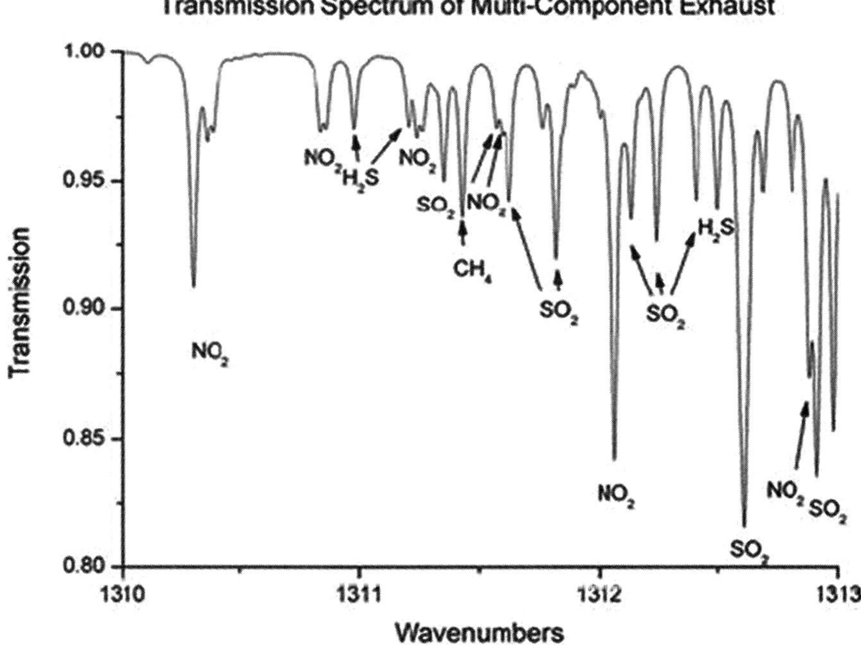

Figure 10.8 A mid-infrared gas spectrum illustrating the analysis of multicomponent smokestack gas mixtures. (Source: [3].)

There are, however, other areas in which such detailed data concerning the actual composition of a particular sample is superfluous and indeed can cause confusion. One example of such an area lies in determining food safety and/or fraudulent food supply. In such cases, a simpler categorization process is far preferable to a detailed chemical analysis. An example of a fiber optic system designed to realize such an analysis [4] is shown in Figure 10.9. The purpose of this system is to collect large amounts of data on the spectral properties of a particular sample, including both transmission and angular scatter, and process this data into useful information.

Consequently, instead of aiming to produce an analysis of the detailed composition of each individual sample, these multiple spectral signatures are combined using one of a number of standard data reduction algorithms, of which principal component analysis is perhaps the best known, into a minimal number, preferably three, of vectors designed so that a particular category sits within a particular sector of the three-dimensional space that these vectors define. Figure 10.10 shows a typical example of this approach applied to a regional analysis of olive oils. The presence of non-extra virgin oil is readily discerned from the signatures, as indeed are also regional variations. While this technique gives no indication of actual content, since the three components are purely the outputs of mathematical analysis, it does provide a reliable guide. This basic idea has been applied to a wide variety of liquid foodstuffs, including not only oils but beers, wines, and whisky. While there are certainly sectors in which these discriminatory capabilities are far better than others, it has proved to be a remarkably versatile approach to solving this particular category of characterization problems. It is a very good example of the evolution from very careful optical fiber sensor physical design into a sensing system that provides easily interpreted output information that hides within it the complex original data sets from the multiple spectroscopic measurements.

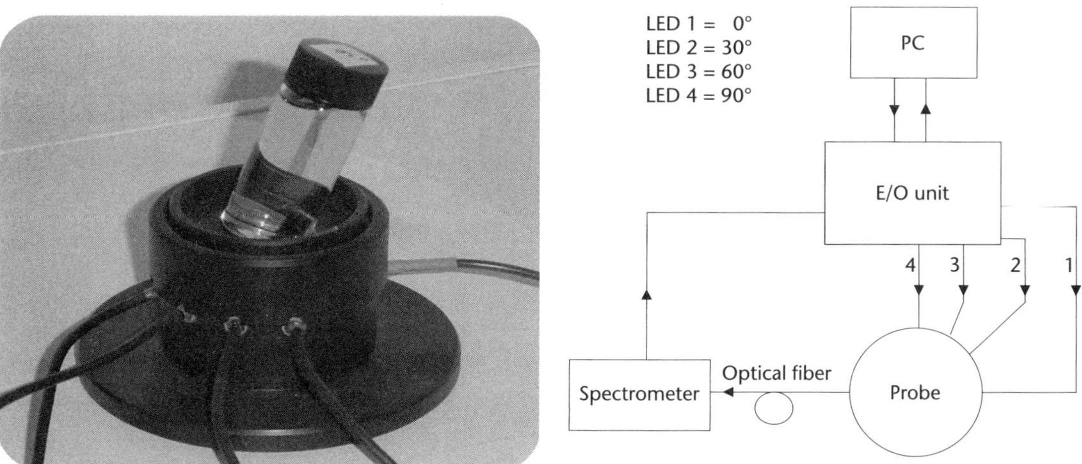

Figure 10.9   A fiber optic linked liquid sample analysis unit for applications involving liquid characterization and discrimination within a particular class of liquids. Multiple carefully controlled white light optical sources from the fibers to the left illuminate the sample in turn to be spectroscopically analyzed via the output fiber at the right. (Courtesy Anna Mignani, CNR IFAC, Florence, Italy.)

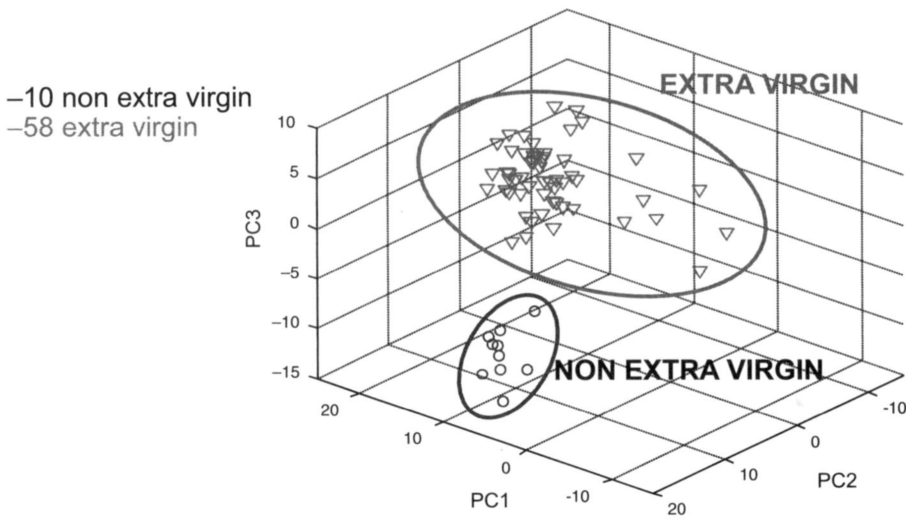

Figure 10.10 Clustering of extra virgin and on extra virgin olive oils production year: 2002/2003—illustrating the discriminating capacity of the system shown in Figure 10.9. The same system with different mathematical analysis tools has also been used to discriminate among extra virgin oils, beers, wines, whiskeys, and hogwash oils. The system can also detect the adulteration of apparently pure samples. (Courtesy of Mignani Anna Grazia, CNR - IFAC, Florence.)

10.2.2 Single Point Sensing: Processing the Fiber Gyroscope

The fiber optic gyroscope (Figure 10.11) is a very powerful example of the application of elegant digital processing to optimize the read-out process in a subtle physical system through careful juxtaposition of digital techniques and deep physical understanding. In principle, the idea is simple. The light traveling in the same direction as the rotation sees the input coupler receding and vice versa for the light proceeding in the opposite direction. Consequently, there is a slight time difference in the optical transit in the two directions, and this is measured as an optical phase. Clearly this phase depends upon a precise knowledge of optical wavelength so regardless of what follows accurate definition of this wavelength is the first fundamental prerequisite [5].

An asymmetrically positioned in-loop phase modulator is the key to extracting the value of this phase difference. In effect, this modulator introduces a differential phase between clockwise and counterclockwise propagating beams within the fiber loop. The simplest version of this modulator introduces a sinusoidal phase modulation at a frequency corresponding to the reciprocal of twice the optical delay around the loop. This implies that the sinusoidal phase modulation is in antiphase on the clockwise and counterclockwise beams resulting in the maximum phase differential. Detection using lock-in techniques at this frequency gives a sinusoidal output as shown in Figure 10.12. The amplitude of the signal detected at the modulation frequency relates to the phase bias and is clearly approximately linear for small phase differences. It is, however, an open loop system and relies on a measurement of optical light intensity. It is consequently subject to detector variations, optical

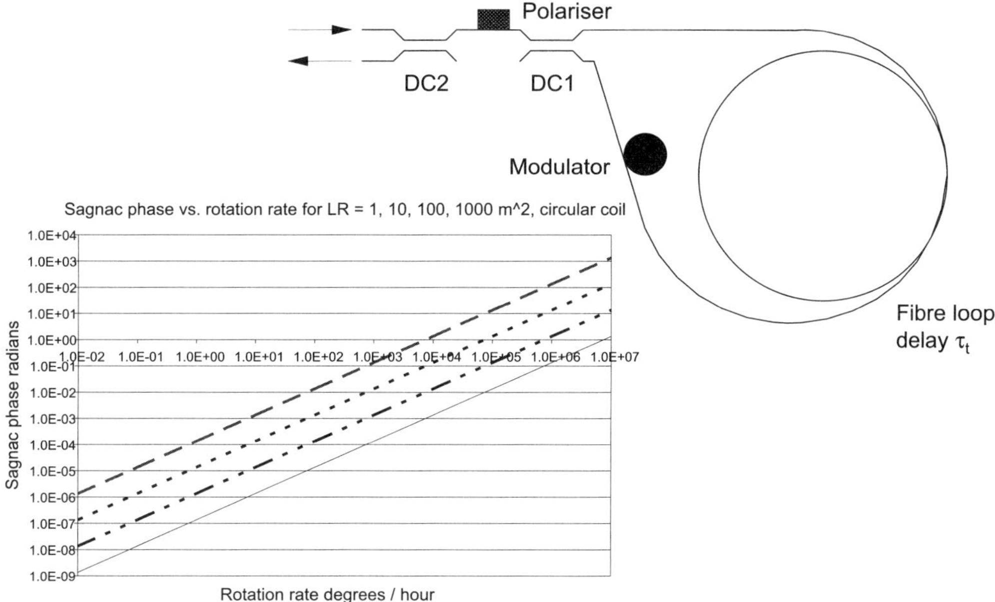

Figure 10.11   The fiber optic gyroscope—a basic schematic diagram is shown at top right. The phase delays for various fiber length to radius products from 1 to 1000 m2 are shown at bottom left. The minimum detected rates from an optimally processed system lie in the region of nanoradians!

input variations, and a host of other interfering phenomena. A few self-referencing analog schemes have been implemented. In particular, the fact that the modulation process also introduces a second harmonic can be exploited and the ratio of the first and second harmonic amplitudes detected to correct for at least some of these errors but at the expense of some quite complex digital circuitry. However, this approach has been successfully used to address some of the relatively low performance (few

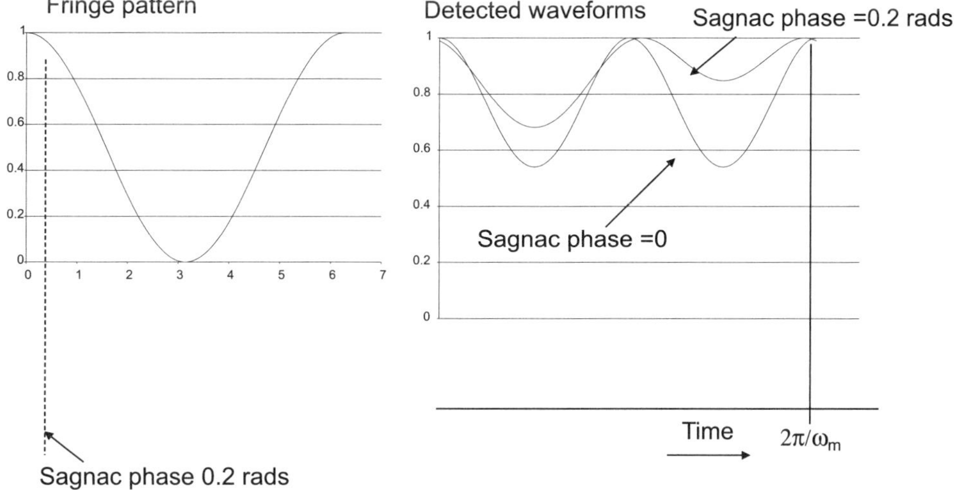

Figure 10.12   Open loop detection using an off center phase modulator (as indicated in Figure 10.11) with detected waveforms given for a Sagnac phase of 0.2 radians and a sinusoidal phase modulation frequency of 1/2tt.

degrees per hour) applications where the fiber gyroscope offers benefits but where there are also highly competitive approaches.

Philosophically, in addition to the open loop approach to processing gyroscope data, we can also close the loop and optimize detection on to the zero phase point. In effect, this requires a known controlled frequency shift between the beams propagating in the two directions. This frequency shift is engineered to ensure that there is *exactly* the same propagation phase in the two directions. One approach to this is to change the phase modulation function from a sinusoid to a phase ramp (sometimes called serrodyne modulation), which in effect introduces a slight change in the optical propagation frequencies in the two directions and this frequency change is in turn proportional to the rotation rate. The trick is, of course, to ensure that the correct slope is applied (Figure 10.13), and this can be done by superimposing the original sinusoidal phase modulation on to the serrodyne ramp and looking at the apparent phase difference between the two directions. This can then, in principle, be effectively set to zero through appropriate changes in the ramp slope. There is, however, an error in this approach in that this feedback loop needs to be closed during the interval between the apparent ramp fly back time in the two directions since, in this region, for a time period corresponding to the transit times thorough the coil, the effective differential phase is incorrect.

This effect can be mitigated by slightly modifying the approach into one that uses the same feedback principles to find the phase difference necessary to sit exactly on the zero fringe point but, instead of utilizing the ramp, implements an abrupt step change in phase which exactly lasts for one loop transit time (Figure 10.14). Thereafter, at the end of the loop transit time it steps up the phase one more time, thereby in effect holding the net effective phase difference between the two directions at zero while removing the step during the fly back. Of course, phase steps cannot go on indefinitely and there is clearly some convenience in terminating these steps just before the 2π has been accumulated and then returning to the zero to start the process one more time. This very brief description conceals a great deal of additional subtlety through which extremely precise fiber optic gyroscopes capable of resolving around 10^{-4} degrees per hour and therefore, for example, suitable for space exploration can be realized. This is another good example of an overall system

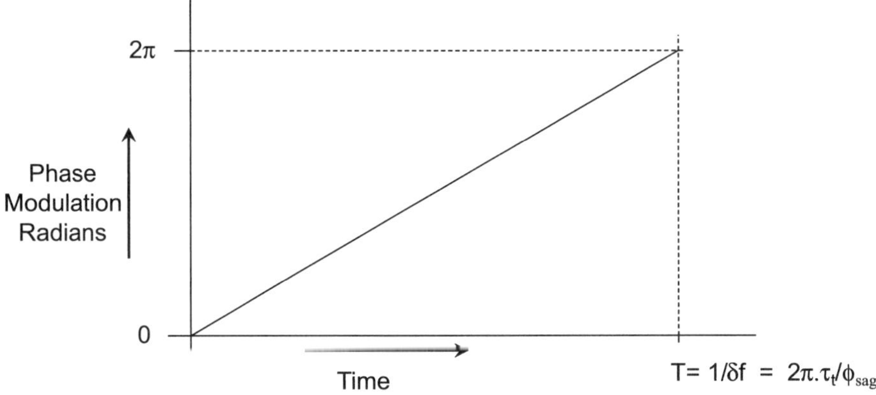

Figure 10.13 The offset phase modulation using the serrodyne principle. The concept is to offset the phase in one direction by the Sagnac induced phase difference fsag during the transit time over the fiber loop.

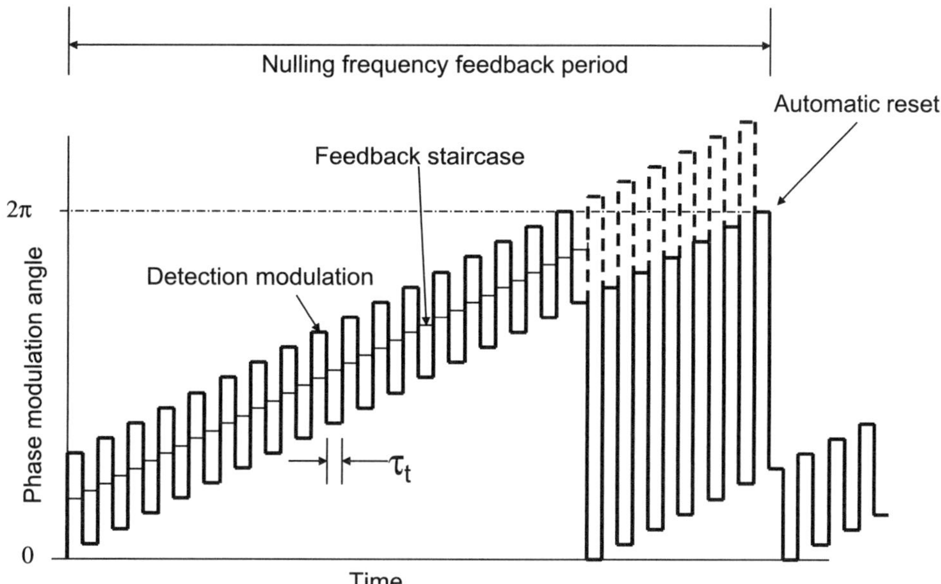

Figure 10.14   Staircase phase difference ramp for closed loop detection showing detection modulation superimposed. (For a much more comprehensive background, see Herve Lefevre's book [5].)

approach that integrates detailed physics with subtle signal processing to facilitate a fiber optic sensor, in this case a gyroscope, with unprecedented performance. It is also an example of a system for which, while the general principles lie in the public domain, the detailed implementation is in fact the result of many years of development. This is discussed in significantly more detail and indeed with much more authority in Hervé le Fèvre's excellent book on the topic [5].

The fiber optic gyroscope uses subtle signal processing in order to optimize the exploitation and interpretation of the underlying physics. There are numerous other similar examples for fiber sensor systems ranging from [6] gas measurement (Figure 10.15) to electrical current measurements on high voltage networks [7] (Figure 10.16). However, at least as important are the signal processing systems applied to extensive multipoint measurement networks based on fiber optics where in the distributed (Figure 10.2) or highly multiplexed (Figure 10.5) architectures. These systems at their most developed form can generate data at the rate of hundreds of megabits per second while the operator typically needs data at the rate of hundreds of bits per day. The complex art of identifying events exemplified in the so-called smart structure is central to making vast quantities of data useful to the user community.

10.2.3 Distributed Sensing

In the context of distributed sensing arguably the most mature technology is that based on using Raman scatter to extract temperature profiles as a function of position along a probing optical fiber. In its raw form the Raman distributed temperature sensor (DTS) produces a map of temperature versus position. In the context of, for example, a fire alarm in a road or rail tunnel, which is a frequent application for

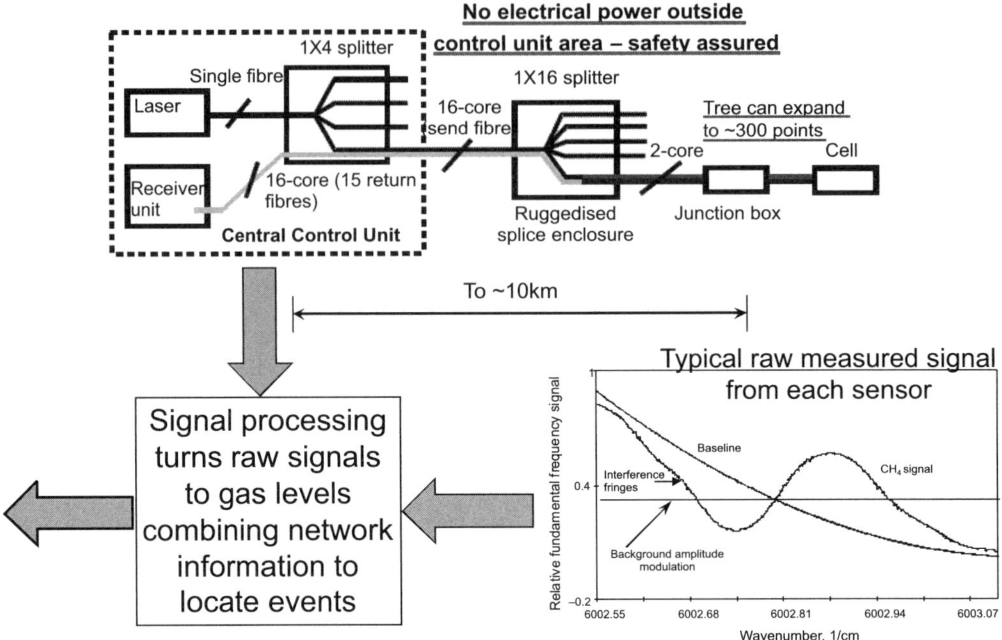

Figure 10.15 An optical fiber based methane gas detection system illustrating again the need for yes: no information rather than detailed raw data in a practical system.

DTS, all that is required is a simple temperature threshold to indicate the presence of a hot zone that can then promptly be investigated. The DTS system, however, finds far more complex application and is extensively used in the oil industry to monitor the behavior of bore holes both during exploration and production. These bore holes form their own complex ecosystem, all of which contributes to their thermal profile. The overriding question then is how to extract from this profile

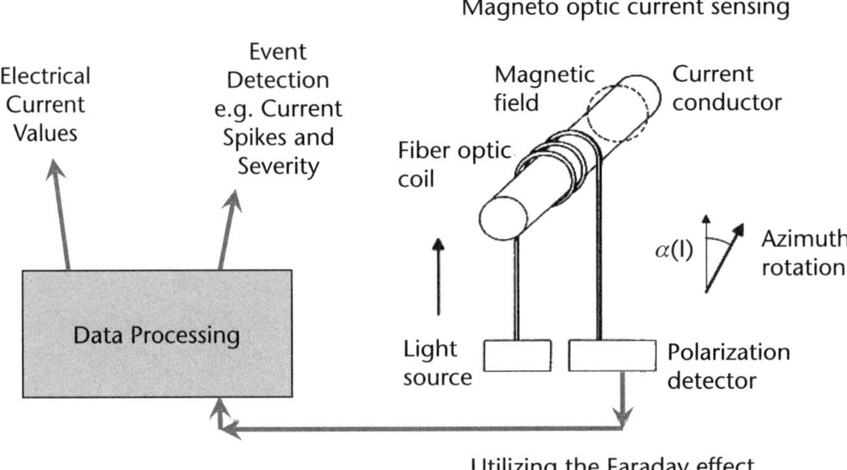

Figure 10.16 Signal processing in the optical fiber current sensor indicating the need for appropriate filtering of the raw current data that the sensor produces.

reliable information that is useful to the operator. Among the many questions that may be asked are, what is the flow rate? Is the flowing liquid multiphase? Are their leaks into or out of the bore hole? What is the state of the reserves remaining in the underground reservoir? Answering these questions requires not just a single shot of temperature versus time along the length of the bore hole—all that is needed for the fire alarm—but a complete history of the evolution of this signature and how to interpret not only the signature itself but its changes with time.

Figure 10.17 shows just a couple of examples of how this extensive data set can be interpreted into usable information [8].

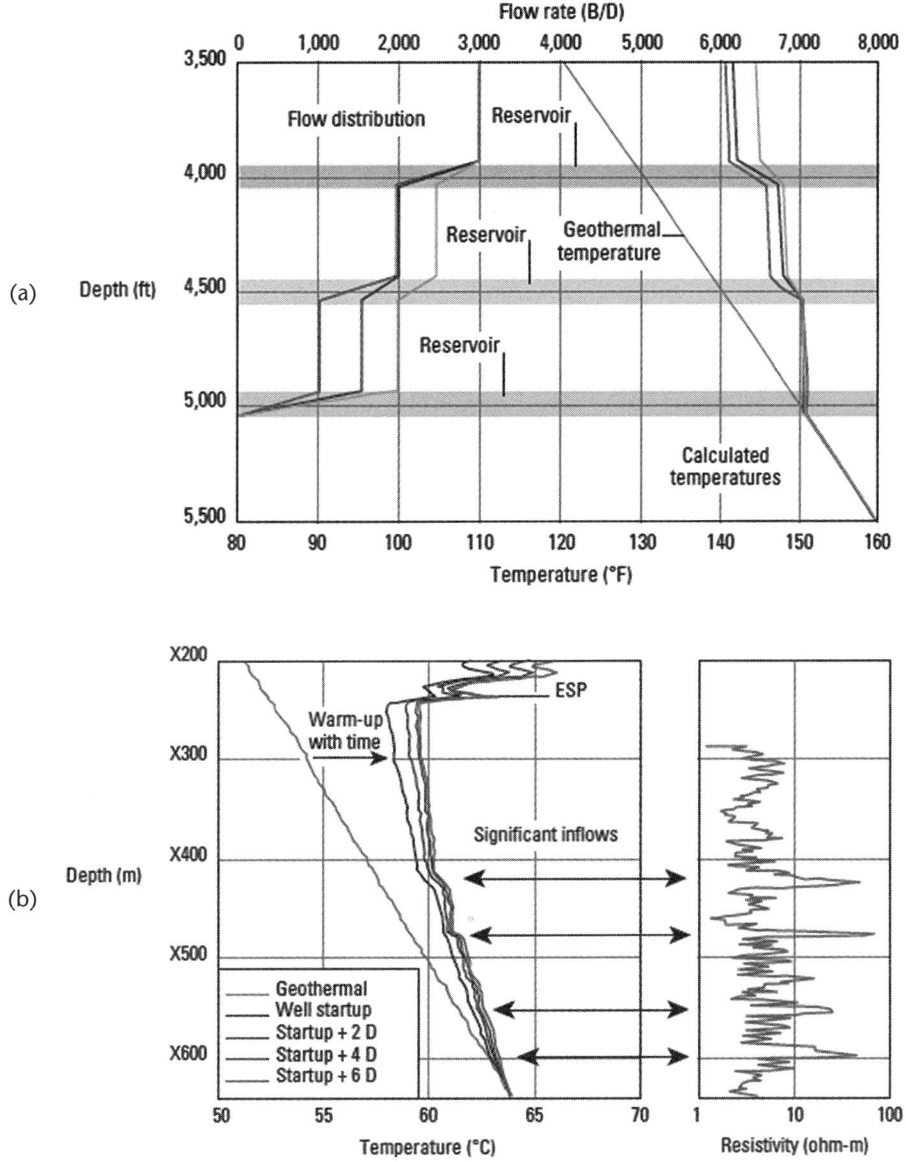

Figure 10.17   Applying data processing to the outputs from a distributed temperature measurement system (DTS) in the context of oil exploration. Left, some idealized modeling results for temperatures and flow rates; right, an example of actual results. (For more details, see [8].)

The distributed acoustic sensor, first demonstrated in principle in 2005, operates as a distributed microphone using coherent detection (Figure 10.18). The so-called DAS has made remarkable progress from first demonstration to extensive application in just a few short years. Its uses include perimeter fence monitoring for unwelcome intruders, leak detection on extended pipelines, and, again, investigations down hole of exploration and production well behavior. The data generation rate is clearly huge since it corresponds to effectively audible bandwidths for each and every meter of a fiber that may extend to tens of kilometres.

Figure 10.19 shows a laboratory demonstration of such a sensing system in operation highlighting the achievable signal-to-noise ratios and the frequency bandwidth. Figure 10.20 shows a time/space evolution plot for another trial system and the corresponding strain versus time distribution [9]. In this particular plot (Figure 10.20(a)), the fiber path has been laid symmetrically about channel 544 so that channels equidistant from 544 should be identical. Figure 10.20(b) indicates temporal outputs for one position corresponding to channels 230 and 858. The traces should be identical, and are very close indeed to being this! These plots give some idea of the capability of DAS and also of the issues involved in selecting appropriate data sets to meet specific requirements.

The filtering processes involved are complex and consequently are jealously guarded by their developers. Furthermore DAS development is still at a relatively early stage. Its unique functionality will undoubtedly find extensive application though will require the emergence of suitable data reduction algorithms targeted at identifying application-specific events. It will also be clear that the same fiber can additionally be interrogated as a DTS so the flexibility implied within a single installation is truly remarkable.

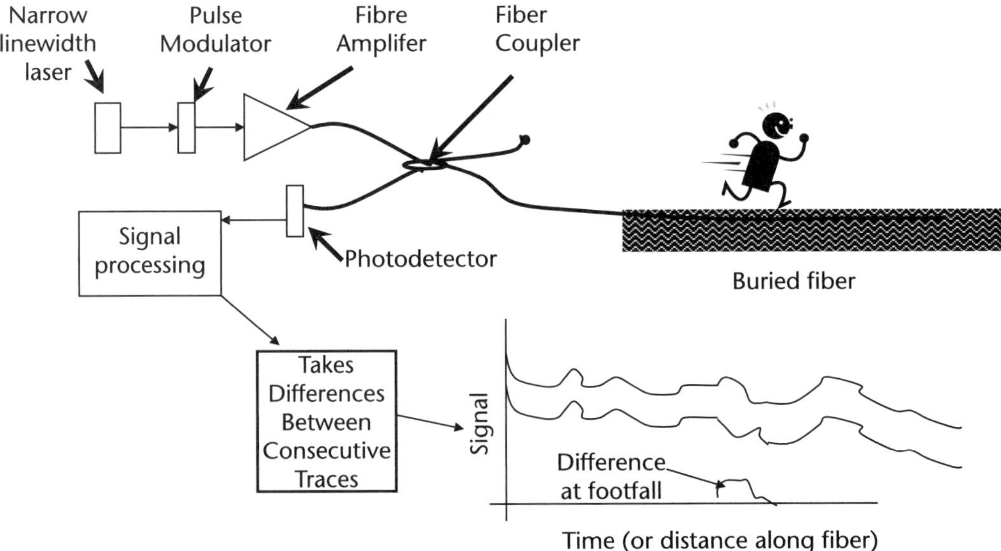

Figure 10.18 The first published demonstration of nonpolarimetric distributed acoustic or vibration sensing using conventional single mode fibers and Rayleigh scatter. The demonstration here is for intrusion sensing. (After Juarez et al, *JLT*, Vol. 23, June 2005, p. 2061.)

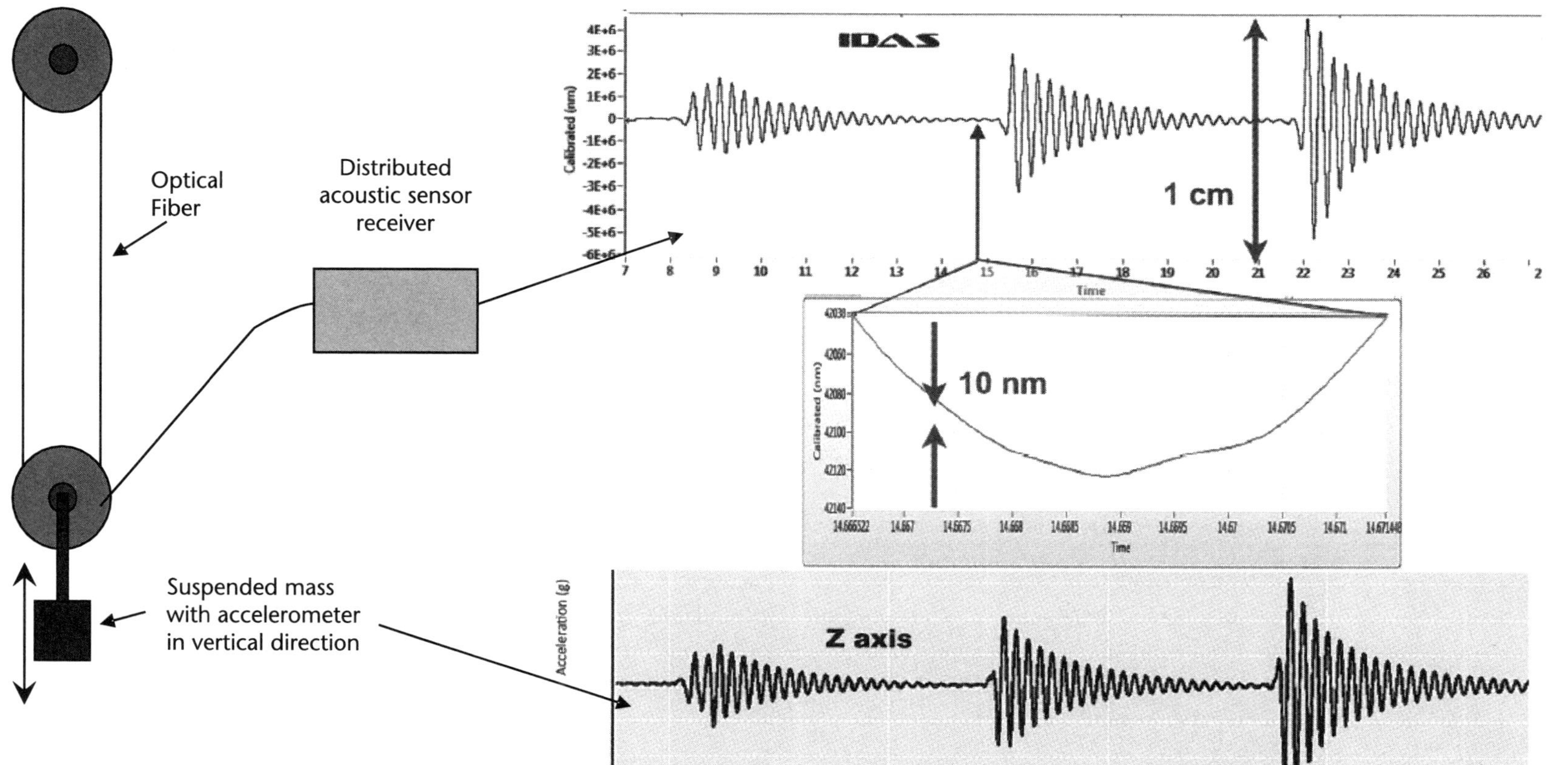

Figure 10.19 A laboratory demonstration of the detectable dynamic range of a distributed acoustic sensor (IDAS) here showing detectability of dynamic signal at the resonance of the pulley structure for strains measured on the z (vertical axis accelerometer from 5 ne to 0.5%. (See [9] for more details.)

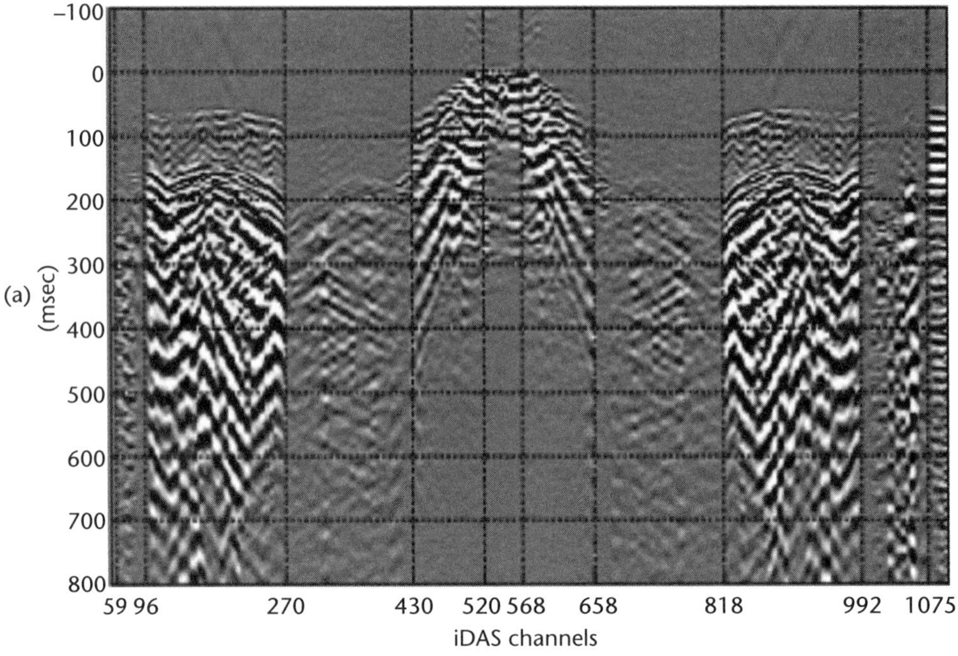

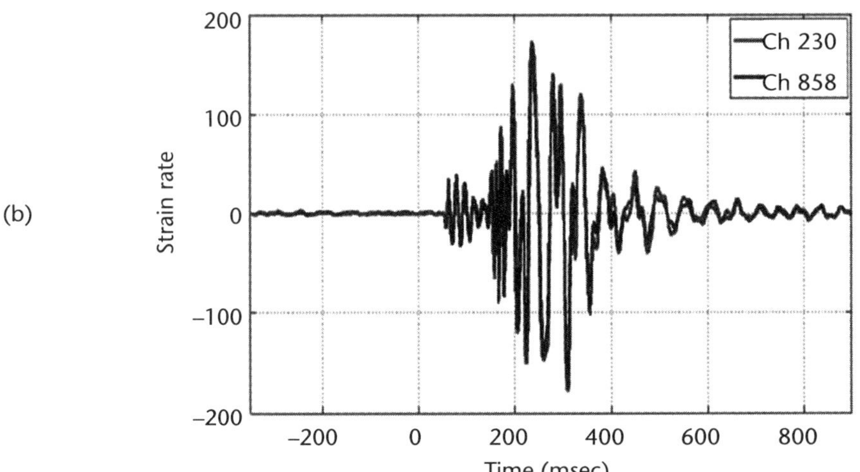

Figure 10.20 A time space plot for a sample set of DAS readings (left, with the DAS signals as brighter for increased amplitude) and the consequent time varying signal for one particular spatial position as read from the two nominally identical symmetrically positioned channels. (See text and also [9] for more details.)

10.2.4 Multiplexed Systems

Of the multiplexed systems, those based on fiber Bragg gratings (Figure 10.4) are the most common. These sensor systems are intrinsically responsive to both the strain and temperature with arguably the simplest implementations addressing periodically predictable dynamic strain fields and interpreting these signals as an indicator of structural loading. Perhaps the most extensive single application to date of the fiber Bragg grating in this context has been in composite wind turbine blades. These are massive structures approaching 100m in length and large enough for a man to stand inside and install grating sensors at appropriate points within the blade itself. The dynamic signals detected are utilized to feed a control system through which the pitch of the turbine into the prevailing wind can be optimized to both minimize the blade stresses during the operation and ensure most efficient electrical power generation. In this particular context, the signal processing algorithms are tuned to the fundamental rotation frequency of the blade assembly, typically a few times per minute, and through this filtering process the influence of interfering variables, most commonly temperature, can be effectively minimized [10].

The fiber Bragg grating has also found its way into an immense range of other structures—including downhole systems where temperature and vibrational fields can be separated and the latter used, for example, as an indicator of structural condition or, indeed, as the basis for the correlation flow meter (Figure 10.21) [11].

There have also been some ventures into incorporating fiber gratings into operational structures such as bridges and high-rise buildings as a prospective indicator of structurally integrity. Here both dynamic and static strain fields are of interest, and while the measurements can be made with temperature compensation where needed there remain two principal issues to be resolved. The first concerns the long-term reliability of such installations and ensuring that a reading made 5, 10, or 20 years ago can be consistently and reliably repeated. This reassurance has yet to be established for any structures, though the particular operational environment of both the wind turbine and the downhill probe provide some opportunities for self-checking procedures. Less closed systems like bridges and buildings are much more difficult to continuously monitor. The second as yet unresolved issue centers, yet again, upon reliable data processing through which the gradual evolution of data sets can be unambiguously interpreted as a means to locate and characterize wear or damage. Again since neither wear nor damage can be intuitively modeled in advance, this remains yet another elusive challenge.

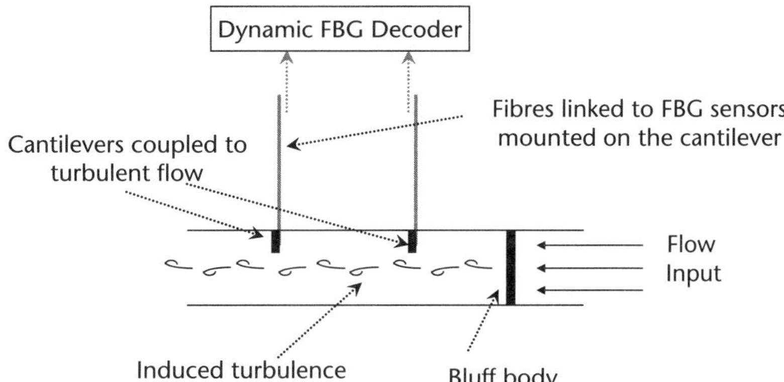

Figure 10.21 FBG utilization in measuring turbulent flow rate. Flow induced vibration on the cantilever arms is detected on the FBG attached to or embedded in the cantilever. The vibration patterns of the two detection arms are correlated and the time delay corresponding to the maximum correlation together with the cantilever spacing gives the flow rate.

10.3 Repeatability, Reliability, and Maintenance

There are many facets of these particular parameters that entail both collaboration among the diverse suppliers and users of particular equipment technology and ingenious approaches toward optimizing longer term performance prospects.

Among these perhaps the most wide-ranging example is the operational characteristics of the fiber Bragg grating. Ideally, gratings and interrogation units from different manufacturers should be readily interchangeable. Any grating designed for a particular interrogation wavelength should be a direct replacement for any other, and the same applies to interrogation units. However, the gratings and the interrogation units can both have different characteristics (Figure 10.22) and can, and often do, give different answers for operation under the same nominal strain field. Similarly there can be differentials in the strain transfer characteristics from one grating to another. Consequently, despite the apparent benefits (Figure 10.23), the necessary interchangeability is currently absent and so for many applications the well-established strain gauge remains the preferred option. Modifying this state of affairs will require significant collaboration among the parties involved and at the time of writing one-to-one user-supplier relationships targeted to specific applications provide satisfactory solutions [12].

In principle, some variation on the theme of digital corrections algorithms could of course be realized to cope with replacement requirements. These could, if necessary, work with the probably reasonable assumption that any system modifications will result in systematic offsets and/or predictable modifications to scale factors that can be modeled and calibrated. Similar comments apply to temperature corrections that are universally felt in all sensors except, of course, those designed to measure temperature.

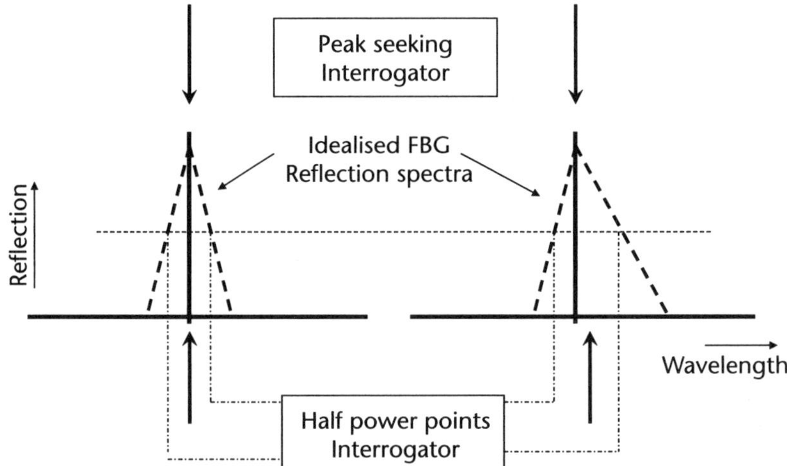

Figure 10.22 Simplified view on how different FBG spectral symmetry and differing FBG interrogator algorithms can produce different outputs for the same grating. This simplistic approach underlies the observation that serious discrepancies can occur for the same grating viewed under identical conditions from different instruments. This problem has yet to be satisfactorily resolved.

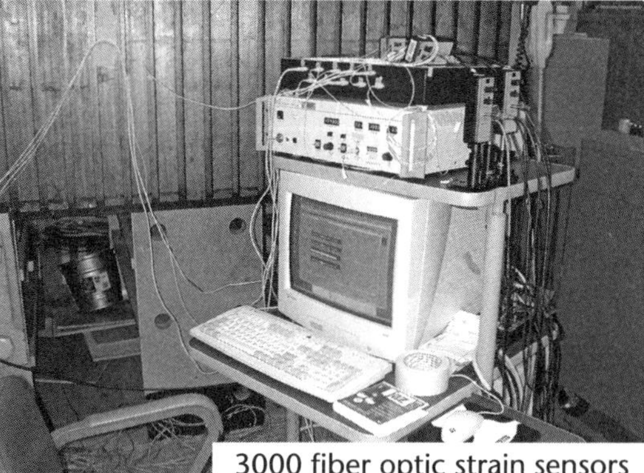

Figure 10.23   Demonstrating the apparent (and in many cases very real!) benefits of the FBG strain sensor compared to conventional electrical wire strain gauges. Not only is the interconnect immune to emi, it is also electrically passive and can extend to kilometers! Even so, many applications continue to demand the strain gauge for which some of the reasons are mentioned in the text. (Source: NASA Langley via Alexis Mendez.)

Numerous other phenomena affect the operation of sensors during their useful life, many of which are unique to the sensing functions, due in part to the much wider range of environmental conditions to which the fiber is subjected. Considerable experience has now been accumulated in operating many of the sensors described earlier, and this experience has facilitated the derivation of correction algorithms to accommodate deterioration and, when appropriate, of alarms to indicate when system performance may be becoming compromised. The environments to which sensors are subjected are immensely varied, including significant temperature cycling, aggressive chemical environments, ionizing radiation, and dynamic changes in relative humidity to mention but a few examples. Consequently, solutions have been developed for specific application contexts through which appropriate corrective action can be taken specific to user interests.

This particular section has been very brief and somewhat superficial. Its objective though has been to indicate that digital processing plays a critical role in the wide ranging possible applications to which fiber sensors have been tasked and that each particular application relies on a combination of digital data analysis and accumulated experience coupled with the ever-necessary physical insight to ensure continued satisfactory operation whether in the context of system repair and replacement or of long-term operational modification to system characteristics. Digital solutions are always necessary when ensuring the necessary reliable measurements.

10.4   What of the Future?

The future evolution of fiber sensors will proceed along two paths. The first lies in enhancing applications [13–15]. While the benefits of fiber sensing are well rehearsed, the momentum of established approaches is enormous within the user community with the exception—most notably, distributed sensing—of approaches that enable entirely new measurement modalities. For these latter situations, the first stage is to excite prospective users about the prospects and thereafter convince them that the excitement is well founded. This has involved exploring long routes toward convincing the user of both technical performance and reliability and thereafter the invariably even greater task of ensuring that the data collected in all its vastness can be conveyed as concise, useful information. We have already explored some examples of this process. The current market optimism for fiber sensors, typically extrapolating a factor of five improvements in market volume [16, 17], from recent figures (referenced to 2013) to 2020, is based upon expanding confidence in the user community coupled with expanding awareness of the need for comprehensive attention to data presentation.

The second area where improvements will inevitably accrue lies in the exploitation of emerging optical technologies. Of these perhaps the most important are squeezed light (Figure 10.24) [18], photonic crystals, nanophotonics [19, 20], and surface plasmons (Figure 10.25) [21, 22] and metamaterials (Figure 10.26) [23]. At a conceptual level these approaches have been with us for 20 or more years; what is changing significantly is the technical implementation of these concepts into interesting innovative devices. The focus of this chapter lies in digital optics rather than innovative physical technologies, but these technologies in turn inevitably rely significantly on digital approaches to their design and computer-controlled precision machining for the realization of these designs. Some initial exploratory steps have been taken including, for example, the realization of a micromechanical device to demonstrate optical squeezing. The demonstrations of these concepts to

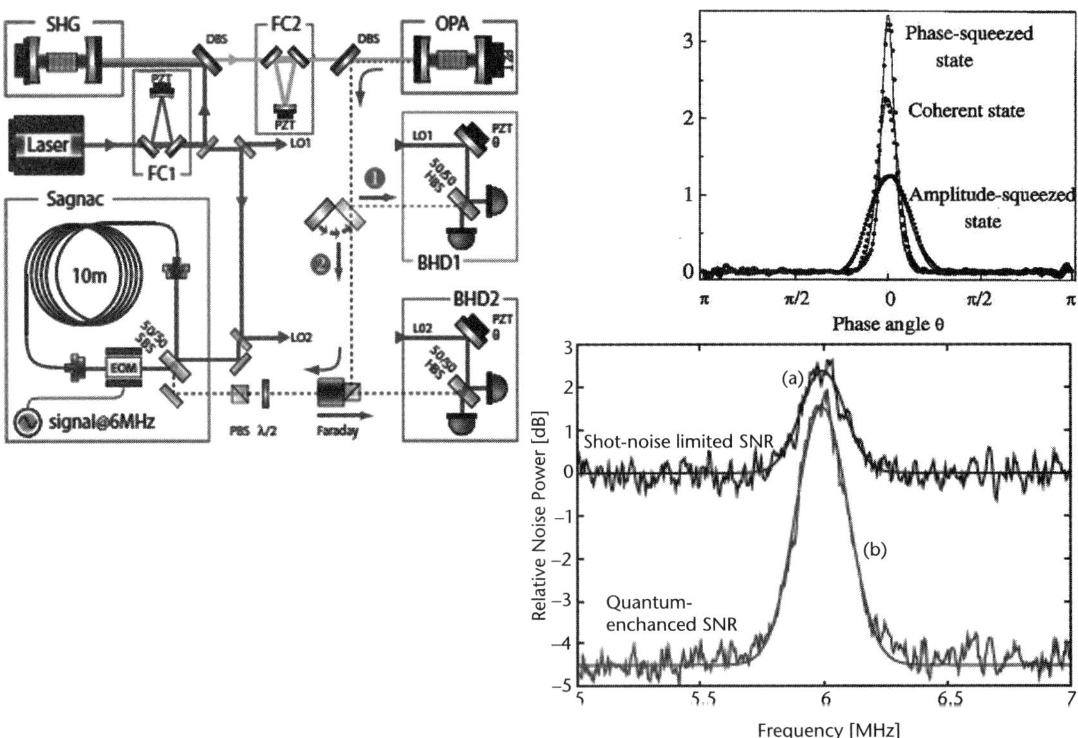

Figure 10.24 Squeezed light: top right, the basic concept and its influence on noise properties; top left, a demonstration of phase squeezing (used on the LIGO telescope) in a Sagnac interferometer; bottom right, the experimental results thereof. (Source: NASA Langley via Alexis Mendez.)

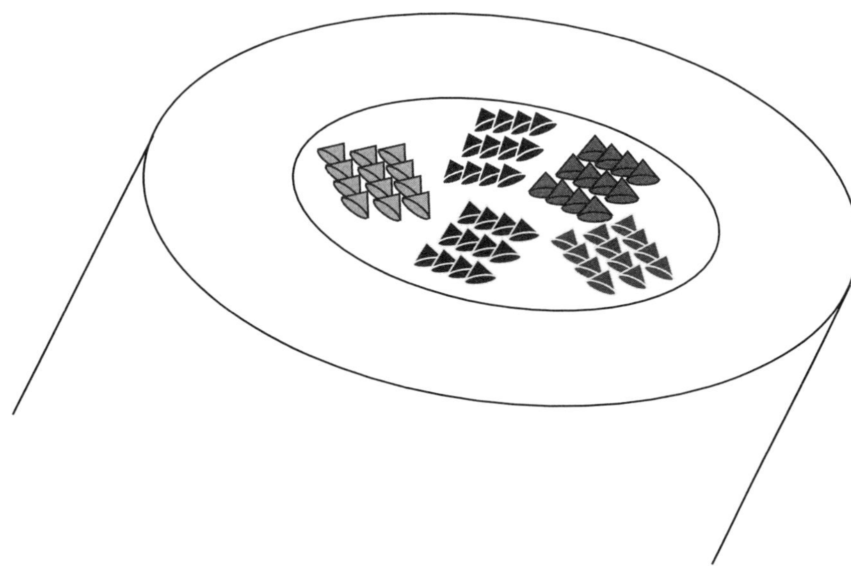

Figure 10.25   A "concept diagram" of a possible surface plasomon nanophotonic based multiparameter (one parameter per color) sensor system fabricated on the end of a fiber. This hypothetical array of nanophotonic "optical wire" cones could be fabricated to be sensitive to either physical and/or (bio)chemical phenomena. There'll be interrogation issues, signal-to-noise ratio considerations, and host of other practicalities to consider, but when nanophotonics becomes optical wiring, much will be achieved.

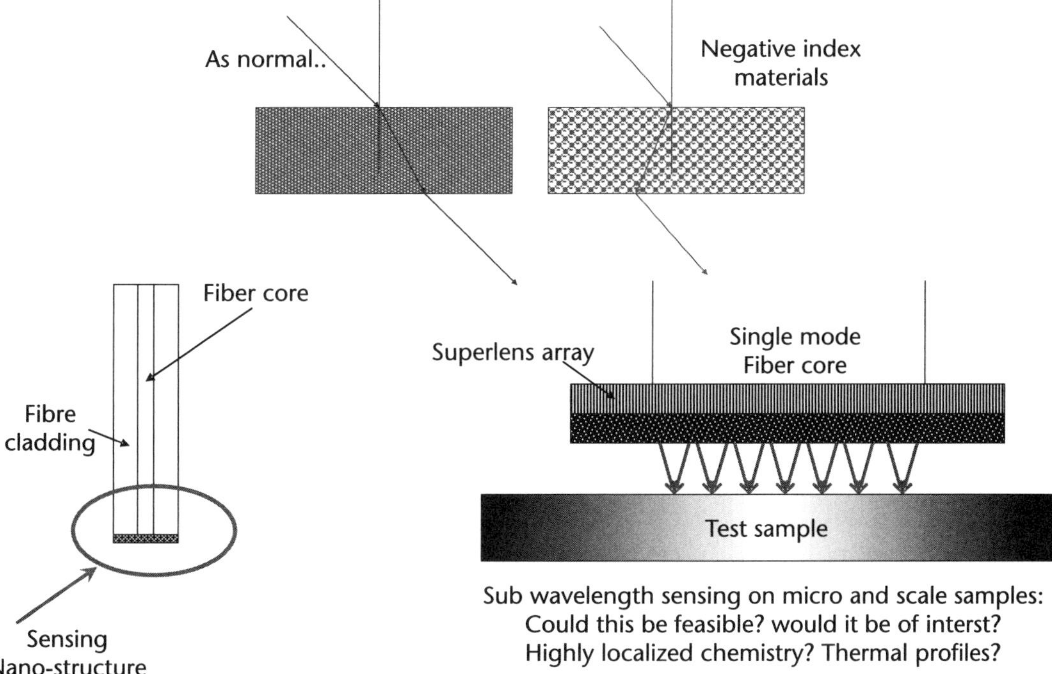

Figure 10.26   Metamaterials have in principle the prospect for perfect focusing beyond the Abbe limit. (top sketch). Could such materials enable a super resolution chemical sensing array using a suitable nanostructure fabricated on the fiber end? Many opportunities are here for digital modeling and for decoding the data set that such a system might produce!

date as sensors have been few and far between, but do include instances of using photonic crystals and photonic crystal fibers and of exploiting squeezed light and surface plasmons.

Taking these concepts through into real application will inevitably involve the processes alluded to earlier in this section—namely, exciting potential users, demonstrating technical performance and reliability, and enhancing data presentation. There will be much to do here for digital techniques.

10.5 Concluding Comments

This chapter has endeavored to highlight the critical role that digital processing has taken in both the successful application of fiber sensor technologies and in their future design and implementation.

Sensor technology is, by its very nature, highly fragmented and consequently the digital solutions applied to each detailed implementation are themselves totally individualistic. The features of each solution are determined by a combination of the physics underlying the sensor technology involved and the needs of the final user. This situation contrasts sharply with the optical fiber communication sector for which uniformity and total interchangeability are dominant requirements.

As fiber sensor technology evolves, a multifaceted approach to its evolution will emerge (Figure 10.27). New technologies will be applied and designed into innovative sensor systems, while current techniques will be improved where possible and practical, standardized for universal adoption, and made more adaptive to idiosyncratic needs. In time, the new technologies will in turn find their niche and probably a decade or more from now will go through similar processes and will settle themselves into user-friendly formats. The total process is, undeniably, time consuming and requires enthusiastic, respectful, and unambiguous communication among the various interests. Concept to product is typically two decades, so

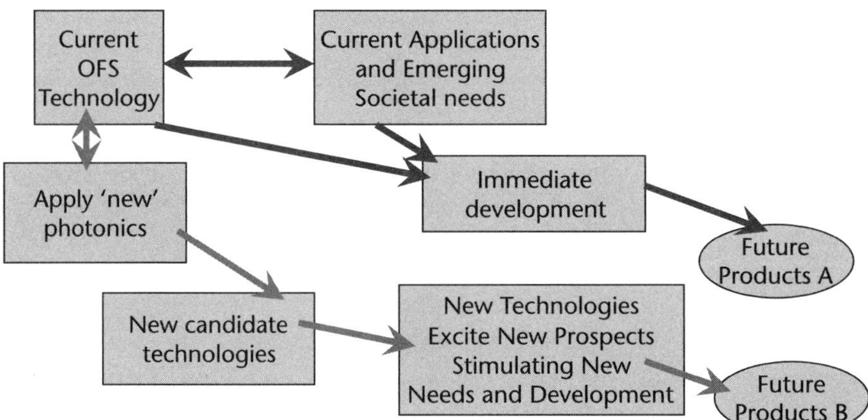

Figure 10.27 An attempt to represent the future prospects of fiber sensors. Developments to meet currently predicted needs (a) will co-exist with innovative photonics becoming ever more available and enabling new, currently inconceivable, measurement modalities and thereafter—in another half century or so—spawning new products; (b) after stimulating currently inconceivable applications.

precision, persistence, and patience are also essential attributes. There is, however, considerable satisfaction to be gained from seeing a long ago lab experiment being put to practical use.

10.6 Acknowledgments

Many people have contributed to formulating this story, among whom former students and fellow workers—too many to name individually—figure largely, as do contributors to the many collaborative projects with which the author has been involved. Most recently the COST TD1001 program has provided an excellent forum through which to evolve the real processes necessary for sensor application, and again the many contributors to lectures and discussions throughout this program most certainly warrant credit.

Notes and References

[1]   The history of fiber optic sensors can be traced through the Proceedings of the International Conference on Optical Fiber Sensors, starting in 1983 and most recently presented in Santander in 2014. Proceedings available from SPIE. See also Culshaw, Brian, and Alan Kersey. "Fiber-optic sensing: A historical perspective." *Journal of lightwave technology* 26.9 (2008): 1064–1078.

[2]   A. Othonos and K. Kalli, *Fiber Bragg Gratings*, Norwood, MA: Artech House, 1999.

[3]   Iain Howieson, "The Application of Quantum Cascade Laser Gas Sensors to Industrial Monitoring," *Gases and Technology*, May 2006.

[4]   Mignani, A. G. et al., "EAT-by-LIGHT: Fiber-Optic and Micro-Optic Devices for Food Quality and Safety Assessment," *Sensors Journal, IEEE*, vol. 8, no.7, pp. 1342, 1354, July 2008 doi: 0.1109/JSEN.2008.926971.

[5]   H. C. Lefvre, *The Fiber Optic Gyroscope*, Norwood, MA: Artech House, 1999, second edition to be released 2014

[6]   Culshaw, B., et al. "Fiber optic techniques for remote spectroscopic methane detection—from concept to system realisation," *Sensors and actuators B: chemical* 51.1 (1998): 25–37. Also a recent field trial in D. Walsh, I. Mauchline, J. Pillans, et al., "Field trial of all-optical, multipoint methane sensing system for coal mine safety & ventilation applications" in *Proceedings 10th IMVC*, Sun City, South Africa, August 2014.

[7]   M. Willsch, T. Bosselmann, M. Villnow, W. Ecke. "Fiber optical sensor trends in the energy field," *Proc. SPIE 8421, OFS2012 22nd International Conference on Optical Fiber Sensors*, 84210R, October 4, 2012, doi:10.1117/12.2007513.

[8]   More information in "The Essentials of Fiber-Optic Distributed Temperature Analysis" published by Schlumberger, Texas, and openly obtainable.

[9]   Tom Parker, Arran Gillies, Sergey V. Shatalin, Mahmoud Farhadiroushan. "The intelligent distributed acoustic sensing," *Proceedings Volume 9157: 23rd International Conference on Optical Fiber Sensors*, May 2014.

[10]   Phil Rhead, "Fiber Optic Sensing Technology and Applications in Wind Energy," Sandia Blade Workshop, 2008 available at http://windpower.sandia.gov/2008BladeWorkshop/PDFs/Wed-06-Rhead.pdf.

[11]   Henrik Krisch, et al., "Advanced photonic vortex flowmeter with interferometer sensor for measurement of wide dynamic range of medium velocity at high temperature and high pressure," *Proceedings Volume 9157: 23rd International Conference on Optical Fiber*

Sensors, May 2014. See also Takashima, Shoichi, Hiroshi Asanuma, and Hiroaki Niitsuma. "A water flowmeter using dual fiber Bragg grating sensors and cross-correlation technique," *Sensors and Actuators A: Physical* 116.1 (2004): 66–74.

[12] Habel, W. R., Schukar, V. G., Kusche, N. "Fiber optic strain sensors are making the leap from lab to industrial use—reliability and validation as a precondition for standards," *Measurement Science and Technology*, 24.9, 094006, 2013.

[13] "Photonics our vision for a key enabling technology of Europe," May 2011. Report available for free download from http://www.photonics21.org.

[14] "Optics and photonics: essential technologies for our nation," 2013.Report available for free download from http://www.nap.edu/catalog/php?record id=13491. See also US National Photonics Initiative https://spie.org/x8893.xml.

[15] The NAE Grand Challenges discourse is at www.engineeringchallenges.org.

[16] *Photonic Sensor Consortium Market Survey Report Fiber Optic Sensors*, June 2013, Information Gatekeepers, Inc.

[17] *Global Fiber Optic Sensor Market 2014–2018*, December 2013, TechNavio inc via reportlinker.com.

[18] Vahlbruch, H., et al. "Observation of squeezed light with 10dB quantum noise reduction" *Physical Review Letters* 100.3, 033602, 2008

[19] Monticone, F., and Alu, A. "Metamaterial enhanced nanophotonics," *Optics and Photonics News, 35*, December 2013.

[20] Vasudevanpillai, B., Tamitake I., Abdulaziz A., Athiyanathil S., and Mitsuru I. "Semiconductor quantum dots and metal nanoparticles: syntheses, optical properties and biological applications," *Analytical and Bioanalytical Chemistry*, 391 (7), 2469–2495, August 2008.

[21] Tudos, A. J., and Schasfoort, R. B. M. (eds), *Handbook of Surface Plasmon Resonance*, London: RSC Publishing, 2008.

[22] Sharma, A. K., Jha, R. and Gupta, B. D. "Fiber optic sensors based on surface plasmon resonance: a comprehensive review," *IEEE Sensors Journal*, 7, 1118, 2007.

[23] Billings L. "Metamaterials World," *Nature*, 500, 138–140, 8 August 2013.

CHAPTER 11

Optical Coherence Tomography: Principles, Implementation, and Applications in Ophthalmology

Yoshiaki Yasuno

11.1 Introduction

11.1.1 What Is Optical Coherence Tomography?

Optical coherence tomography (OCT) is an interferometric modality that provides noninvasive tomography of in vivo human tissues [1]. OCT is essentially a scanning low-coherence interferometer that utilizes coherence gating to resolve the depth structure of a sample. During the measurement, OCT illuminates a sample with a probe beam focused on it. The transversal structure is then obtained by transversally scanning the probe beam using a rotating mirror, typically a galvanometric scanning mirror. If the transversal scanning is one-dimensional, a two-dimensional (2D) cross-sectional image is obtained. Similarly, 2D transversal scanning, typically achieved by a pair of galvanometric scanning mirrors, provides a three-dimensional (3D) volumetric tomography.

Among several medical tomographic modalities, OCT is characterized by the following properties. First, OCT is noninvasive. Since OCT uses a weak near-infrared (NIR) light as a probe, the sample remains free of photochemical and photothermal damage. Second, the OCT has higher resolution than other clinical tomographic methods such as X-ray computed tomography, magnetic resonance imaging, and ultrasound tomography [2]. OCT typically provides a depth resolution of less than 10 μm. Third, the measurement speed of OCT is fast. The first generation of OCT, known as time-domain OCT (TD-OCT), had an imaging speed between several hundreds to several thousands of depth lines (A-lines) per second [3]. A more recent OCT method, Fourier-domain OCT (FD-OCT) provides an even faster speed of around several hundred-thousand A-lines per second [4, 5]. This high speed enables 3D investigation of in vivo organs within a realistic measurement time. Another characteristic property of OCT is its maximum measurement depth or penetration depth. Since OCT uses NIR light, the probe beam is relatively highly scattered by the sample to be measured. This high scattering unfortunately limits the penetration depth of the OCT to within a few millimeters. Considering these properties of

OCT, it is clear that OCT is not an all-purpose tomographic modality. However, there are some organs that are extremely suitable for OCT.

11.1.2 OCT and Ophthalmology

11.1.2.1 The Epoch of OCT

OCT has been utilized for variety of clinical fields such as dermatology [6], dentistry [7], gastrointestinal medicine [8], oncology [9], cardiology [10], and ophthalmology [11]. Among these clinical fields, ophthalmology is perfectly suited to OCT.

The first demonstration of OCT was described in a research paper published in 1991 [1]. In this seminal paper, it is clear that OCT was intended to be utilized in ophthalmology from the beginning. This intention can be inferred from three facts. First, the samples measured in this paper were a macular and an optic nervehead of an ex vivo porcine retina. Structural disorder of the macula is associated with several severe eye diseases such as age-related macular degeneration, diabetic retinopathy, and central serous chorioretinopathy, while the abnormality of the optic nervehead is associated with glaucoma. This fact suggests that the authors (i.e., the inventors of OCT) already intended to use OCT for clinical practice in ophthalmology even at its epoch.

The second fact implicitly suggesting the authors' intention is the authorship of the paper that comprises laser specialists, including James G. Fujimoto from the Massachusetts Institute of Technology, and ophthalmology specialists, including Joel S. Schuman and Carmen A. Puliafito from the Department of Ophthalmology, Harvard Medical School. In addition, the first author, David Huang, was a PhD student studying both engineering and medicine.

The third fact concerns the series of papers that were quickly published by the authors after this first publication. In 1992, the same authors published a high-speed low-coherence reflectometry method [12]. It should be noted that a high imaging speed is crucial for in vivo OCT imaging of an eye. The submission date of this paper was September 1991, suggesting that the authors were preparing for in vivo OCT eye imaging even when they were working on their first 1991 publication. Furthermore, in 1993, the authors demonstrated the first in vivo human retinal imaging [13].

These facts suggest that OCT was, from the beginning, developed for use in clinical ophthalmology. As proof of this speculation, OCT was commercialized as a clinical ophthalmic imaging device only five years after the first demonstration.

Regarding this point, the question arises as to why they were interested in ophthalmology.

11.1.2.2 Eye Is a Perfect Target of OCT

The eye is a perfect organ for investigation by optical modalities. This is mainly because the eye itself is an optical instrument. Because the eye is an imaging system that images an object on the retina, it is also easy to build an optical system that images the retina on an arbitrary imaging plane using the eye optics.

Another important characteristic of the eye is transparency. As discussed earlier, OCT penetration is limited to a few millimeters, mainly because of scattering in tissue. However, we can ignore the scattering in the optical media of the eye including the cornea, aqueous humor (that fills the anterior eye chamber), crystalline lens, and vitreous. Hence, several-millimeter penetration of retinal OCT is defined not from the surface of the eye but from the surface of the retina, located around 24-mm under the eye surface. Because the thickness of the retina is about a few hundred micrometers, the limited penetration of OCT is not a serious issue for retinal imaging.

The direction of the tomographic cross section also enhances the value of OCT. Almost all eye imaging modalities such as ophthalmoscope, color fundus photography, and angiographies are en-face imaging modalities, while OCT is the only one that is an in vivo modality that provides a depth-oriented cross section of the retina. Because the retina has a fine multilayered structure and its destruction is highly associated with eye diseases, the in vivo cross-sectional tomography provided by OCT is extremely important for ophthalmic diagnosis.

The noninvasiveness of OCT is also particularly important in ophthalmology. This is because the eye is one of a few organs that we cannot biopsy. Since all portions of the retina are associated with visual function, it is impossible to excise even a small portion of a tissue for diagnosis purposes.

Considering these properties of the eye, it is natural to conclude that the eye is a perfect organ for investigation by OCT. Furthermore, it is natural that the eye was selected as the first target for OCT.

11.2   OCT Technologies

11.2.1   Requirement for Eye Imaging

The first generation of OCT, TD-OCT, is a variant of low-coherence interferometry. Hence, TD-OCT is a relatively slow imaging modality. TD-OCT scans the optical path-length difference for depth sectioning and scans the probe beam transversally to obtain transversal structural information. In particular, a 2D scan is required for 2D cross-sectional imaging, and a 3D scan is required for volumetric tomography. This multidimensional mechanical scanning limits the measurement speed. For example, one widely utilized retinal OCT device (Stratus OCT, Carl Zeiss Meditec, CA, USA) requires 1.5s to obtain a single cross-sectional scan (B-scan). The A-line (depth scan) rate of this device is only approximately 340 A-lines/s.

Measurement speed is particularly important for in vivo eye imaging because the eye is always moving. Involuntary eye motions including drift, tremor, saccade, and micro-saccade are normal physiological actions, and hence inevitable. The OCT measurement speed needs to be fast enough to render eye motion negligible. The previously mentioned speed of 340 A-lines/s was somewhat acceptable for 2D cross-sectional imaging but not fast enough for volumetric measurement. In fact, the retinal cross-section taken by clinical TD-OCT was significantly distorted by eye motion. Namely, TD-OCT was not sufficiently fast enough, even for 2D imaging. A state-of-the-art TD-OCT has demonstrated a higher A-line rate of 4,000 A-lines/s [3], but this speed was still not sufficiently fast for volumetric imaging. In order to obtain a volumetric tomography of the eye, a measurement speed on the order of tens of thousands of A-lines/s is required.

Another important requirement for ophthalmic OCT is high sensitivity. Because the reflectivity of retina can be as low as 10^{-5}–10^{-6}, a high system sensitivity and the resulting high signal-to-noise ratio (SNR) are required. In general, the SNR of OCT is roughly proportional to the measurement time and optical power of the probe beam. As discussed earlier, the OCT measurement time should be short for eye investigation. Hence, it is not a good strategy to use long measurement times to increase the SNR. Increasing the probe beam power is also not an optimal solution. Because the eye is a photosensitive organ, it is also very sensitive to photodamage. Hence, the probe beam power of retinal OCT is strictly limited by safety standards such as ANSI [14] and ISO [15]. For example, the maximum allowable probe power for OCT is around 700 μw for a probe wavelength of 840 nm and around 1.2–1.7 mW for 1060 nm. To overcome these inherent sensitivity limitations, a new OCT method that had a higher probe beam throughput for image formation was required.

These two main requirements (i.e., high speed and high sensitivity) have driven researchers to develop a new OCT variation: FD-OCT.

11.2.2 TD-OCT Versus FD-OCT

11.2.2.1 Advantages of FD-OCT over TD-OCT

Although FD-OCT is a variant of OCT, it is based on a completely different interferometric configuration. Namely, FD-OCT detects the interference signal in the spectral domain. This spectral domain detection is typically achieved by a high-speed spectrometer or by a high-speed wavelength sweeping laser that is equipped with an intracavity monochrometer. FD-OCT based on the former scheme is called spectral domain OCT (SD-OCT) or spectral radar [16, 17], and that based on the latter scheme is swept-source OCT (SS-OCT) or optical frequency domain imaging [18].

Because SD-OCT provides an A-scan without mechanical scanning, it achieves a faster measurement speed than TD-OCT. The first 2D in vivo retinal imaging using SD-OCT was performed with a measurement speed of 1,000 A-lines/s in 2002 [19] (Figure 11.1(a), reprinted from [19]) and the first 3D in vivo retinal image was achieved in 2004 with a measurement speed of 29,000 A-lines/s [20] (Figure 11.1(b), reprinted from [20]). Although SS-OCT requires an intracavity mechanical wavelength scanning mechanism such as a piezoelectric Fabry-Perot filter, a polygon-mirror-based wavelength tuner, or a MEMS wavelength tuner, it provides a measurement speed of several ten- to hundred-thousand A-lines/s [18].

FD-OCT first attracted researchers because of its fast measurement speed. However, soon it was recognized that it had a higher sensitivity than TD-OCT [21–23]. This higher sensitivity can be seen as a result of the higher throughput of the probe beam to the image. TD-OCT is a low-coherence interferometer in which the reference and probe beams generate an interference signal only when the optical-path-length difference (OPLD) is less than the coherence length. Although this is the cause of the depth resolution of TD-OCT, it also means that the probe beam that is outside of the coherence length cannot contribute to image formation. On the other hand, FD-OCT is considered to be a bundle of monochromatic interferometers. Namely, in the case of SD-OCT, the signal output at each wavelength channel of the spectrometer provides monochromatic interference between the reference and probe beams. Hence, the beams generate an interference signal even when the OPLD is

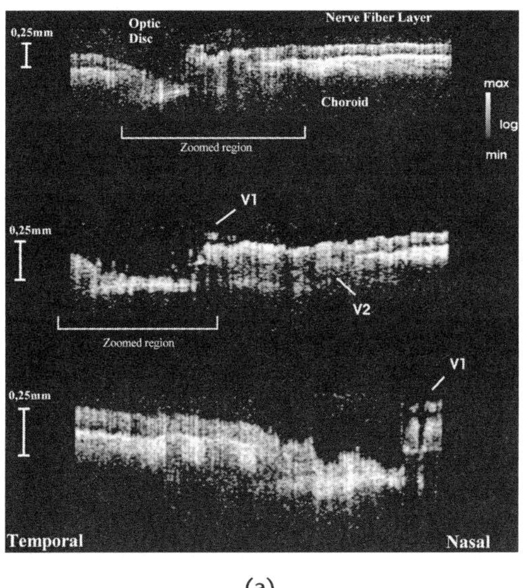

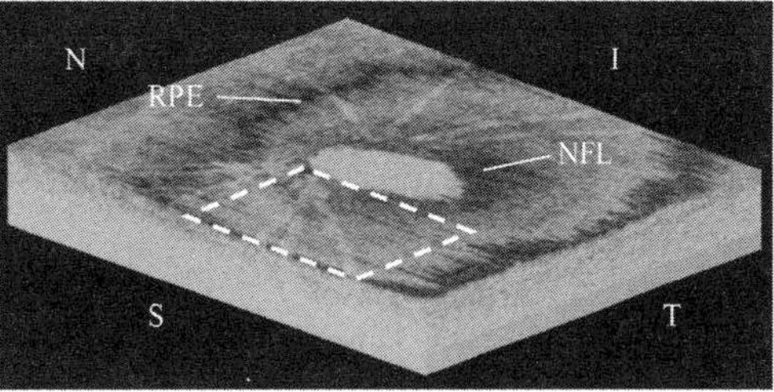

(a)   (b)

Figure 11.1   (a) The first in vivo SD-OCT image of a human retina. V1 and V2 indicate blood vessels. (b) Example of an en-face resection of a 3D volume of a human retina obtained by a high-speed SD-OCT. The labels I, S, N, and T, respectively represent the inferior, superior, nasal, and temporal directions. RPE and NFL indicate the retinal pigment epithelium and nerve fiber layer, respectively.

several millimeters long. The broadband interference (i.e., low-coherence interference, which provides the depth resolution) is then numerically performed later in a computer. Because of this interference scheme, almost all portions of the probe beam power contribute to image formation. The detailed theory that supports this high sensitivity is described elsewhere [21–23].

Because of these two major advantages of FD-OCT (i.e., high speed and high sensitivity), it has quickly become the de facto standard for tomographic investigation of the eye.

11.2.3   Principles of FD-OCT

11.2.3.1   Hardware Scheme of SD-OCT

In this section, SD-OCT is described as a representative FD-OCT method. An SD-OCT system comprises two subsystems, as depicted in Figure 11.2. The first subsystem is the OCT interferometer including the retinal scanner. The second subsystem is a high-speed spectrometer that typically consists of collimation optics, an optical grating, a focusing lens (Fourier transform lens), and a high-speed line sensor. The line sensor is typically a 1-D CCD or CMOS camera.

The light source in the interferometer is a broadband light source that can be a broadband pulse laser such as Ti:Sapphire laser or a superluminescent diode (SLD). A typical center wavelength is around 830 nm, and the bandwidth is from about 50 nm to more than 100 nm. The reasons for this typical center wavelength and bandwidth are described in Section 11.2.6. The light from the light source is split by a fiber coupler. A portion of the light is introduced into a probe arm that is equipped

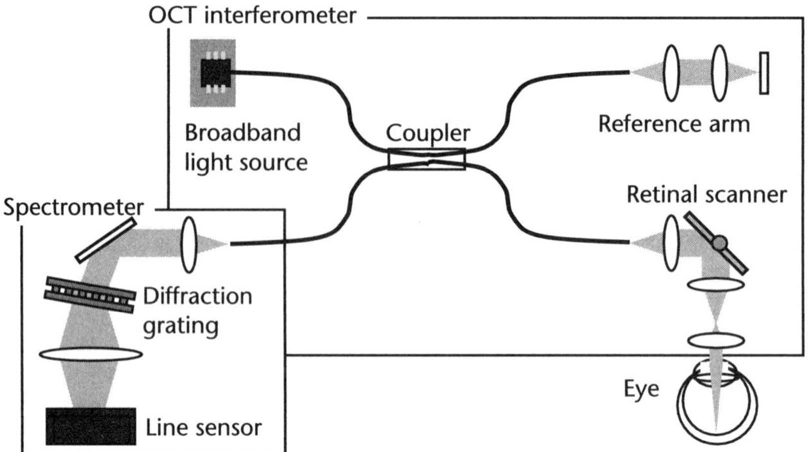

Figure 11.2 Simplified SD-OCT scheme consisting of two subsystems (i.e., the OCT interferometer that is almost equivalent to that of TD-OCT and a high-speed spectrometer).

with a retinal scanner. This portion is called the probe beam. The residual portion of the beam is introduced into the reference arm of the interferometer and becomes the reference beam. The backscattered probe beam and reflected reference beam are recombined by the coupler and then introduced into the high-speed spectrometer. These two beams interfere with each other and create an interferogram, not in the time domain but in the spectral domain. The spectral interference signal is then detected by a line sensor and numerically processed in a PC to form an OCT A-line. A pair of galvanometric scanning mirrors in the retinal scanner is additionally utilized to obtain a 2D or 3D OCT tomography.

11.2.3.2 Tomography Reconstruction

The numerical process to reconstruct an OCT A-line from the spectral interferogram consists of three phases: rescaling, inverse-Fourier transform, and logarithmic compression.

The first phase of the numerical process is rescaling. It mainly rescales the spectral interferogram into the wavenumber domain. The raw spectral interferogram detected by the high-speed spectrometer is almost linearly spaced in wavelength. In the rescaling process, the spectral interferogram is numerically resampled to be linear in wavenumber.

The rescaling process is further divided into two steps: calibration and resampling. In the calibration step, the wavelength spacing property of the spectrometer is characterized. The required accuracy of the wavelength calibration is significantly higher than that of standard spectrometers and can be as high as 0.1 nm or more. Thus, the standard calibration methods of a spectrometer cannot be utilized. On the other hand, some fringe analysis based methods can be used [24, 25]. This calibration is required only once per spectrometer. In the resampling process, the measured spectral interferogram is numerically resampled at a set of new sampling points that are evenly spaced in wavenumber. In this numerical resampling, the spectral interferogram is interpolated by a polynomial/spline function. It is also common to apply zero-padding interpolation before the polynomial/spline interpolation [20]. The purpose of the zero-padding is to extend the inherent cut-off frequency of the polynomial/spline interpolation [26]. The resampling is applied to all spectral interferograms obtained by the spectrometer.

The second phase of OCT A-line reconstruction is the inverse-Fourier transform. This part is mathematically described as follows. The electric field output from the OCT interferometer, and similarly that on the line sensor in the spectrometer, is described in the wavenumber domain as

$$\tilde{E}_d(k) = \sqrt{p_p \rho_p}\, \tilde{s}(k)\tilde{\alpha}(k)\tilde{\gamma}_p(k) + \sqrt{p_r \rho_r}\, \tilde{s}(k)\tilde{\gamma}_r(k), \tag{11.1}$$

where k is the angular wavenumber, p_p and p_r are the optical powers of the probe and reference beams, respectively, ρ_p and ρ_r are the intensity throughput of the probe and reference arms, respectively, and $\tilde{s}(k)$ is the complex spectrum of the light source. $\tilde{\gamma}_p(k)$ and $\tilde{\gamma}_r(k)$ are phase-only functions representing the group delay and the dispersion at the probe and sample arms, respectively. The complex spectral reflectivity of the sample is $\tilde{\alpha}(k)$, and is defined using the complex reflectivity of the sample along depth $\alpha(z)$ as $\tilde{\alpha}(k) \equiv \mathfrak{F}[\alpha(z)]$ where $\mathfrak{F}[\ ]$ represents Fourier transform and z is a variable representing the round trip optical path length (and is the Fourier counterpart of k). By assuming a refractive index of n, the depth position in the probe and reference arms z' is expressed as $z' = z/2n$. Note that because $\alpha(z)$ can be regarded as the structure of the sample along the depth, this is what we wish to measure. Thus, the following processes aim to retrieve $\alpha(z)$ from $\tilde{E}_d(k)$.

Although the spectrometer can perform spectrally resolved detection, the electric field $\tilde{E}_d(k)$ cannot be measured directly. This is because the line sensor is not sensitive to the electric field but to light intensity. Hence, we measure the power spectrum as

$$\begin{aligned}\tilde{I}_d(k) = \left|\tilde{E}_d(k)\right|^2 &= p'_p\left|\tilde{s}(k)\right|^2\left|\tilde{\alpha}(k)\right|^2 + p'_r\left|\tilde{s}(k)\right|^2 \\ &+ \sqrt{p'_p p'_r}\left|\tilde{s}(k)\right|^2 \tilde{\alpha}(k)\tilde{\gamma}_p(k)\tilde{\gamma}_r^*(k) + c.c.\end{aligned} \tag{11.2}$$

where $p'_p \equiv p_p\rho_p$ and $p'_r \equiv p_r\rho_r$, $\tilde{I}_d(k)$ is the power spectrum, the superscript $*$ represents the complex conjugate, and $c.c.$ represents the complex conjugate of the third term. Here, we have utilized $\left|\tilde{\gamma}_p(k)\right|^2 = \left|\tilde{\gamma}_r(k)\right|^2 = 1$ (i.e., $\tilde{\gamma}_p(k)$ and $\tilde{\gamma}_r(k)$ are phase-only functions). It should be noted that the raw spectrum spread on the line sensor is not evenly spaced in wavenumber, but it is rescaled to the wavenumber domain by the previously described rescaling process.

The measured spectrum is then inverse-Fourier transformed as

$$\begin{aligned}I_d(z) \equiv \mathfrak{F}^{-1}\left[\tilde{I}_d(k)\right] &= p'_p\Gamma(z) * \mathfrak{A}[\alpha(z)] + p'_r\Gamma(z) \\ &+ \sqrt{p'_p p'_r}\,\Gamma(z) * \alpha(z) * \mathfrak{F}^{-1}\left[\tilde{\gamma}_p(k)\tilde{\gamma}_r^*(k)\right] \\ &+ \sqrt{p'_p p'_r}\,\Gamma(z) * \alpha^*(-z) * \mathfrak{F}^{-1}\left[\tilde{\gamma}_p^*(k)\tilde{\gamma}_r(k)\right]\end{aligned} \tag{11.3}$$

where $\mathfrak{F}^{-1}[\ ]$ represents inverse-Fourier transform, the operator $*$ represents convolution, $\Gamma(z)$ is the coherence function of the light source, defined as $\mathfrak{F}^{-1}[|\tilde{s}(k)|^2]$, and is the auto-correlation of $\alpha(z)$, which is, according to the Wiener Khinchin's theorem, equivalent to $\mathfrak{F}^{-1}[|\tilde{\alpha}(k)|^2]$. Assuming the probe and reference arms have the same high-order phase dispersion, $\phi(k)$, the phase-only functions become $\tilde{\gamma}_p(k) = e^{i\phi(k)}e^{-ikz_p}$ and $\tilde{\gamma}_r(k) = e^{i\phi(k)}e^{-ikz_r}$, where z_p and z_e are the round-trip path length of the probe and reference arms, respectively. By substituting these phase terms into (11.3), the Fourier-transformed power spectrum becomes

$$
\begin{aligned}
I_d(z) = {}& p'_p\Gamma(z) * \mathfrak{A}[\alpha(z)] + p'_r\Gamma(z) \\
& + \sqrt{p'_p p'_r}\,\Gamma(z) * \alpha(z) * \delta\!\left(z - z_d\right) + \sqrt{p'_p p'_r}\,\Gamma(z) * \alpha^*(-z) * \delta\!\left(z + z_d\right)
\end{aligned}
\tag{11.4}
$$

where $z_d \equiv z_p - z_r$. This signal is schematically depicted in Figure 11.3. In this figure, the horizontal axis represents the double pass depth position z and the vertical axis represents the squared power of $I_d(z)$.

The first and second terms of the equation appear close to the origin of z, where the optical paths of the probe and reference beams match each other. On the other hand, the third term appears at around $z = z_d$, as indicated by convolution with delta-function $\delta(z - z_d)$.

Observing the third term, it is evident that the shape of this term is the sample structure $\alpha(z)$ convolved with the coherence function $\Gamma(z)$. In other words, this term provides us the sample structure blurred by a point spread function $\Gamma(z)$. Namely, this term is the OCT signal that we wish to measure. Note that (11.4) represents a numerical signal obtained by the process described earlier. Namely, the sample structure along the depth has been obtained as numerical data.

The fourth term is the mirrored conjugate of the third term (OCT signal) and is frequently denoted as a mirror signal. Since the OCT signal and the mirror signal carry identical information, one can arbitrarily choose one of the signals. In addition, the four signals are well separated if Δz is properly selected. Hence, it is easy to separate the OCT signal from the other terms.

This process is intuitively depicted in Figure 11.4. The process flow starts with the complex temporal output from the interferometer, in which the sample structure is encoded in a time-of-flight manner in the probe beam. The subsequent process extracts the sample structure from this signal. First, the signal is physically Fourier transformed by the spectrometer. The spectrometer first yields a joint complex spectrum of the probe and reference beams that is represented by (11.1). The joint power

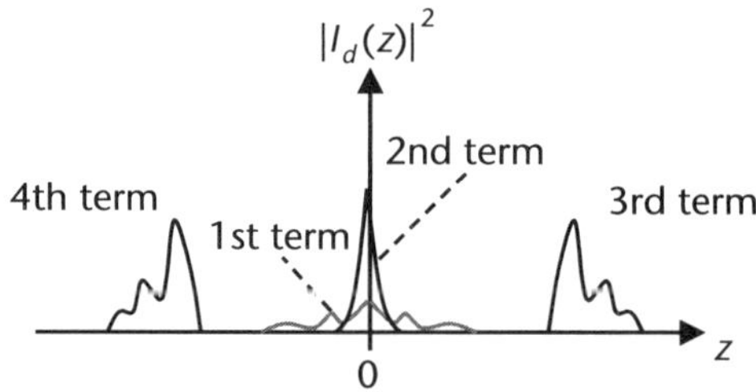

Figure 11.3 Schematic of the Fourier-transformed power spectrum.

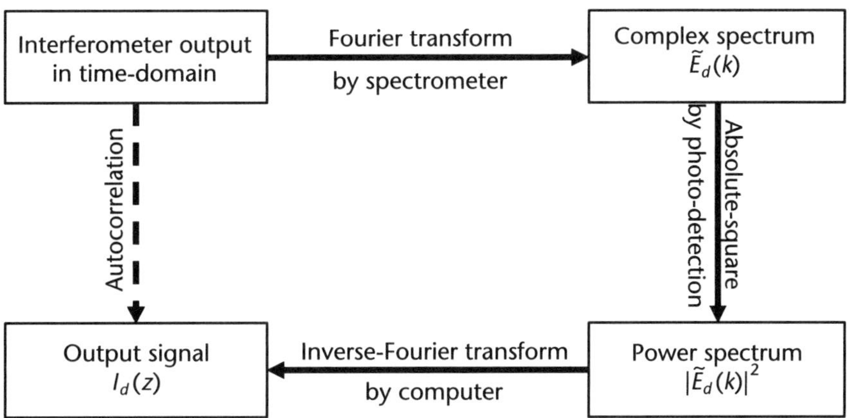

Figure 11.4   Processing flow of OCT signal reconstruction.

spectrum is then converted into its squared power during the detection process. Hence, from the line sensor we obtain the joint power spectrum of the probe and reference beams. This joint power spectrum is digitized and numerically inverse-Fourier transformed into $I_d(z)$.

This process flow can be summarized as follows: an input signal is first Fourier transformed, squared, and then inverse-Fourier transformed. Summarized this way, the process flow of SD-OCT is found to be equivalent to the calculation flow of the Wiener-Khinchin's theorem. This suggests that the final output of this flow $I_d(z)$ is the autocorrelation of the input signal. Since the input signal consists of the probe and reference beams, the output signal includes not only the autocorrelations of the probe beam and the reference beam, but also the crosscorrelation between them. These cross-correlation terms are represented by the third and fourth terms of (11.4) (i.e., the OCT and mirror signals).

The third and final part of the OCT reconstruction is logarithmic compression. This process is needed because of the sensitivity of OCT, and hence the SNR of an OCT image is significantly higher than the color resolution of an OCT monitor (computer display) or the contrast sensitivity of a human eye (observer). For instance, an SD-OCT system with a standard configuration possesses a sensitivity of around 98 dB [27, 28]. Assuming the reflectivity of the eye is -50 dB, the SNR of an OCT image is around 48 dB. On the other hand, the typical contrast resolution of a computer display is 256 grayscales, which is only around 24 dB.

BOX 11.1   Logarithmic Compression of Signal Range

To properly display the OCT signal on a computer screen, the OCT signal is commonly compressed into logarithmic space using $10\log|I_d(z)|^2$. Here, the base of the logarithm is 10, and $I_d(z)$ is absolute-squared in order to convert it to signal energy. Because $|I_d(z)|^2$ is an energy, $10\log|I_d(z)|^2$ represents the OCT signal in dB scale. This definition of dB scale is conventionally utilized for OCT.

11.2.4 SD-OCT and SS-OCT

As mentioned in previous sections, there are two implementations of FD-OCT: SD-OCT and SS-OCT [18, 24, 29]. SD-OCT uses a broadband light source and the spectrally resolved detection is achieved by a high-speed spectrometer. On the other hand, SS-OCT uses a nonspectrally resolved point detector. The spectral resolution of SS-OCT is achieved by a high-speed wavelength scanning light source, a laser with a high-speed wavelength tuner in its cavity [5, 18, 29, 30]. Because of the wavelength scanning, a spectrally resolved interferogram is obtained as a time-resolved signal measured by the point photodetector. SS-OCT is regarded as a monochrometer-version of FD-OCT.

The decision to use SD-OCT or SS-OCT is mainly based on the probe wavelength. As described in Section 11.2.3, SS-OCT uses a line sensor to detect the spectral interferogram. This line sensor is typically a silicon device that is only sensitive to wavelengths shorter than 1 μm. Hence, FD-OCT with a short probe-wavelength is commonly implemented as SD-OCT. It should be noted that longer-wavelength FD-OCT can also be implemented as SD-OCT [25, 31] if an InGaAs line sensor is used. However, the relatively high price of InGaAs line sensors has impeded the popularization of SD-OCT with long-wavelength probes.

BOX 11.2 Different Noise Properties between SD- and SS-OCT

FD-OCT using a long probe wavelength such as 1060 or 1310 nm is commonly implemented as SS-OCT. One potential drawback of SS-OCT is its wide detection bandwidth. For instance, if the acquisition time of a single spectral interferogram is τ, the detection bandwidth of SD-OCT is $1/\tau$. On the other hand, the bandwidth of SS-OCT is N/τ, where N is the number of sampling points in a single spectral interferogram. Here we have assumed a 100% detection cycle duty. In a typical FD-OCT implementation, N is about 1,000–2,000. Namely, the detection bandwidth is 1,000–2,000 times wider in SS-OCT than in SD-OCT. This wider detection bandwidth results in fewer rejections of relative intensity noise (RIN) of the light source. Hence, SS-OCT frequently operates at the RIN limit, while SD-OCT easily operates close to the shot-noise limit (see also Section 11.2.8). This drawback of SS-OCT can be partially compensated for by using a balanced photodetection scheme. Because of the RIN rejection capability of balanced detection, recent SS-OCT systems can operate close to the shot-noise limit (i.e., only a few dB below it). This difference in the detection scheme (i.e., balanced or non-balanced), results in different interferometer designs for SD- and SS-OCT.

11.2.5 Interferometer Design of FD-OCT

FD-OCT interferometers can be classified as balanced or nonbalanced. The non-balanced interferometer is commonly configured as a Michelson interferometer and is used for SD-OCT. On the other hand, the balanced interferometer is used for SS-OCT and can be implemented as either a Mach-Zehnder or Michelson interferometer.

11.2.5.1 Nonbalanced Michelson Interferometer

As discussed in Section 11.2.4, SD-OCT does not require balanced photodetection because of its narrow detection bandwidth. Hence, SD-OCT is typically implemented with a nonbalanced Michelson interferometer. An example of a non-balanced Michelson interferometer is depicted in Figure 11.5(a).

The advantage of the nonbalanced Michelson interferometer is its simplicity. The minimally required fiber component for this type of interferometer is a coupler. An isolator and several polarization controllers are also frequently used. The former is utilized to reject back reflection light to the light source and consequently avoid photodamaging it. The latter is to maximize the OCT signal, achieved when the polarization states of the probe and reference beams are identical at the detector.

A disadvantage of the Michelson interferometer is the inevitable loss of the probe beam. Because the total throughput of the probe beam as well as the collection rate of the back scattered probe beam are defined by the splitting ratio of the coupler, the throughput cannot be 100%. The maximum total throughput is obtained with a 50/50 coupler and is 25%.

However, for eye imaging, the probe collection rate is more important than the total throughput. The probe power on the eye is limited by a safety standard. Thus, the collection rate is the primary factor that limits the sensitivity. One typical example of the coupler for retinal imaging is a 20/80 coupler. The 20% portion of the beam is utilized as a probe beam, and the 80% portion of the back scattered beam from the sample is collected for imaging. Although the total throughput of this configuration is only 16%, this configuration provides better sensitivity than the 50/50 coupler when the probe power on the eye reaches the safety limit.

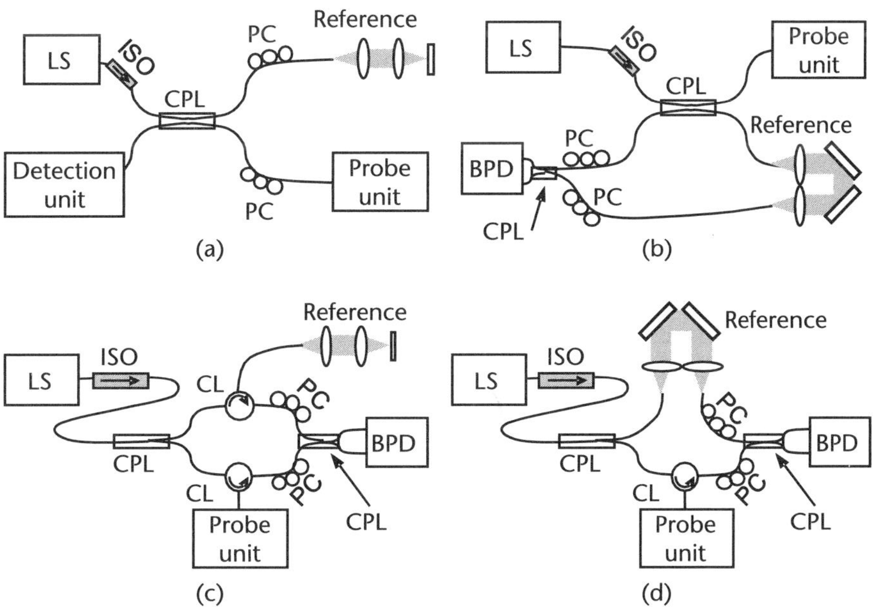

Figure 11.5 Examples of interferometer configurations: (a) nonbalanced Michelson interferometer, (b) balanced Michelson interferometer, and balanced Mach-Zehnder interferometers are depicted in (c) and (d). Here, LS is a light source, ISO is an isolator, CPL is a coupler, CL is a circulator, PC is a polarization controller, and BPD is a balanced photo-detector.

A nonbalanced Michelson interferometer is typically utilized for SD-OCT at any wavelength.

11.2.5.2 Balanced Mach-Zehnde r Interferometer

Typical configurations of the balanced Mach–Zehnder interferometer are depicted in Figure 11.5(c) and (d). In these configurations, the light is first split into probe and reference beams by a coupler. These beams are reflected and back scattered by a reference mirror and a sample and then the recombined by another coupler.

The biggest advantage of the Mach-Zehnder interferometer is its high probe beam throughput. Because it uses a circulator, all the optical power of the back scattered probe beam contributes to image formation. Since the OCT sensitivity is proportional to the probe beam power at the detector [21–23], this advantage directly results in high system sensitivity.

A possible demerit of the Mach-Zehnder interferometer is the circulator. Although a circulator is a well-established device for 1310-nm wavelengths, it is not well developed for 1050 nm. Hence, the circulator for 1050 nm is more expensive than that for 1310 nm, and its performance is not excellent. For example, circulators are known to have polarization mode dispersion (PMD). Although, the PMD of the circulator is negligible for standard 1310-nm OCT, it significantly affects a polarization sensitive extension of OCT known as polarization sensitive OCT (PS-OCT) [32–34]. Because the circulator performance is generally worse at 1050 nm, circulator PMD is not negligible, even for a standard nonpolarization-sensitive OCT at 1050 nm.

In summary, the balanced Mach-Zehnder interferometer is suitable for 1310-nm SS-OCT.

11.2.5.3 Balanced Michelson Interferometer

Another interferometer for SS-OCT is the balanced Michelson interferometer. A typical configuration is depicted in Figure 11.5(b).

The main advantage of this type of interferometer is balanced photodetection without the need for circulators. The balanced detection scheme means that this interferometer is suitable for SS-OCT. In addition, the absence of a circulator makes it suitable for 1050-nm wavelengths. Hence, this interferometer is frequently utilized for 1050-nm SS-OCT [4, 35].

Potential drawbacks of this interferometer are its limited total throughput and limited collection ratio of the probe beam. These limitations are identical to that of a nonbalanced Michelson interferometer. Hence, a similar selection strategy for the coupler splitting ratio can be utilized to optimize the interferometer for eye imaging.

11.2.6 Wavelength Selection

Because water is the main light absorption source in living tissue, the wavelength of the OCT probe beam is selected to match one of local minima of water absorption. Figure 11.6 shows the light absorption coefficients of water in the NIR region,

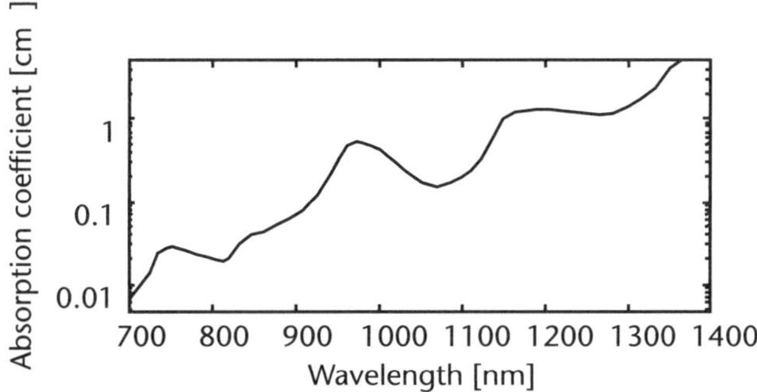

Figure 11.6 Absorption coefficient of water in the NIR region.

where the plot was created using a dataset from [36]. In this absorption curve, there are three local absorption minimums at approximately 830, 1050, and 1300 nm. These three wavelengths are the most common OCT probing wavelengths.

The selection of one of these three wavelengths is roughly based on two general properties of light. The first property is that the shorter wavelength is less absorbed by water. This property leads to the first rule of penetration: the shorter wavelength probe penetrates the tissue more deeply. The second property is that, in general, longer wavelengths are less scattered by the tissue. The first rule is therefore in conflict with that of the second, namely, that the longer wavelength has higher penetration. In practice, tissue penetration is defined to balance these two rules. Hence, the wavelength for the best penetration greatly depends on the scattering and absorption properties of the sample to be measured. The probe wavelength is then selected mainly to best penetrate a specific sample.

11.2.6.1 The 830-nm Band

In the NIR region, water has the lowest absorption at 830 nm. On the other hand, this wavelength is the most highly scattered by the tissues. Hence, the 830-nm probe is optimal for tissue that has fewer scatterers and comprises large amounts of water. This suggests that the retina is the perfect target for 830 nm.

First, the retina possesses relatively fewer scatterers. The photoreceptor cells that convert optical signals to electric signals are located at the outer (posterior) part of the retina. Hence, when we see objects, the light that carries the visual information stimulates the photoreceptor cells after passing through the inner part of the retina. This suggests that the scattering of the inner retina is low; otherwise, the human visual function would be poor. Because of this low scattering, the 830-nm probe can achieve relatively high penetration into the retina, even though it is more highly affected by scattering than other, longer wavelengths.

Another characteristic condition of retinal imaging concerns the optical media of the eye. The retina is positioned beneath the eye optics that includes the cornea, aqueous humor, crystalline lens, and vitreous that forms a total length of around 24 mm (i.e., a 48-mm round trip for the probe beam). These tissues mainly consist of water and have nearly no scattering. Hence, for retinal imaging, the probe-beam should be less absorptive by water.

These two conditions make the 830-nm probe beam particularly suitable for retinal imaging. It is also noteworthy that OCT with an 830-nm probe generally has better depth resolution than with other wavelengths. This is because the depth resolution of OCT is proportional to the square of the center wavelength (see Section 11.2.7). Because a retina consists of a fine-layered structure, this high resolution provides particular advantages for clinical examination. Hence, almost all commercial retinal OCT is implemented with an 830-nm probe.

A Ti:Sapphire mode-locked laser and an SLD are two of the most common light sources for 830-nm OCT. In practice, the former has a center wavelength of about 800 nm, while that of the latter is between 830–840 nm. For clinical retinal OCT, SLD is more commonly utilized than the Ti:Sapphire laser because of its stability, cost effectiveness, and compactness. A typical SLD utilized for retinal SD-OCT has a center wavelength of 840 nm and a bandwidth of 50 nm, providing a depth resolution of around 5 μm in tissue.

11.2.6.2 The 1050-nm Band

The 1050-nm band is an emerging wavelength band for retinal imaging. Because water absorption has a second minimum at 1050 nm, this wavelength can pass through the ocular optical medium. In addition, this relatively long wavelength suppresses scattering by the retinal tissue and provides higher penetration than the 830-nm probe. This high penetration enables investigation of the deep posterior eye (i.e., the choroid).

Retinal imaging at the 1050-nm band was demonstrated with TD-OCT on ex vivo porcine eyes in 2003 [37] and for in vivo human eyes in 2004 [38]. Since 2006, the 1050-nm band has been commonly utilized for retinal SS-OCT [39–41]. Retinal SD-OCT at 1050 nm has also been demonstrated [25, 42, 43]. Although only one model of retinal 1050-nm OCT is commercially available as of 2014, several other commercial prototypes have also been demonstrated. Hence, 1050 nm is expected to be the next generation of clinical retinal OCT, replacing 840-nm SD-OCT.

BOX 11.3   Optical Components for 1050-nm Band

In general, optical and fiber components for 1050-nm beams are more expensive than other wavelengths and of relatively low quality. However, recent developments in Ytterbium fiber lasers with a wavelength of 1060 nm have coincidentally driven the development of cost-effective and high-quality optical and fiber components at this wavelength. The availability of these components further motivates the use of 1050 nm for the next generation of clinical retinal OCT.

11.2.6.3 The 1310-nm Band

The 1310-nm band is a telecom wavelength band, and hence the optical and fiber components at this wavelength are generally cost effective and high quality. Hence, this band has been widely utilized for OCT.

Compared with the other OCT wavelengths, 1310-nm light is minimally scattered by tissue. Hence, it has very high tissue penetration. As a result, this wavelength band is the most common wavelength for several clinical applications such as dermatology, cardiology, and gastrointestinal imaging. For ophthalmology, this wavelength band is utilized for anterior eye imaging [24, 44]. On the other hand, 1310-nm light cannot be utilized for retinal OCT because it is highly absorbed by water.

11.2.7 Resolution and Dispersion

11.2.7.1 Transversal Resolution

Since OCT is a point-scanning imaging modality and the signal intensity is proportional to the field amplitude instead of the squared power of the probe, the OCT transversal resolution is defined as the spot size on the sample. Namely, assuming a Gaussian incident beam with diameter, the transversal resolution Δx becomes

$$\Delta x = \frac{4\lambda_c}{\pi}\left(\frac{f}{d}\right) \tag{11.5}$$

where λ_c is the center wavelength of a probe beam and f is the focal length of the focusing lens [45]. This equation suggests that the transversal resolution can be improved by increasing the incident beam diameter or decreasing the focal length.

However, the depth of focus, computed as $\pi\Delta x^2/2\lambda$, decreases as the transversal resolution increases. Hence, the transversal resolution should be properly selected to maintain a sufficient depth of focus.

For retinal imaging, the focal length is that of the eye optics and hence is not selectable. On the other hand, the beam diameter can be increased to achieve better transversal resolution. One specific issue for retinal imaging is eye aberrations. Because the human eye has a significant amount of aberrations, over-increasing the beam diameter sometimes worsens the transversal resolution. Typically, the best achievable transversal resolution of retinal OCT is around 20 μm.

It is worth noting that high transversal resolution, as high as transform-limited resolution, can be achieved by a retinal OCT equipped with adaptive optics [46, 47].

EXERCISE 11.1

Calculate the transversal resolution of the following example of a typical OCT configuration. The focal length of the lens is 40 mm, the center wavelength is 1.3 μm, and the beam diameter is 1.8 mm.

Solution

By substituting $f = 40 \times 10^{-3}$ m, $\lambda_x = 1.3 \times 10^{-6}$ m, and $d = 1.8 \times 10^{-3}$ into (11.5), the transversal resolution is obtained as $\Delta x = 3.7$ μm.

11.2.7.2 Depth Resolution

Equation (1.4) suggests that the OCT axial point spread function (PSF) is defined by the coherence function $\Gamma(z)$, which is a Fourier transform of the intensity spectrum of the light source. Hence the depth resolution of OCT is defined by the width of the coherence function.

Assuming a Gaussian intensity spectrum with a full width at half maximum (FWHM) of Δk in wavenumber, the full width of -6 dB of the maximum of $\Gamma(z)$ is obtained as $\Delta z = 4\ln2/\Delta k$. This equation can be approximated using the center wavelength of the probe beam λ_c and the FWHM of wavelength of $\Delta\lambda$ as

$$\Delta z = \frac{2\ln2}{\pi}\frac{\lambda_c^2}{\Delta\lambda} \tag{11.6}$$

This Δz is commonly utilized as the depth resolution of OCT.

Note that this resolution was derived assuming a Gaussian spectrum. Recent wide-band SLDs frequently have non-Gaussian spectra, and this equation provides only a rough measure of depth resolution for such light sources.

EXERCISE 11.2

The specifications of an example of commonly utilized SLD light sources for SD-OCT are following. Calculate the in-air and in-tissue depth resolutions of OCT utilizing this lightsource. The lightsource has a Gaussian-shaped spectrum with FWHM of 55 nm and its center wavelength is 840 nm. The refractive index of the air is 1.0 and that of tissue can be assumed to be 1.38.

Solution

By substituting $\lambda = 840 \times 10^{-9}$ m and $\Delta\lambda = 55 \times 10^{-9}$ m into (11.6), the depth resolution is obtained as $\Delta z = 5.7 \ \mu$m in air. By dividing the in-air resolution by the refractive index of the tissue (1.38), the in-tissue resolution is obtained as $\Delta z = 4.1 \ \mu$m.

11.2.7.3 Dispersion and Dispersion-Correction

Recall that (11.4) was derived by assuming the probe and reference arms have the same high order dispersion. To consider the dispersion effect in OCT imaging, we need to remove this assumption.

The OCT equation that does not make this assumption (11.3) shows that the OCT signal is blurred by the function $\Gamma(z) * \mathfrak{F}\left[\tilde{\gamma}_p(k)\tilde{\gamma}_r^*(k)\right]$. To consider different high order dispersions, here we let $\tilde{\gamma}_p(k) = e^{i\phi_p(k)}e^{-ikz_p}$ and $\tilde{\gamma}_r(k) = e^{i\phi_r(k)}e^{-ikz_r}$, where $\phi_p(k)$ and $\phi_r(k)$ represent high order dispersions in the probe and reference arms, respectively. The previous function then becomes $\Gamma(z) * \mathfrak{F}^{-1}\left[e^{i\phi_d(k)}\right] * \delta(z - z_d)$, where $\phi_d(k) = \phi_p(k) - \phi_r(k)$ and represents an unbalanced dispersion between the two arms. Ignoring the shifting, the axial PSF under unbalanced dispersion then becomes

$$PSF = \Gamma(z) * \mathfrak{F}^{-1}\left[e^{i\phi_d(k)}\right] \tag{11.7}$$

Namely, the PSF is additionally broadened by the unbalanced dispersion. Another remarkable point is that the PSF under dispersion is asymmetric, while PSF without dispersion is symmetric, because the inverse-Fourier transform of an arbitrary function becomes asymmetric unless the function is a real function. This fact can be utilized to identify the cause of unexpected PSF broadening when building an OCT system.

Dispersion unbalance comes not only from the OCT system but also from the sample itself. Particularly for retinal imaging, the ocular media causes dispersion unbalance. In addition, the ocular dispersion varies subject by subject. Hence, dispersion correction is essential for retinal imaging.

Unbalanced dispersion can be roughly cancelled by designing the OCT optics to have minimally unbalanced dispersion. However, this hardware cancellation cannot cancel the residual dispersion that varies among subjects. This residual dispersion is commonly canceled using numerical methods.

Using the same $\phi_p(k)$ and $\phi_r(k)$ as in (11.7), the third term of (11.2) (i.e., the component of spectral interferogram corresponding to the OCT signal) becomes $\sqrt{p'_p p'_r}\left|\tilde{s}(k)\right|^2 \tilde{\alpha}(k)e^{i\phi_d(k)}e^{-ikz_d}$, where $e^{-i\phi_d(k)}$ is the unbalanced dispersion. Because the interferogram has been digitized, the unbalanced dispersion can be canceled by numerically multiplying the interferogram by the complex conjugate of $e^{-i\phi_d(k)}$ (i.e., $e^{i\phi_d(k)}$), called the counter dispersion.

Unfortunately, $e^{-i\phi_d(k)}$ is normally unknown. Hence, iterative optimization algorithms are utilized to cancel the unbalanced dispersion. In such algorithms, an estimated counter dispersion is applied to the spectral interferogram before the Fourier transform. Subsequently, a sharpness metric is computed from a reconstructed OCT image. The counter dispersion is then iteratively optimized to maximize the sharpness metric. Several types of sharpness metrics have been utilized for retinal imaging, including the number of high-intensity pixels [27] and image information entropy [40].

11.2.8   SNR and Sensitivity

11.2.8.1   Definition and Measurement of SNR and Sensitivity

Sensitivity and SNR are commonly utilized to evaluate the imaging performance of OCT. The sensitivity is defined as the maximum measurable attenuation of the probe beam. Thus, sensitivity is a specification of the system. On the other hand, SNR is the ratio of the signal energy (at a certain point in the image) to the noise energy and hence is a specification of an image. Although the sensitivity and SNR are specifications of two different things, they are tightly coupled. Namely, the SNR is proportional to the sensitivity when a same sample is measured at a same position.

The sensitivity of an OCT system is commonly measured in the following manner. An OCT signal is acquired using a perfectly reflecting mirror with a neutral density (ND) filter placed before the mirror as a sample. This ND filter mimics the attenuation of a target tissue. In the case of retinal imaging, an ND filter with an optical density of 2.5–3.0 is utilized. This gives a round-trip attenuation of −50 to −60 dB and mimics the reflectivity of the eye. Absorption-type ND filters are not optimal for this purpose because they sometimes have a large dispersion. For this measurement, transversal scanning should not be performed to avoid inessential signal attenuation. The signal intensity at the mirror is then determined from the OCT image. Multiple A-lines are then obtained with the mirror sample removed. A series of complex OCT signals are taken at the position where the mirror sample was located, and the variances of the real and imaginary parts of the OCT signal, σ_r^2 and σ_i^2, are obtained. The noise energy can then be determined as a summation of the variances (i.e., $\sigma_r^2 + \sigma_i^2$). The SNR of the OCT image of the mirror is then determined as the ratio between the signal and noise energies. Finally, the sensitivity is determined in dB scale by subtracting the SNR from the attenuation of the ND filter. For example, if the attenuation of the ND filter is −60 dB and the mirror signal has an SNR of 40 dB, the sensitivity becomes −60 dB − 40 dB = −100 dB.

According to its literal definition, the sensitivity is a negative value in dB. However, it is conventionally presented as a positive value by omitting the negative sign. We also use this convention in this chapter. With this convention, the SNR becomes proportional to the sensitivity in linear scale, or a constantly biased value in dB. Particularly for a sample with 100% of reflectivity, the SNR is identical to the sensitivity.

11.2.8.2 Noise Source and Sensitivity Optimization

There are three major noise sources in OCT. The first one is shot noise that originates from statistical fluctuations in counting photons and/or electrons. Because it is a fluctuation in counting, shot noise is proportional to the square-root of the optical power, and its energy is proportional to the optical power. Namely, short noise energy σ_{shot}^2 is proportional to $p_p' + p_r'$. In practical OCT imaging, we can assume $p_r' \gg p_p'$, hence $\sigma_{\text{shot}}^2 \propto p_r'$.

The second source of noise is the RIN of the light source. Since RIN is a fluctuation in light intensity, its energy σ_{RIN}^2 is proportional to $(p_p' + p_r')^2$. With the same assumptions as for shot-noise, we can say $\sigma_{\text{RIN}}^2 \propto p_r'^2$.

The third source of noise is detection noise. This is an aggregation of nonoptical noises such as the thermal noise of the photodetector and electrical noise of a signal digitizer. As evident from its definition, the energy of detection noise is independent to p_p' and p_r'. Here, detection noise energy is denoted as σ_{det}^2.

The energy (i.e., squared power) of the OCT signal ε_s is proportional to $p_p' p_r'$, as indicated by (11.3) and (11.4). Using the energies of the signal and various noises, the SNR is expressed as

$$\text{SNR} \equiv \frac{\varepsilon_s}{\sigma_{\text{shot}}^2 + \sigma_{\text{RIN}}^2 + \sigma_{\text{det}}^2} = \frac{c_a p_p' p_r'}{c_b p_r' + c_c p_r'^2 + \sigma_{\text{det}}^2}, \tag{11.8}$$

where c_a, c_b, and c_c are proportional constants that are defined by hardware specifications. Assuming a mirror sample with 100% reflectivity, this equation also represents the sensitivity.

Although we omit discussion of the practical values of the proportional constants, this equation provides some important insights into OCT sensitivity. Most importantly, OCT sensitivity is proportional to probe power. Therefore, it is a common strategy to increase the probe power in order to improve sensitivity. Second, sensitivity is a function of the power of a reference beam. From this perspective, the OCT operation can be classified into three regimes: shot-noise limited, RIN limited, and detection-noise limited. In each regime, the sensitivity has a different dependency on the reference power.

The regime where $\sigma_{shot}^2 \gg \sigma_{RIN}^2, \sigma_{det}^2$ is the shot-noise limited regime. In this regime, the sensitivity is approximated by $(c_a/c_b)p'_p$ and is evidently independent from the reference power. Similarly, in the RIN limited regime, the sensitivity is proportional to $(c_a/c_c)(p'_p/p'_r)$. Namely, the sensitivity is inversely proportional to the reference power. In the detection-noise limited regime, the sensitivity is proportional to $(c_a/\sigma_{det}^2)(p'_p p'_r)$ and hence proportional to the reference power.

These facts are summarized in Figure 11.7(a), where the logarithmic sensitivity is plotted as a function of the logarithm of the reference power on the detector. The solid curve depicted in this figure is called a sensitivity curve. The dashed curves (i)–(iii) are the sensitivity curves assuming only detection noise, shot-noise, or RIN exists, respectively. The dashed-dotted lines divide the three noise-regimes. As the reference power increases, first the sensitivity and subsequently the SNR increase. In this regime, the sensitivity is dominated by the detection noise (the detection noise limited regime). Once the sensitivity reaches the shot-noise limited regime, it becomes constant. Finally, it begins to decrease in the RIN regime. Hence, the OCT reference power should be optimized such that OCT works in the shot-noise limited regime.

Another situation regarding the sensitivity is depicted in Figure 11.7(b), where OCT can never operate at the shot-noise limit. In this case, the reference power can be suboptimally optimized to have the maximum achievable sensitivity. In addition, modifications of the OCT system should be considered, such as employing a light source with a lower RIN, using detector/detection electronics with lower detection noise, or both.

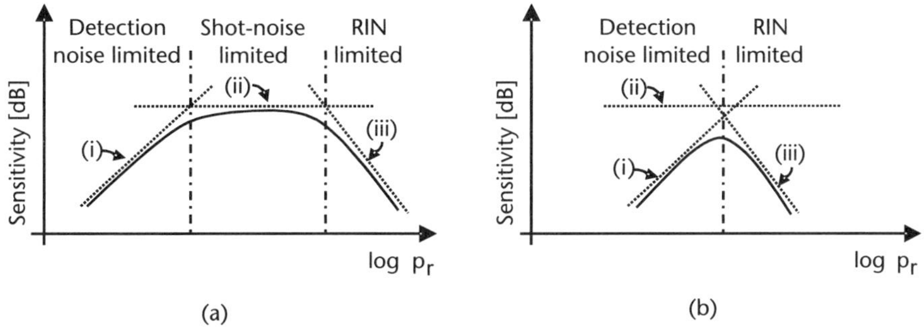

Figure 11.7 Examples of OCT sensitivity curves, where the sensitivity is plotted in a dB-scale as a function of logarithm of the power of a reference beam on a detector. The solid curves are the sensitivity curve, the dashed lines (i)–(iii) are the sensitivity curves assuming only detection noise, shot-noise, or RIN exists, respectively. The OCT can operate as a shot-noise limited regime in (a), but cannot in (b).

To roughly design an OCT system, shot-noise limited sensitivity is particularly important. This is because shot-noise is inevitable noise and hence is the true theoretical limit of OCT sensitivity. The shot-noise limited sensitivity of FD-OCT is given as [18, 22, 48]

$$\text{Sensitivity} = \frac{\eta \tau}{h \nu_0} p_p', \tag{11.9}$$

where p_p' is the optical power of the probe on the detector when the sample is a mirror with 100% reflectivity, h is the Plank constant, ν_0 is the center frequency of the light source, η is the quantum efficiency of the photodetector, and τ is the integration time of a line sensor (for SD-OCT) or the inverse of the light source wavelength scanning frequency (for SS-OCT).

EXERCISE 11.3

Calculate the shot-noise limited sensitivity of an SD-OCT with the following specifications. The light source is an SLD with 840-nm center wavelength. The interferometer is a nonbalanced Michelson interferometer. A 30/70 coupler splits the beam into reference and probe arms and 30% portion of the beam is lead to the probe arm. The probe power on the sample is 750 μW. The diffraction efficiency of a grating utilized in a spectrometer is 90%. The quantum efficiency of a line CCD sensor utilized in the spectrometer is 40%. The line rate of the sensor is 70 kHz, and its duty is 100%.

Solution

The coupler leads a 70% portion of the probe beam reflected from a sample. By considering this collection efficiency of the coupler and the diffraction efficiency of the grating, the probe power on the sensor becomes $p_p' = 750 \times 10^{-6} \times 0.7 \times 0.9$ W. The exposure time of the sensor is obtained from the line rate and duty as $\tau = 1.0/70000$ s. The center frequency $\nu_0 = c/\lambda_c$ with c as the speed of light. And $\eta = 0.4$.

By substituting these values into (11.9), the shot-noise limited sensitivity is estimated as 1.14×10^{10}, which is in 100.6 dB in logarithmic scale.

11.3 OCT in Ophthalmology

In this section, we discuss the utility of OCT in ophthalmology. The modalities for ophthalmic diagnosis can be characterized by three factors. The first factor is whether the modality is objective or subjective; the second factor is whether it is structural or functional; and the third factor is invasiveness. For example, visual acuity and field tests are subjective, functional, and noninvasive examinations. Similarly, angiographies, including fluorescein angiography and indocyanine green angiography (ICGA), are objective, functional, and invasive modalities. Among other modalities, OCT is characterized as an objective, structural, and noninvasive modality. Color fundus photography and fundus autofluorescence imaging are also objective, structural, and noninvasive, but these modalities only provide 2D en-face images, while OCT can provide a 3D volumetric tomography.

Thus far, we have not particularly differentiated OCT systems for posterior and anterior eye examination. However, the optimal OCT is not the same for these two parts of the eye. For example, the probe beam for posterior eye imaging should not be highly absorbed by the ocular media, as discussed in Section 11.2.6, while this condition is not significant for anterior imaging. Additionally, the information needed for the diagnosis of posterior and anterior eye diseases differs, and hence the optimal design of an anterior or posterior OCT also differs. In this section, we independently discuss the features and utilities of posterior and anterior OCT.

11.3.1 Posterior Eye Imaging

11.3.1.1 Structure of the Posterior Eye

The posterior part of the eye consists of three major layers, the retina, choroid, and sclera, in anterior (inner) to posterior (outer) order. The retina is the sensory part of the eye and consists of a further 10 layers, from the anterior to posterior: inner limiting membrane (ILM), nerve fiber layer (NFL), ganglion cell layer (GCL), inner plexiform layer (IPL), inner nuclear layer (INL), outer plexiform layer (OPL), outer nuclear layer (ONL), external limiting membrane (ELM), photo-receptors, retinal pigment epithelium (RPE), and Bruch's membrane. The photo-receptor layer is further divided into two segments: inner and outer. The choroid is a vascular-rich tissue beneath the retina that nourishes the posterior part of the retina. The retina and the choroid are separated by the RPE, a melanin rich cellular monolayer, and Bruch's membrane. The sclera is a further outer layer and provides mechanical support for the eye. A narrow definition of the retina refers to only the retina, while the wider sense of the word refers to the retina and choroid.

11.3.1.2 Posterior Imaging at 830 nm

Figure 11.8 shows an example OCT cross-section of an in vivo normal human macular, which is the visual center of the posterior eye. The image was taken by an ultra-high-resolution SD-OCT [49] at the 830-nm band (a center wavelength of 870 nm and a bandwidth of 170 nm) possessing a depth resolution of 2.9 μm in tissue. The top and bottom of the image represents the anterior and posterior directions, respectively, and this image orientation is the convention for displaying posterior OCT images. In this image, the OCT signals are displayed with an inverted grayscale (i.e., black represents hyperscattering and white represents hyposcattering). Because the ILM is too thin to be visualized, the topmost hyperreflective layer appearing in the OCT image is the NFL. From the NFL to the posterior, the retinal layers shown are the GCL (hyposcattering), IPL (hyperscattering), INL (hyposcattering), OPL (hyperscattering), ONL (hyposcattering), and ELM (thin hyperscattering layer). The strong hyperscattering layer beneath the ELM is the outermost part of the inner segment of the photoreceptor, called the ellipsoid region (ELS). Although it is currently believed to a part of the inner segment, it was once believed to be the junction between the inner and outer segments of the photoreceptor and hence

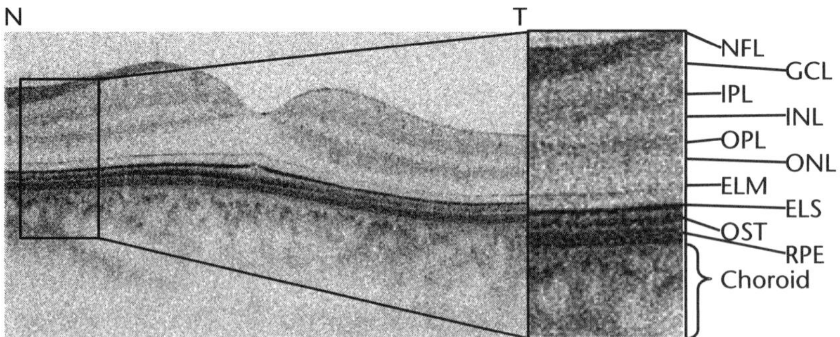

Figure 11.8 An example of an ultra-high-resolution OCT of in vivo normal macula. Nasal and temporal directions are represented by N and T, respectively. NFL: nerve fiber layer, GCL: ganglion cell layer, IPL: inner plexiform layer, INP: inner nuclear layer, OPL: outer plexiform layer, ONL: outer nuclear layer, ELM: external limiting membrane, ELS: ellipsoid region, OST: the tip of photoreceptor outer segment, and RPE.

is still sometimes denoted as the IS/OS. Another hyperscattering line beneath the ELS is the outer tip of the outer segment of the photoreceptor (OST). The final strong-scattering line beneath the OST is the RPE. Bruch's membrane cannot be identified with the normal eye because it is closely attached to the RPE. However, it sometimes appears as a faint hyperscattering line in some pathologic cases. The ELS, OST, and RPE layers are sometimes denoted as the RPE complex for historical reasons. Namely, these layers were hard to separate using early generations of posterior OCT because of their relatively low depth resolution.

The choroid can be roughly classified into three layers. The inner-most layer is the choriocapillaris (CC) that is just beneath Bruch's membrane and appears as a thin hyperscattering layer in an OCT image. Sattler's layer is located to the posterior of the CC and consists of medium-diameter blood vessels. Finally, Haller's layer is located at the outer-most part of the choroid and consists of large-diameter blood vessels. The blood vessels in Sattler's and Haller's layers appear as hyposcattering regions in the OCT image.

The RPE and choroid contain certain amounts of melanin that highly absorb the probe beam at the 830-nm band [38, 40]. It limits the image penetration, and the image contrast of the choroid is very low at 830-nm. Additionally, the sclera is almost invisible at this wavelength.

Despite the limited penetration, 830-nm SD-OCT is very widely used for posterior diagnosis in eye clinics. An example of a clinical case using 830-nm SD-OCT is shown in Figure 11.9. The case concerns a macular hole examined by an early ophthalmic SD-OCT prototype with a probe wavelength of 840 nm and an axial resolution of 4.5 μm in tissue (see [50] for details). In the cross-sectional image (Figure 11.9(a)), the detachment of the retina from the RPE is clearly visible. Additionally, the traction of the retina by detached vitreous is clearly visualized (arrowheads). The volume rendering of the same case (Figure 11.9(b)) allows a more intuitive understanding of the pathology.

In addition to the qualitative observation of structural abnormalities in retinal pathology, OCT has been used to quantify them. The quantification is mainly done by measuring the thickness of the retinal layers.

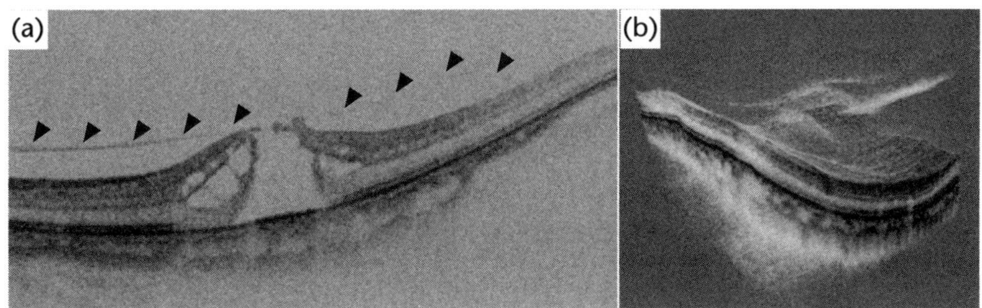

Figure 11.9 Cross-sectional 830-nm SD-OCT image of a macular hole (a) and the volume rendered retina of the same case (b).

The peripapillary NFL thickness is known as a good indicator of glaucoma. Almost all clinical posterior OCT devices can automatically measure this thickness and provide a risk score for glaucoma. Recently, the thicknesses of the inner retinal layers (i.e., NFL, GCL, and IPL) have been used as more sensitive measures of glaucoma [51].

11.3.1.3 High-Penetration Posterior OCT at 1050 nm

As discussed earlier, the early generations of posterior OCT that use a 830-nm probe do not provide high-contrast images at the choroid. This issue has been gradually resolved by 1050-nm OCT.

Figure 11.10 shows a comparison of OCT images of a normal macula taken by 1050-nm SS-OCT (Figure 11.10(a)) and 830-nm SD-OCT (Figure 11.10(b)). (The figures are modified and reprinted from [37].) The 1050-nm SS-OCT and 830-nm SD-OCT respectively possess depth resolutions of 10.4 and 4.5 μm in tissue. Although the depth resolution of 1050-nm OCT is lower, it clearly shows the retinal layers. Additionally, it visualizes the choroid with high contrast. In 1050-nm OCT image, the hyperscattering CC layer beneath the RPE and large hyposcattering structures that are large blood vessels in the Haller's layer are visible. The interface between the choroid and sclera is also visible as a hyperscattering line.

OCT at 1050 nm can be used for both qualitative observation and quantitative examinations. Figure 11.11 shows examples of OCT cross-sections of a case of polypoidal choroidal vasculopathy. The images were modified and reprinted from [52]. The images are displayed in grayscale (i.e., bright pixels represent hyperscattering

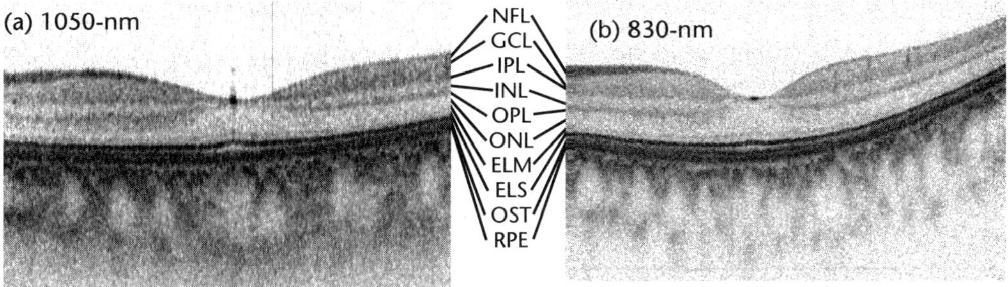

Figure 11.10 Retinal OCT images taken by (a) 1050-nm SS-OCT and (b) 830-nm SD-OCT. The images are of the same macular region of a normal in vivo retina.

and dark pixels represent hyposcattering). Figure 11.11(a) and (b) were obtained by 1050-nm SS-OCT, while Figure 11.11(c) and (d) were obtained by 830-nm SD-OCT. Figure 11.11(a) and (c) and Figure 11.11(b) and (d) were taken at the same locations on the retina. In all images, a severe detachment of the RPE is visible. In the 830-nm images, the structure beneath the RPE detachment is unclear. However, the 1050-nm images have imaged two types of regions beneath the RPE detachment: hyposcattering and hyperscattering regions. A clinical study showed that the hyperscattering regions correspond to the polypoidal hyperfluorescence in ICGA [52]. Another remarkable feature of the 1050-nm images are their high penetration into the choroid and sclera.

Another clinical utility of 1050-nm OCT is its ability to quantify choroidal thickness because of the full-depth penetration into the choroid. Because the choroid is known to be associated with several retinal diseases, a quantitative choroidal thickness measurement is expected to be a powerful tool for quantitative diagnosis. Currently, the choroidal thickness has been found to be associated with several posterior eye diseases, including age-related macular degeneration and polypoidal choroidal vasculopathy [53], central serous chorioretinopathy [54, 55], and some whole-eye diseases including glaucoma [56] and myopia [57, 58].

Although almost all commercial OCT systems for posterior eye imaging are 830-nm SD-OCT, a 1050-nm posterior OCT device is already commercially available and some other semi-commercial research prototypes were demonstrated in 2014. These commercial and semicommercial devices have shown high clinical utilities and are expected to be the next generation of clinical posterior OCT, replacing 830-nm SD-OCT.

11.3.2 Anterior Eye Imaging

The three targets of anterior eye imaging by OCT are the anterior chamber, tissues at the eye surface, and the cornea. For the anterior chamber and the eye surface,

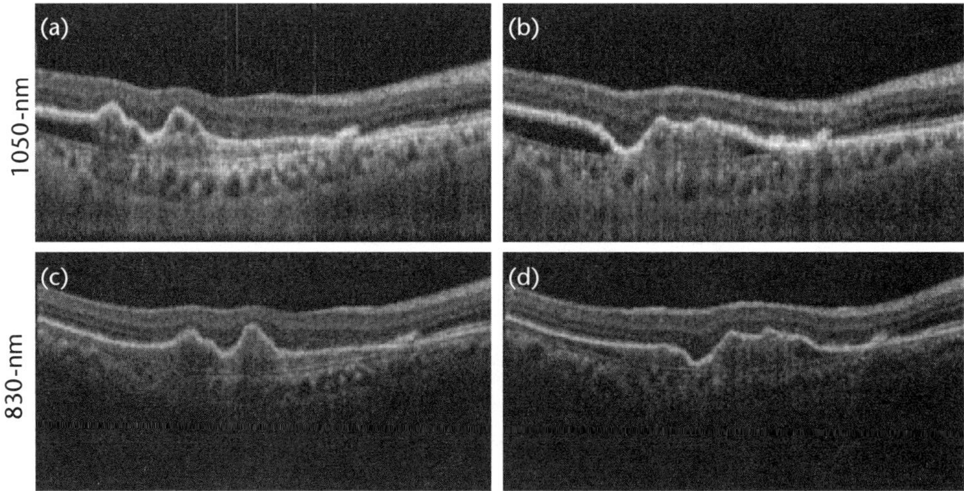

Figure 11.11 OCT images of polypoidal choroidal vasculopathy: (a) and (b) were taken by 1050-nm SS-OCT, while (c) and (d) were taken by 830-nm SD-OCT.

1310-nm OCT is mainly used because it has higher penetration than other wavelengths. The purpose of corneal imaging can be further classified into visualization of tissue abnormalities and morphometric investigation. Because of the two distinct purposes, both 830- and 1310-nm are used for this target.

11.3.2.1 Anterior Eye Chamber Imaging

When OCT imaging the anterior eye chamber, some parts are imaged in a transscleral manner, and hence the anterior OCT should have a high penetration. Conversely, the main interest of anterior OCT imaging is the gross morphology in the eye chamber. Hence, the required resolution is lower than that of posterior OCT. For these reasons, anterior eye chamber imaging is mainly performed with 1310-nm OCT.

Among several clinically significant targets in anterior chamber imaging, quantitative evaluation of anterior angle, the angle between the iris and cornea, is known to be useful for assessing the risk of angle closure glaucoma. In more detail, the anterior eye chamber is filled with fluid called the aqueous humor. The aqueous humor is generated at the ciliary body located posterior to the iris and drained out from the chamber through a channel located just anterior to the anterior angle called the trabecular meshwork. If the trabecular meshwork is morphologically blocked by the iris, it causes an acute elevation of intraocular pressure that damages the optic nerve and results in angle closure glaucoma. To prevent such an acute glaucoma attack, it is important to assess the risk of angle closure. Quantitative assessment of the anterior angle using OCT aims to find eyes at high risk of glaucoma attack.

To quantitatively assess the anterior angle, a parameter that correctly reflects the morphology of the anterior eye chamber is required. Some morphometric parameters originally developed for the analysis of the ultrasound images of anterior chamber, such as angle-opening distance at 500 μm (AOD500) [59], and angle recess area (ARA) [60] are commonly used for the morphological analysis of OCT.

The first generation of clinical anterior eye OCT was 1310-nm TD-OCT, and hence it provides only 2D cross-sections of the anterior eye chamber. However, the recent generation of anterior eye OCT is 1310-nm SS-OCT and provides a full 3D morphology of the anterior eye. Because of this 3D capability, anterior SS-OCT provides whole circumference visualization of the anterior angle, called a virtual gonioscopy in a single scan, as shown in Figure 11.12. The full 3D tomography also enables quantitative assessment of the angle for all circumferences [61].

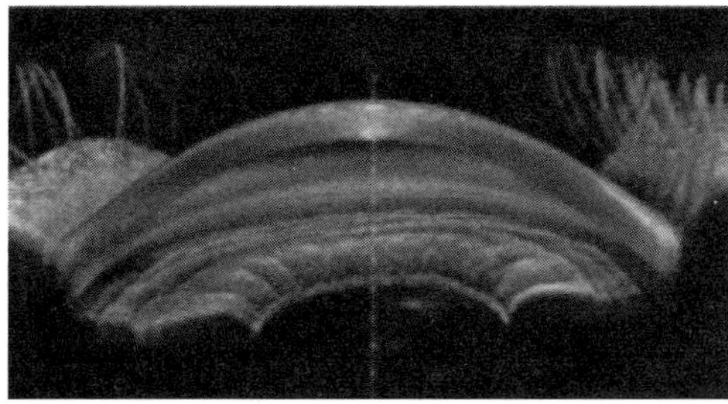

Figure 11.12 Example of whole circumference imaging of the anterior eye chamber. The image was provided from Tomey Corporation.

11.3.2.2 Ocular Surface Investigation

Among several clinically significant targets at the ocular surface, a filtering bleb structure created by trabeculectomy surgery is considered to be one of the most suitable targets for OCT. A trabeculectomy is a glaucoma surgery in which the trabecular meshwork is excised in part and a filtering structure for the aqueous humor, the bleb, is created beneath the conjunctiva. The aqueous humor is then drained into the bleb, and this finally reduces the intraocular pressure.

Because the bleb is an artificially created wound in the sclera, it will naturally heal with time. This undesirable wound healing and the associated scarring results in malfunctioning of the bleb, causing the intraocular pressure to rise again. Although the anti–wound healing drug, mitomycin C, is frequently used to prevent the wound from healing, it does not always successfully stop tissue scarring. Additional intervention, called bleb revision, including secondary surgery could be necessary if scarring occurs. To optimize the secondary intervention, we need a proper modality to assess the filtering bleb.

Several studies have shown that anterior eye segment OCT at 1310 nm is a suitable modality to assess the filtering bleb [62–64]. Figure 11.13 (modified and reprinted from [63]) shows an example of a trabeculectomy bleb examined by 3D anterior OCT. Here, the anterior OCT is a 3D SS-OCT with a 1310-nm probe beam and a depth resolution of 11 μm. Figure 11.13(a), (b), and (c) respectively

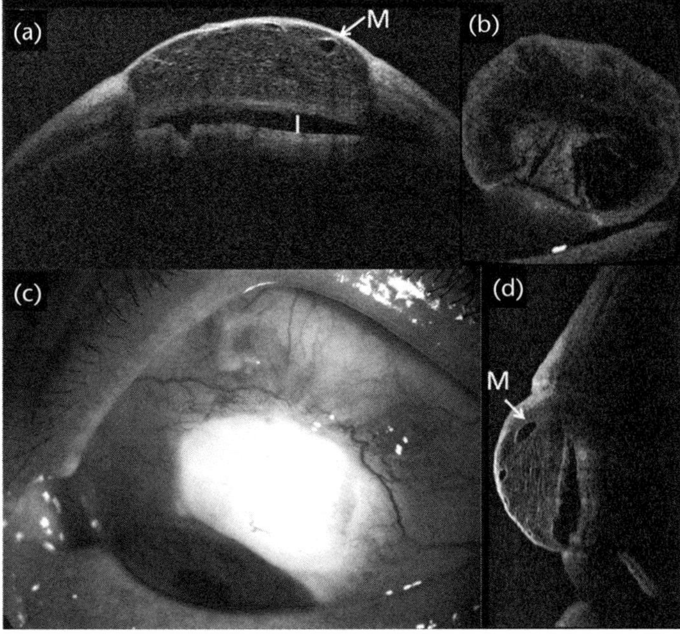

Figure 11.13 Example of a trabeculectomy bleb visualized by 3D anterior OCT. (c) is a color picture of the bleb, and (a), (b), and (d) are OCT cross-sections in horizontal, vertical, and en face directions, respectively. These OCT cross-sections were extracted from a single volumetric OCT. Labels M and I indicate a microcyst and an internal fluid-filled cavity, respectively.

show horizontal, vertical, and en face OCT cross-sections that were created by a single OCT volume. A microcyst (M) and fluid-filled cavity (I) are visible in the bleb. Additionally, the image clearly shows an internal reticulate structure in the bleb that cannot be noninvasively visualized by modalities other than OCT.

11.3.2.3 Corneal Evaluation by OCT

Corneal evaluation by OCT has two major purposes: observation of microscopic disorders and quantification of the optical quality (i.e., aberrations in a wide sense). For the first purpose, both 1310- and 830-nm OCT are used. Currently, there is no commercial 830-nm OCT device that specializes in anterior investigation. However, most commercially available posterior 830-nm SD-OCT devices have a measurement mode for the anterior eye. Because 830-nm OCT, in general, possesses higher resolution than 1310-nm OCT, it provides fine detail of corneal structural disorders.

Some 3D anterior eye-segment OCT can quantitatively assess the morphological (aberration) properties of the cornea. This assessment is particularly useful for the diagnosis of keratoconus, a bilateral, progressive, and noninflammatory disease that is characterized by ectasia and conic deformation of the cornea. The full 3D capability of OCT has enabled the automatic morphological analysis of cornea and is known to have a high sensitivity for keratoconus detection [65].

11.4 Conclusion

OCT was, from the beginning, intended for clinical investigation. From a technological perspective, OCT is just a variation of low-coherence interferometry. If so, it is natural to ask what the main innovation of OCT is. The answer is in the name: OCT. In its etymology, interferometry is a measurement method to observe light interference. Namely, the purpose of interferometry is to measure an interference signal. Conversely, tomography originates from the Greek word *tomos* that means to slice, and *-graphy* means recording. As evident in this etymology, the purpose of tomography is to obtain a cross-sectional slice of a subject. Namely, the low-coherence interferometer was given a clear clinical purpose when it was renamed OCT.

As discussed throughout the manuscript, current OCT is no longer based on a standard low-coherence interferometer (TD-OCT) but on a spectral interferometer. However, it is still called OCT. Additionally, the research field of OCT is no longer limited to the field of optics. It frequently covers photonics, electronics, signal and image processing, and basic and clinical medicine. More recently, the research field also covers informatics such as machine learning and computer aided diagnosis. This expansion of OCT can be understood if OCT is not viewed as the name of a technology but the name of an application-oriented research field.

From this point of view, we believe there are still many important research topics in OCT that have not yet been recognized. Further, there is still room for many researchers to contribute to this field.

References

[1] D. Huang, E. A. Swanson, C. P. Lin, J. S. Schuman, W. G. Stinson, W. Chang, M. R. Hee, T. Flotte, K. Gregory, C. A. Puliafito, and J. G. Fujimoto, "Optical coherence tomography," *Science* 254, 1178–1181 (1991).

[2] J. G. Fujimoto and W. Drexler, "Introduction to Optical Coherence Tomography," in *Optical Coherence Tomography Technology and Applications*, J. G. Fujimoto and W. Drexler, eds. (Springer, 2008), pp. 1–45.

[3] A. Rollins, S. Yazdanfar, M. Kulkarni, R. Ung-Arunyawee, and J. Izatt, "In vivo video rate optical coherence tomography," *Opt. Express* 3, 219–229 (1998).

[4] B. Potsaid, B. Baumann, D. Huang, S. Barry, A. E. Cable, J. S. Schuman, J. S. Duker, and J. G. Fujimoto, "Ultrahigh speed 1050nm swept source/Fourier domain OCT retinal and anterior segment imaging at 100,000 to 400,000 axial scans per second," *Opt. Express* 18, 20029–20048 (2010).

[5] T. Klein, W. Wieser, L. Reznicek, A. Neubauer, A. Kampik, and R. Huber, "Multi-MHz retinal OCT," *Biomed. Opt. Express* 4, 1890–1908 (2013).

[6] M. C. Pierce, J. Strasswimmer, B. H. Park, B. Cense, and J. F. de Boer, "Advances in Optical Coherence Tomography Imaging for Dermatology," *J. Invest. Dermatol.* 123, 458–463 (2004).

[7] L. L. Otis, M. J. Everett, U. S. Sathyam, and B. W. Colston, "Optical Coherence Tomography: A New Imaging Technology for Dentistry," *Journal of the American Dental Association* 131, 511–514 (2000).

[8] Michalina J. Gora, J. S. Sauk, R. W. Carruth, K. A. Gallagher, M. J. Suter, N. S. Nishioka, L. E. Kava, M. Rosenberg, B. E. Bouma, and G. J. Tearney, "Tethered capsule endomicroscopy enables less invasive imaging of gastrointestinal tract microstructure," *Nat. Med.* 19, 238–240 (2013).

[9] B. J. Vakoc, D. Fukumura, R. K. Jain, and B. E. Bouma, "Cancer imaging by optical coherence tomography: preclinical progress and clinical potential," *Nat Rev Cancer* 12, 363–368 (2012).

[10] G. J. Tearney, S. Waxman, M. Shishkov, B. J. Vakoc, M. J. Suter, M. I. Freilich, A. E. Desjardins, W.-Y. Oh, L. A. Bartlett, M. Rosenberg, and B. E. Bouma, "Three-Dimensional Coronary Artery Microscopy by Intracoronary Optical Frequency Domain Imaging," *J Am Coll Cardiol Img* 1, 752–761 (2008).

[11] L. M. Sakata, J. DeLeon-Ortega, V. Sakata, and C. A. Girkin, "Optical coherence tomography of the retina and optic nerve—a review," *Clin. Exp. Ophthalmol.* 37, 90–99 (2009).

[12] E. A. Swanson, D. Huang, M. R. Hee, J. G. Fujimoto, C. P. Lin, and C. A. Puliafito, "High-speed optical coherence domain reflectometry," *Opt. Lett.* 17, 151–153 (1992).

[13] E. A. Swanson, J. A. Izatt, M. R. Hee, D. Huang, C. P. Lin, J. S. Schuman, C. A. Puliafito, and J. G. Fujimoto, "In vivo retinal imaging by optical coherence tomography," *Opt. Lett.* 18, 1864–1866 (1993).

[14] American National Standard Institute, *American National Standard for the Safe Use of Lasers ANSIZ 136.1—2014* (American National Standards Institute, 2014).

[15] CH/172/6, *Ophthalmic Instruments. Fundamental Requirements and Test Methods. General Requirements Applicable to All Ophthalmic Instruments, BS EN ISO 15004-1:2009* (International Organization for Standardization, 2009).

[16] A. F. Fercher, C. K. Hitzenberger, G. Kamp, and S. Y. El-Zaiat, "Measurement of intraocular distances by backscattering spectral interferometry," *Opt. Commun.* 117, 43–48 (1995).

[17] G. Häusler, "'Coherence Radar' and 'Spectral Radar'—New Tools for Dermatological Diagnosis," *J. Biomed. Opt.* 3, 21 (1998).

[18] S. Yun, G. Tearney, J. de Boer, N. Iftimia, and B. Bouma, "High-speed optical frequency-domain imaging," *Opt. Express* 11, 2953–2963 (2003).

[19] M. Wojtkowski, R. Leitgeb, A. Kowalczyk, T. Bajraszewski, and A. F. Fercher, "In vivo human retinal imaging by Fourier domain optical coherence tomography," *J. Biomed. Opt.* 7, 457 (2002).

[20] N. Nassif, B. Cense, B. Park, M. Pierce, S. Yun, B. Bouma, G. Tearney, T. Chen, and J. de Boer, "In vivo high-resolution video-rate spectral-domain optical coherence tomography of the human retina and optic nerve," *Opt. Express* 12, 367–376 (2004).

[21] R. Leitgeb, C. K. Hitzenberger, and A. F. Fercher, "Performance of fourier domain vs. time domain optical coherence tomography," *Optics Express* 11, 889–894 (2003).

[22] J. F. de Boer, B. Cense, B. H. Park, M. C. Pierce, G. J. Tearney, and B. E. Bouma, "Improved signal-to-noise ratio in spectral-domain compared with time-domain optical coherence tomography," *Opt. Lett.* 28, 2067–2069 (2003).

[23] M. Choma, M. Sarunic, C. Yang, and J. Izatt, "Sensitivity advantage of swept source and Fourier domain optical coherence tomography," *Opt. Express* 11, 2183 (2003).

[24] Y. Yasuno, V. D. Madjarova, S. Makita, M. Akiba, A. Morosawa, C. Chong, T. Sakai, K.-P. Chan, M. Itoh, and T. Yatagai, "Three-dimensional and high-speed swept-source optical coherence tomography for in vivo investigation of human anterior eye segments," *Opt. Express* 13, 10652–10664 (2005).

[25] S. Makita, T. Fabritius, and Y. Yasuno, "Full-range, high-speed, high-resolution 1-μm spectral-domain optical coherence tomography using BM-scan for volumetric imaging of the human posterior eye," *Opt. Express* 16, 8406–8420 (2008).

[26] C. Dorrer, N. Belabas, J.-P. Likforman, and M. Joffre, "Spectral resolution and sampling issues in Fourier-transform spectral interferometry," *J. Opt. Soc. Am. B* 17, 1795–1802 (2000).

[27] M. Wojtkowski, V. Srinivasan, T. Ko, J. Fujimoto, A. Kowalczyk, and J. Duker, "Ultrahigh-resolution, high-speed, Fourier domain optical coherence tomography and methods for dispersion compensation," *Opt. Express* 12, 2404–2422 (2004).

[28] B. Cense, N. Nassif, T. Chen, M. Pierce, S.-H. Yun, B. Park, B. Bouma, G. Tearney, and J. de Boer, "Ultrahigh-resolution high-speed retinal imaging using spectral-domain optical coherence tomography," *Opt. Express* 12, 2435–2447 (2004).

[29] R. Huber, M. Wojtkowski, K. Taira, J. Fujimoto, and K. Hsu, "Amplified, frequency swept lasers for frequency domain reflectometry and OCT imaging: design and scaling principles," *Opt. Express* 13, 3513–3528 (2005).

[30] W. Y. Oh, S. H. Yun, G. J. Tearney, and B. E. Bouma, "115 kHz tuning repetition rate ultrahigh-speed wavelength-swept semiconductor laser," *Opt. Lett.* 30, 3159–3161 (2005).

[31] S. Yun, G. Tearney, B. Bouma, B. Park, and J. de Boer, "High-speed spectral-domain optical coherence tomography at 1.3 μm wavelength," *Opt. Express* 11, 3598–3604 (2003).

[32] E. Z. Zhang and B. J. Vakoc, "Polarimetry noise in fiber-based optical coherence tomography instrumentation," *Opt. Express* 19, 16830–16842 (2011).

[33] E. Z. Zhang, W.-Y. Oh, M. L. Villiger, L. Chen, B. E. Bouma, and B. J. Vakoc, "Numerical compensation of system polarization mode dispersion in polarization-sensitive optical coherence tomography," *Opt. Express* 21, 1163–1180 (2013).

[34] M. Villiger, E. Z. Zhang, S. Nadkarni, W.-Y. Oh, B. E. Bouma, and B. J. Vakoc, "Artifacts in polarization-sensitive optical coherence tomography caused by polarization mode dispersion," *Opt. Lett.* 38, 923–925 (2013).

[35] Boy Braaf, K. A. Vermeer, V. A. D. P. Sicam, E. van Zeeburg, J. C. van Meurs, and J. F. de Boer, "Phase-stabilized optical frequency domain imaging at 1-μm for the measurement of blood flow in the human choroid," *Opt. Express* 19, 20886–20903 (2011).

[36] G. M. Hale and M. R. Querry, "Optical Constants of Water in the 200-nm to 200-μm Wavelength Region," *Appl. Opt.* 12, 555–563 (1973).

[37] B. Považay, K. Bizheva, B. Hermann, A. Unterhuber, H. Sattmann, A. Fercher, W. Drexler, C. Schubert, P. Ahnelt, M. Mei, R. Holzwarth, W. Wadsworth, J. Knight, and P. S. J.

Russell, "Enhanced visualization of choroidal vessels using ultrahigh resolution ophthalmic OCT at 1050 nm," *Opt. Express* 11, 1980–1986 (2003).

[38] A. Unterhuber, B. Považay, B. Hermann, H. Sattmann, A. Chavez-Pirson, and W. Drexler, "In vivo retinal optical coherence tomography at 1040 nm-enhanced penetration into the choroid," *Opt. Express* 13, 3252–3258 (2005).

[39] E. C. Lee, J. F. de Boer, M. Mujat, H. Lim, and S. H. Yun, "In vivo optical frequency domain imaging of human retina and choroid," *Opt. Express* 14, 4403–4411 (2006).

[40] Y. Yasuno, Y. Hong, S. Makita, M. Yamanari, M. Akiba, M. Miura, and T. Yatagai, "In vivo high-contrast imaging of deep posterior eye by 1-μm swept source optical coherence tomography and scattering optical coherence angiography," *Opt. Express* 15, 6121–6139 (2007).

[41] V. J. Srinivasan, D. C. Adler, Y. Chen, I. Gorczynska, R. Huber, J. S. Duker, J. S. Schuman, and J. G. Fujimoto, "Ultrahigh-Speed Optical Coherence Tomography for Three-Dimensional and En Face Imaging of the Retina and Optic Nerve Head," *Invest. Ophthalmol. Vis. Sci.* 49, 5103–5110 (2008).

[42] B. Považay, B. Hermann, A. Unterhuber, B. Hofer, H. Sattmann, F. Zeiler, J. E. Morgan, C. Falkner-Radler, C. Glittenberg, S. Blinder, and W. Drexler, "Three-dimensional optical coherence tomography at 1050nm versus 800nm in retinal pathologies: enhanced performance and choroidal penetration in cataract patients," *J. Biomed. Opt.* 12, 041211–041211-7 (2007).

[43] P. Puvanathasan, P. Forbes, Z. Ren, D. Malchow, S. Boyd, and K. Bizheva, "High-speed, high-resolution Fourier-domain optical coherence tomography system for retinal imaging in the 1060 nm wavelength region," *Opt. Lett.* 33, 2479–2481 (2008).

[44] S. Radhakrishnan, A. M. Rollins, J. E. Roth, S. Yazdanfar, V. Westphal, D. S. Bardenstein, and J. A. Izatt, "Real-Time Optical Coherence Tomography of the Anterior Segment at 1310 nm," *Arch. Ophthalmol.* 119, 1179–1185 (2001).

[45] B. Bouma, ed., *Handbook of Optical Coherence Tomography*, 1st ed. (Informa Healthcare, 2001).

[46] R. Zawadzki, S. Jones, S. Olivier, M. Zhao, B. Bower, J. Izatt, S. Choi, S. Laut, and J. Werner, "Adaptive-optics optical coherence tomography for high-resolution and high-speed 3D retinal in vivo imaging," *Opt. Express* 13, 8532–8546 (2005).

[47] F. Felberer, J.-S. Kroisamer, B. Baumann, S. Zotter, U. Schmidt-Erfurth, C. K. Hitzenberger, and M. Pircher, "Adaptive optics SLO/OCT for 3D imaging of human photoreceptors in vivo," *Biomed. Opt. Express* 5, 439–456 (2014).

[48] R. A. Leitgeb, C. K. Hitzenberger, A. F. Fercher, and T. Bajraszewski, "Phase-shifting algorithm to achieve high-speed long-depth-range probing by frequency-domain optical coherence tomography," *Opt. Lett.* 28, 2201–2203 (2003).

[49] Y. Hong, S. Makita, M. Yamanari, M. Miura, S. Kim, T. Yatagai, and Y. Yasuno, "Three-dimensional visualization of choroidal vessels by using standard and ultra-high resolution scattering optical coherence angiography," *Opt. Express* 15, 7538–7550 (2007).

[50] S. Makita, Y. Hong, M. Yamanari, T. Yatagai, and Y. Yasuno, "Optical coherence angiography," *Opt. Express* 14, 7821–7840 (2006).

[51] O. Tan, G. Li, A. T.-H. Lu, R. Varma, and D. Huang, "Mapping of Macular Substructures with Optical Coherence Tomography for Glaucoma Diagnosis," *Ophthalmology* 115, 949–956 (2008).

[52] Y. Yasuno, M. Miura, K. Kawana, S. Makita, M. Sato, F. Okamoto, M. Yamanari, T. Iwasaki, T. Yatagai, and T. Oshika, "Visualization of Sub-retinal Pigment Epithelium Morphologies of Exudative Macular Diseases by High-Penetration Optical Coherence Tomography," *Invest. Ophthalmol. Vis. Sci.* 50, 405–413 (2009).

[53] S. E. Chung, S. W. Kang, J. H. Lee, and Y. T. Kim, "Choroidal Thickness in Polypoidal Choroidal Vasculopathy and Exudative Age-related Macular Degeneration," *Ophthalmology* 118, 840–845 (2011).

[54] I. Maruko, T. Iida, Y. Sugano, A. Ojima, and T. Sekiryu, "Subfoveal choroidal thickness in fellow eyes of patients with central serous chorioretinopathy," *Retina* (Philadelphia, Pa.) 31, 1603–1608 (2011).

[55] S. Kuroda, Y. Ikuno, Y. Yasuno, K. Nakai, S. Usui, M. Sawa, M. Tsujikawa, F. Gomi, and K. Nishida, "Choroidal thickness in central serous chorioretinopathy," *Retina* 33, 302–308 (2013).

[56] S. Usui, Y. Ikuno, A. Miki, K. Matsushita, Y. Yasuno, and K. Nishida, "Evaluation of the Choroidal Thickness Using High-Penetration Optical Coherence Tomography With Long Wavelength in Highly Myopic Normal-Tension Glaucoma," *Am. J. Ophthalmol.* 153, 10–16.e1 (2012).

[57] T. Fujiwara, Y. Imamura, R. Margolis, J. S. Slakter, and R. F. Spaide, "Enhanced Depth Imaging Optical Coherence Tomography of the Choroid in Highly Myopic Eyes," *Am. J. Ophthalmol.* 148, 445–450 (2009).

[58] Y. Ikuno, S. Fujimoto, Y. Jo, T. Asai, and K. Nishida, "Choroidal thinning in high myopia measured by optical coherence tomography," *Clin. Ophthalmol.* 7, 889–893 (2013).

[59] C. J. Pavlin, K. Harasiewicz, and F. S. Foster, "Ultrasound biomicroscopy of anterior segment structures in normal and glaucomatous eyes," *Am. J. Ophthalmol.* 113, 381–389 (1992).

[60] H. Ishikawa, J. M. Liebmann, and R. Ritch, "Quantitative assessment of the anterior segment using ultrasound biomicroscopy," *Curr Opin Ophthalmol* 11, 133–139 (2000).

[61] M. Baskaran, S.-W. Ho, T. A. Tun, A. C. How, S. A. Perera, D. S. Friedman, and T. Aung, "Assessment of Circumferential Angle-Closure by the Iris–Trabecular Contact Index with Swept-Source Optical Coherence Tomography," *Ophthalmology* 120, 2226–2231 (2013).

[62] M. Singh, P. T. K. Chew, D. S. Friedman, W. P. Nolan, J. L. See, S. D. Smith, C. Zheng, P. J. Foster, and T. Aung, "Imaging of Trabeculectomy Blebs Using Anterior Segment Optical Coherence Tomography," *Ophthalmology* 114, 47–53 (2007).

[63] K. Kawana, T. Kiuchi, Y. Yasuno, and T. Oshika, "Evaluation of Trabeculectomy Blebs Using 3-Dimensional Cornea and Anterior Segment Optical Coherence Tomography," *Ophthalmology* 116, 848–855 (2009).

[64] S. Kojima, T. Inoue, T. Kawaji, and H. Tanihara, "Filtration Bleb Revision Guided by 3-Dimensional Anterior Segment Optical Coherence Tomography," *Journal of Glaucoma* 23, 312–315 (2014).

[65] S. Fukuda, S. Beheregaray, S. Hoshi, M. Yamanari, Y. Lim, T. Hiraoka, Y. Yasuno, and T. Oshika, "Comparison of three-dimensional optical coherence tomography and combining a rotating Scheimpflug camera with a Placido topography system for forme fruste keratoconus diagnosis," *Br. J. Ophthalmol.* 97, 1554–1559 (2013).

About the Authors

Jim Burge is professor of optical sciences, astronomy, and mechanical engineering at the University of Arizona in Tucson, where he leads a group of researchers, engineers, and students that push the state of the art in fabrication and testing technologies and apply them for manufacturing challenging mirrors and optical systems. He directs the Large Optics Fabrication and Testing Laboratory that supports a range of telescope projects from small telescopes used in space to the 25m Giant Magellan Telescope. He has authored over 320 technical publications and has cofounded two companies, Arizona Optical Metrology and Arizona Optical Systems.

Lujie Chen received his B.Eng. from Nanjing University of Technology, China, in 2001 and his Ph.D. from the National University of Singapore, Singapore, in 2006. From 2007 to 2010, he was a research associate at the University of Cambridge, UK. He joined Singapore University of Technology and Design (SUTD) as a faculty in 2010. In 2011, he had a one-year academic visit at Massachusetts Institute of Technology, where he studied quantum dots for biomedical imaging. His research interests include optical metrology, medical imaging, and rapid prototyping. He is the developer of UU and Fig, a general-purpose image processing system with a focus on optical metrology and rapid prototyping.

Brian Culshaw graduated from University College London with a First Class Honours degree in physics in 1966 and a Ph.D. in electrical engineering in 1970. He has held appointments at Cornell University (1970), Bell Northern Research, Ottawa, Canada (1971–1973), University College London (1974–1983), Stanford University (1982), and joined the Strathclyde University in 1983 where he is professor emeritus. He has served as a director of SPIE, the International Society for Optical Engineering based in Bellingham, WA, and was the 2007 president of the society. He has published in excess of 300 technical papers, several conference proceedings, seven textbooks, and a dozen patents.

Rishikesh Kulkarni received his MTech from the Indian institute of Sciences, Bangalore, India. He has industrial experience in the fields of biomedical and automobile electronics. He is currently pursuing his Ph.D. in photonics from the Ecolé Polytechnique Federale de Lausanne (EPFL). His interest lies in digital signal processing and its applications in optical metrology.

Bing Pan is with the School of Aerospace Science & Engineering at Beihang University (BUAA), China. He received his Ph.D. in mechanical engineering from Tsinghua

University in 2008. After working as a research fellow for one year in Nanyang Technological University, Singapore, he joined the Institute of Solid Mechanics, BUAA, in 2009. His current research interests mainly focus on advanced optical techniques and their applications in experimental mechanics. He has published 60 peer-reviewed articles in international journals. He won the National Natural Science Funds for Excellent Young Scholar in 2013, and was selected for the Program for New Century Excellent Talents in University by the Ministry of Education, China in 2012.

Chenggen Quan graduated in 1982 from Harbin Institute of Technology (HIT), China, with a B.Eng. in mechanical engineering. He received his M.Eng. in optical engineering from HIT in 1988 and his Ph.D. from Warwick University, UK, in 1992. Before joining the National University of Singapore (NUS), he worked as a senior engineer at National Measurement Centre, Singapore Productivity and Standards Board. He is currently an associate professor in the Department of Mechanical Engineering, NUS. His research interests include optical nondestructive testing, experimental mechanics, digital holography, optical image encryption, and automatic fringe-pattern analysis. He has authored or coauthored over 150 international refereed journal papers.

K. Ramesh is currently a professor in the Department of Applied Mechanics, IIT Madras, was its chairman from 2005–2009, and was formerly a professor in the Department of Mechanical Engineering, IIT Kanpur. He has made significant contributions to the advancement of digital photoelasticity. He has authored a monograph, *Digital Photoelasticity—Advanced Techniques and Applications*, and contributed a chapter titled "Photoelasticity" in the *Springer Handbook of Experimental Solid Mechanics*. He has pioneered a new paradigm in engineering education by writing innovative e-books such as *Engineering Fracture Mechanics* and *Experimental Stress Analysis*. He is a Fellow of the Indian National Academy of Engineering since 2006 and has received the Zandman Award from the Society of Experimental Mechanics in the United States (2012) for his contributions to the advancement of photoelastic coatings.

Julio Soria was awarded a B.E. (Hons) in 1983 and a Ph.D. in mechanical engineering in 1989 from the University of Western Australia. After a postdoctoral fellowship at Stanford University and NASA Ames Research Centre in 1990, he joined CSIRO in 1991. In 1993 he joined Monash University where he is currently a professor with a personal chair in mechanical engineering. He is director of the Laboratory for Turbulence Research in Aerospace & Combustion (LTRAC), which he established in 1994. He is a Fellow of the Australasian Fluid Mechanics Society. His current research interests include turbulent boundary layer flows, subsonic and supersonic transitional and turbulent jet flows, and the development of nonintrusive optical experimental measurement methods (PIV, Stereo-PIV, Tomo-PIV, HPIV, Tomo-HPIV). He has authored more than 420 papers.

Cho Jui Tay received his B.Sc. and Ph.D. from the University of Strathclyde, UK. He is currently an associate professor at the National University of Singapore. His research interests include experimental stress analysis and optical interferometry and its applications to MEMS and NEMS devices. He has published widely in international journals and has coauthored several books and book chapters. He serves on the editorial boards of several international journals and committees of various grant award agencies. He holds a number of patents and has won several research and best paper awards including the David Cargill Prize, UK.

Matias R. Viotti studied mechanical engineering at the National University of Rosario, Argentina, where he received his diploma in 1999. His doctoral degree was obtained at the Physics Institute of Rosario, Argentina, in 2005. He has been working since 2005 as an associate researcher at the Metrology and Automation Laboratory linked to the National University of Santa Catarina, Brazil. He is a senior member of SPIE. His research interests include digital speckle pattern interferometry, measurement of residual stresses, and the development of coherent techniques for strain analysis and nondestructive testing. He has coauthored more than 50 papers published in international journals and conference proceedings and one book chapter.

Richard Vollmerhausen is a founding member and chief engineer of St. Johns Optical Systems. He received his Ph.D. in electrical engineering from the University of Delaware. He has experience as an instrumentation engineer at Douglas Aircraft and as a staff physicist at the Naval Weapons Center at China Lake. At the U.S. Army Night Vision Lab, he developed the Target Task Performance metric that is currently used for the design of all U.S. and most of NATO electro-optical weapon sights. He also performed extensive flight experiments to establish the psychophysical requirements for helicopter pilotage. He has published close to 100 research papers, chapters on electro-optical system performance, and two books on imaging system performance.

Lionel Watkins was a research fellow, then lecturer, at the School of Electronic Engineering Science at the University of Wales, Bangor, UK, obtaining his Ph.D. from the same institution in 1991. In 1996 he joined the physics department at the University of Auckland, New Zealand. His research interests are in optical metrology, principally ellipsometry, and interferometry. In the former, he has designed a number of novel ellipsometers that borrow concepts from interferometry while in the latter, his interest has been focused on wavelet methods for phase recovery, and more recently, on novel techniques for finding best-fit ellipses to the Lissajous figures generated by heterodyne interferometers.

Yoshiaki Yasuno leads the Computational Optics Group at the University of Tsukuba. He obtained his Ph.D. in spatiotemporal optical computing in 2000, and extended the theory of optical computing to optical measurement. Since 2003 he has been studying an application of optical computing to high-speed medical imaging, which is currently recognized as Fourier-domain optical coherence tomography (FD-OCT). His current research interest is in the application of FD-OCT to ophthalmology, and he is also actively involved in the extension of FD-OCT based on the perspective of optical computing.

Chunyu Zhao is cofounder and president of Arizona Optical Metrology LLC, a company that provides computer-generated holograms for testing aspheric surfaces. He received a B.S. in physics and a B.E. in mechanical engineering in 1993 at Tsinghua University, Beijing, China, and an M.S. in 1999 and Ph.D. in 2002, both in optical engineering at the University of Arizona. He has over 15 years of experience in developing and building a variety of interferometric systems for testing large aspheric optical surfaces. He has authored/coauthored about 70 technical papers and presented dozens of talks at international conferences.

Ping Zhou is an assistant research professor at the College of Optical Sciences, University of Arizona. Her research is in the area of precision optical fabrication and testing, especially for large optics. Her research activities include optical system design, high-precision metrology, and error analysis and data processing for various measurements. She received her doctoral degree at the University of Arizona in 2009. Since then, she has been working in the large optics fabrication and testing group at the University of Arizona.

About the Editor

Professor Pramod Rastogi received his MTech degree from the Indian Institute of Technology Delhi, and his doctorate from the University of Franche Comté in France. He joined the Ecolé Polytechnique Federale de Lausanne (EPFL) in Switzerland 1978. He is the author or coauthor of over 150 scientific papers published in peer-reviewed archival journals. Professor Rastogi is also the author of encyclopedia articles, and has published several books in the field of optical metrology.

Professor Rastogi is the 2014 recipient of the SPIE *Dennis Gabor Award* and is a member of the Swiss Academy of Engineering Sciences. He is a Fellow of the Society of the Photo-Optical Instrumentation Engineers (1995) and a Fellow of the Optical Society of America (1993). He is also a recipient of the *Hetényi Award* for the most significant research paper published in Experimental Mechanics in 1982. Professor Rastogi is the co-editor-in-chief of the *International Journal of Optics and Lasers in Engineering*, Elsevier.

Index

Optical FDM Network Technologies, Kiyoshi Nosu

Optical Fiber Amplifiers: Materials, Devices, and Applications, Shoichi Sudo, editor

Optical Fiber Communication Systems, Leonid Kazovsky, Sergio Benedetto and Alan Willner

Optical Fiber Sensors, Volume Three: Components and Subsystems, John Dakin and Brian Culshaw, editors

Optical Fiber Sensors, Volume Four: Applications, Analysis, and Future Trends, John Dakin and Brian Culshaw, editors

Optical Measurement Techniques and Applications, Pramod Rastogi

Optical Transmission Systems Engineering, Milorad Cvijetic

Optoelectronic Techniques for Microwave and Millimeter-Wave Engineering, William M. Robertson

Reliability and Degradation of III-V Optical Devices, Osamu Ueda

Signal Processing and Performance Analysis for Imaging Systems, S. Susan Young, Ronald G. Driggers, and Eddie L. Jacobs

Smart Structures and Materials, Brian Culshaw

Substrate Surface Preparation Handbook, Max Robertson

Surveillance and Reconnaissance Imaging Systems: Modeling and Performance Prediction, Jon C. Leachtenauer and Ronald G. Driggers

Wavelength Division Multiple Access Optical Networks, Andrea Borella, Giovanni Cancellieri, and Franco Chiaraluce

For further information on these and other Artech House titles, including previously considered out-of-print books now available through our In-Print-Forever[®] (IPF[®]) program, contact:

Artech House	Artech House
685 Canton Street	16 Sussex Street
Norwood, MA 02062	London SW1V 4RW UK
Phone: 781-769-9750	Phone: +44 (0)20 7596-8750
Fax: 781-769-6334	Fax: +44 (0)20 7630-0166
e-mail: artech@artechhouse.com	e-mail: artech-uk@artechhouse.com

Find us on the World Wide Web at: www.artechhouse.com